Advanced Techniques against Food Adulteration

Waleed Alahmad • Mohammad Ashfaq
Editors

Advanced Techniques against Food Adulteration

Springer

Editors
Waleed Alahmad
Department of Chemistry
Chulalongkorn University
Bangkok, Thailand

Mohammad Ashfaq
School of Sciences
Woxsen University
Hyderabad, Telangana, India

ISBN 978-3-031-99453-1 ISBN 978-3-031-99454-8 (eBook)
https://doi.org/10.1007/978-3-031-99454-8

This Springer imprint is published by the registered company Springer Nature Switzerland AG
The registered company address is: Gewerbestrasse 11, 6330 Cham, Switzerland

Preface

Food fraud poses a significant global challenge, raising concerns about the safety and authenticity of food products. Addressing this issue requires continuous research and the development of robust food authentication methods. Over the years, various analytical techniques and quality assurance frameworks have been created, each with distinct advantages, to ensure the genuineness of food. Among these, isotopic methods have emerged as powerful tools for combating deceptive claims, particularly those of geographical origin.

Isotopic methods provide a reliable means of verifying the authenticity of food by analyzing stable isotopes present in the samples. These methods help establish the geographic origin of food products, making them invaluable in detecting fraudulent practices. Additionally, targeted and untargeted omics approaches enable the rapid acquisition of comprehensive chemical profiles for food samples. This aids in identifying potential adulterants and ensuring the overall integrity of the food supply chain.

PCR-powered bioassays have become the gold standard for species identification, offering a precise and efficient means of detecting mislabeled or substituted ingredients. Spectral analyses, such as those using infrared or Raman spectroscopy, have gained prominence as non-destructive methods for effectively distinguishing adulterated foods. These techniques provide valuable insights into the composition of food products without altering their integrity.

Real-time rapid detection methods, including electronic sensors and immunoassays, have proven to be essential tools for addressing adulteration in an affordable yet resilient manner. These methods enable quick and on-the-spot analysis, contributing to the prevention of fraudulent activities. Moreover, recent advancements in the miniaturization of spectrometers and other sensing devices have paved the way for point-of-need food analysis, offering convenience and efficiency. The pillars of quality assurance against food fraud are advancing rapidly, encompassing a multidisciplinary approach that combines various analytical methods and technologies. To capture these advancements and facilitate knowledge dissemination, there is a need to compile them into a comprehensive book. This book is a timely resource, combining the latest research findings, technological breakthroughs, and practical applications in the fight against food fraud. By integrating these diverse methodologies and staying abreast of emerging technologies, the food industry can strengthen its defenses against fraudulent activities, ultimately ensuring the safety and authenticity of the global food supply. This proposed book aims to be a valuable guide for researchers, industry professionals, and policymakers engaged in safeguarding the integrity of the food we consume.

Food adulteration remains one of the most pressing concerns in the global food industry, with serious implications for consumer health, regulatory compliance, and market integrity. In an era marked by complex food supply chains and increased consumer awareness, the demand for advanced, reliable, and accessible detection strategies has never been greater. The continued evolution of adulteration methods calls for equally dynamic solutions—ones that integrate scientific rigor, technological innovation, and practical applicability.

This book, *Advanced Techniques Against Food Adulteration*, has been conceived as a timely response to this challenge. It offers a multidisciplinary overview of current and next-generation analytical methods, from well-established techniques to emerging technologies that hold great promise in reshaping food authentication and safety assurance.

The volume opens with discussions on the detection of synthetic dyes and meat adulterants, two areas of persistent concern across global markets. It progresses into chapters that explore chemical fingerprinting through gas chromatography and novel low-cost platforms like microfluidic paper-based analytical devices. These methods illustrate how traditional and portable technologies can coexist and complement one another in real-world applications.

Further chapters delve into advanced sensor technologies, including electrochemical sensors, optical arrays, and electronic tongue and nose systems—each offering new avenues for rapid, sensitive, and often non-destructive analysis. The integration of nanotechnology, liquid chromatography, and surface-enhanced Raman spectroscopy highlights the frontier of detection science, particularly in identifying trace adulterants with high specificity.

In addition to detection, this book recognizes the growing relevance of green chemistry and sustainable solutions. The use of deep eutectic solvents for extraction represents a move toward more environmentally friendly practices in analytical chemistry. Capillary electrophoresis and enzyme-linked immunosorbent assays (ELISA) further underscore the breadth of available techniques, ranging from high-resolution separations to immunological screenings.

Of equal importance is the chapter on risk mitigation strategies and the holistic integration of technologies into next-generation food safety systems. These insights aim to support not only the detection of adulteration but also its prevention through informed, technology-driven policies and practices.

As editors of this volume, I am deeply grateful to the contributing authors, reviewers, and collaborators whose expertise and dedication have brought this project to life. Their collective efforts reflect a shared commitment to safeguarding the authenticity, quality, and trustworthiness of the global food supply.

It is our hope that this book will serve as a valuable reference for researchers, industry professionals, regulatory authorities, and academic institutions. By bridging scientific innovation and practical implementation, it aspires to contribute meaningfully to the global fight against food fraud.

Bangkok, Thailand Waleed Alahmad
Hyderabad, India Mohammad Ashfaq

Contents

Analytical Methods for Dye Food Adulteration

Thidarat Samkumpim, Niluh Indria Wardani, Pakorn Varanusupakul, and Waleed Alahmad

Contents

Abstract This chapter comprehensively reviews analytical methods for detecting dye adulteration in food, emphasizing their crucial role in ensuring food safety and public health. The discussion begins by examining the widespread use of synthetic dyes in the food industry, driven by their esthetic and economic advantages despite

Thidarat Samkumpim and Niluh Indria Wardani contributed equally with all other contributors.

T. Samkumpim · N. I. Wardani · P. Varanusupakul · W. Alahmad (✉)
Faculty of Science, Department of Chemistry, Chulalongkorn University, Bangkok, Thailand
e-mail: waleed.al@chula.ac.th

W. Alahmad, M. Ashfaq (eds.), *Advanced Techniques against Food Adulteration*,
https://doi.org/10.1007/978-3-031-99454-8_1

their associated health risks. Key analytical techniques, including spectrophotometry, chromatography, and mass spectrometry, are highlighted for their accuracy and reliability in identifying harmful adulterants. Advances in sample preparation methods, such as microextraction technologies and green solvent systems, are showcased as environmentally sustainable solutions to traditional approaches. Additionally, this chapter explores cutting-edge developments in electrochemical sensors and smartphone-based colorimetric assays, recognized for their high sensitivity, cost-efficiency, and adaptability for field applications. Case studies from regions illustrate the health consequences of dye adulteration and underline regulatory enforcement challenges. This chapter concludes by emphasizing future directions, advocating for eco-friendly, scalable detection technologies, and harmonizing global food safety regulations to protect consumers and enhance food quality standards.

Keywords Food dye adulteration · Analytical detection methods · Spectrophotometry and chromatography · Electrochemical sensors · Food safety and regulations

Introduction

It is essential to consume food to maintain existence. Food consumption patterns are also changing as lifestyles change globally. To enhance the allure of instant and processed foods (commonly referred to as "junk foods"), manufacturers frequently implement a variety of synthetic food colors. The color of food is a critical factor that influences our perception of its quality and flavor [1]. Food coloring, or color additive, is any dye, pigment, or substance that imparts color when added to food or drink. Color additives, or food coloring, are any dye, pigment, or substance that imparts color when applied to foods or beverages. These color additives have the potential to conceal imperfections and enhance the appeal of lower-quality foods. They are available in diverse forms, such as liquids, granules, gels, and pastes.

Food adulteration refers to the intentional or unintentional addition of inferior quality, harmful, or non-permitted substances to food products or the removal of valuable ingredients. This practice is typically carried out to increase the quantity and enhance the appearance or shelf-life of the food, often at the expense of its quality, safety, and nutritional value [2]. Food adulteration can be defined as the practice of adulterating food or contamination of food materials by adding a few substances, which are collectively called adulterants. Food adulteration is the process of contaminating food by introducing substances collectively referred to as adulterants. Food adulteration is the deliberate or unintentional introduction of harmful, unauthorized, or inferior-quality substances into food products, or the removal of valued ingredients [3]. This practice is typically intended to enhance the appearance or shelf-life of food and increase its supply, often at the expense of its quality, safety, and nutritional value. The adulteration of food pigments is particularly hazardous.

The increasing diversity of chemical compounds utilized in dyes presents a challenge for researchers in maintaining up-to-date safety evaluations, and the presence of hazardous, unregulated synthetic compounds can have adverse effects on consumers. As health awareness increases, more consumers are reviewing labels to avoid food dyes and synthetic additives. Labels that assert that a product is additive-free must be verified through appropriate testing.

The practice of adding adulterants is quite prevalent in developing countries despite the existence of a variety of penalties and actions. Synthetic food colors are extensively employed in a variety of culinary products. Synthetic colorants can be categorized as water-soluble or fat-soluble, depending on their solubility [4]. Water-soluble colorants, including caramel color and anthocyanins, are compatible with recipes incorporating water-based ingredients, including sponge cake, cream, custard, gelatin, almond paste, and marzipan. These colorants dissolve in water. Fat-soluble colorants, such as Sudan I, Sudan II, Sudan III, and Sudan IV, are azo compounds that disintegrate in fat. They are appropriate for high-fat recipes, including chocolate, ganache, almond paste, sugar paste, and parfaits. Nowadays, one of the most significant challenges in society is the adulteration or contamination of natural food products. Natural products are adulterated through a variety of methods. For example, Sudan dyes, fat-soluble azo dyes frequently employed in the textile industry, are known to be added to palm oil. Sudan dyes are a synthetic family that has been extensively utilized in the coloring and enhancing of industrial products [5]. Palm oil has been frequently substituted into other oils to increase volume and reduce production costs, despite its low price, lustrous color, and wide availability on the global market. The International Agency for Research on Cancer (IARC) has classified these dyes as group 3 carcinogens, and studies have demonstrated their potential association with certain bladder and liver-related malignancies. Consequently, the IARC has prohibited their use as food additives [6].

Foodborne diseases are typically infectious or toxic and are caused by entering bacteria, viruses, parasites, or chemical substances into the body through contaminated food. Dye adulteration is a severe public health issue that impacts millions of individuals annually (Fig. 1). According to the World Health Organization (WHO), approximately 600 million individuals (approximately one in ten) fall ill, and 420,000 individuals perish annually as a consequence of consuming unsafe food [7]. Consumers' health may be adversely affected by contamination and adulteration in food, which reduces its nutritional quality and may compromise its safety [8]. The impact on health can be defined as:

1. Acute toxicity: Certain adulterants can be toxic, resulting in acute or chronic health conditions. For instance, including non-edible dyes can result in long-term health, organ harm, and poisoning.
2. Nutrient deficiency: The food's overall nutritional value is often diminished due to the dilution or eradication of nutritious components during adulteration.
3. Allergic reactions: Certain adulterants have the potential to induce hypersensitivity reactions or allergies in susceptible individuals.

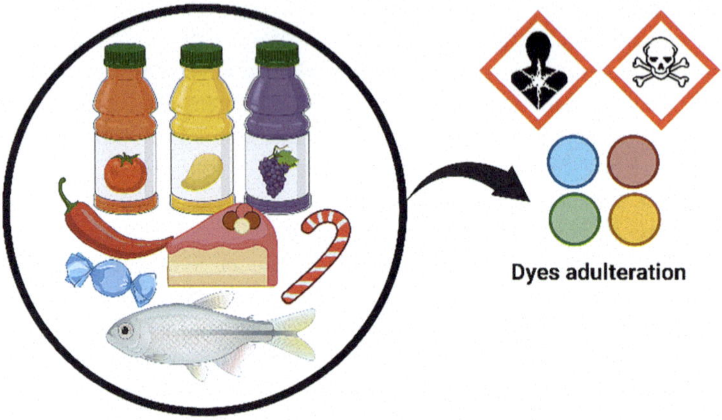

Fig. 1 Illustration of toxic and carcinogenic dye adulteration in food and beverages. BioRender software was used to create this figure under an academic license

4. Chronic diseases: Certain adulterants may induce hypersensitivity reactions or allergies in susceptible individuals.

Commonly Used Dyes in Food Adulteration

The most prevalent additives in food are non-permitted colors. Dyes are frequently employed in various food industries, including adult nutrition, animal nutrition, processed foods, and herbal products. The most frequently adulterated food products are cereal and bakery products, edible oils and fats, milk and dairy products, meat and fish, fruit and fruit beverages, honey, coffee, tea, wine, and organic products, among others [9]. In particular, it is a frequently employed adulterant in chili to enhance its red and yellow hues. The color of chili impacts consumers' emotions when they are making a purchasing decision. Using dyes as adulterants in food products is an unethical and illegal practice intended to improve the visual allure of food, thereby presenting it as a fresher or of higher quality than it is. There are some examples of commonly used dyes in food adulteration.

The International Agency for Research on Cancer (IARC) has classified Sudan dyes I–IV as class 3 carcinogens, a family of synthetic azo dyes. During the early 2000s, they were discovered in certain food products in certain European countries. The European Commission decided on emergency measures for chili, chili products, Curcuma, and palm oil in response to the resulting alarm. In recent years, notifications regarding commodities adulterated with Sudan dyes have decreased significantly. Consequently, European countries have only maintained oversight of the Sudan I and IV regulations. Some products, such as chili, seasonings, and spice mixtures, frequently contain Sudan I and IV [10].

Calcium carbide, sodium cyclamate, cyanide, and formalin are hazardous chemicals that are frequently employed to ripen verdant tropical fruits, maintain their freshness, and preserve them until they are sold. Vegetables, fruits, popular sweetmeats, soft drinks, beverages, and confectioneries are colored with low-cost textile dyes to attract consumers [11]. Many toxic compounds were added to food, while consumers were unaware of toxic contamination.

Food must be safe and nutritious as a fundamental requirement for human consumers. Different factors must be considered because food dye regulations differ by country. The Food and Drug Administration (FDA) in the United States has approved a list of specific synthetic and natural food dyes that are considered safe for ingestion, provided they are used within specified limits. The following synthetic dyes are frequently approved: FD&C Red No. 40 (Allura Red), FD&C Yellow No. 5 (Tartrazine), FD&C Yellow No. 6 (Sunset Yellow), FD&C Blue No. 1 (Brilliant Blue), FD&C Blue No. 2 (Indigo Carmine), FD&C Green No. 3 (Fast Green), and FD&C Red No. 3 (Erythrosine). The list of ingredients for any food product containing synthetic dyes must specify the specific dye used. A precautionary statement must be included with any pigment that has the potential to cause allergic reactions or other health issues. In the European Union, the use of food pigments and other food additives is subject to specific laws. These laws stipulate the reported categories of food products, maximum usable quantities, chemical characterization, and purity.

Analytical Techniques for Dye Detection

Dyes are widely used as additives in food production to enhance visual appeal. However, their use raises significant concerns due to their environmental impact and potential health risks, including cancer, attention-deficit disorder, and hyperactivity in children. Synthetic dyes are often implicated in food safety violations due to cross-contamination or adulteration. Although regulations limit their allowable concentrations, the illicit use of these dyes persists, underscoring the urgent need for sensitive and reliable detection methods to safeguard public health. Current analytical methods for dye detection include spectrophotometry, fluorescence-based sensors, high-performance liquid chromatography (HPLC), and mass spectrometry (MS). While these techniques offer high sensitivity, they are often expensive, time-consuming, and require specialized expertise. Spectrophotometry is particularly effective for detecting high concentrations of dyes due to their strong UV and visible light absorption properties. However, detecting trace amounts necessitates advanced methods that are less accessible because of their complexity and cost. This highlights the need to develop innovative, cost-effective, and efficient detection techniques. Emerging methods, such as electrochemical and colorimetry, show great potential by simplifying detection processes without compromising sensitivity. Advancing such techniques is vital to identify harmful dyes, prevent fraudulent practices, and protect consumer health [12–15].

Sample Preparation for Dye Analysis in Foods

Adequate sample preparation is required to remove interfering substances and enrich analytes, ensuring the concentration meets the method's detection limit. Appropriate instrumental technology, robust pre-treatment protocols, interference clean-up, and analyte enrichment are critical for overcoming matrix effects that can compromise quantitative analysis and method reproducibility, particularly at trace levels. Extraction is commonly used to determine dye content in foods, followed by techniques such as spectrophotometry, chromatography, or electrophoresis (Fig. 2). However, these methods face significant challenges, such as lengthy sample preparation, low extraction efficiency, and reliance on hazardous organic solvents. Furthermore, the diverse solubility characteristics of dyes, which range from water-soluble to water-insoluble, necessitate the careful selection or combination of extractants tailored to each dye. This variability complicates the sample preparation process, making it more complicated and inefficient altogether [16–18].

Sample preparation is critical in analytical processes because it affects almost every subsequent step, ensuring accurate analyte identification, confirmation, and quantification. A well-prepared sample improves separation and detection, whereas inadequate preparation can result in longer instrument maintenance times and higher analysis costs. There are two main types of sample preparation methods: liquid partitioning (liquid–liquid extraction (LLE)) and adsorptive techniques (solid-phase extraction (SPE)). Furthermore, miniaturizing the sample preparation process is generally more effective for analytical purposes than scaling it up because it increases efficiency. Liquid-phase microextraction (LPME) and solid-phase microextraction (SPME) are two well-known miniaturized techniques that are commonly used for efficient sample preparation [19].

Significant advancements and innovations in sample preparation techniques have emerged in recent years, with many undergoing extensive refinement to address the challenges of analytes in complex matrices [19]. This section focuses on these developments, with a particular emphasis on LPME and SPME (Table 4). The development of LPME and SPME has progressed in two key directions:

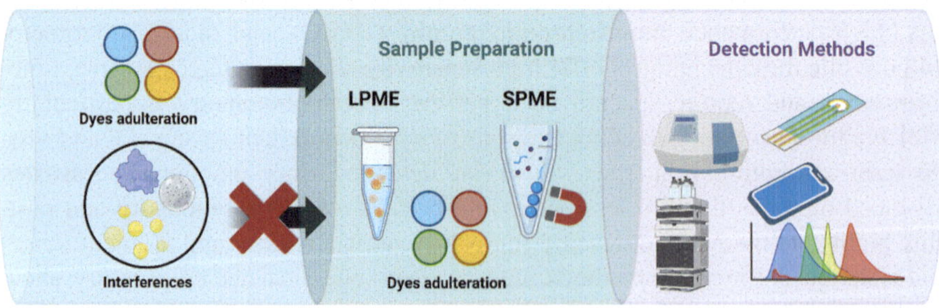

Fig. 2 Illustration of analytical techniques for dye detection. BioRender software was used to create this figure under an academic license

optimization of extractants (solvents and sorbents) and design of extraction device configurations to accommodate efficient miniaturized systems.

LPME

LPME, a scaled-down version of traditional LLE, is renowned for its minimal solvent requirements. However, choosing the right organic solvent for dye extraction from food samples remains challenging, as it must align with its constituents' matrix composition and chemical properties. A thorough assessment of these factors is vital to optimize extraction efficiency and effectiveness. Single solvents are frequently adequate as extractants for synthetic dyes with similar polarity, whereas mixed solvents are required for dyes with a wider polarity range. The solvent used is determined by the compound's properties, such as solubility in water or fat and molecular polarity. With its high solubility and permeability, acetonitrile is widely used as a dispersion or extraction solvent due to its ability to precipitate proteins and carbohydrates, making it ideal for food matrix extractions. Zhang et al. found that acetonitrile alone is less effective for extracting acid dyes with high log P values, strong lipophilicity, and containing sulfonic acid groups because it does not interact well with these compounds. However, methanol can interact with sulfonic acid groups, providing improved selectivity for acidic or electronegative compounds. However, the interaction of methanol with hydroxyl groups in fat-soluble neutral dyes can reduce extraction efficiency. As a result, using a bifunctional solvent mixture, such as methanol-acetonitrile, provides a balanced method for extracting various synthetic dyes from complex food matrices while minimizing matrix interference due to high sugar, fat, and protein content. Additionally, ion-pairing agents enhance the hydrophobic properties of anionic dyes, aiding their transfer from aqueous solutions into the organic extraction phase and improving overall extraction efficiency [16, 20, 31, 32].

In the analysis of dye adulteration in food, innovative solvents like deep eutectic solvents (DES), supramolecular solvents (SUPRAS), and aqueous two-phase systems (ATPSs) have emerged as sustainable alternatives to traditional organic solvents, offering low toxicity, controlled polarity, and environmental benefits. DES are formed by components that establish stable hydrogen bonds, significantly reducing their melting point and resulting in a liquid solvent at room temperature. These solvents dissolve polar dyes through electrostatic interactions, hydrogen bonding, and acid–base reactions. In contrast, non-polar dyes are extracted via interactions with hydrophobic components like hydrocarbon chains or aromatic structures. Muradymov et al. developed quasi-hydrophobic DES composed of both polar and non-polar components, achieving efficient simultaneous extraction of polar dyes (e.g., Sunset Yellow FCF, Azorubine, and Ponceau 4R) and non-polar dyes (e.g., Sudan I–III) from food products. Additionally, Ahmed et al. developed a supramolecular solvent-based liquid–liquid phase microextraction (SUPRAS-LLME) method for determining Sudan I–IV dyes. This approach is characterized by rapid extraction, minimal solvent usage, excellent selectivity, simplicity, cost-effectiveness, and strong adherence to green chemistry principles. The SUPRAS solvents are easy to prepare and effectively recover dyes with low detection limits, further emphasizing their environmental advantages. These advancements showcase the potential of innovative solvent systems to enhance dye analysis in complex food matrices while promoting sustainability. Moreover, aqueous two-phase systems (ATPSs) are made up of cationic and anionic surfactant mixtures that are effective for the extraction of synthetic food dye. These systems prioritize environmental sustainability by using biodegradable, renewable, and non-toxic ingredients. Smirnova and Apyari created a benzethonium-based ATPS microextraction method combined with colorimetry to efficiently analyze dyes, such as Allura Red AC, Azorubine, Sunset Yellow, Tartrazine, and Fast Green FCF. This approach is known for its simplicity, affordability, speed, and environmental friendliness. Dye extraction occurred concurrently with ATPS formation by mixing benzethonium chloride in equimolar proportions with sodium N-lauroylsarcosinate or sodium dihexyl sulfosuccinate, allowing for efficient sample preparation [18, 21, 22, 33].

LPME development focuses on solvent generation and system design improvements to enhance extraction efficiency and simplicity. One such method is dispersive liquid–liquid microextraction (DLLME), which provides excellent pre-concentration for sample preparation. The extractant phase is dispersed using a magnetic stirring syringe [23], ultrasonication [24], vortex [25] system, and a specially designed container with a narrow space. This method is quick, sensitive, and simple, removing the need for centrifugation. However, DLLME faces challenges in collecting the extractant. To address this limitation, solidifying the floating organic drop provides an alternative system setup, making the collection of the final acceptor phase easier and eliminating the need for a specialized chamber [20]. Another type of LPME, gel electromembrane extraction (G-EME), is distinguished by its ability to separate anionic and cationic dyes of the same color from food samples without the use of chemometric or derivative methods. This distinct advantage distinguishes G-EME from techniques such as deep eutectic or switchable solvent extraction coupled with spectrophotometry, which cannot achieve similar separations [14].

SPME

SPME has several advantages, including non-exhaustive extraction, fast processing, improved sample cleaning accuracy, and fewer organic solvents. The technique is based on analyte partitioning between the sample solution and the sorbent material through various interactions. Current SPME research prioritizes the advancement of highly selective and multifunctional sorbents to enhance extraction performance. Multifunctional sorbents are designed by integrating diverse functional moieties into a single material, enabling mixed adsorption mechanisms and significantly improving sensitivity. Nanocomposites exemplify these multifunctional sorbents, offering synergistic benefits by combining their unique properties to provide superior adsorption capacity. Additionally, highly selective sorbents, including molecularly imprinted polymers (MIPs) [26] and restricted-access materials (RAMs) [27], are being developed to target specific analytes among interference molecules in food samples [19, 28].

Inorganic–organic nanocomposites combine the advantages of both material types, effectively overcoming challenges, such as low stability, limited surface area, and adsorption efficiency. Haghdout et al. developed an innovative nanocomposite designed for pre-concentration of trace amounts of Sunset Yellow. This material, derived from maleic acid-based compounds, is highly effective for selective solid-phase extraction (SPE), especially for basic compounds. Moreover, Salamat and Soylak had developed a novel nanocomposite consisting of Ni-MOF nanoparticles integrated onto a multimetallic MnCoFe-LDH nano-flower structure (designated as MnCoFe-LDH/Ni-MOF hierarchically nanostructured), which was used as a coating on a hollow fiber membrane (HFM). This composite, with high sensitivity and effectiveness, was an excellent adsorbent for amaranth dye extraction. Electrostatic forces and π–π stacking facilitated strong interactions between the negatively charged amaranth dye and the positively charged surface of the adsorbent [28, 29, 34].

Traditional extraction methods frequently use hazardous chemicals and lack specificity, whereas metal-organic frameworks (MOFs) offer a more environmentally friendly and precise alternative for sample preparation. De Jesus et al. created a metal-organic framework (MOF), namely [Ni(BDC)]*n*, as a solid adsorbent for detecting crystal violet and Congo red in fish samples. Their adsorption studies revealed the MOF's high affinity for the target analytes. Combined with matrix-solid phase dispersion-UV-Vis spectrophotometry, the technique demonstrated exceptional analytical performance, with great sensitivity and selectivity. Moreover, the Arata-Possetto method is another environmentally friendly approach that uses natural wool as a green sorbent for dye extraction. This method involves boiling defatted wool with potassium bisulfate or hydrochloric acid, followed by alkaline retro-extraction for further investigation. Antela et al. improved this method by extracting azo dyes from natural wool and analyzing the results using a smartphone camera. This advancement allows for the simultaneous preparation of multiple samples, lowering costs and analysis time while promoting environmental sustainability [12, 30].

Dispersive solid-phase extraction (d-SPE) is a popular variant of solid-phase extraction (SPE) that is particularly useful for dye extraction from food due to its efficiency and simplicity. This technique allows for quick processing and effective interference removal, making it ideal for complex food samples. By dispersing a solid sorbent into the sample solution, d-SPE ensures efficient interaction with the target analytes, providing a quick and dependable method for sample preparation, especially for extracting synthetic dyes from food. Furthermore, when combined with magnetic separation and affinity techniques, SPME is a powerful tool for extracting target compounds. In this procedure, affinity reagents are immobilized on magnetic particles, allowing analytes to be separated from the solution by applying a magnetic field. The method is more cost-effective and space-efficient when compact, high-strength magnets are used. Furthermore, automated configuration innovations have improved SPME's flexibility and performance, making it a versatile and efficient tool for various analytical applications [13, 19].

Zarghampour et al. developed a microfluidic chip that combines micro solid-phase extraction and an RGB detection system to detect trace amounts of acidic–basic dyes like erythrosine and crystal violet. The procedure starts with on-chip electromembrane extraction (CEME) to separate analytes from an aqueous solution. These analytes are then transferred to a micro-packed column containing strong cation and anion exchange sorbents and adsorbed there. The resulting colors are quickly measured using RGB colorimetry on a smartphone, allowing for precise dye analyses. This method can be used as a model for dye detection in food matrices [35].

Spectroscopic Methods

Spectrophotometry uses specific wavelengths of light to measure the quantity of a class of compounds (which have particular absorption and transmission wavelength ranges). Ultraviolet-visible (UV-Vis) spectroscopy is a common analytical tool to detect chromophores in compounds by monitoring absorbance spectral wavelengths [36]. The amount of light transmitted through a sample is used to quantify the amount of a compound available to absorb light, which refers to fingerprints for specific dyes. It is crucial to prepare the sample in a clear solution that is free of particles, as the particles will disperse the incident light and produce inaccurate results. As with single-beam instruments, a reference solution is always necessary, although it may not be utilized after the initial background scan has been recorded. Reference cells are necessary during the scanning process for double-beam instruments. UV-Vis spectrophotometric approaches are straightforward and user-friendly, and their low costs and quick results make them popular for colorant testing. However, there are limitations in specificity, meaning the specific compounds responsible for absorbance are unknown, and synthetic and natural dyes can rarely be distinguished. Food analysis has also been incorporated into the application of UV-Vis. This method is advantageous for analyzing food dyes due to the increased sensitivity of highly pigmented compounds to light (Table 1). For instance, the detection of adulterants such as rhodamine B and commercial red textile dye in chili commercial powder at 600 nm was facilitated by UV-Vis spectroscopy and principal component analysis (PCA) modeling [44].

Infrared spectroscopy (IR) has been applied to make numerous determinations from various food matrices and analytes [43]. The wavelengths of infrared (IR) photon spectra, which are optical, electromagnetic radiation components, range from 700 nm to 30 μm [45]. This spectral range is located between microwave radiation and visible light, and the photon energy of these electromagnetic radiations can be mathematically expressed using the well-known Planck-Einstein eq. IR photons are employed in infrared spectroscopy to investigate the spectral absorption

Table 1 Spectroscopic studies for the determination of food dyes

Analyte	Method	Linear range	LOD	Sample	Recovery (%)	References
Sudan I, II	UV–visible spectroscopy and chemometric	0–12 mg g^{-1}	1.5 mg g^{-1} • For amaranth and Sudan I 0.0 mg g^{-1} • For amaranth and Sudan II	Paprika	–	[37]
Para red dye	UV–visible spectroscopy and TLC spectroscopy	5–35 mg L^{-1}	[a]	Chili powder, tomato sauce	[a]	[38]
Sudan I	Fluorescence spectroscopy combined with second-order calibration	0.4–3 mg g^{-1} and 0.5–3 mg g^{-1}	0.4–3 mg g^{-1} and 0.5–3 mg g^{-1}	Paprika powder	–	[39]
Sudan I–IV	Nanosensor with fluorescence spectroscopy	0.05–40, 0.01–20, 0.01–40, 0.05–40 μg mL^{-1} • For or Sudan I–IV, respectively	15.1, 2.83, 3.52, and 16.7 ng mL^{-1} • For or Sudan I–IV, respectively	Chili powder		[40]
Sunset yellow	FTIR spectroscopy With machine learning tools			Tea powder	–	[41]
Sudan dye	NIRS		0.25%	Chili powder	–	[42]
Sudan dye	FTIR spectroscopy		0.016%	Paprika powder	89.5–106.2	[43]

[a] Estimation method

behavior of the molecules under investigation. Absorption is one potential mode of light-matter interaction that may occur when infrared radiation photons interact with molecules. The selectivity is derived from the specific spectral absorption behaviors of the various molecules, which can be used as a determining signature for molecule recognition (molecule's spectral fingerprint). This technique can be used to detect dye adulteration in food products. Fourier transform infrared spectroscopy (FT-IR) spectroscopy is used for scans based on a sample's molecular composition [46]. FTIR is a chemical profile-generating analytical technique that monitors molecules' fundamental vibrational and rotational stretching modes in the mid-IR region (4000–400 cm^{-1}) of the electromagnetic spectrum [47]. FTIR spectroscopy offers a more comprehensive understanding of the scanned sample than NIR spectroscopy, as it measures the fundamental vibrations rather than the harmonics and combination bands that are measured in the near infrared (NIR) region [48]. In contrast to high-performance liquid chromatography (HPLC) procedures, infrared (IR) spectroscopy is a non-destructive technique. HPLC methods are time-consuming, destructive, and cannot be considered high throughput, despite their high sensitivity and low LODs (ng kg^{-1}). Recently, a variety of vibrational spectroscopic techniques have demonstrated the potential to be sensitive and rapid methods for the authentication and quality analysis of a wide range of agro-food samples when used in conjunction with chemometrics (Table 1). These methods are advantageous in that they are non-destructive and relatively inexpensive. For sample preparation, infrared spectroscopy (IR) is a method that can detect solid samples, so it is easy to prepare real samples such as chili powder, tea powders, etc. IR spectroscopy is in contrast with UV-VIS spectroscopy. The common sample for IR is solid. A solid sample will be ground with mineral oil, some solvents, or KBr before being compressed in the pellet.

Fluorescence spectroscopy is highly sensitive and specific, allowing for detecting very low concentrations of fluorescent molecules and providing detailed information about the molecular environment and interactions. Fluorescence spectroscopy is a straightforward and rapid approach for evaluating food quality, requiring no sample preprocessing. Food samples contain fluorescent compounds, or fluorophores, that are exceedingly susceptible to their surroundings. Food samples contain fluorescent compounds, or fluorophores, that are exceedingly susceptible to their surroundings. Fluorescence is a spectroscopic method that is highly sensitive to its environment, as fluorescent compounds—fluorophores—are more environmentally sensitive than absorption techniques. The principle of fluorescence spectroscopy relies on molecules' excitation and subsequent emission of light. Molecules are returned to the ground state by emitting photons with an energy that differs from that absorbed when they are promoted to an electronically excited state through UV or visible (vis) radiation absorption. Fluorescence and phosphorescence are the two forms of photoluminescence [49]. Each fluorophore can be identified by its emission and excitation spectrums. A complex dietary sample's steady state fluorescence emission spectrum may consist of two or more individual components that correspond to a variety of fluorophores. Determining the number and emission profiles of components in an integral spectrum is necessary for gaining a specific understanding of their structure and is complementary to other analytical techniques. Measuring a series of emission spectra at various excitation wavelengths within a wavelength range can accomplish this [50]. The adulterated samples exhibited distinct emission profiles in their fluorescence spectra. Vitamins, amino acids, and dye adulterants are among the products in the market that are inherently fluorescent (Table 1).

Chromatographic Methods

Chromatography is a separation technique that separates the analyses from the matrix by utilizing the partition coefficient of the molecules in the column during extraction. The results will be displayed in a chromatogram, with each molecule having a specific retention time (the time that was detected). Traditional thin-layer chromatography (TLC) is a simple, versatile, and cost-effective analytical technique used for separating and identifying compounds in a mixture. Thin-layer chromatography is an improvement on paper chromatography because the more uniform particles used generally make the method more reproducible. Samples are prepared and plated on a thin silica gel plate and various solvents are used to separate compounds. TLC separates compounds based on their weight, size, and affinity to a solvent [51]. TLC can be used as a rapid screening testing for quick assessment of whether a sample has been adulterated. It offers a fast and straightforward approach, which is especially useful in field testing and preliminary quality control processes. TLC can also be combined with other techniques to improve its performance, such as determining various water-soluble synthetic dyes [52].

High-performance liquid chromatography (HPLC) is a sophisticated liquid chromatography technique used to quantify, identify, and separate components in a mixture. HPLC employs the same principles as TLC, which involves the separation of compounds according to their weight and affinity for a solvent. The differential distribution of a stationary phase (a solid or liquid-coated particle) and a mobile phase (a liquid solvent) within a column is the basic concept of HPLC [31]. The reverse phase is the most frequently employed mode in liquid chromatography, in which the stationary phase is relatively non-polar, and the mobile phase is polar. After separation, the sample fragments are directed to a detector, such as a diode array detector, which measures the amount of light absorbed during the elution of each chemical species from the chromatographic column or UV light to detect the presence of compounds that absorb specific wavelengths. In various industries, HPLC is a frequently employed analytical instrument for detecting and analyzing dye adulteration. HPLC-UV is a simple method for detecting adulteration and can detect low levels of synthetic compounds. HPLC is implemented to detect and quantify synthetic pigments or adulterants not authorized for inclusion in food products. This can guarantee adherence to food safety regulations and protect public health by identifying potentially hazardous additives (Table 2). HPLC-UV can quantify the concentration of anticipated pigments and detect their presence. This can be employed as a tool to standardize the concentration of natural colorants, such as anthocyanins, where the color intensity of a fruit or vegetable concentrate may vary naturally.

Gas chromatography (GC) is an analytical method employed to quantify, identify, and separate volatile substances in a sample. The fundamental principle of gas chromatography is the differential partitioning of compounds between a stationary phase and a mobile phase, which is a carrier gas, such as helium, nitrogen, or hydrogen. The gas chromatograph, a ubiquitous instrument in the majority of analytical laboratories worldwide, can be effectively employed in the authentication and fraud detection processes of a wide range of food and beverage products, including olive oil and other edible vegetable oils, honey and other bee products, milk and dairy products, cereals and bakery products, meat, fish, and seafood, as well as coffee and tea [57]. GC is frequently employed to analyze volatile compounds; however, it can be especially advantageous for identifying volatile adulterants or degradation products linked to specific pigments. GC can detect solvents or other compounds that are indicators of dye adulteration even though dyes themselves may be non-volatile. Complex matrices with a variety of components are frequently present in food products. GC can assist in the overall profiling of food products by separating volatile components that may indicate adulteration or contamination.

Table 2 Chromatographic studies for the determination of food dyes

Analyte	Method	Linear range	LOD	Sample	Recovery (%)	References
Amaranth Brilliant blue Indigo carmine Ponceau 4R Sunset yellow Tartrazine	HPLC-UV-vis • Sample preparation with solid-phase extraction	–	0.15 mg L^{-1}	Lemon flavor sport drink	66–115	[53]
Amaranth Ponceu 4R Allura red Azorubine Erythrosine	HPLC • Sample preparation with DLLME	0.5–500 μg L^{-1}	0.05 μg L^{-1} • For amaranth 0.08 μg L^{-1} • For ponceu 4R 0.03 μg L^{-1} • For allura red 0.02 μg L^{-1} • For azorubine 0.01 μg L^{-1} • For erythrosine	Cherry jelly, carbonated beverage, chocolate dragee	95.8–98.7 • For amaranth 96.6–98.6 • For ponceu 4R 98.0–100.8 • For allura red 94.2–95.4 • For azorubine 95.4–96.8 • For erythrosine	[54]
Butter yellow, Sudan orange G, Para red, Sudan I, Sudan II, Sudan III, Sudan IV, Sudan red 7B, scarlet 808	HPLC–DAD	0.5–25 mg kg^{-1}	0.01–0.04 mg kg^{-1}	Turmeric	96.0–102.6	[55]
Magenta III Rhodamine B	TLC and MS	[a]	[a]	Counterfeit saffron	[a]	[56]

[a] Classification method

Occasionally, the analytes in actual samples are present in low concentrations. Additionally, the signal is disrupted or the column is destroyed in chromatography due to the presence of numerous matrices in real samples. The extraction can be employed to prepare the sample prior to injecting the solution into the instrument. The extraction process encompasses a variety of techniques, including simple filtration and centrifugation of liquids, solid-phase extraction, LLE, and microwave-assisted extraction. Based on the characteristics of the analytes and sample, the user has a choice to select.

Mass Spectrometry (MS)

Mass spectrometry (MS) is an analytical method that is employed to quantify the mass-to-charge ratio of ions. It gives both qualitative and quantitative data regarding the molecular composition of a sample, such as the relative abundances, structural characteristics, and mass of compounds. MS is capable of detecting and identifying unauthorized synthetic pigments in food products. MS can verify the presence of prohibited or restricted dyes by comparing the mass spectra of sample components to known standards. MS is capable of overcoming the constraints that have been observed in other techniques by providing high sensitivity, selectivity, and structural information for the identification, confirmation, and characterization of analytes in complex matrices at low concentration levels [31]. The necessity of sample preparation, which is necessitated by the ionization sources, is indicative of the limitations of MS. This necessitates the samples to be in either a gaseous or liquid state. Furthermore, mass spectrometric equipment necessitates the operation of highly experienced individuals and is expensive. LC-MS and GC-MS for dye detection.

High-performance liquid chromatography and gas chromatography are mostly combined with the most powerful detector mass spectrometer to achieve trace analysis. MS analysis is generally performed by a mass analyzer attached to a separation instrument, such as HPLC or ultra-HPLC (UHPLC or GC for more volatile compounds). Following chromatography separation, compounds are ionized, and the mass of ionized form is measured and further fragmented into additional ions [58]. This provides structural information to identify individual compounds, improving the resolving power and identification potential. The ability to identify individual compounds allows researchers to not only determine if there is food dye adulteration but also notify companies of the exact dye in their products. LC-MS and GC-MS are the most frequently employed MS techniques in the analysis of food, including spices and plants (Table 3).

Electrochemical Methods

Electrochemical techniques are highly promising for dye adulteration analysis in food due to their simplicity, rapid response, high sensitivity, low cost, broad detection range, repeatability, stability, and resistance to interference. Electrochemical

Table 3 Mass spectrometry studies for the determination of food dyes

Analyte	Method	Linear range	LOD	Sample	Recovery (%)	References
Water soluble dye; tartrazine, amaranth, ponceau 4R, sunset yellow FCF, and fat-soluble dye; Sudan (I–IV)	HPLC/DAD/ESI-MS	0.40–40 μg mL^{-1} • For water soluble dye 0.10–10 μg mL^{-1} • For fat soluble dye	2.0–3.5 ng • For water soluble dye 5.4–10.5 ng • For water soluble dye	Soft drink, ginger, chili powders, chili spices	97.9–102.8 • For tartrazine 94.2–104.1 • For amaranth 93.7–102.6 • For ponceau 4R 94.5–104.5 • For sunset yellow FCF 93.9–107.0 • For Sudan (I–IV)	[4]
Sudan I	HPLC with LC/MS	24–5591 μg kg^{-1}	7 μg kg^{-1}	Chili powder and chili-containing food products	–	[59]
Sudan I–IV, Sudan Orange, Sudan red 7B, Para red, rhodamine B	UHPLC–MS/MS	15–10,000 μg L^{-1}	1–48 μg kg^{-1}	Turmeric, curry, chili products	–	[60]
Azod dyes	GC-MS	0.6–2.4 μg mL^{-1}	10.6–84.4 ng mL^{-1}	Paprika	71.2 ± 3.5–118.9 ± 2.5	[61]

nanosensors, which integrate the benefits of electrochemistry and nanotechnology, have proven effective for dye adulteration determinations in food and beverage samples. High-performance electrochemical sensors demand materials with excellent catalytic activity, high conductivity, and large specific surface areas. While metal nanoparticles excel in catalytic properties, their high cost and sustainability issues have spurred interest in non-metallic alternatives like carboxylated multi-walled carbon nanotubes (COOH-MWCNTs). These nanotubes offer superior conductivity and catalytic properties but tend to aggregate due to Van der Waals forces, a limitation mitigated by effective carriers. Deng et al. addressed this by developing an MWCNTs-CONH-66-UiO/PEDOT:PSS composite for detecting Sunset Yellow in food, utilizing-NHCO-linkages formed through amidation between UiO-66-NH2 and COOH-MWCNTs to enhance sensitivity and detection limits. Similarly, modified carbon paste electrodes (CPEs), prepared from graphite and pasting liquids, provide broad potential windows, low background currents, and modifiable surfaces, as exemplified by Mohammadi et al.'s CPE incorporating nickel-iron layered double hydroxide (Ni-Fe LDH) and ionic liquid (IL), which achieved high sensitivity for Sudan I detection with lower oxidation potentials. Advances in nanotechnology have further refined sensors with materials like layered double hydroxides (LDHs) and ionic liquids, offering properties such as high porosity and stability. In the realm of sustainability, De Faria et al. developed a 3D-printed electrode using a composite of graphite (G) and carbon nitride (C_3N_4) embedded in polylactic acid (PLA), combining mechanical stability with eco-friendly materials to minimize sample preparation and support green chemistry. These developments demonstrate the convergence of sensitivity, efficiency, and environmental responsibility in current electrochemical sensor design [62–66].

Multifunctional hydrogels and MIPs are another way to improve the sensitivity of electrochemical sensors. Chen et al. created a multifunctional hydrogel, especially a dual-network magnetically conductive hydrogel. N-methylene bisacrylamide was used to cross-link cationic guar gum (CGG) and β-cyclodextrin (β-CD), forming the major three-dimensional network in MCHG. Hydrogen bonding and electrostatic interactions were used in the secondary network to connect CGG, β-CD, and magnetite@carboxylate-terminated carbon nanotubes. When added to a glassy carbon electrode, this MCHG performed well as an electrochemical sensor for detecting Ponceau 4 R. The MCHG's porous structure, high Ponceau 4 R adsorption capacity, and superior conductivity make it an ideal signal amplifier for sensitive electrochemical detection of this dye. George et al. also created a dual-template molecularly imprinted polymer-based electrochemical sensor that can analyze two food additive dyes simultaneously: dazzling blue FCF and tartrazine. The monomer for the MIP layer was 3-aminophenyl boronic acid (3-APBA), which was electropolymerized on a carbon fiber paper electrode. This approach, which combined electrochemical analysis with molecular imprinting technology, enabled the simple, selective, and cost-effective detection of vivid blue FCF and tartrazine in actual food samples [67, 68].

Moreover, screen-printed graphite electrodes (SPGE) have emerged as a suitable technology for electrochemical detection due to their simplicity and effectiveness.

Electrosensing with SPGEs may be done with just one drop of sample, avoiding the requirement for lengthy pre-treatment. Furthermore, a simple cleaning method allows for reusing a normally disposable sensor, increasing its cost-effectiveness. Muriqi et al. proved this potential by establishing an inexpensive electrochemical approach for rapidly monitoring carminic acid, a food colorant. This method streamlines the detection procedure by putting a tiny liquid sample straight to the SPGE, with no extra preparation necessary. This screening assay's practical relevance for food safety management was effectively confirmed by analyzing sweet candies, soft drinks, and energy beverages, demonstrating its dependability and adaptability in real-world food safety monitoring. Furthermore, to lower the cost of electrode manufacturing, Alves created a disposable sensor employing a low-cost conductive graphite ink supported on cardboard paper to detect Sunset Yellow and tartrazine using differential pulse voltammetry simultaneously. The proposed method did not require a complex sample treatment step (only simple dilution), and it was used to determine both colorants in isotonic, juice, powdered juice, freeze pop, and gelatin samples, with satisfactory selectivity in comparison to other food ingredients found in the analyzed samples [69, 70].

Emerging Technologies

Emerging analytical chemistry technologies are transforming how we identify and evaluate food adulteration, making procedures more efficient, accessible, and accurate. Smartphone-based colorimetry, chemometric-assisted analysis, and automated analysis have attracted attention for their capacity to shorten operations and increase the precision of food safety evaluations.

Smartphone-Based Colorimetry

Smartphone-based colorimetry is a versatile and widely available tool that allows quick and straightforward image analysis in most work environments. Compared to other analytical techniques, it stands out for its simplicity, affordability, speed, and lower environmental impact due to the use of minimal reagents. Jacinto et al., for example, created the RadesPhone, a device that measures tartrazine in carbonated beverages using MIPs. Their method converts digital images to RGB color values and analyzes the data using partial least squares (PLS) regression, resulting in high precision and minimal prediction errors. This method does not require solvents or specialized laboratory equipment, making it suitable for people with no technical training and consistent with point-of-care testing principles. Smartphone-based colorimetry, while less sensitive and with a narrower detection range than spectrophotometry, has several advantages. These include eliminating the need for cuvettes and other devices and being cost-effective, user-friendly, portable, and capable of analyzing multiple samples at once. Furthermore, its compatibility with Internet-based systems increases its usefulness for real-time monitoring and data sharing [22, 26, 71]. Smartphone-based colorimetry platforms can be designed using versatile formats, such as paper strips [72], innovative droplet-shaped digital image colorimetry applied to wetted paper substrates [73], and wool sorbent [12], combining simplicity with innovative analytical precision (Table 4).

Smartphone-based detection systems provide a viable alternative for on-the-spot analysis, stressing mobility, cost, and usability, particularly in food safety applications requiring artificial food colorants (AFCs). For example, Kotian et al. created a novel "3D Bio-Pen cassette," which includes a cotton strand-based ink cartridge for detecting malachite green, a forbidden chemical used to preserve crops and aquaculture goods. The 3D Bio-Pen detects Okra using two label-free mechanisms: the bleaching activity of citric acid ink and the frictional resistance caused by rubbing the pen against the Okra. Citric acid's triprotic nature allows the conversion of water-insoluble leuco-malachite green into water-soluble malachite green via an acid–base reaction, allowing for quick and easy identification. Similarly, Wang et al. developed a point-of-care test based on filter paper and a smartphone for detecting artificial food colorants in drinks with the naked eye. Filter paper functions as both a sample and responding medium, while the smartphone analyzes the RGB and grayscale values of stained indicator papers and colorimetric cards. This digital colorimetry technique decreases mistakes caused by subjective color assessment, increasing the trustworthiness of the results. Such systems, with their mobility, speed, cost-effectiveness, and user-friendliness, provide a solid foundation for AFC-related food safety monitoring [74, 75].

Chemometric-Assisted Analysis

Dong et al. created a chemometric-assisted HPLC-DAD approach for effectively and concurrently screening and quantifying 12 azo dyes unlawfully added to food goods. This method improved the separation performance of classic HPLC-DAD procedures by using "mathematical separation." The approach accurately identified individual analytes despite spectrum overlaps and complicated backgrounds using second-order calibration methods, sometimes known as the "second-order advantage." This decreased the need for full chromatographic resolution, resulting in easier sample preparation. The approach shows outstanding speed and reliability in detecting azo dyes in a variety of food matrices with a short elution time. Its simplicity, efficiency, environmental sustainability, and global application make it an excellent choice for food safety analysis. In a similar development, de Matos applied the CIELab color parameters to dye screening and chemometric categorization. The approach identified the presence and kind of colors in cassava flour by developing a classification tree. This novel technique provides a practical and effective alternative to chemometric analysis, simplifying the screening procedure and making it more accessible for ordinary applications. These approaches demonstrate the value

Table 4 Sample preparation for the determination of food dyes

Analyte	Method	Linear range	LOD	Sample	Recovery (%)	References
Allura red Tartrazine Fast green	IP-SA-DLLME-SFOD[a]	15.0–1500.0 μg L^{-1} 35.0–2000.0 μg L^{-1} 3.0–1200.0 μg L^{-1}	3.7 μg L^{-1} 9.5 μg L^{-1} 0.83 μg L^{-1}	Candy, soft drinks, and jellies	95.3–98.6 96.0–100 96.6–97.5	[20]
Sudan I Sudan II Sudan III Sudan IV	SUPRAS-LLME[b]	5–100 μg L^{-1} (Sudan I–IV)	0.23 μg L^{-1} 0.51 μg L^{-1} 0.42 μg L^{-1} 0.39 μg L^{-1}	Red sweet, yellow sweet, orange sweet, red jelly sweets, red chili pepper, red sweet pepper, cherry juice powder, turmeric, tomato sweet sauce, tomato hot sauce, tomato paste, isot pepper, curry powder, sumac	82–109	[21]
Sunset yellow FCF Azorubine Ponceau 4R Sudan I Sudan II Sudan III	Automated quasi-hydrophobic deep eutectic based microextraction	1–100 mg L^{-1} kg^{-1} (all dyes)	0.13 mg L^{-1} kg^{-1} 0.12 mg L^{-1} kg^{-1} 0.07 mg L^{-1} kg^{-1} 0.03 mg L^{-1} kg^{-1} 0.05 mg L^{-1} kg^{-1} 0.07 mg L^{-1} kg^{-1}	Paprika powder, candied fruit, churchkhela, cotton candy, confectionery sprinkles	>85	[18]
Allura red AC Azorubine Sunset yellow Tartrazine Fast green FCF	Aqueous two-phase system based microextraction	0.09–2.1 mg L^{-1} 0.02–1.6 mg L^{-1} 0.04–1.5 mg L^{-1} 0.04–0.7 mg L^{-1} 0.03–0.4 mg L^{-1}	0.03 mg L^{-1} 0.005 mg L^{-1} 0.01 mg L^{-1} 0.01 mg L^{-1} 0.008 mg L^{-1}	Beverages	85–112	[22]
Tartrazine	MSIS-DLLME[c]	0.03 μg L^{-1}	0.1–1 μg L^{-1}	Saffron powder, fruit juice, pineapples powder, dessert bananas, powder oranges syrup	98–102	[23]
Sudan I Sudan II Sudan III Sudan IV	UA-SUPRAS-ME[d]	1–100 ng mL^{-1}	0.48–0.63 ng mL^{-1}	Red wine, powder beverage, chili sauce, red pepper	91–102	[24]

Brilliant blue Rhodamine B	VA-DES-SLPME[d]	0.068–4.5 μg mL^{-1} 0.042–3 μg mL^{-1} (dynamic range)	20 μg L^{-1} 13 μg L^{-1}	Rose sherbet, energy drink, aphrodisiac effective energy drink, chili spice sauce, candy, sumac, worm jelly, red hot pepper, smurf marshmallow, chili pepper, and powdered sweet pepper	92–102	[25]
Tartrazine Auramine O	Gel electromembrane extraction	0.05–2.0 μg mL^{-1} (both)	0.0256 μg mL^{-1} 0.0310 μg mL^{-1}	Chewing gum, smarties, and candy	85.0–111 91.7–112	[14]
Tartrazine	MIP-SPE[e]	1.2 mg L^{-1}	0–20 mg L^{-1}	Soda drinks	N.A.[f]	[26]
Methylene blue Basic fuchsin	RAM-SPE[g]	0.2–50 mg L^{-1} 0.5–50 mg L^{-1}	0.1212–0.1424 mg L^{-1} 0.3789–0.4409 mg L^{-1}	Fish and shrimp	88–97	[27]
Amaranth	Nanocomposite hollow fiber membrane-SPE	20–1000 μg L^{-1}	5.6 μg L^{-1}	Pomegranate juice, cherry juice, fruit pastille, smarties	94–99	[28]
Sunset yellow	Inorganic–organic nanocomposite-DSPME[h]	50–4800 ng mL^{-1}	12.45 ng mL^{-1}	Candy and pastilles	96–103	[29]
Crystal violet and Congo red	MOF-MSPD[i]	100–1000 μg L^{-1}	12 μg L^{-1}	Fish	81–103	[30]
Fat-soluble synthetic dyes	Dispersive solid-phase extraction	1.0–10 ng mL^{-1} 10–100 ng mL^{-1}	2.5–25 μg kg^{-1}	Chili powder, tomato sauce, egg, and dried pork	61–119	[13]

[a] Ion pair-based surfactant-assisted liquid–liquid microextraction with solidification of floating organic drops
[b] Supramolecular solvent-based microextraction
[c] Magnetic stirring in syringe DLLME
[d] Vortex-assisted deep eutectic solvent-based supported liquid-phase microextraction
[e] Molecularly imprinted polymer-based solid-phase extraction
[f] Not available
[g] Restricted-access material-based solid-phase extraction
[h] Dispersive solid-phase microextraction
[i] Metal-organic framework-based matrix solid-phase dispersion

of integrating chemometric tools with analytical procedures to enhance dye detection and classification in complicated food samples [76, 77].

Automated Analysis

Automated colorimetric systems with 3D-printed modules and incorporated color sensors provide a low-cost analysis method. These devices provide consistent, reproducible illumination while showing raw data directly on a screen, avoiding the need for picture processing. Compact and efficient, they use fewer reagents, make less waste, and execute tests faster than traditional equipment, making them perfect for in-place applications. Antela et al. created a low-cost system that includes a 3D-printed carousel sampler and a Raspberry Pi 4-controlled module with a TC34725 RGBC sensor capable of assessing tartrazine, Allura Red AC, and Brilliant Blue FCF. Another type of analysis automation is flow analysis systems, which address the issues associated with employing toxic, volatile, and flammable solvents in food analysis. Traditional procedures, which require considerable amounts of these solvents for sample dissolution, analyte separation, and derivatization, produce significant toxic waste during mass analysis. To lessen environmental effect, automation and downsizing of these processes, as well as the use of ecologically friendly solvents, have become a priority. Shishov et al. used an automated hydrophobic deep eutectic solvent (DES)-based microextraction approach to separate and concentrate Sudan dyes (I, II, and III) from chili-based sauces. The low viscosity and high hydrophobicity of DES consisting of terpenes (thymol and menthol) and fatty acids (hexanoic to decanoic acid) allow for facile extraction in a lab-in-syringe system, providing a more sustainable and effective method to food analysis [78, 79].

Applications and Case Studies

In developed countries, there has been an annual increase in analytical methods for food detection. India serves as an illustration of the case study. India is an expansive nation with a diverse array of cultures. People in India consume herb products and seasonings daily. Adding food colorants to food is intended to influence consumers' purchasing decisions (Table 5).

Regarding food flavor, smell, and color, India is believed to be a center of culinary diversity. The market for culinary colorants was expanding at a rapid pace. The rate of food adulteration in India has nearly doubled in the past 5 years. The natural colorant industry's expansion is a direct result of the growing awareness of the severe health consequences of the extensive use of synthetic dyes. Nevertheless, the industry's rapid expansion is hindered by the exorbitant cost of natural colorants compared to their synthetic counterparts. However, despite the discrepancies regarding the health impacts of long-term use, the perception of food vendors about the

Table 5 Electrochemical detection method for the determination of food dyes

Analyte	Method	Linear range	LOD	Sample	Recovery (%)	References
Sunset yellow Tartrazine	ePAD-DPV[a]	4.0–60.0 μmol L^{-1} (both)	2.38 μmol L^{-1} 2.09 μmol L^{-1}	Isotonic, juice, powdered juice, freeze pop, and gelatin	95–108 99–106	[70]
Carminic acid	SPCE-SWV[b]	0.0001–0.5 mg mL^{-1}	0.00004 mg mL^{-1}	Beverage, energy drink, and candy	103	[69]
Amaranth	3D-printed electrode-SWV	0.2–4.2 μmol L^{-1}	0.06 μmol L^{-1}	Jelly, popsicles, isotonic drinks, and food flavoring	87–115	[64]
Sunset yellow	Composite electrode-DPV	0.01–70 μM	3.3 nM	Plain cake and marinating meat	98–106	[62]
Sudan I	Modified electrode based electrochemical sensor	0.03–510 μM	0.01 μM	Ketchup and chili powder	96–104	[63]
Sunset yellow Sudan I	Modified electrode-DPV	0.1–1000 μM 0.1–800 μM	0.1 μM 0.1 μM	Soft drink and chili powder	93–110	[65]
Ponceau 4R	MCHG/GCE[c] electrochemical sensor	0.01–200 μM	1.8 nM	Honey and liqueur	88–107	[67]
Tartrazine Brilliant blue FCF	MIP-DPV[d]	0.02–0.34 μM 0.02–0.34 μM	0.010 μM 0.0083 μM	Nutrient-enhanced sport drink beverages, dry fruit, and frozen green peas	90–106 90–110	[68]

[a] Electrochemical paper based analytical device with differential pulse voltammetry
[b] Screen-printed carbon electrode-square-wave voltammetry
[c] Magnetic conductive hydrogel/glassy carbon electrode
[d] Molecularly imprinted polymer-based differential pulse voltammetry

permitted level of use, and the flow of non-permitted food colors on black markets, many of these prohibited dyes are permitted to be used in India within certain limits. In various regions of India, the widespread use of non-permitted carcinogenic, neurotoxic colorants is a prevalent practice. Metanil yellow, a potent carcinogenic dye that the Government of India has prohibited, was detected in high concentrations in turmeric, ladoo, and besan in the unorganized food sectors of West Bengal. This finding supports the notion that there is a lack of quality control and awareness of food regulations. Candyfloss, sugar toys, beverages, mouth fresheners, ice candy, and confectionary product samples exceeded the limit of colorants, according to an analytical study conducted to investigate the use of synthetic food colorants in various states of India. The detrimental effects of colorants on health have been demonstrated in prior research. Synthetic colorants have a profound impact on young children, resulting in symptoms, such as irritability, sleep disturbances, hyperactivity, impulsivity, and a lack of attention [a clinical condition known as "attention-deficit/hyperactivity disorder" (ADHD)] [80]. Identifying adulterated dyes in a variety of foods is a difficult task. Nevertheless, it is labeled as containing harmless adulterants. The producer may defraud the consumer.

Future Trends and Conclusion

Ensuring food safety through effectively detecting synthetic dyes is a continuous problem that necessitates new and adaptive solutions. While existing procedures are successful in controlled laboratory environments, they sometimes rely on expensive equipment and need specialist knowledge. Future efforts should prioritize the creation of accessible, scalable, and robust analytical procedures to meet the issues provided by complex food matrices, fat-soluble pigments, and uncontrolled synthetic additives. Improving sample preparation and streamlining analytical procedures will also increase technique sensitivity and consistency. By harmonizing international rules and encouraging standard testing techniques, synthetic dyes in food can be monitored more consistently and efficiently. Through sustained innovation and collaboration, the sector can help to make the world's food supply safer and more sustainable.

Due to regulatory discrepancies and a lack of comprehensive detection tools, analyzing synthetic dyes in food items simultaneously is a significant challenge. Current food colorant law varies greatly between nations, resulting in coverage gaps and inconsistencies. To solve these issues, future research must concentrate on developing analytical methods that are efficient, user-friendly, and capable of detecting a large range of dye types across a variety of food matrices. Such developments will allow for improved monitoring and enforcement of food safety regulations. Rapid detection techniques, such as multiplexed colorimetric systems, are one promising area for advancement. These techniques, which employ optimized procedures such as the utilization of sorbent indicator materials, provide a less expensive and more portable alternative to standard instrumentation. While these approaches have shown promise for on-site use, their accuracy, sensitivity, and capacity to analyze complicated samples require further improvement to satisfy practical requirements.

Another rising trend in synthetic dye detection is the development of eco-friendly or "green" analytical procedures. These techniques focus on lowering the usage of hazardous chemicals, reducing waste, and increasing energy efficiency while retaining high sensitivity and precision. Green analytical methods are consistent with worldwide initiatives to implement sustainable practices across all scientific domains, including food safety. Solvent-free extraction, miniaturized systems, and using biodegradable or recyclable materials are gaining popularity in analytical processes. Using green chemistry principles in food dye analysis not only solves environmental issues but also lowers operating costs and improves technique accessibility. For example, integrating green extraction processes with portable detection equipment can result in long-term and efficient method for on-site analysis.

Acknowledgments This work is supported by the Second Century Fund (C2F), Chulalongkorn University, Thailand. Waleed Alahmad gratefully acknowledges the Department of Chemistry, Faculty of Science, Chulalongkorn University, for the provision of research facilities, resources, and technical support.

References

1. Randhawa J, Meehnian H, Gupta P, Verma M (2020) Food colours. CRC Press, Boca Raton, pp 79–101. https://doi.org/10.1201/9780429354557-4
2. Hong E, Lee SY, Jeong JY, Park JM, Kim BH, Kwon K, Chun HS (2017) Modern analytical methods for the detection of food fraud and adulteration by food category. J Sci Food Agric 97(12):3877–3896. https://doi.org/10.1002/jsfa.8364
3. Anagaw YK, Ayenew W, Limenh LW, Geremew DT, Worku MC, Tessema TA, Simegn W, Mitku ML (2024) Food adulteration: causes, risks, and detection techniques-review. SAGE Open Med 12:20503121241250184. https://doi.org/10.1177/20503121241250184
4. Ma M, Luo X, Chen B, Su S, Yao S (2006) Simultaneous determination of water-soluble and fat-soluble synthetic colorants in foodstuff by high-performance liquid chromatography–diode array detection–electrospray mass spectrometry. J Chromatogr A 1103(1):170–176. https://doi.org/10.1016/j.chroma.2005.11.061
5. Zhou W, Wang X, Liu Y, Zhang W, Di X (2023) Novel Cu^{2+}-based immobilized metal affinity magnetic nanoparticles for fast magnetic solid-phase extraction of trace Sudan dyes in food samples. Food Chem 404:134432
6. Rebane R, Leito I, Yurchenko S, Herodes K (2010) A review of analytical techniques for determination of Sudan I–IV dyes in food matrixes. J Chromatogr A 1217(17):2747–2757. https://doi.org/10.1016/j.chroma.2010.02.038
7. WHO (2022) WHO global strategy for food safety 2022–2030: towards stronger food safety systems and global cooperation. World Health Organization, Geneva
8. Attrey DP (2017) Chapter 10 - Detection of food adulterants/contaminants. In: Gupta RK, Dudeja MS (eds) Food safety in the 21st century. Academic Press, San Diego, pp 129–143. https://doi.org/10.1016/B978-0-12-801773-9.00010-8
9. Danezis GP, Tsagkaris AS, Brusic V, Georgiou CA (2016) Food authentication: state of the art and prospects. Curr Opin Food Sci 10:22–31

10. Vera DN, Ruisánchez I, Callao MP (2018) Establishing time stability for multivariate qualitative methods. Case study: Sudan I and IV adulteration in food spices. Food Control 92:341–347. https://doi.org/10.1016/j.foodcont.2018.04.057
11. Al-Mamun M, Chowdhury T, Biswas B, Absar N (2018) Chapter 11 - Food poisoning and intoxication: a global leading concern for human health. In: Grumezescu AM, Holban AM (eds) Food safety and preservation. Academic Press, San Diego, pp 307–352. https://doi.org/10.1016/B978-0-12-814956-0.00011-1
12. Antela KU, Sáez-Hernández R, Morales-Rubio Á, Cervera ML, Luque MJ (2024) Smartphone-based procedure to determine content of single synthetic dyes in food using the Arata-Possetto extraction method. Talanta 270:125537
13. Qi P, Zhou Q, Chen G, Lin Z, Zhao J, Xu H, Gao H, Liu D, Mao X (2023) Simultaneous qualitative and quantitative determination of 104 fat-soluble synthetic dyes in foods using disperse solid-phase extraction and UHPLC-Q-Orbitrap HRMS analysis. Food Chem 427:136665
14. Saeidi F, Shokrollahi A, Zarghampour F, Navaser M (2024) Simultaneous extraction and preconcentration of two dyes with the same color from food samples by dual gel electromembrane and determination by spectrophotometric method. Microchem J 202:110777
15. Sedighi V, Faridbod F (2024) Synthesis of polymer dots as fluorescent nanoprobe for the detection of Ponceau 4R, an additive color abuse in food. Food Chem 434:137402
16. Zhang T, Cheng J, Huang X, Liang Y, Yao X, Chen S, Wei X, Lei H (2025) Development of an external standard method for the high-throughput determination of broadly polar multiclass synthetic dyes in high-sugar, high-fat, and high-protein foods. Food Chem 464:141820
17. Mohammadi Z, Sabzehmeidani MM, Ghaedi M, Dashtian K, Abbasi-Asl H (2024) Dispersive micro-solid phase extraction coupled with spectrophotometric using (MgFe CLDH)/GO magnetically separable sorbent for pre-concentration of anionic food dyes in water samples. Emerg Contamin 10(4):100347
18. Muradymov R, Paul N, Das NK, Banerjee T, Shishov A (2024) Quasi-hydrophobic deep eutectic solvents for simultaneous automated determination of polar and non-polar dyes in food products. Microchem J 206:111510
19. Chen Y, Guo Z, Wang X, Qiu C (2008) Sample preparation. J Chromatogr A 1184(1–2):191–219
20. Elbakri AM, Eltabey RM, El-Defrawy MM, Mortada WI (2025) Rapid ion pair-based surfactant-assisted liquid-liquid microextraction with solidification of floating organic drop for simultaneous spectrophotometric determination of selected anionic dyes in food samples. Food Chem 463:141060
21. Ahmed HEH, Kori AH, Gumus ZP, Soylak M (2024) Supramolecular solvent-based liquid–liquid microextraction (SUPRAS-LLME) of Sudan dyes from food and water samples with HPLC. Microchem J 201:110682
22. Smirnova SV, Apyari VV (2023) Aqueous two-phase systems based on cationic and anionic surfactants mixture for rapid extraction and colorimetric determination of synthetic food dyes. Sensors 23(7):3519
23. Mohammadzadeh A, Amiri M (2024) Magnetic stirring in syringe dispersive liquid-liquid microextraction as an effective method for preconcentration of tartrazine dye from food samples: a multivariate analysis approach. Heliyon 10(5):e27611
24. Dalmaz A, Özak SS (2023) Environmentally-friendly supramolecular solvent microextraction method for rapid identification of Sudan I–IV from food and beverages. Food Chem 414:135713
25. Namlı H, Bişgin AT (2024) Simultaneous deep eutectic solvent based microextraction for monitoring brilliant blue and rhodamine B in foodstuff and industrial samples. J Food Compos Anal 127:105980
26. Jacinto C, Maza Mejía I, Khan S, López R, Sotomayor MD, Picasso G (2023) Using a smartphone-based colorimetric device with molecularly imprinted polymer for the quantification of tartrazine in soda drinks. Biosensors 13(6):639

27. Bo C, Tang X, Jia Z, Li Y, Gong B, Ma X, Dai X (2024) Preparation of copolymer brush-type restricted access microspheres with adjustable adsorption selectivity to variety of dyes. J Appl Polym Sci 141(5):e54884
28. Salamat Q, Soylak M (2024) MnCoFe-LDH/Ni-MOF nanocomposite-coated hollow fiber membrane for solid phase extraction of amaranth dye from water and food samples followed by spectrophotometric analysis. J Mol Liq 396:123987
29. Haghdoust S, Ghaedi M, Arabkhani P, Asfaram A, Ghaderi S (2024) Efficient pre-concentration of trace levels of the synthetic food dye in foodstuff samples through an adsorption extraction technique using a novel inorganic–organic nanocomposite. Microchem J 199:109941
30. de Jesus JR, Ribeiro IS, de Carvalho JP, de Oliveira KLA, Moreira RPL, da Silva RC (2024) Successful synthesis of eco-friendly metal-organic framework ([Ni (BDC)] n) allows efficient extraction of multiresidues pesticides and dyes from fish samples. Microchem J 201:110592
31. Yamjala K, Nainar MS, Ramisetti NR (2016) Methods for the analysis of azo dyes employed in food industry–a review. Food Chem 192:813–824
32. Rutkowska M, Płotka-Wasylka J, Sajid M, Andruch V (2019) Liquid–phase microextraction: a review of reviews. Microchem J 149:103989. https://doi.org/10.1016/j.microc.2019.103989
33. Pereira JFB, Coutinho JAP (2020) Chapter 5 - Aqueous two-phase systems. In: Poole CF (ed) Liquid-phase extraction. Elsevier, Amsterdam, pp 157–182. https://doi.org/10.1016/B978-0-12-816911-7.00005-0
34. Hou Y, Jin M, Sun D, Li J, Bi R, Zhang L, Zhu S (2023) Multifunctional inorganic/organic nanocomposite microspheres-reinventing eggs for bone repair applications. Compos Part B 255:110644
35. Zarghampour F, Yamini Y, Dil EA, Shokrollahi A, Javadian G (2023) A new microfluidic-chip device followed by sensitive image analysis of smart phone for simultaneous determination of dyes with different acidic–basic properties. Talanta 254:124168
36. Farag MA, Sheashea M, Zhao C, Maamoun AA (2022) UV fingerprinting approaches for quality control analyses of food and functional food coupled to chemometrics: a comprehensive analysis of novel trends and applications. Foods 11(18):2867. https://doi.org/10.3390/foods11182867
37. Márquez C, Ruisánchez I, Callao MP (2019) Qualitative and quantitative multivariate strategies for determining paprika adulteration with Sudan I and II dyes. Microchem J 145:686–692. https://doi.org/10.1016/j.microc.2018.11.034
38. Mustafa S, Khan R, Sultana I, Nasir N, Tariq M (2013) Estimation of Para red dye in chilli powder and tomato sauces by a simple spectrophotometric method followed by thin layer chromatography. J Appl Sci Environ Manag 17:177–184. https://doi.org/10.4314/jasem.v17i2.2
39. Monago-Maraña O, Eskildsen CE, Muñoz de la Peña A, Galeano-Díaz T, Wold JP (2020) Nondestructive fluorescence spectroscopy combined with second-order calibration as a new strategy for the analysis of the illegal Sudan I dye in paprika powder. Microchem J 154:104539. https://doi.org/10.1016/j.microc.2019.104539
40. Fang A, Long Q, Wu Q, Li H, Zhang Y, Yao S (2016) Upconversion nanosensor for sensitive fluorescence detection of Sudan I–IV based on inner filter effect. Talanta 148:129–134. https://doi.org/10.1016/j.talanta.2015.10.048
41. Amsaraj R, Mutturi S (2023) Rapid detection of sunset yellow adulteration in tea powder with variable selection coupled to machine learning tools using spectral data. J Food Sci Technol 60(5):1530–1540. https://doi.org/10.1007/s13197-023-05694-3
42. Haughey SA, Galvin-King P, Ho Y-C, Bell SEJ, Elliott CT (2015) The feasibility of using near infrared and Raman spectroscopic techniques to detect fraudulent adulteration of chili powders with Sudan dye. Food Control 48:75–83. https://doi.org/10.1016/j.foodcont.2014.03.047
43. Lohumi S, Joshi R, Kandpal LM, Lee H, Kim MS, Cho H, Mo C, Seo Y-W, Rahman A, Cho B-K (2017) Quantitative analysis of Sudan dye adulteration in paprika powder using FTIR spectroscopy. Food Add Contamin Part A 34(5):678–686. https://doi.org/10.1080/19440049.2017.1290828

44. Rohaeti E, Muzayanah K, Septaningsih DA, Rafi M (2019) Fast analytical method for authentication of chili powder from synthetic dyes using UV-vis spectroscopy in combination with chemometrics. Indonesian J Chem 19(3):668–674
45. Aulia MTN, Hery S (2022) Infrared spectroscopy for detecting adulterants in food and traditional indonesian herbal medicine. In: Marwa E-A, Khalid A-S, Ahmed SE-S (eds) Infrared spectroscopy. IntechOpen, Rijeka. https://doi.org/10.5772/intechopen.106803
46. Bhunia AK (2008) Chapter 1 - Biosensors and bio-based methods for the separation and detection of foodborne pathogens. In: Taylor SL (ed) Advances in food and nutrition research. Academic Press, San Diego, pp 1–44. https://doi.org/10.1016/S1043-4526(07)00001-0
47. Sun D-W (2009) Infrared spectroscopy for food quality analysis and control. Academic Press, San Diego
48. Lohumi S, Lee S, Lee H, Cho B-K (2015) A review of vibrational spectroscopic techniques for the detection of food authenticity and adulteration. Trends Food Sci Technol 46(1):85–98. https://doi.org/10.1016/j.tifs.2015.08.003
49. Sikorska E, Khmelinskii I, Sikorski M (2019) Fluorescence spectroscopy and imaging instruments for food quality evaluation. In: Zhong J, Wang X (eds) Evaluation technologies for food quality. Woodhead Publishing, Sawston, pp 491–533. https://doi.org/10.1016/B978-0-12-814217-2.00019-6
50. Radotić K, Stanković M, Bartolić D, Natić M (2023) Intrinsic fluorescence markers for food characteristics, shelf life, and safety estimation: advanced analytical approach. Foods 12(16):3023. https://doi.org/10.3390/foods12163023
51. Gold IL, Imoisi OB, Akpose OI (2023) Thin layer chromatographic techniques for detection of Sudan dyes in palm oil. J Chem Soc Nigeria 48(2):877. https://doi.org/10.46602/jcsn.v48i2.877
52. Oka H, Ikaia Y, Ohno T, Kawamura N, Hayakawa J, Harada K-I, Suzuki M (1994) Identification of unlawful food dyes by thin-layer chromatography-fast atom bombardment mass spectrometry. J Chromatogr A 674(1):301–307. https://doi.org/10.1016/0021-9673(94)85235-9
53. Floriano L, Ribeiro LC, Saibt N, Bandeira NMG, Prestes OD, Zanella R (2018) Determination of six synthetic dyes in sports drinks by dispersive solid-phase extraction and HPLC-UV-vis. J Braz Chem Soc 29(3):602–608. https://doi.org/10.21577/0103-5053.20170173
54. Faraji M (2019) Determination of some red dyes in food samples using a hydrophobic deep eutectic solvent-based vortex assisted dispersive liquid-liquid microextraction coupled with high performance liquid chromatography. J Chromatogr A 1591:15–23. https://doi.org/10.1016/j.chroma.2019.01.022
55. Yang H-J, Lim S, Lee D-H, Yun C-I, Kim Y-J (2024) Validation and measurement uncertainty of analytical methods for various azo dye adulterants in Curcuma longa L. J Food Sci 89(9):5993–6004. https://doi.org/10.1111/1750-3841.17251
56. Bhooma V, Nagasathiya K, Vairamani M, Parani M (2020) Identification of synthetic dyes magenta III (new fuchsin) and rhodamine B as common adulterants in commercial saffron. Food Chem 309:125793. https://doi.org/10.1016/j.foodchem.2019.125793
57. Kristian P, Marijana A, Djura V (2019) Gas chromatography in food authentication. In: Peter K (ed) Gas chromatography. IntechOpen, Rijeka, p 6. https://doi.org/10.5772/intechopen.88512
58. Osman AG, Raman V, Haider S, Ali Z, Chittiboyina AG, Khan IA (2019) Overview of analytical tools for the identification of adulterants in commonly traded herbs and spices. J AOAC Int 102(2):376–385. https://doi.org/10.5740/jaoacint.18-0389
59. Mazzetti M, Fascioli R, Mazzoncini I, Spinelli G, Morelli I, Bertoli A (2004) Determination of 1-phenylazo-2-naphthol (Sudan I) in chilli powder and in chilli-containing food products by GPC clean-up and HPLC with LC/MS confirmation. Food Add Contamin 21(10):935–941. https://doi.org/10.1080/02652030400007252
60. Arrizabalaga-Larrañaga A, Epigmenio-Chamú S, Santos FJ, Moyano E (2021) Determination of banned dyes in red spices by ultra-high-performance liquid chromatography-atmospheric pressure ionization-tandem mass spectrometry. Anal Chim Acta 1164:338519. https://doi.org/10.1016/j.aca.2021.338519

61. Otero P, Saha SK, Hussein A, Barron J, Murray P (2017) Simultaneous determination of 23 azo dyes in paprika by gas chromatography-mass spectrometry. Food Anal Methods 10(4):876–884. https://doi.org/10.1007/s12161-016-0648-6
62. Deng L, Li J, Yue R, Xu J, Duan X (2024) Fabrication of an electrochemical sensing platform based on MWCNT-CONH-66-UiO and PEDOT: PSS for highly sensitive and rapid detection of sunset yellow. Microchem J 202:110722
63. Mohammadi SZ, Tajik S, Hamidian H, Mousazadeh F, Emambakhsh F, Farzan A (2025) Efficient electrochemical determination of Sudan I in food samples based on modified electrode by using Ni-Fe layered double hydroxide nanosheets and ionic liquid. Inorg Chem Commun 171:113500
64. de Faria LV, Villafuerte LM, do Nascimento SF, de Sá IC, Peixoto DA, Ribeiro RSDA, Nossol E, Lima TDM, Semaan FS, Pacheco WF (2024) 3D-printed electrodes using graphite/carbon nitride/polylactic acid composite material: a greener platform for detection of amaranth dye in food samples. Food Chem 442:138497
65. Sun R, Lv R, Li Y, Du T, Chen L, Zhang Y, Zhang X, Zhang L, Ma H, Sun H (2023) Simple and sensitive electrochemical detection of sunset yellow and Sudan I in food based on AuNPs/Zr-MOF-graphene. Food Control 145:109491
66. Kaya SI, Cetinkaya A, Ozkan SA (2021) Latest advances on the nanomaterials-based electrochemical analysis of azo toxic dyes sunset yellow and Tartrazine in food samples. Food Chem Toxicol 156:112524
67. Chen Y, Sun Y, Pang X, Wang R, Waterhouse GI, Xu Z (2023) Three-dimensional dual-network magnetic conductive hydrogel for the highly sensitive electrochemical detection of ponceau 4R in foods. Biosens Bioelectron 241:115698
68. George A, Bharath M, Ghosh M, Varghese A (2024) Single-monomer dual templated MIP based electrochemical sensor for tartrazine and brilliant blue FCF. Electrochim Acta 475:143682
69. Muriqi S, Patočka J, Červenka L, Sýs M (2024) A simple and universal electrochemical assay for rapid field monitoring of E120 food additive. J Food Compos Anal 136:106709
70. Alves GF, de Faria LV, Lisboa TP, de Souza CC, Fernandes BLM, Matos MAC, Matos RC (2022) A portable and affordable paper electrochemical platform for the simultaneous detection of sunset yellow and tartrazine in food beverages and desserts. Microchem J 181:107799
71. Saadati M (2021) Smartphone-based digital image analysis for determination of some food dyes in commercial products. Food Anal Methods 14(11):2367–2374. https://doi.org/10.1007/s12161-021-02059-4
72. Wang S, Li H, Li T, Xu R, Yang C, Geng T, Yan T, Zhang Q, Li D, Yan J (2025) Paper strips-based narrowing colorimetry: an all-in-one strategy for rapid visual detection and adulteration identification of food dyes. Food Control 168:110885
73. Wardani NI, Alahmad W, Tabani H, Varanusupakul P (2023) Emulsification liquid-liquid microextraction coupled with droplet digital image colorimetric detection for determination of rhodamine B in food and beverage samples. Adv Sample Prep 8:100091
74. Kotian A, Prabhu D, Sundarrajan B, Prabhu A, Mani NK (2024) 3D pen with cotton cartridge for on-site detection of malachite green dye coating on okra. J Food Compos Anal 128:106021
75. Wang S, Wang H, Ding Y, Li W, Gao H, Ding Z, Lin P, Gu J, Ye M, Yan T, Chen H, Ye J (2022) Filter paper- and smartphone-based point-of-care tests for rapid and reliable detection of artificial food colorants. Microchem J 183:108088. https://doi.org/10.1016/j.microc.2022.108088
76. Dong M-Y, Wu H-L, Long W-J, Wang T, Yu R-Q (2021) Simultaneous and rapid screening and determination of twelve azo dyes illegally added into food products by using chemometrics-assisted HPLC-DAD strategy. Microchem J 171:106775
77. de Matos MFR, Bezerra PQM, Correia LCA, Viola DN, de Oliveira Rios A, Druzian JI, Nunes IL (2021) Innovative methodological approach using CIELab and dye screening for chemometric classification and HPLC for the confirmation of dyes in cassava flour: a contribution to product quality control. Food Chem 365:130446. https://doi.org/10.1016/j.foodchem.2021.130446

78. Antela KU, Sáez-Hernández R, Cervera ML, Morales-Rubio Á, Luque MJ (2023) Development of an automated colorimeter controlled by raspberry Pi4. Anal Methods 15(4):512–518
79. Shishov A, Pochivalov A, Dubrovsky I, Bulatov A (2023) Deep eutectic solvents with low viscosity for automation of liquid-phase microextraction based on lab-in-syringe system: separation of Sudan dyes. Talanta 255:124243
80. Varghese R, Ramamoorthy S (2023) Status of food colorants in India: conflicts and prospects. J Consum Prot Food Saf 18(2):107–118. https://doi.org/10.1007/s00003-023-01427-y

Advanced Techniques Against Meat Adulteration

Saeid Jafari, Anat Denyingyhot, Mahdi Ebrahimi, Dharmendra K. Mishra, and Kitipong Assatarakul

Contents

S. Jafari
Faculty of Science, Department of Food Technology, Chulalongkorn University, Bangkok, Thailand

Department of Food Science, Purdue University, West Lafayette, IN, USA

A. Denyingyhot
The Halal Science Center, Chulalongkorn University, Bangkok, Thailand

M. Ebrahimi
Faculty of Veterinary Medicine, Department of Veterinary Preclinical Sciences, University Putra Malaysia, Serdang, Selangor, Malaysia

D. K. Mishra (✉)
Department of Food Science, Purdue University, West Lafayette, IN, USA
e-mail: mishra67@purdue.edu

K. Assatarakul (✉)
Faculty of Science, Department of Food Technology, Chulalongkorn University, Bangkok, Thailand
e-mail: Kitipong.A@chula.ac.th

W. Alahmad, M. Ashfaq (eds.), *Advanced Techniques against Food Adulteration*,
https://doi.org/10.1007/978-3-031-99454-8_2

Abstract As the world's meat consumption grows, it is more important than ever to guarantee the integrity and authenticity of meat products. An overview of cutting-edge methods and recently developed technology used in the identification and avoidance of meat adulteration is given in this book chapter. The sensitivity, specificity, and scalability of conventional meat adulteration detection techniques, such as visual examination and chemical analysis, are all limited. Advanced strategies utilizing state-of-the-art analytical approaches and technology advancements have surfaced to solve these issues. The identification of species in meat products has been transformed by DNA barcoding, which also makes it possible to verify the origin of the meat and identify species substitution. High-throughput sequencing of DNA samples is made possible by next-generation sequencing (NGS) technologies, which also make it easier to identify adulterants at the molecular level and provide comprehensive insights into the genetic makeup of meat products. By analyzing complex mixtures of compounds in meat samples, mass spectrometry techniques—such as liquid chromatography-mass spectrometry (LC-MS) and gas chromatography-mass spectrometry (GC-MS)—make it possible to detect contaminants, additives, and residues used for adulteration. Meat samples can be non-destructively analyzed using nuclear magnetic resonance (NMR) spectroscopy, which can reveal information on the chemical makeup of the meat and identify minute alterations that may be signs of adulteration. Using imaging methods like hyperspectral imaging and microscopy, meat samples can be visually inspected for contamination or adulteration. New technologies provide innovative ways to fight meat adulteration, such as artificial intelligence (AI) devices. Large datasets can be analyzed by AI algorithms to identify possible patterns of adulteration. In summary, cutting-edge ways for preventing meat adulteration encompass a variety of strategies that include interdisciplinary cooperation, technical advancements, and cutting-edge analytical methodologies to protect consumer health and trust in the meat supply chain.

Keywords Detection methods · Food safety · Food fraud prevention · Halal food · Innovative solutions · Meat adulteration

Introduction

Meat adulteration (Fig. 1) has emerged as a critical concern in the global food supply chain, posing significant risks to public health, consumer trust, and food safety standards. The term "meat adulteration" encompasses various practices, including intentionally misrepresenting meat species, non-meat substances, and prohibited additives, all aimed at reducing production costs or deceiving consumers to increase profitability [2]. These unethical practices compromise the nutritional quality of meat products and have profound implications for safety, allergen exposure, religious and cultural dietary restrictions, and legal compliance.

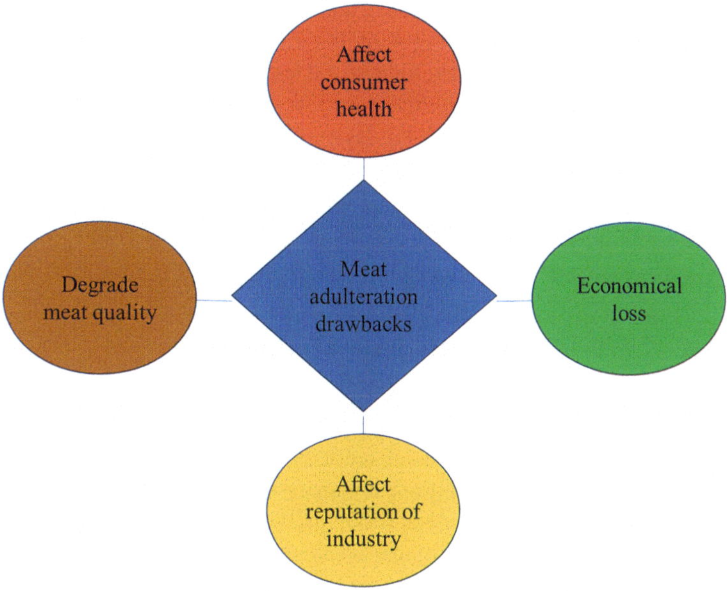

Fig. 1 Various drawbacks of meat adulteration (inspired by Vinothkanna et al. [1])

The complexity of meat adulteration lies in its multifaceted nature and the diversity of fraudulent practices. Common forms include substituting high-value meat species with lower-value alternatives, partial or complete replacement with plant- or animal-derived adulterants, and extending meat products with water, starches, or synthetic compounds [3]. In addition, there is increasing concern over using undeclared or harmful substances, such as dyes and preservatives, to enhance appearance or extend shelf life.

The detection of meat adulteration has become increasingly sophisticated, leveraging advancements in molecular biology, proteomics, and analytical chemistry. Techniques such as DNA barcoding, mass spectrometry, and spectroscopic analysis have been instrumental in identifying adulterants with high precision [4]. Despite these technological advancements, the global nature of food supply chains introduces challenges, including regulatory gaps, lack of standardization, and the need for cross-border cooperation. From a public health perspective, the consequences of meat adulteration can be severe. The consumption of adulterated meat has been associated with outbreaks of foodborne illnesses, exposure to harmful contaminants, and the ingestion of substances that may trigger severe allergic reactions. Furthermore, economic fraud in the meat industry undermines consumer confidence and disrupts market integrity, causing financial losses for legitimate producers and traders [5].

Addressing meat adulteration requires a multidisciplinary approach involving regulatory enforcement, technological innovation, and public awareness. International organizations such as the World Health Organization (WHO), the

Food and Agriculture Organization (FAO), and various national food safety authorities play a pivotal role in establishing guidelines, conducting surveillance, and fostering collaboration among stakeholders. This chapter aims to provide a comprehensive analysis of meat adulteration, covering its historical context, current trends, and advanced detection methods. By synthesizing recent advancements and identifying gaps in existing knowledge, this article seeks to contribute to the development of more effective strategies for combating meat adulteration and ensuring the integrity of the global food supply.

Overview of Traditional Detection Methods

Detecting adulteration is critical to ensure the authenticity, safety, and nutritional integrity of meat products. Traditional detection methods, which have been the cornerstone of food analysis for decades, play an essential role in identifying fraudulent practices such as species substitution, including non-meat ingredients, and using inferior or illegal additives. Though increasingly supplemented or replaced by modern techniques, these methods remain valuable due to their simplicity, cost-effectiveness, and accessibility [4].

Sensory evaluation is one of the oldest and most basic methods for detecting meat adulteration. It involves assessing meat products' appearance, texture, odor, and taste. For instance, a visual inspection can reveal irregularities in color or texture that might indicate the presence of non-meat fillers or inferior meat cuts. However, sensory analysis is subjective and heavily reliant on the evaluator's expertise, making it less reliable for subtle or chemically complex adulterations [3]. Physical methods such as microscopic examination have been employed to detect adulteration in meat products. Microscopy can identify non-meat components like starch granules, plant fibers, or tissue fragments from other animal species. Additionally, meat particle size, connective tissue structure, and fat distribution can be scrutinized to reveal discrepancies indicative of adulteration. While these methods visually confirm adulteration, they are often labor-intensive and require skilled technicians [2].

Chemical methods are widely used to detect specific adulterants or verify meat product composition. Techniques such as determining protein content using the Kjeldahl method, fat content analysis through Soxhlet extraction, and detecting starch or other fillers using iodine tests are common. These methods are relatively straightforward and cost-effective, making them suitable for routine quality control. However, their specificity and sensitivity are limited, and they may not distinguish between adulterants of similar chemical composition.

Enzyme-based assays, such as those for detecting peroxidase activity or specific protein markers, have been employed to identify species-specific adulteration. For instance, identifying hemoglobin or myoglobin from non-target species can confirm the presence of unauthorized meat sources. Biochemical tests are relatively rapid and easy to perform, but they often require careful sample preparation and are prone

Table 1 Comparison of traditional vs. advanced detection methods for meat adulteration

Aspect	Traditional methods	Advanced methods
Sensitivity	Low; struggles to detect low levels of adulteration	High; detects minute amounts of adulterants
Specificity	Limited; cannot distinguish closely related species	High; can identify specific species accurately
Effectiveness on processed products	Ineffective for highly processed meats	Effective even with processed and degraded samples
Speed	Rapid for initial screening	Time-consuming, depending on the technique used
Cost	Generally low	High; requires expensive equipment and reagents
Subjectivity	Subject to human interpretation errors	Objective: Producing quantifiable and reproducible data
Regulatory compliance	Often does not meet modern legal standards	Meets global regulatory standards for food fraud cases
Ease of use	Simple to implement, minimal training required	Requires specialized training and skilled personnel
Applications	Limited to basic adulteration detection	Broad range, including contamination, substitution, and fillers
Equipment requirements	Basic lab equipment	Advanced instruments like PCR machines and sequencers

to interference from food processing conditions [6]. Despite their utility, traditional detection methods have significant limitations (Table 1). They often lack the precision and sensitivity required to detect low levels of adulteration. Many of these methods are labor-intensive and time-consuming, and their reliance on human expertise introduces variability in results. Additionally, the growing complexity of food adulteration practices, including using advanced technologies to mask fraudulent activities, has challenged the effectiveness of traditional approaches. Table 1 highlights the trade-offs between cost and effectiveness in choosing the appropriate method for detecting meat adulteration. As shown, advanced methods offer superior precision and reliability despite higher costs and complexity.

Overview of Advanced Detection Methods

The affordability and straightforwardness of traditional detection methods make them particularly valuable in resource-constrained settings or as preliminary screening tools. However, the limitations of these methods underscore the need for integration with advanced techniques, such as molecular and spectroscopic analyses, to ensure comprehensive and reliable detection. As food fraud becomes increasingly sophisticated, a multidisciplinary approach combining traditional and modern methodologies will be essential to safeguard the authenticity and safety of meat products [7]. Advanced detection methods have emerged as critical tools for ensuring the authenticity and integrity of meat products in the face of increasingly sophisticated adulteration techniques. These methods leverage cutting-edge molecular biology, spectroscopy, and machine learning technologies, offering high sensitivity, specificity, and rapid results.

DNA Barcoding: Principles and Applications

DNA barcoding is a cornerstone technology in the fight against meat adulteration, offering unparalleled accuracy and reliability in species identification. As meat adulteration practices become more sophisticated, DNA barcoding will continue to evolve as a critical tool for ensuring meat products' authenticity, safety, and integrity in global food systems. It involves using short, standardized DNA sequences—known as "barcodes"—to identify and authenticate the species origin of meat products [7]. This method is particularly effective in detecting mislabeled or fraudulent meat products, even in processed or mixed samples where traditional morphological or biochemical techniques often fail. However, there are some key aspects of DNA barcoding in meat adulteration detection, which will be discussed as follows [8]: (1) Species identification: DNA barcoding enables precise identification of the species present in meat products by comparing the DNA sequence of a sample to a reference database. Common barcode regions include cytochrome c oxidase I (COI), which is widely used in animals for species identification, as well as 16S rRNA and 12S rRNA. Effective in distinguishing between closely related species and nuclear DNA markers: Used in some cases for higher resolution in species differentiation. (2) Detection of adulteration in complex samples: Meat adulteration often involves mixing lower-value or unauthorized species with premium meat. DNA barcoding can identify multiple species within a single sample, including those present in trace amounts, making it invaluable for detecting adulteration in processed or ground meat. (3) Verification of label claims: DNA barcoding is used to verify claims about meat products, such as the declared species on product labels. It is beneficial in identifying cases of fraudulent labeling, such as substituting one species for another (e.g., pork for beef or horse meat for lamb). (4) Overcoming challenges in processed meat: High-temperature processing, grinding, and chemical treatments can degrade DNA in meat products, making identification difficult. DNA barcoding can still amplify short and highly conserved DNA regions, ensuring accurate detection even in heavily processed meat.

There are many advantages of DNA Barcoding, such as (1) High specificity and sensitivity: It can detect species-specific DNA sequences even in small quantities. (2) Universality: Using standardized barcode regions allows broad application across various species. (3) Scalability: High-throughput sequencing technologies can simultaneously analyze large numbers of samples and global databases. Access to comprehensive DNA barcode libraries (e.g., BOLD and GenBank) enables robust identification. However, despite its advantages, DNA barcoding faces challenges such as the need for comprehensive reference databases, the possibility of DNA degradation in certain samples, and costs associated with high-throughput sequencing. Advances in molecular biology and bioinformatics, coupled with collaborative efforts to expand barcode libraries, are expected to enhance its effectiveness further. Nehal et al. [8] emphasized DNA barcoding's role in addressing challenges in food supply chains, including adulteration, substitution, and mislabeling of products, such as legumes, seafood, oils, herbal products, and spices (Table 2). By facilitating accurate species identification, DNA barcoding enhances quality control, food safety, and trade supervision while serving as a valuable tool in combating food

Table 2 Application of advanced detection methods (DNA barcoding and next -generation sequencing) for meat adulteration in the published research papers

No.	Title	Type of study	Technology	Key findings	References
1	DNA barcoding: a modern age tool for detection of adulteration in food	Review	DNA barcoding	• DNA barcoding is a biotechnological innovation that utilizes standardized, unique DNA sequences, typically 400–800 base pairs long, from mitochondrial (e.g., COI), plastidial (e.g., rbcL), or nuclear (e.g., ITS) regions to identify and categorize food products	Nehal et al. [8]
2	Application of DNA barcoding for ensuring food safety and quality	Review	DNA barcoding	• Despite some limitations, DNA barcoding is a reliable tool for food traceability, regulatory compliance, and consumer protection	Dawan and Ahn [7]
3	Identification of meat and poultry species in food products using DNA barcoding	Research	DNA barcoding	• Full barcoding proved to be a reliable technique for identifying species in meat and poultry products, except for canned items • Mini-barcoding holds significant promise for species identification in processed products, but it requires an enhanced primer set for optimal application	Hellberg et al. [9]

(continued)

Table 2 (continued)

No.	Title	Type of study	Technology	Key findings	References
4	DNA barcoding as a molecular tool to track down mislabeling and food piracy	Review	DNA barcoding	• DNA barcoding for genetic traceability is an effective and reliable method for identifying and authenticating food products	Barcaccia et al. [10]
5	Occurrence of mislabeling in meat products using DNA-based assay	Research	DNA barcoding	• A high substitution rate was observed among the meat products, revealing a mislabeling rate of 57%, leading to substantial discrepancies with the label information. This raises significant concerns regarding food safety and consumer protection	Di Pinto et al. [11]
6	DNA barcoding and mini-barcoding in authenticating processed animal-derived food: A case study involving the Chinese market	Research	DNA barcoding	• 23% (12 of 52) of animal-derived products were mislabeled and, in most cases, contained undeclared species	Xing et al. [12]
7	Application of DNA barcoding and metabarcoding for species identification in Salmon products	Research	DNA barcoding	• DNA barcoding and metabarcoding of commercial salmon food products identified mislabeling in 16 out of 32 samples (50%)	Wang et al. [13]
8	Authentication of origin of meat species processed under various Indian culinary procedures using DNA barcoding	Research	DNA barcoding	• Indian culinary practices, which involve extensive processing and the use of strong spices in popular meat recipes, do not affect the quality of meat DNA for subsequent species identification through DNA barcoding using the COI gene	Ahmed et al. [14]

(continued)

Table 2 (continued)

No.	Title	Type of study	Technology	Key findings	References
9	Authentication technologies using DNA-based approaches for meats and halal meats determination	Review	DNA barcoding	• DNA barcoding technology shows great potential as a forensic and authenticity tool for verifying halal animal and meat products	El Sheikha et al. [15]
10	DNA barcoding reveals mislabeling of game meat species on the U.S. commercial market	Research	DNA barcoding	• The enforcement of mandatory inspection of game meats and the verification of species labeling in the United States	Quinto et al. [16]
11	Evaluation and application of a next generation sequencing approach for meat species identification	Research	Next-generation sequencing	• A total of 45 ground meat samples obtained from local markets in Europe and Asia were examined, revealing that 18% contained cases of adulteration involving undeclared meat species	Cottenet et al. [17]
12	Application of next generation sequencing for species identification in meat and poultry products: a DNA metabarcoding approach	Research	Next-generation sequencing	• The presence of mislabeling of processed meat and poultry products were confirmed	Xing et al. [18]
13	Preliminary study on species authentication in poultry meat products by next-generation sequencing	Research	Next-generation sequencing	• Undeclared species, such as pig and cattle, were highlighted in 8/13 of the samples • Metabarcoding is a promising tool to identify meat species in mixtures	Mottola et al. [19]
14	A 16S next generation sequencing based molecular and bioinformatics pipeline to identify processed meat products contamination and mislabeling	Research	Next-generation sequencing	• The majority of the meat samples were primarily contaminated with pork • The bioinformatics pipeline proved its specificity by accurately identifying species and quantifying contamination levels in all samples	Chaora et al. [20]

(continued)

Table 2 (continued)

No.	Title	Type of study	Technology	Key findings	References
15	The future of NGS (next generation sequencing) analysis in testing food authenticity	Review	Next-generation sequencing	• NGS has the potential to emerge as the benchmark technology for food authenticity testing, offering significant advantages to consumers, regulators, and industry stakeholders	Haynes et al. [21]
16	Simultaneous identification of four meat species (cattle, chicken, fish, and pig) using next generation sequencing (NGS)	Research	Next-generation sequencing	• Next-generation sequencing (NGS) combined with ribosomal RNA mitochondrial DNA genes can be utilized to analyze mixtures of meat samples • 16S-KH showed better detection effect in all species comparing with 12S-Ki and 16S-Ki	Mahama et al. [22]
17	Application of next generation semiconductor-based sequencing for species identification and meat derived products authentication	Research	Next-generation sequencing	• Six commercial animal products were found to contain other species inconsistent with their product labels	He et al. [23]
18	Detection of adulterated meat products by a next-generation sequencing-based metabarcoding analysis within the framework of the operation OPSON X: a cooperative project of the German National Reference Centre for authentic food (NRZ-Authent) and the competent German food control authorities	Research	Next-generation sequencing	• The 16S rDNA metabarcoding method is comparable to the techniques currently employed by German food control laboratories. As a result, this method shows significant potential as a screening tool for authenticating mammal and poultry species in meat products	Kappel et al. [24]

(continued)

Table 2 (continued)

No.	Title	Type of study	Technology	Key findings	References
19	Using a next-generation sequencing approach to DNA metabarcoding for identification of adulteration and potential sources of mercury in commercial cat and dog foods	Research	Next-generation sequencing	• The 16sUniF_16sUniR primer set produced a relatively higher number of reads across the widest range of taxa, although several other primer sets were also effective in identifying common animal- and plant-based ingredients in commercial pet food products	Dunham-Cheatham et al. [25]
20	Combining a COI mini-barcode with next-generation sequencing for animal origin ingredients identification in processed meat product	Research	Cytochrome oxidase I (*COI*) mini-barcode with next-generation sequencing (NGS)	• The NGS method outperforms the clone sequencing method in terms of accuracy, sensitivity, and detection efficiency	Pan et al. [26]

fraud. The review also advocated collaboration between research laboratories and industries to establish global standardized protocols for food authentication. Dawan and Ahn [7] demonstrated that DNA barcoding is widely applied to detect fraud in meat, seafood, and plant-based products, ensuring accurate labeling and reducing health risks from allergens or contaminants. For meat and poultry, it identified undeclared species, addressing public health and ethical concerns. In seafood, it detected species substitution and prevented illegal fishing practices. It authenticated herbs and medicinal plants for plant-based products, protecting consumers from harmful substitutions. According to them, advances in next-generation sequencing (NGS) and mini-barcodes enhance the method's speed, sensitivity, and cost-efficiency. Challenges such as DNA degradation in processed foods and designing universal primers for all species should be carefully addressed. Hellberg et al. [9] evaluated the effectiveness of DNA barcoding in identifying species in meat and poultry products, comparing the results of full-length barcoding (658 bp) and mini-barcoding (127 bp). Sixty samples were analyzed, including deli meats, ground meats, dried meats, and canned meats. Each sample underwent complete and mini-barcoding of the cytochrome c oxidase subunit I (COI) gene, and the resulting sequences were matched against the Barcode of Life Data System (BOLD) and GenBank for species identification. Mini -barcoding showed a high potential to be used for species identification in processed products. Future studies were proposed to combine mini-barcoding with NGS to facilitate the sequencing-based identification of species mixtures in food products. Barcaccia et al. [10] described potential challenges in using mtDNA and cpDNA to determine species boundaries, including hybridity and polyploidy in the genomes. While DNA barcoding with one or more genes may be suitable for animals, it is less effective for plants due to their distinct life cycle traits, evolutionary backgrounds, and the occurrence of hybridization and polyploidization events. Di Pinto et al. [11] examined processed meat products from Italian markets and supermarkets using the mitochondrial cytochrome b gene qualitative PCR identification system to check for species substitution or mislabeling. Strengthening European legislative guidelines would involve developing and implementing reliable labeling practices and traceability systems to ensure an effective food safety framework. Additionally, ongoing monitoring, enhanced detection methods, and stricter penalties for violators could help reduce authentication issues in the future. Xing et al. [12] applied DNA barcoding and mini-barcoding to test animal-derived food products, including meat, poultry, and fish. They found that the 16S rRNA mini-length barcode was a valuable alternative to the full-length COI barcode for species identification in heavily processed products that typically contain degraded DNA. The combination of mini-length barcode and NGS was proposed to overcome this limitation. Wang et al. [13] explored the use of NG for identifying mixed salmon products. Salmon samples containing up to eight species were amplified using 16S rRNA mini-barcode primers and sequenced on an Illumina HiSeq2500 platform. The study showed that species identification of salmon mixtures using the Illumina HiSeq NGS method with the 16S rRNA mini -barcode was effective. Additionally, combining DNA barcoding and metabarcoding proved helpful in identifying different types of salmon products.

Ahmed et al. [14] evaluated using DNA barcoding to authenticate different meat species processed according to Indian culinary practices. Their findings indicated that DNA barcoding effectively confirmed species authenticity and geographical origin. However, pickled products could not be tracked for species authentication, as the culinary processes compromised DNA quality, making further applications difficult. El Sheikha et al. [15] reviewed all DNA-based techniques developed for the authenticity of meat derived from halal and non-halal animals and their derivatives. Denyingyhot et al. [27] developed an advanced on-site tool for detecting contamination in halal meat from five non-halal animals (monkeys, dogs, rats, pigs, and cats) using a multiplex polymerase chain reaction combined with a DNA strip. This method was used to analyze 115 meat product samples available in Thailand. Their findings revealed that 16 samples were contaminated with pig DNA, while no contamination from other non-halal animals was detected. They concluded that this technique offers significant potential for on-site monitoring of non-halal animal contamination in halal products in Thailand.

DNA barcoding technology, an innovative method, has advanced due to the decreasing cost of DNA sequencing equipment and the widespread adoption of various other DNA-based identification techniques. Quinto et al. [16] found that mislabeling occurred in commercially sold game meat products in the United States, with 18.5% of the analyzed game meats identified as potentially mislabeled. They attributed the mislabeling to factors such as economic motives and product mishandling. Based on the mentioned studies, DNA barcoding has emerged as a transformative technology in combating meat adulteration, offering unparalleled accuracy and reliability in species identification, even in processed and mixed samples. Despite challenges such as DNA degradation, the need for comprehensive reference databases, and cost considerations, advancements in NGS, mini-barcoding, and bioinformatics continue to enhance its efficiency and applicability. Collaborative efforts among researchers, industries, and regulatory bodies are crucial to overcoming these challenges and standardizing protocols globally, ensuring food authenticity, safety, and consumer trust in the fight against food fraud.

Next-Generation Sequencing (NGS) for Meat Authentication

Next-generation sequencing is a cornerstone technology in the fight against meat adulteration, providing unparalleled precision and scalability for authenticating meat products. Its ability to detect adulteration in complex mixtures and processed foods ensures consumer trust, regulatory compliance, and the integrity of global food systems. NGS is a transformative technology that has revolutionized meat authentication by enabling high-throughput and comprehensive DNA analysis in meat products. Its ability to identify multiple species simultaneously, even in complex or processed mixtures, makes it a powerful tool against meat adulteration, which is a significant issue in food safety, public health, and consumer trust. Some key features of NGS in meat authentication are discussed as follows [17]: (1) Comprehensive species identification: NGS facilitates the identification of all species present in a sample, even when multiple species are mixed. NGS can detect species with high sensitivity and specificity, including trace amounts of adulterants, by sequencing the entire DNA content (metagenomics) or specific barcode regions (amplicon sequencing). (2) Detection of complex adulteration: Traditional techniques like PCR are limited to simultaneously detecting one or a few species. In contrast, NGS can simultaneously analyze DNA from all species in a sample, making it ideal for uncovering sophisticated adulteration practices involving multiple species. (3) Authentication of processed meat: Meat adulteration often occurs in processed products like sausages, burgers, or minced meat, where DNA may be fragmented or degraded. NGS excels in amplifying and analyzing short, degraded DNA fragments, ensuring accurate species identification even in highly processed foods. (4) Verification of label claims: NGS is widely used to verify labeling claims, such as species authenticity, organic certification, or geographical origin. It can identify undeclared species or contaminants, ensuring that consumers receive products that align with their expectations and regulatory standards. (5) Building comprehensive DNA databases: By sequencing and cataloguing species-specific DNA, NGS expands reference databases like BOLD (Barcode of Life Data System) and GenBank. These resources improve the accuracy and reliability of meat authentication efforts globally [18].

The main applications of NGS in meat adulteration detection are as follows: (1) Identification of non-declared or unauthorized species: NGS can detect species substitutions, such as the inclusion of pork in beef products, horsemeat in lamb, or non-meat fillers in meat products, ensuring the authenticity of premium meat products. (2) Detection of contaminants and allergens: NGS helps identify potential contaminants, including undeclared animal species, that may pose consumer allergenic or cultural/religious concerns. (3) Tracing geographical origin and supply chain integrity: By analyzing specific genetic markers, NGS can trace the geographical origin of meat, ensuring compliance with claims related to local or region-specific production. (4) High-throughput testing for large-scale monitoring: NGS allows simultaneous analysis of hundreds of samples, making it a cost-effective option for large-scale meat authentication programs in regulatory or industrial settings [19].

The main advantages of NGS in meat authentication are high sensitivity and specificity, which detects even trace amounts of adulterants; unbiased analysis, which provides a complete picture of species composition without prior assumptions; scalability, which is suitable for high-throughput analysis of multiple samples; and flexibility, which analyzes raw, cooked, or processed meat products. However, several challenges need to be addressed, including the high cost and complexity of NGS and the need for advanced bioinformatics tools and expertise to manage and interpret large volumes of sequencing data. Additionally, identification accuracy depends on the completeness and quality of reference DNA databases. Integrating NGS with machine learning and advanced bioinformatics will further improve its effectiveness in meat authentication. Moreover, reducing costs and streamlining workflows will enhance the accessibility of NGS for regulatory authorities and industries worldwide.

Cottenet et al. [17] evaluated the effectiveness of a commercial NGS solution as a non-targeted method for identifying meat species. Their findings demonstrated that closely related species were accurately distinguished, and exotic species were correctly identified. The NGS Food Authenticity Workflow also offered kits specifically designed for fish and plant identification, providing suitable options for detecting potential adulteration in these high-risk food categories. The development of a dedicated kit for insect species identification is anticipated to become a necessity in the near future. Xing et al. [18] investigated the ability of high-throughput DNA metabarcoding to identify mammalian and avian species in mixed products. NGS of a short segment of the 16S ribosomal RNA (16S rRNA) mitochondrial gene was performed. DNA metabarcoding is not yet routinely used for food authenticity verification. However, sequencing costs are decreasing quickly with the rapid advancements and expansion of NGS technology. Mottola et al. [19] utilized DNA metabarcoding to assess the labeling accuracy of commercial multispecies poultry meat products. The molecular analysis verified that all products contained the species listed on their labels, with any discrepancies involving the presence of one or more undeclared meat species. This innovative DNA metabarcoding approach shows promise as an effective method for authenticating animal species in mixed meat products. Chaora et al. [20] developed a diagnostic pipeline combining molecular and bioinformatics methods, leveraging the mitochondrial 16S ribosomal RNA (rRNA) gene to detect mislabeling in processed meat products using NGS. A total of 155 processed meat products, including minced meat, biltong, burger patties, and sausages, were collected from various locations across South Africa. Sausage samples exhibited the highest contamination rate, with 46% containing beef, pork, or mutton mixtures within a single sample. Haynes et al. [21] reviewed the current state of NGS applications in food authenticity and explored potential future technological advancements. They highlighted the capabilities of NGS platforms like Illumina and PacBio, the ability to analyze complex food matrices and simultaneously identify multiple ingredients, and the potential to detect unknown adulterants and screen for genetically modified organisms (GMOs) without prior knowledge. The authors predicted that with continued advancements, NGS could establish itself as the benchmark technology for food authenticity testing, offering significant benefits to consumers, regulators, and industry stakeholders. Mahama et al. [22] employed NGS in conjunction with the rRNA gene to identify four meat species (cattle, chicken, fish, and pig) using three primer sets (12S-Ki, 16S-KH, and 16S-Ki). The primers 12S-Ki and 16S-Ki could not amplify DNA from fish and chicken, likely due to mismatches between the primer sequences and the annealing regions of these species. These findings showed that NGS paired with the rRNA mtDNA gene could be effectively used as a reliable routine testing method. This technique can be applied to monitor and control meat adulteration in halal meat production and the industry.

He et al. [23] used next-generation semiconductor-based sequencing technology to identify species in highly processed and complex meat products. Their findings showed precise identification of all animal components in mixed samples, with a detection threshold of 1%. The bioinformatics pipeline they developed effectively analyzed sequencing data, validating a library of 17 species for screening purposes. Kappel et al. [24] gathered further data on validating the 16S rDNA metabarcoding method using representative samples. This method successfully detected species that were not targeted by the standard analytical methods employed, but, as noted previously, it failed to detect fallow deer. The NGS method was effective in identifying mammal and poultry species, particularly in food samples with complex species compositions. Dunham-Cheatham et al. [25] measured total mercury (Hg) and methylmercury levels in a range of commercial pet food products ($n = 127$) and used genetic tools to identify if specific ingredients contributed to elevated Hg concentrations in the final product. They also employed next-generation amplicon sequencing with ten barcode primers to target distinct taxa and assessed whether one primer set was more effective than the others in amplifying the degraded DNA found in pet food products. Their findings showed that NGS techniques could be used to evaluate the prevalence of adulteration in pet food products. Pan et al. [26] introduced a method that combines a novel cytochrome oxidase I (COI) mini-barcode with NGS to accurately and efficiently identify various animal species (swine, bovine, Caprinae, as well as some fish, shrimp, and poultry) in processed meat products. Their findings suggested that to differentiate between closely related species and minimize the impact of competitive efficiency, it is advisable to use a combination of different DNA markers. In summary, next-generation sequencing

(NGS) is a transformative and versatile technology for meat authentication, offering unparalleled sensitivity, specificity, and scalability in detecting adulteration across complex and processed food matrices. NGS strengthens consumer trust and food safety by enabling comprehensive species identification, verifying labeling claims, and enhancing regulatory compliance. Despite challenges like high costs, data complexity, and reliance on robust reference databases, continued advancements in bioinformatics, machine learning, and sequencing methodologies are poised to make NGS the benchmark tool for ensuring the integrity and authenticity of global meat supply chains.

Mass Spectrometry in the Analysis of Meat Adulterants

Mass spectrometry is critical in detecting meat adulterants by providing high-precision analysis of species-specific proteins and chemical contaminants. Its versatility and reliability make it a cornerstone technology in safeguarding meat authenticity, protecting consumer trust, and ensuring regulatory compliance in the food industry. Mass spectrometry (MS) is a powerful analytical technique widely employed in detecting and identifying meat adulterants. Its ability to precisely analyze the molecular composition of samples makes it an essential tool for ensuring the authenticity and safety of meat products. MS is particularly effective in identifying both species-specific proteins and chemical adulterants in meat, even in highly processed forms [28].

The key role of MS in meat adulteration detection is protein-based species identification, which can detect species-specific proteins or peptides in meat products, distinguishing between different animal species. Techniques such as tandem mass spectrometry are particularly effective in identifying unique peptide biomarkers associated with specific species. MS can identify heat-stable or denatured proteins even when DNA is degraded during processing, ensuring reliable species identification. In addition, MS is highly effective in detecting chemical adulterants, such as preservatives (e.g., formaldehyde), illegal additives (e.g., synthetic dyes), and extenders (e.g., starch or plant-based fillers). Coupled with chromatographic techniques like gas chromatography (GC-MS) or liquid chromatography (LC-MS), MS provides precise quantification and structural characterization of these compounds. In terms of quantitative analysis, MS offers unparalleled sensitivity and specificity for quantifying adulterants at trace levels. This capability is crucial for detecting low concentrations of undeclared or illegal additives that could compromise food safety. For fat and lipid profiling, MS is used to analyze the lipid composition of meat, which can indicate adulteration with non-meat fats or oils. Fatty acid profiles derived through MS indicate meat purity and quality. To detect non-meat additives, MS can identify and quantify plant-based adulterants, such as soy proteins or starches, that are used to dilute or mimic meat. These analyses ensure the authenticity of premium meat products. MS is also used to verify the absence of particular species, such as pork or beef, in products labeled as halal or kosher, helping ensure compliance with religious and cultural dietary requirements [29] (Table 3).

Table 3 Application of advanced detection methods (mass spectrometry and nuclear magnetic resonance) for meat adulteration in the published papers

No.	Title	Type of study	Technology	Key findings	References
1	Detection of meat species adulteration using high-resolution mass spectrometry and a proteogenomics strategy	Research	Mass spectrometry and proteogenomics	• The method effectively identified 1% (w/w) of pork or horse meat in a mixture, both prior to and after cooking to an internal temperature of 71 °C	Ruiz Orduna et al. [28]
2	Liquid chromatography–mass spectrometry bottom-up proteomic methods in animal species analysis of processed meat for food authentication and the detection of adulterations	Review	Liquid chromatography-mass spectrometry (LC-MS)	• The high potential of both LC-HRMS and LC-MS/MS MRM methods in identification of new species-specific heat -stable peptide markers in processed foodstuffs was confirmed	Stachniuk et al. [29]
3	Identification of marker proteins for the adulteration of meat products with soybean proteins by multidimensional liquid chromatography-tandem mass spectrometry	Review	Liquid chromatography-tandem mass spectrometry	• Integrating prefractionation and mass spectrometry for the accurate detection of food adulterants	Leitner et al. [30]
4	Ambient mass spectrometry based on REIMS for the rapid detection of adulteration of minced meats by the use of a range of additives	Research	Rapid evaporative mass spectrometry (REIMS)	• REIMS is a rapid screening technique suitable for use as a frontline testing tool to verify the quality and authenticity of meat products	Kosek et al. [31]

(continued)

Table 3 (continued)

No.	Title	Type of study	Technology	Key findings	References
5	Untargeted metabolomics and proteomics approach using liquid chromatography-orbitrap high resolution mass spectrometry to detect pork adulteration in *Pangasius hypophthalmus* meat	Research	Liquid chromatography-Orbitrap high -resolution mass spectrometry (HRMS)	• Both metabolomics and proteomics using LC-HRMS could detect pork at the lowest concentration level (0.5% w/w)	Windarsih et al. [32]
6	The current and potential applications of ambient mass spectrometry in detecting food fraud	Review	Ambient mass spectrometry	• Ambient mass spectrometry techniques deliver rapid and accurate results that are notably comparable to those achieved with traditional methods	Black et al. [5]
7	Making complex measurements of meat composition fast: Application of rapid evaporative ionization mass spectrometry to measuring meat quality and fraud	Review	Rapid evaporative ionization mass spectrometry (REIMS)	• REIMS produces a comprehensive mass spectral fingerprint that accurately represents a meat sample without requiring any sample preparation	Ross et al. [33]
8	Comparative review and the recent progress in detection technologies of meat product adulteration	Review	Mass spectrometry	• Mass spectrometry technologies are capable of identifying species by simultaneously detecting several distinct peptides	Li et al. [34]
9	Rapid analysis and authentication of meat using the MasSpec pen technology	Research	Mass spectrometry	• MasSpec pen technology offers a robust alternative analytical method for analyzing meat and detecting meat fraud	Gatmaitan et al. [35]

(continued)

Table 3 (continued)

No.	Title	Type of study	Technology	Key findings	References
10	Identification and quantification of fox meat in meat products by liquid chromatography–tandem mass spectrometry	Research	Liquid chromatography–tandem mass spectrometry	• LC-MS/MS method can be used for the accurate identification and quantification of fox meat products	Zhang et al. [36]
11	H-NMR-based metabolomics: An integrated approach for the detection of the adulteration in chicken, Chevon, beef and donkey meat	Research	Nuclear magnetic resonance (NMR) spectroscopy	• The NMR spectral data yielded valuable insights, leading to the identification of 37 metabolites across chicken, goat, beef, and donkey meat • H-NMR-based metabolomics is a reliable method with potential applications in the meat industry for ensuring quality control	Akhtar et al. [37]
12	High resolution nuclear magnetic resonance spectroscopy (NMR) studies on meat components: Potentialities and prospects	Research	HR-nuclear magnetic resonance (NMR) spectroscopy	• The 1H HR-MAS NMR technique enables the direct acquisition of high-resolution NMR spectra from meat samples	Brescia et al. [38]
13	Nuclear magnetic resonance (NMR) spectroscopy in food science	Review	Nuclear magnetic resonance (NMR) spectroscopy	• The combination of NMR with chemometrics (multivariate statistical analysis) is emphasized as a powerful tool for food classification and biomarker identification	Hatzakis [39]

(continued)

Table 3 (continued)

No.	Title	Type of study	Technology	Key findings	References
14	Authentication of butter from lard adulteration using high-resolution of nuclear magnetic resonance spectroscopy and high-performance liquid chromatography	Research	High-resolution of nuclear magnetic resonance spectroscopy and high-performance liquid chromatography	• The combination of 1H-NMR and HPLC offers a robust alternative for detecting lard adulteration in butter	Fadzillah et al. [40]
15	Detection and quantification of Milk adulteration using time domain nuclear magnetic resonance (TD-NMR)	Research	Time domain nuclear magnetic resonance (TD-NMR)	• 1H TD-NMR can serve as a fast and effective alternative method for detecting and quantifying adulteration in milk	Santos et al. [41]
16	When detection of dairy food fraud fails: an alternative approach through proton nuclear magnetic resonance spectroscopy	Research	Nuclear magnetic resonance	• Proton nuclear magnetic resonance (1H NMR) descriptors, derived from various integral ratios of signals related to the CH_2 group compared to those associated with butyric and n-3 fatty acids, were suggested as a means to detect adulterations	Hanganu and Chira [42]
17	Rapid identification and quantitation of pork and duck meat of binary and ternary adulteration in minced beef by 1H NMR combined with multivariate data fusion	Research	1H NMR combined with multivariate data fusion	• 1H NMR with orthogonal partial least squares-discriminant analysis and partial least squares are effective in the quality control of meat products	Leng et al. [43]
18	Rapid and non-destructive detection of water-injected pork using low-field nuclear magnetic resonance (LF-NMR) and magnetic resonance imaging (MRI)	Research	Low-field nuclear magnetic resonance (LF-NMR) and magnetic resonance imaging (MRI)	• LF-NMR combined with MRI offers an effective method for the detection of water-injected pork	Gai et al. [44]

(continued)

Table 3 (continued)

No.	Title	Type of study	Technology	Key findings	References
19	Advances in nuclear magnetic resonance spectroscopy for food authenticity testing	Review	Nuclear magnetic resonance (NMR) spectroscopy	• NMR spectroscopy is not only a reliable quantitative analytical method but also provides effective tools for identifying compounds, even without the need for standards	Sobolev and Mannina [45]
20	Recent developments in the food quality detected by non-invasive nuclear magnetic resonance technology	Review	Nuclear magnetic resonance (NMR)	• NMR technology is a powerful and non-invasive method for assessing the authenticity, composition, and quality of food, including fruits, vegetables, meat, and seafood, during processing and storage • When combined with multivariate analysis, NMR can quickly identify and categorize foods	Fan and Zhang [46]

There are three main techniques in MS for meat adulteration, which will be listed as follows: (1) Matrix-assisted laser desorption/ionization (MALDI-MS), which is ideal for protein and peptide analysis, MALDI-MS is commonly used to identify species-specific biomarkers in meat adulteration studies. (2) Isotope ratio mass spectrometry (IRMS) analyzes the isotopic composition of meat to verify the geographical origin or production methods, providing evidence of fraudulent labeling. (3) High-resolution mass spectrometry (HRMS), which offers unmatched precision in identifying unknown or complex adulterants, making it valuable for forensic-level analysis of adulteration [30].

The key advantages of MS include its ability to detect trace amounts of adulterants, its applicability to a diverse range of adulterants such as proteins, lipids, and chemicals, its effectiveness in analyzing processed or mixed meat products where other techniques may fall short, and its suitability for large-scale monitoring of meat products. Despite its advantages, MS faces challenges, including the high cost of instruments and operations, the requirement for skilled personnel, the complexity of sample preparation, which can reduce throughput, and the dependence on comprehensive databases of protein and chemical profiles for reliable analysis. Advances such as the development of miniaturized MS instruments, streamlined workflows, and the incorporation of artificial intelligence for data interpretation are expected to improve the accessibility and efficiency of MS in meat authentication. Ruiz Orduna et al. [28] investigated meat adulteration using proteogenomic annotation, strategically selected surrogate tryptic peptides, and high-resolution MS. The mass spectrometer operated in full-scan mode with high resolution and accurate mass, achieving resolving powers of 140,000 and 17,500 (FWHM) for full-scan MS and MS/MS, respectively. Their method detected and identified highly specific tryptic peptides for all four target proteins in each animal species, with observed m/z values deviating by less than 3 ppm from the theoretical m/z. The analyses were successfully conducted on both raw and cooked meat. Stachniuk et al. [29] demonstrated that high-performance liquid chromatography combined with tandem MS and in-solution protein digestion is emerging as the preferred method for detecting, identifying, and monitoring heat-stable species-specific peptides in highly processed meat products. However, extensive validation is still required. They anticipate that sufficient data should be gathered within a few years to enable widespread acceptance of this methodology by regulatory bodies and the food industry. Leitner et al. [30] investigated the detection of soybean protein adulteration in processed meat products by identifying marker proteins through advanced analytical techniques. The research integrated multidimensional liquid chromatography with tandem mass spectrometry (nanoLC-MS/MS) to target specific proteins. Variants and subunits of glycinin and β-conglycinin, the primary soybean seed storage proteins, were successfully identified. The method demonstrated sufficient sensitivity to differentiate closely related protein variants within complex food matrices. Additionally, the study laid the groundwork for creating mass spectrometry-based diagnostic tools to ensure food quality control. Kosek et al. [31] evaluated the potential of rapid evaporative ionization MS (REIMS) for detecting undeclared additives in pork and chicken meat products, including sausages and burgers. The study demonstrated that REIMS, combined with multivariate analysis, could serve as a promising and rapid analytical method for identifying unusual patterns in pork and chicken meat indicative of adulteration with protein-based bulking agents. The REIMS technique discovered such adulterants with high confidence when more than 2.5% of these substances were added. The researchers concluded that REIMS is well-suited for high-throughput screening of large sample volumes, efficiently identifying suspicious cases that warrant further investigation using techniques like PCR or LC-MS. Windarsih et al. [32] successfully employed untargeted metabolomics and proteomics using liquid chromatography-high-resolution mass spectrometry (LC-HRMS) combined with chemometric techniques, including principal component analysis (PCA) and partial least square-discriminant analysis (PLS-DA), to distinguish authentic *Pangasius hypothalamus* meat from adulterated samples. The approach demonstrated strong model fitness ($R > 0.95$) and predictivity ($Q > 0.5$). This study contributed to advancing analytical methods for halal authentication of fish meat products, aiming to establish robust, standardized, and reliable techniques for detecting pork adulteration in fish. The researchers also recommended further investigating lipidomics, as lipid compounds were identified as potential metabolite markers. Black et al. [5] explored various aspects of food adulteration, including its economic and public health impacts, as well as the role of ambient MS in detecting such adulteration. They compared ambient MS with traditional methods like LC-MS, NMR, ELISA, PCR, and spectroscopic techniques commonly used for investigating food adulteration, authenticity, traceability, and safety. Unlike these conventional methods, ambient MS required no sample preparation and minimal sampling time, providing rapid and accurate results that are on par with those of established techniques. In a review study, Ross et al. [33] highlighted that rapid evaporative ionization MS (REIMS) had the potential to bridge the gap between sophisticated laboratory analysis and technology capable of screening quality, safety, and authenticity parameters in a single test. REIMS was successfully used to detect species fraud and hormone use in livestock, monitor meat processing, and identify off-flavors like boar taint. Li et al. [34] reviewed both destructive meat adulteration technologies, which rely on DNA, protein, and metabolite analyses, and non-destructive techniques based on spectroscopy, providing a comparative analysis. They emphasized that developing convenient and effective chemometric models is essential for non-destructive methods, and the creation of portable devices for on-site monitoring is a future direction. Additionally, omics technologies, particularly proteomics, play a vital role in laboratory detection by enabling multispecies identification and screening of unknown targets through mass spectrometry databases. Gatmaitan et al. [35] conducted MS Pen analysis directly on various types of meat and fish, including grain-fed beef, grass-fed beef, venison, cod, halibut, Atlantic salmon, sockeye salmon, and steelhead trout, with each sample analyzed in just 15 s. Statistical models created using the Lasso method and a training set of samples resulted in accuracies of 95% for the beef model, 100% for the beef versus venison model, and 84% for the multiclass fish model. Additionally, MS Pen testing for identifying venison and grass-fed beef samples adulterated with grain-fed beef achieved 100% prediction accuracy for both classifiers using test sets. Zhang et al. [36] developed a liquid chromatography-tandem MS (LC-MS/MS) method for precise identification and quantitative analysis of fox meat products. They found that the LC-MS/MS method is straightforward, accurate, and reliable, making it suitable for species identification and quantitative analysis of everyday meat products.

Despite challenges like high costs and complex workflows, MS advances, such as miniaturized instruments and AI integration, enhance its efficiency and accessibility. The previously mentioned studies demonstrated its effectiveness in identifying species-specific biomarkers and adulterants in processed meats and real-time screening using innovative methods like REIMS. As MS continues to evolve, it plays a pivotal role in safeguarding the integrity of global meat products.

Nuclear Magnetic Resonance (NMR) Spectroscopy

Nuclear magnetic resonance (NMR) spectroscopy is a powerful analytical tool for identifying and combating meat adulteration. Its role can be categorized into the following aspects: Akhtar et al. [37]: (1) Detection of species adulteration: NMR spectroscopy can identify the specific metabolic fingerprints of different meat species. This makes it possible to detect the substitution of high-value meats (e.g., beef or lamb) with cheaper alternatives (e.g., pork or poultry), ensuring authenticity. (2) Quantification of adulterants: NMR provides quantitative data on the chemical composition of meat. It can accurately detect and measure adulterants like added water, non-meat proteins, or plant-based fillers, offering precise quality control. (3) Authentication of geographic origin: By analyzing isotopic and metabolic profiles, NMR can verify meat's geographical origin, helping combat fraudulent claims about premium or region-specific meat products. (4) Assessment of meat quality: NMR effectively determines meat freshness and detects spoilage indicators. It can also identify the presence of prohibited additives, like chemical preservatives or dyes, used to mask poor quality. (5) Rapid and non-destructive analysis: One of the significant advantages of NMR is its non-destructive nature, allowing for rapid screening without altering the sample. This is particularly useful in high-throughput testing scenarios. (6) Detection of processed meat adulteration: In processed meat products like sausages and burgers, NMR can analyze the lipid and protein profiles to detect the addition of non-declared ingredients or the presence of contaminants. (7) High sensitivity and specificity: The high resolution of NMR ensures the detection of even trace amounts of adulterants. Its ability to provide detailed molecular-level insights surpasses many traditional methods [38].

The advantages of NMR in analyzing meat adulteration include its exceptional precision in differentiating authentic meat from adulterated samples, its ability to provide comprehensive chemical and molecular insights, consistent and reproducible results ideal for regulatory applications, and minimal sample preparation, which lowers the risk of contamination or errors during testing. Akhtar et al. [37] explored metabolic variations across different meat samples using NMR spectroscopy combined with multivariate data analysis techniques, such as principal component analysis (PCA) and orthogonal partial least squares-discriminant analysis (OPLS-DA). Through 1H-NMR spectroscopy, they identified 37 metabolites in the gluteal muscle tissues of cows, goats, donkeys, and chickens. The findings highlighted that NMR-based metabolomics is a robust approach for identifying novel signatures (potential biomarkers) to differentiate meat sources and holds promise for quality control applications to distinguish various types of meat. Brescia et al. [38] introduced the innovative use of proton high-resolution magic angle spinning (1H HR-MAS) NMR for analyzing meat characteristics. Their research demonstrated that this technique allows for the rapid acquisition of both qualitative and quantitative data on chemical composition without requiring extensive sample preparation. Additionally, the study emphasized the method's potential in determining meat origin and detecting meat adulteration. Hatzakis [39] addressed the knowledge gap in applying NMR techniques to food science by providing a comprehensive overview. The review detailed the fundamental principles of NMR spectroscopy and its significance in analyzing the complexity and heterogeneity of food matrices. Quantitative NMR (qNMR) was emphasized for its accuracy in measuring food components, enhancing safety and authenticity verification. Nonetheless, challenges such as high operational costs and specialized expertise requirements were also recognized. Fadzillah et al. [40] employed proton nuclear magnetic resonance (1H-NMR) spectroscopy combined with high-performance liquid chromatography (HPLC) to authenticate butter and detect lard adulteration. High-resolution NMR spectroscopy effectively identified the triacylglycerol composition of lard as a chemical marker for Halal authentication. Integrating 1H-NMR and HPLC proved to be a high-performance method for comprehensively detecting lard adulteration in butter, encompassing all proton-bearing components. Santos et al. [41] investigated the potential use of 1H time domain NMR (1H TD-NMR) as a quick method for evaluating milk quality (Table 3). To conduct this study, they added whey, urea, hydrogen peroxide, synthetic urine, and synthetic milk to the milk samples at 5, 15, 25, 35, and 50% v/v concentrations. The researchers concluded that 1H TD-NMR, combined with chemometric analysis, could be utilized for the online automation of milk analysis, allowing for a higher sampling frequency and portable equipment. Hanganu and Chira [42] explored the limitations of the current method for determining the fatty acid profile of milk fats using proton NMR (1H-NMR) data. They demonstrated that contrary to a previously reported approach, the butyric and linolenic groups in dairy fat samples cannot be accurately identified from 1H-NMR spectra due to the overlapping methyl signal resonances at 0.96 ppm. As a result, while 1H-NMR spectroscopy can be applied to detect dairy fraud, it is ineffective for determining the fatty acid profile. Leng et al. [43] successfully used 1H NMR spectra combined with chemometrics to rapidly identify and quantify pork and duck meat adulterated in ground beef. The 1H NMR spectra revealed that the main components of pure beef, pork, and duck were similar, consisting primarily of lactic acid, creatine, and carnosine, along with small amounts of amino acids, vitamins, and nucleic acids. However, the characteristic peaks for each type of meat were distinct. The method developed, which combines 1H NMR with orthogonal partial least squares-discriminant analysis and partial least squares, holds potential for use in the quality control of meat products. Gai et al. [44] explored the potential of using LF-NMR and MRI to detect water-injected pork. The *Longissimus dorsi* muscles were injected with deionized water at concentrations of 0%, 5%, 10%, 15%, 20%, and 25%. LF-NMR T2 relaxation tests on the adulterated pork revealed the presence of injected water at the molecular level. They concluded that LF-NMR, combined with MRI, was an effective method for detecting water-injected pork. In a review by Sobolev and Mannina [45], it was shown that NMR spectroscopy has been effectively used to detect adulteration of pure beef muscle tissue with offal, such as beef liver or kidney, using methods like PCA, PLS, and PLS-DA. PCA analysis of 97 samples identified clear separations between beef muscle, kidney, and liver, with 90% of the data captured in the first two principal components. PLS calibration accurately predicted the amounts of added liver and kidney, yielding high correlation coefficients (0.990 and 0.986) and low standard errors (3.51% and 3.93%). Additionally, 1H NMR was used to distinguish beef from horse meat based on triglyceride signatures, achieving near-perfect classification accuracy. NMR combined with pattern recognition also successfully determined the geographical origins of beef from New Zealand, Australia, the United States, and Korea, highlighting NMR's potential in verifying meat authenticity and origin. Fan and Zhang [46] reviewed recent advancements in NMR technologies for assessing food quality during processing and storage over the past decade. They highlighted that NMR is an effective, non-invasive tool for detecting the authenticity, composition, and quality of meat and other food products. They also proposed several future trends: (1) NMR should be combined with other spectroscopy methods like NIR spectroscopy, size exclusion chromatography, ultrasound, mass spectrometry, scattering techniques, and rheometry to enhance accuracy, sensitivity, and stability in quality detection. (2) Improvements in NMR hardware and user-friendly software will simplify equipment operation and ensure data reliability for food researchers. (3) A comprehensive NMR database should be developed for specific food products, reflecting the variety and richness of food types. (4) New NMR sequences and automated data processing procedures should be designed to enable high throughput.

Nuclear magnetic resonance (NMR) spectroscopy is a versatile and non-invasive analytical tool that offers unparalleled precision in detecting meat adulteration, ensuring authenticity, quality, and compliance. At the same time, future advancements promise greater accessibility and efficiency in food analysis.

Multispectral and Hyperspectral Imaging

Multispectral and hyperspectral imaging (MSI and HSI) are advanced, non-destructive technologies that play a critical role in detecting and preventing meat adulteration. These techniques combine imaging and spectroscopy, capturing spatial and spectral information to analyze meat's chemical and physical properties. Their roles can be outlined as follows [4]: (1) Detection of species adulteration: (a) Species identification: MSI and HSI can differentiate between various meat species by analyzing their unique spectral signatures. For example, they can detect the substitution of high-value meat like beef with cheaper alternatives such as pork or poultry. (b) Non-invasive analysis: These techniques do not require destructive sampling, preserving the meat for further use. (2) Identification of non-meat additives: MSI and HSI can identify and quantify the presence of non-meat adulterants such as plant-based fillers, starch, and other additives that are often mixed with meat to increase volume and also enable the detection of adulterants based on differences in chemical composition. (3) Assessment of meat quality: HSI, in particular, can assess meat quality by detecting changes in color, texture, and moisture content. It can also identify spoilage-related compounds and provide spatial distribution maps of fat and protein, which can help verify whether the nutritional composition aligns with labeling claims. (4) Detection of contaminants and additives: MSI and HSI can detect the use of harmful chemical preservatives, synthetic dyes, or other illegal additives, ensuring food safety and compliance with regulations and can also identify physical contaminants, such as foreign objects in processed meat products. (5) Authentication of geographic origin: Spectral analysis can differentiate meat based on geographic or production-related attributes, such as organic versus conventionally raised meat or specific regional breeds. (6) Rapid, non-destructive screening provides real-time results without damaging the sample, making them ideal for high-throughput screening in industrial and regulatory settings. (7) Hyperspectral imaging for processed meat is particularly effective for processed meat products, where visual cues are less reliable. It can detect unauthorized ingredients and verify the presence of declared components. (8) Advanced machine learning integration: When paired with machine learning algorithms, MSI and HSI can enhance classification accuracy, automate detection processes, and handle large datasets efficiently. Overall, the advantages of MSI and HSI in meat adulteration analysis primarily include their high sensitivity and specificity, broad applicability across various meat products, long-term cost-effectiveness by minimizing the need for chemical reagents and labor-intensive processes, and real-time monitoring capabilities, making them well-suited for inline quality control in meat production facilities. Ropodi et al. [6] explored using MSI combined with multivariate data analysis to detect fraudulent substitution between minced beef and pork for the first time. The study demonstrated excellent discrimination between pure and adulterated samples, achieving a correct classification rate of 98.48% using PLS-DA and LDA. Furthermore, 98.48% of samples were accurately classified within a ±10% adulteration range using LDA, while PLS-DA achieved a classification rate of 96.97% within the same range. Kamruzzaman et al. [4] introduced and optimized a novel rapid analytical technique utilizing near-infrared (NIR) HIS to detect adulteration levels in minced lamb for the first time. Minced lamb samples were adulterated with minced pork in increments of approximately 2%, ranging from 2 to 40% (w/w). The findings demonstrated that NIR hyperspectral imaging, combined with suitable multivariate methods, could effectively detect adulteration in minced lamb. The partial least squares regression (PLSR)-based prediction model exhibited a strong correlation ($R^2cv = 0.99$) with the adulteration level. Additionally, they highlighted that HIS offers the advantage of spatial concentration mapping, a feature unavailable in conventional methods. Ropodi et al. [47] examined the potential of MSI combined with data analysis techniques to detect minced beef adulteration with horsemeat and assess model performance during refrigerated storage (Table 3). The findings demonstrated that MSI could distinguish pure and freshly ground samples with a maximum error of 20%. Although the classification models were less effective than those for pork adulteration, the study highlighted the importance of independent sample validation and the notable color changes during storage. Feng et al. [48] presented a comprehensive review of recent advancements in HSI and MSI technologies for assessing the quality and safety of various meat products and their potential for widespread application. The study suggested that selecting key wavelengths and developing models based on these wavelengths significantly simplified the process while maintaining high accuracy. Utilizing HSI or MSI to monitor chemical parameters or quality attributes during meat processing provides valuable insights for better process control in manufacturing. The study also concluded that HSI and MSI are versatile techniques applicable to both raw and processed meat products. Al-Sarayreh et al. [49] conducted an in-depth evaluation of HSI's effectiveness in detecting adulteration in red meat products. They collected a dataset of line-scanning images of lamb, beef, and pork muscles, considering various meat states, including fresh, frozen, thawed, and samples packed or unpacked in transparent bags. The study found that the convolutional neural network (CNN) model delivered the highest performance, achieving an overall classification accuracy of 94.4%, regardless of the product's state. Additionally, the research demonstrated that HIS systems are powerful tools for rapid, reliable, and non-destructive detection of adulteration in red meat products. Jiang et al. [50] studied using HSI to detect and visualize beef adulteration with duck meat. Minced beef was adulterated with duck meat in varying amounts (0–100% w/w) in 10% increments. The results showed that HSI measurements could accurately determine the level of duck meat adulteration in beef. They also demonstrated that visible and near-infrared (VNIR) HSI could serve as a rapid, non-destructive, and effective method for accurately screening minced beef adulterated with minced duck. Jiang et al. [51] suggested using HSI in combination with chemometrics and a wavelength selection algorithm to detect and visualize adulteration of minced pork jowl meat in ground pork. The multispectral RC-PLSR model showed the best results in the prediction set, with an Rp^2 of 0.9063, an RPD of 2.30, and a limit of detection (LOD) of 6.50%. They also concluded that HSI offers more detailed visualization than conventional imaging and NIRS in identifying adulteration levels. Achata et al. [52] assessed the use of HSI in the 400–1000 nm spectral range, combined with multivariate analysis, to explore the feasibility of creating generic models for detecting minced beef adulteration with other meats. The spectral data from different minced meat adulterants (such as chicken and turkey) successfully predicted the adulteration levels of other meats (pork, lamb, and duck) by estimating the beef concentration in each sample type. While the models showed good results across the entire spectral range, applying a band selection method enhanced the performance of the prediction models. The authors highlighted the potential of HSI and multivariate analysis for predicting adulteration levels of various adulterants in minced beef and/or determining the beef content in adulterated samples. Kamruzzaman et al. [3] were the first to identify pork adulteration in minced beef using an HSI technique. The minced beef samples were mixed with minced pork at concentrations ranging from 2% to 50% (w/w) in increments of about 2%. Images were captured using a visible near-infrared HSI (VNIR-HSI) system, and the spectral data were analyzed. The authors concluded that, with proper adjustment and calibration, HSI techniques could be widely adopted for both laboratory and industrial applications. Kamruzzaman et al. [2] investigated the effectiveness of visible near-infrared (VNIR) HSI (400–1000 nm) combined with machine learning (ML) to detect chicken adulteration in fresh minced beef. The minced beef samples were adulterated with minced chicken at levels ranging from 0% to 50% (w/w) in approximately 2% increments. The authors concluded that HSI paired with ML could serve as a rapid, non-invasive method for accurately detecting adulteration in minced meat. They also highlighted the advantage of this technology's non-destructive nature, making it ideal for fast analysis in online, offline, or at-line inspections.

Detection Using Artificial Intelligence

Artificial intelligence (AI) is transformative in addressing meat adulteration by enhancing detection accuracy, streamlining analytical processes, and enabling predictive quality control. By leveraging ML, deep learning (DL), and data analytics, AI empowers researchers, producers, and regulators to safeguard the integrity of meat products. Its contributions are detailed as follows [53]: (1) Enhanced detection and classification: (a) AI algorithms can analyze complex datasets, such as spectral or chemical signatures, to identify adulterants in meat with high accuracy and (b) AI models trained on unique features of different meat species can differentiate between authentic and adulterated products, even in complex mixtures. (2) Integration with analytical techniques: (a) MSI/HSI data analysis: AI is often used alongside multispectral and hyperspectral imaging to process vast amounts of spectral data, enabling faster and more accurate identification of adulteration and (b) NMR and FTIR spectroscopy: AI enhances the interpretation of spectral data from NMR and Fourier transform infrared (FTIR) spectroscopy, identifying adulterants and contaminants with minimal human input. (3) Detection of non-meat adulterants, which identifies chemical and biological markers associated with non-meat adulterants such as plant fillers, water, or starches, ensures the purity of meat products. (4) Real-time monitoring can be integrated into production lines to detect adulteration in real-time, providing immediate feedback and reducing the risk of adulterated products reaching the market. (5) Traceability and supply chain integrity: (a) Blockchain and AI: By combining AI with blockchain technology, it is possible to create transparent and tamper-proof systems that track meat products throughout the supply chain, ensuring authenticity and origin and (b) Anomaly detection: AI can flag irregularities in supply chain data, such as unusual changes in product composition or origin, which may indicate adulteration. (6) Predictive quality control, which predicts potential adulteration risks based on historical data and environmental factors, helps manufacturers take proactive measures. (7) Consumer protection and labeling: (a) AI-powered tools can verify claims on meat labels, such as "organic" or "grass-fed," by analyzing key markers associated with those attributes and (b) Smartphone applications utilizing AI can empower consumers to test meat quality and authenticity using portable devices. (8) Cost and time efficiency reduce reliance on traditional, time-consuming laboratory methods by automating data processing and analysis, significantly speeding up adulteration detection while lowering operational costs. In a nutshell, AI is a game-changer in the fight against meat adulteration. By enhancing the efficiency, accuracy, and speed of detection methods, AI ensures product authenticity and safety and fosters trust among consumers and stakeholders in the food industry. Kazemi et al. [53] provided a comprehensive review of advancements in AI applications for detecting meat adulteration, focusing on spectral-based techniques like near infrared (NIR) and HSI. They also discussed practical challenges, such as environmental variations and the need for high-quality calibration datasets, which impact the implementation of AI-based solutions in industrial settings. According to them, future outlooks should emphasize the integration of AI with real-time and online detection systems to revolutionize meat safety assessment. The findings underscored the transformative potential of AI-driven spectroscopy techniques in the food industry, paving the way for innovative, non-destructive authentication solutions. Shi et al. [54] emphasized the necessity for further studies to enhance the robustness, refine algorithms, and validate these technologies in real-world production settings. The study examined advancements in non-destructive methods powered by AI for assessing meat quality. AI-based approaches (e.g., NIR and HSI) evaluate characteristics, such as texture, freshness, nutritional composition, and adulteration. Each method offers distinct advantages, such as HSI for integrating internal and external attributes. Furthermore, incorporating AI models like support vector machines (SVM) and partial least squares regression (PLSR) has significantly enhanced the precision and effectiveness of meat quality evaluations. Goyal et al. [55] highlighted certain limitations and the significant amount of work needed to enhance the accuracy of food adulteration detection systems. One major challenge is the unavailability of datasets online, as researchers often do not share links to their datasets (Table 4). Additionally, the lack of annotated datasets and the time-intensive nature of creating labeled datasets for various food items pose significant barriers. In the future, providing such resources could enable other researchers to focus on improving system efficiency by developing innovative methods for detecting food adulteration. Alvarez-Garcia et al. [56] reviewed the latest advancements in this field, emphasizing the critical role of computer vision, AI, and ultrasonography in assessing quality and efficiency in meat production and monitoring processes. Integrating sensors and AI techniques has notably enhanced precision and efficiency in evaluating key parameters, such as intramuscular fat, water-holding capacity, moisture, color, shear force, pH, and more. This technology enables real-time monitoring, early detection of quality issues, and improved traceability throughout the supply chain. To simulate adulteration, Rady and Adedeji [57] tested several plant- and animal-based materials as potential adulterants in minced beef and pork. Using selected wavelengths from HSI data, classification models improved by 100% for pork, 20% for chicken, and 4.8% for wheat gluten, with similar 100% performance for beef. Beef adulterated with textured vegetable protein showed a 13.6% improvement, while chicken showed a 5% boost. This study's findings, involving a broad range of adulterants and adulteration ratios, represent a novel contribution to the field.

Table 4 Application of advanced detection methods (multispectral and hyperspectral imaging and AI) for meat adulteration in the published papers

No.	Title	Type of study	Technology	Key findings	References
1	Multispectral image analysis approach to detect adulteration of beef and pork in raw meats	Research	Multispectral images	• Multispectral imaging offers a fast and non-invasive approach to detect adulteration	Ropodi et al. [47]
2	Fast detection and visualization of minced lamb meat adulteration using NIR hyperspectral imaging and multivariate image analysis	Research	NIR hyperspectral imaging and multivariate image analysis	• Traditional analytical methods, which are often labor-intensive and time-consuming, could be replaced by spectral data to enable rapid, cost-effective, and non-destructive testing for detecting adulteration in minced lamb meat	Kamruzzaman et al. [4]
4	Hyperspectral imaging and multispectral imaging as the novel techniques for detecting defects in raw and processed meat products: current state-of-the-art research advances	Review	Hyperspectral imaging and multispectral imaging	• Hyperspectral imaging, a promising tool for developing rapid and non-invasive methods, can effectively detect defects in both raw and processed meat products	Feng et al. [48]
5	Detection of red-meat adulteration by deep spectral–spatial features in hyperspectral images	Research	Deep spectral–spatial features in hyperspectral images	• Deep learning techniques, such as convolutional neural networks, offer powerful features for effectively classifying hyperspectral data of meat products	Al-Sarayreh et al. [49]
6	Rapid identification and visualization of jowl meat adulteration in pork using hyperspectral imaging	Research	Hyperspectral imaging	• HIS is a promising method for detecting and visualizing minced jowl meat in pork	Jiang et al. [50]

(continued)

Table 4 (continued)

No.	Title	Type of study	Technology	Key findings	References
7	Multivariate optimization of hyperspectral imaging for adulteration detection of ground beef: Towards the development of generic algorithms to predict adulterated ground beef and for digital sorting	Research	Hyperspectral imaging	• The potential of using HSI and multivariate analysis to predict adulteration levels of various adulterants in minced beef and/or to determine the proportion of beef in an adulterated sample	Achata et al. [52]
8	Hyperspectral imaging for a rapid detection and visualization of duck meat adulteration in beef	Research	Hyperspectral imaging	• HSI detects minced beef adulteration with duck meat	Jiang et al. [51]
9	Hyperspectral imaging in tandem with multivariate analysis and image processing for non-invasive detection and visualization of pork adulteration in minced beef	Research	Hyperspectral imaging in tandem with multivariate analysis	• When combined with suitable data analysis, VNIR-HSI can be reliably and accurately used to detect and quantify the amount of adulterant in minced beef	Kamruzzaman et al. [52]
10	Rapid and non-destructive detection of chicken adulteration in minced beef using visible near-infrared hyperspectral imaging and machine learning	Research	Visible near-infrared hyperspectral imaging and machine learning	• Hyperspectral imaging combined with machine learning can be employed to identify, measure, and visualize the level of chicken adulterant in minced beef	Kamruzzaman et al. [2]
11	Detection of adulteration of ground meat by spectral-based techniques and artificial intelligence (2020–2024)	Review	Spectral-based techniques and artificial intelligence	• AI-driven is non-destructive authentication solutions for food adulteration	Kazemi et al. [53]

(continued)

Table 4 (continued)

No.	Title	Type of study	Technology	Key findings	References
12	A review on meat quality evaluation methods based on non-destructive computer vision and artificial intelligence technologies	Review	Artificial intelligence technologies	• The study emphasized on combining multiple detection methods and data fusion to enhance predictive accuracy and achieve comprehensive quality assessments	Shi et al. [54]
13	Food adulteration detection using artificial intelligence: a systematic review	Review	Artificial intelligence technologies	• Researchers have predominantly utilized machine learning and deep learning approaches, accounting for approximately 75%, compared to other sensor-based and hybrid methods • There is a need to build online systems which can perform food adulteration detection	Goyal et al. [55]
14	Implementing artificial intelligence to measure meat quality parameters in local market traceability processes	Review	Artificial intelligence	• The integration of artificial intelligence with sensor development holds great promise for the meat industry • The Peruvian meat industry is progressing significantly, with artificial intelligence technologies enhancing quality control, traceability, safety, and the consumption of high-value meat products	Alvarez-Garcia et al. [56]

(continued)

Table 4 (continued)

No.	Title	Type of study	Technology	Key findings	References
15	Application of hyperspectral imaging and machine learning methods to detect and quantify adulterants in minced meats	Research	Hyperspectral imaging (HIS) and machine learning (ML)	• The study showed effectiveness of hyperspectral imaging (HSI) (400–1000 nm) as a non-destructive method for detecting meat adulteration • Increasing the sample size could enhance the applicability of HSI in food adulteration detection	Rady and Adedeji [57]

Conclusions and Future Directions

Advancements in analytical techniques such as DNA barcoding, next-generation sequencing (NGS), mass spectrometry, nuclear magnetic resonance (NMR) spectroscopy, hyperspectral imaging (HSI), and artificial intelligence (AI) have revolutionized the detection of meat adulteration. These methods provide unparalleled accuracy, sensitivity, and scalability, addressing the limitations of traditional approaches. Their ability to identify species substitution, quantify non-meat additives, verify geographical origins, and detect contaminants enhances meat products' authenticity, safety, and quality. Despite challenges such as high costs, complex workflows, and the need for robust databases, these technologies continue to foster consumer trust, regulatory compliance, and food supply chain integrity. Collaborative efforts among researchers, industries, and regulatory bodies remain essential to overcome barriers and standardize protocols globally.

Acknowledgments This study was supported by the Second Century Fund (C2F), Chulalongkorn University. The Department of Food Science at Purdue University is also acknowledged.

References

1. Vinothkanna A, Dar OI, Liu Z, Jia AQ (2024) Advanced detection tools in food fraud: a systematic review for holistic and rational detection method based on research and patents. Food Chem 446:138893
2. Kamruzzaman M, Makino Y, Oshita S (2016) Rapid and non-destructive detection of chicken adulteration in minced beef using visible near-infrared hyperspectral imaging and machine learning. J Food Eng 170:8–15

3. Kamruzzaman M, Makino Y, Oshita S (2015) Hyperspectral imaging in tandem with multivariate analysis and image processing for non-invasive detection and visualization of pork adulteration in minced beef. Anal Methods 7(18):7496–7502
4. Kamruzzaman M, Sun DW, ElMasry G, Allen P (2013) Fast detection and visualization of minced lamb meat adulteration using NIR hyperspectral imaging and multivariate image analysis. Talanta 103:130–136
5. Black C, Chevallier OP, Elliott CT (2016) The current and potential applications of ambient mass spectrometry in detecting food fraud. TrAC Trends Anal Chem 82:268–278
6. Ropodi AI, Pavlidis DE, Mohareb F, Panagou EZ, Nychas GJ (2015) Multispectral image analysis approach to detect adulteration of beef and pork in raw meats. Food Res Int 67:12–18
7. Dawan J, Ahn J (2022) Application of DNA barcoding for ensuring food safety and quality. Food Sci Biotechnol 31(11):1355–1364
8. Nehal N, Choudhary B, Nagpure A, Gupta RK (2021) DNA barcoding: a modern age tool for detection of adulteration in food. Crit Rev Biotechnol 41(5):767–791
9. Hellberg RS, Hernandez BC, Hernandez EL (2017) Identification of meat and poultry species in food products using DNA barcoding. Food Control 80:23–28
10. Barcaccia G, Lucchin M, Cassandro M (2015) DNA barcoding as a molecular tool to track down mislabeling and food piracy. Diversity 8(1):2
11. Di Pinto A, Bottaro M, Bonerba E, Bozzo G, Ceci E, Marchetti P, Tantillo G (2015) Occurrence of mislabeling in meat products using DNA-based assay. J Food Sci Technol 52:2479–2484
12. Xing RR, Hu RR, Han JX, Deng TT, Chen Y (2020) DNA barcoding and mini-barcoding in authenticating processed animal-derived food: a case study involving the Chinese market. Food Chem 309:125653
13. Wang N, Xing RR, Zhou MY, Sun RX, Han JX, Zhang JK, Chen Y (2021) Application of DNA barcoding and metabarcoding for species identification in salmon products. Food Add Contamin Part A 38(5):754–768
14. Ahmed N, Sangale D, Tiknaik A, Prakash B, Hange R, Sanil R, Khedkar G (2018) Authentication of origin of meat species processed under various Indian culinary procedures using DNA barcoding. Food Control 90:259–265
15. El Sheikha AF, Mokhtar NFK, Amie C, Lamasudin DU, Isa NM, Mustafa S (2017) Authentication technologies using DNA-based approaches for meats and halal meats determination. Food Biotechnol 31(4):281–315
16. Quinto CA, Tinoco R, Hellberg RS (2016) DNA barcoding reveals mislabeling of game meat species on the US commercial market. Food Control 59:386–392
17. Cottenet G, Blancpain C, Chuah PF, Cavin C (2020) Evaluation and application of a next generation sequencing approach for meat species identification. Food Control 110:107003
18. Xing RR, Wang N, Hu RR, Zhang JK, Han JX, Chen Y (2019) Application of next generation sequencing for species identification in meat and poultry products: a DNA metabarcoding approach. Food Control 101:173–179
19. Mottola A, Piredda R, Lorusso L, Armani A, Di Pinto A (2023) Preliminary study on species authentication in poultry meat products by next-generation sequencing. Food Control 145:109459
20. Chaora NS, Khanyile KS, Magwedere K, Pierneef R, Tabit FT, Muchadeyi FC (2022) A 16S next generation sequencing based molecular and bioinformatics pipeline to identify processed meat products contamination and mislabelling. Animals 12(4):416
21. Haynes E, Jimenez E, Pardo MA, Helyar SJ (2019) The future of NGS (next generation sequencing) analysis in testing food authenticity. Food Control 101:134–143
22. Mahama S, Chebako H, Sirikwanpong S, Mahamad P, Santiworakul NY, Suksuwan A, Nopponpunth V (2022) Simultaneous identification of four meat species (cattle, chicken, fish, and pig) using next generation sequencing (NGS). In: Proceedings of the international halal science and technology conference, vol 14. IHSATEC, Bangkok, pp 182–193

23. He M, Xing S, Yao G, Yao M, Wen X, Ling Y, Guo W (2024) Application of next generation semiconductor-based sequencing for species identification and meat derived products authentication. Food Control 165:110639
24. Kappel K, Gadelmeier A, Denay G, Gerdes L, Graff A, Hagen M, Haase L (2023) Detection of adulterated meat products by a next-generation sequencing-based metabarcoding analysis within the framework of the operation OPSON X: a cooperative project of the German National Reference Centre for authentic food (NRZ-Authent) and the competent German food control authorities. J Consum Prot Food Saf 18(4):375–391
25. Dunham-Cheatham SM, Klingler KB, Estrada MV, Gustin MS (2021) Using a next-generation sequencing approach to DNA metabarcoding for identification of adulteration and potential sources of mercury in commercial cat and dog foods. Sci Total Environ 778:146102
26. Pan Y, Qiu D, Chen J, Yue Q (2020) Combining a COI mini-barcode with next-generation sequencing for animal origin ingredients identification in processed meat product. J Food Qual 1:2907670
27. Denyingyhot A, Srinulgray T, Mahamad P, Ruangprach A, Sa S, Saerae T, Keeratipibul S (2022) Modern on-site tool for monitoring contamination of halal meat with products from five non-halal animals using multiplex polymerase chain reaction coupled with DNA strip. Food Control 132:108540
28. Ruiz Orduna A, Husby E, Yang CT, Ghosh D, Beaudry F (2017) Detection of meat species adulteration using high-resolution mass spectrometry and a proteogenomics strategy. Food Add Contamin Part A 34(7):1110–1120
29. Stachniuk A, Sumara A, Montowska M, Fornal E (2021) Liquid chromatography–mass spectrometry bottom-up proteomic methods in animal species analysis of processed meat for food authentication and the detection of adulterations. Mass Spectrom Rev 40(1):3–30
30. Leitner A, Castro-Rubio F, Marina ML, Lindner W (2006) Identification of marker proteins for the adulteration of meat products with soybean proteins by multidimensional liquid Chromatography–Tandem mass spectrometry. J Proteome Res 5(9):2424–2430
31. Kosek V, Uttl L, Jírů M, Black C, Chevallier O, Tomaniová M, Hajšlová J (2019) Ambient mass spectrometry based on REIMS for the rapid detection of adulteration of minced meats by the use of a range of additives. Food Control 104:50–56
32. Windarsih A, Warmiko HD, Indrianingsih AW, Rohman A, Ulumuddin YI (2022) Untargeted metabolomics and proteomics approach using liquid chromatography-Orbitrap high resolution mass spectrometry to detect pork adulteration in Pangasius hypophthalmus meat. Food Chem 386:132856
33. Ross A, Brunius C, Chevallier O, Dervilly G, Elliott C, Guitton Y, Vanhaecke L (2021) Making complex measurements of meat composition fast: application of rapid evaporative ionisation mass spectrometry to measuring meat quality and fraud. Meat Sci 181:108333
34. Li YC, Liu SY, Meng FB, Liu DY, Zhang Y, Wang W, Zhang JM (2020) Comparative review and the recent progress in detection technologies of meat product adulteration. Compr Rev Food Sci Food Saf 19(4):2256–2296
35. Gatmaitan AN, Lin JQ, Zhang J, Eberlin LS (2021) Rapid analysis and authentication of meat using the MasSpec pen technology. J Agric Food Chem 69(11):3527–3536
36. Zhang Y, Liu M, Wang S, Kang C, Zhang M, Li Y (2022) Identification and quantification of fox meat in meat products by liquid chromatography–tandem mass spectrometry. Food Chem 372:131336
37. Akhtar MT, Samar M, Shami AA, Mumtaz MW, Mukhtar H, Tahir A, Kaka U (2021) 1H-NMR-based metabolomics: an integrated approach for the detection of the adulteration in chicken, chevon, beef and donkey meat. Molecules 26(15):4643
38. Brescia MA, Jambrenghi AC, Martino VD, Sacco D, Giannico F, Vonghia G, Sacco A (2002) High resolution nuclear magnetic resonance spectroscopy (NMR) studies on meat components: potentialities and prospects. Ital J Anim Sci 1(2):151–158
39. Hatzakis E (2019) Nuclear magnetic resonance (NMR) spectroscopy in food science: a comprehensive review. Compr Rev Food Sci Food Saf 18(1):189–220

40. Fadzillah NA, Rohman A, Salleh RA, Amin I, Shuhaimi M, Farahwahida MY, Khatib A (2017) Authentication of butter from lard adulteration using high-resolution of nuclear magnetic resonance spectroscopy and high-performance liquid chromatography. Int J Food Prop 20(9):2147–2156
41. Santos PM, Pereira-Filho ER, Colnago LA (2016) Detection and quantification of milk adulteration using time domain nuclear magnetic resonance (TD-NMR). Microchem J 124:15–19
42. Hanganu A, Chira NA (2021) When detection of dairy food fraud fails: an alternative approach through proton nuclear magnetic resonance spectroscopy. J Dairy Sci 104(8):8454–8466
43. Leng T, Hu X, Chen Y, Gan B, Xie J, Yu Q (2023) Rapid identification and quantitation of pork and duck meat of binary and ternary adulteration in minced beef by 1H NMR combined with multivariate data fusion. Food Control 154:110018
44. Gai S, Zhang Z, Zou Y, Liu D (2019) Rapid and non-destructive detection of water-injected pork using low-field nuclear magnetic resonance (LF-NMR) and magnetic resonance imaging (MRI). Int J Food Eng 15(8):20180313
45. Sobolev AP, Circi SIMONE, Mannina LUISA (2016) Advances in nuclear magnetic resonance spectroscopy for food authenticity testing. In: Advances in food authenticity testing. Woodhead Publishing, Sawston, pp 147–170
46. Fan K, Zhang M (2019) Recent developments in the food quality detected by non-invasive nuclear magnetic resonance technology. Crit Rev Food Sci Nutr 59(14):2202–2213
47. Ropodi AI, Panagou EZ, Nychas GJE (2017) Multispectral imaging (MSI): a promising method for the detection of minced beef adulteration with horsemeat. Food Control 73:57–63
48. Feng CH, Makino Y, Oshita S, Martín JFG (2018) Hyperspectral imaging and multispectral imaging as the novel techniques for detecting defects in raw and processed meat products: current state-of-the-art research advances. Food Control 84:165–176
49. Al-Sarayreh MM, Reis M, Qi Yan W, Klette R (2018) Detection of red-meat adulteration by deep spectral–spatial features in hyperspectral images. J Imaging 4(5):63
50. Jiang H, Wang W, Zhuang H, Yoon SC, Yang Y, Zhao X (2019) Hyperspectral imaging for a rapid detection and visualization of duck meat adulteration in beef. Food Anal Methods 12:2205–2215
51. Jiang H, Cheng F, Shi M (2020) Rapid identification and visualization of jowl meat adulteration in pork using hyperspectral imaging. Foods 9(2):154
52. Achata EM, Mousa MA, Al-Qurashi AD, Ibrahim OH, Abo-Elyousr KA, Aal AMA, Kamruzzaman M (2023) Multivariate optimization of hyperspectral imaging for adulteration detection of ground beef: towards the development of generic algorithms to predict adulterated ground beef and for digital sorting. Food Control 153:109907
53. Kazemi A, Mahmoudi A, Khojasteh Najand M (2024) Detection of adulteration of ground meat by spectral-based techniques and artificial intelligence (2020-2024). Iranian Food Sci Technol Res J 20:90
54. Shi Y, Wang X, Borhan MS, Young J, Newman D, Berg E, Sun X (2021) A review on meat quality evaluation methods based on non-destructive computer vision and artificial intelligence technologies. Food Sci Anim Resour 41(4):563
55. Goyal K, Kumar P, Verma K (2022) Food adulteration detection using artificial intelligence: a systematic review. Arch Comput Methods Eng 29(1):397–426
56. Alvarez-García WY, Mendoza L, Muñoz-Vílchez Y, Nuñez-Melgar DC, Quilcate C (2024) Implementing artificial intelligence to measure meat quality parameters in local market traceability processes. Int J Food Sci Technol 59(11):8058–8068
57. Rady A, Adedeji AA (2020) Application of hyperspectral imaging and machine learning methods to detect and quantify adulterants in minced meats. Food Anal Methods 13:970–981

Gas Chromatography for Chemical Fingerprinting of Volatile Compounds and Adulteration Tests in Food

Nuttanee Tungkijanansin and Chadin Kulsing

Contents

Abstract Chemical fingerprinting is a structured analytical approach to identify and differentiate samples by analyzing their chemical profiles. These profiles can be generated using various techniques depending on the types of samples. Multicomponent samples are typically analyzed using chromatographic methods, and advanced systems of two-dimensional gas chromatography (2D-GC) can be used to obtain chemical fingerprints of highly complex samples, such as petrochemicals or

N. Tungkijanansin
Faculty of Science, Department of Chemistry, Chulalongkorn University, Bangkok, Thailand

C. Kulsing (✉)
Faculty of Science, Department of Chemistry, Chulalongkorn University, Bangkok, Thailand

Metabolomics for Life Sciences Research Unit, Chulalongkorn University, Bangkok, Thailand
e-mail: chadin.k@chula.ac.th

W. Alahmad, M. Ashfaq (eds.), *Advanced Techniques against Food Adulteration*,
https://doi.org/10.1007/978-3-031-99454-8_3

biologicals. The reliability of fingerprinting depends on experimental procedures and the ability to separate relevant compounds from the interferences. Without a well-optimized method, the results may be insufficient to detect differences, such as those found in adulterated food samples. Selecting suitable experimental conditions is thus essential to obtaining high-quality fingerprinting results. Since this often involves multiple experiments, the process can be time-consuming. However, applying experimental design techniques can reduce the number of experiments relying on predictive models of outcomes based on a limited dataset. This chapter discusses the core principles of chemical fingerprinting, the range of available analytical techniques, and the importance of experimental design in optimizing fingerprinting procedures.

Keywords Gas chromatography · Chemical fingerprint · Volatile compounds · Food

Introduction to Chemical Fingerprinting and the Related Analytical Techniques

Chemical fingerprinting is an approach to obtaining unique chemical patterns of samples using specialized analytical techniques. Like conventional fingerprints that could be used to inform people's identities or barcodes showing item prices in a shopping center, chemical fingerprinting of samples could reveal their identities. For example, fingerprints based on chemical components of blood, food, perfume, and fuel could be used to screen for disease and analyze the ingredients. The approach could also be effective for adulteration tests where the fingerprints of the authentic food samples differ from those of the adulterated ones. Analytical chemists could obtain chemical fingerprints using various analytical techniques and instruments, as shown in Fig. 1.

Fingerprinting of atoms could be obtained using an atomic emission spectrometer, which results in the characteristic spectrums of individual atoms. The fingerprints of molecules can be obtained using infrared spectrometry or nuclear magnetic resonance. Mixtures could be fingerprinted using chromatography with a suitable detector, resulting in different chromatograms. In the case of more complex samples, more complicated/expensive instruments with higher separation performance, such as two-dimensional gas chromatography (2D-GC) for analysis of petrochemical samples, can be used. A challenge in fingerprinting is that the results depend on the experimental approaches.

Regarding separation science, without a good fingerprinting approach, correct information about samples may not be obtained or it may be obtained with an ambiguous result or incapability to differentiate adulterated food. Good fingerprinting generally requires separating more compounds from the target compounds to improve sample identification. Besides the data analysis and interpretation of the

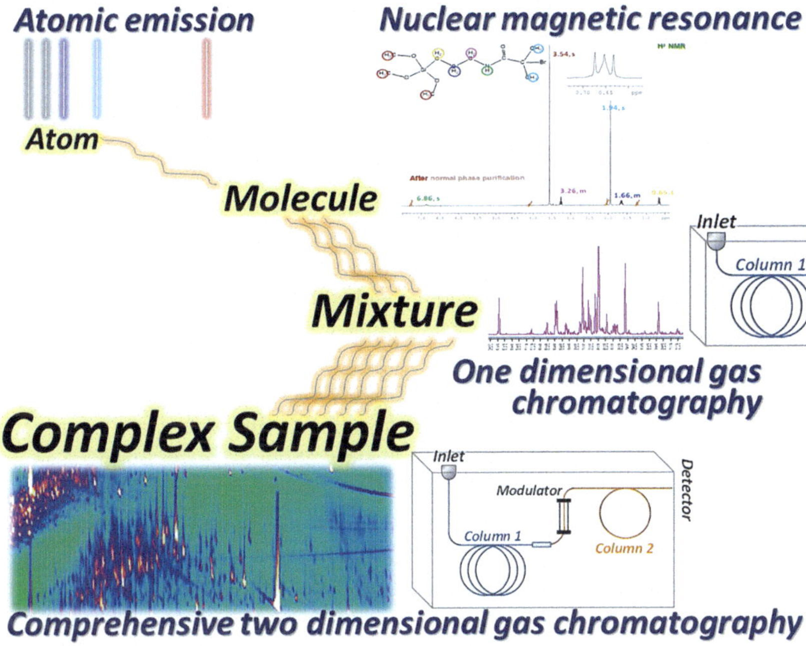

Fig. 1 Chemical fingerprinting of different species using different analytical techniques

fingerprinting results, suitable experimental conditions should be selected to get a good fingerprinting result in chromatography. These are wear-and-tear processes, especially in optimization, to obtain the best experimental condition in chemical fingerprinting. According to the possibility of experimental variation, many experimental results are required to analyze each sample. The number of experiments could be reduced using experimental design concepts where prediction of the results at any conditions of interest could be performed based on a suitable set of experimental data.

Gas Chromatography and Fingerprinting of Food

This chapter focuses mainly on the analysis of food volatile compounds. Among different approaches, gas chromatography (GC) is a high-performance separation technique that provides volatile chemical fingerprinting (unique chemical patterns of samples based on specialized analytical techniques) according to the separation of chemical components of samples. In addition to the general optimization of experimental conditions such as column dimension, carrier flow rate, and temperature program, column selection should be performed to separate a particular sample with suitable selectivity, resulting in the maximum number of separated peaks.

Conventionally, poly(siloxane) and poly(ethylene glycol) stationary phases are applied to separate samples according to vapor pressure and polarity difference of compounds. Fingerprinting of aromatic compounds or unsaturated fatty acid methyl esters (FAME) in derivatized milk can be effectively performed using stationary phases with high contents of cyano or phenyl groups. Alternative stationary phases can also be applied, such as applying a wide range of ionic liquid (IL) stationary phases with desirable properties, such as tunable selectivity and stability at high temperatures. Phosphonium- and imidazolium-based IL columns have been commercialized and effectively applied, e.g., for improved separation of FAME in derivatized milk and pesticides in vegetable and sulfur compounds in mushrooms.

GC is often hyphenated with mass spectrometry, which is the technique abbreviated as GC-MS. This combines the benefits of two different analytical methods. GC separates volatile compounds in a sample, and MS detects the separated components. This results in a chromatogram showing a fingerprint of separated compound peaks. Electron ionization (EI) is the most widely applied ionization interface for GC-MS, providing different MS fingerprints for various compounds. This hyphenation thus offers the GC and MS fingerprinting results useful in untargeted analyses, such as analyses of food, medical, environmental, and biological samples. Apart from this qualitative aspect, quantitative analysis can also be performed, e.g., by injecting standard compounds with different concentrations to construct their calibration curves (targeted analysis). As in conventional GC with other hyphenated detection, compounds in a sample mixture analyzed with GC-MS should be volatile and thermally stable.

An important concept in general separation/detection hyphenation is that the detector acquisition frequency should be fast enough to obtain information on all the chromatographic peaks. However, a detector requires some processing time for sufficient signal quality. A compromise between acquisition time and frequency should thus be made for different sample analyses. Good hyphenation is also expected, along with good connection and compatibility (flow rate and temperature) between the combined analytical techniques. The system performance should also be stable, providing repeatable analysis results with sufficient sensitivity toward different analytical goals. With adequate data acquisition frequency (e.g., >20 Hz), GC-MS delivers the result of a total ion chromatogram showing separated peaks with varying times of retention (GC fingerprint). EI MS spectrum (MS fingerprint) can also be obtained for each separated peak, which can be used to propose compound identity for each separated peak (untargeted analysis). This can be performed by comparing the experimental mass fingerprint with that from a library with the example GC-MS analysis of Tom Yum (a traditional Thai dish with a sour and spicy taste) provided in Fig. 2. A peak of β-bisabolene could be identified based on the experimental MS spectrum showing the top match score with the databased MS spectrum. This compound is odor active with the smell of sweet, lightly woody and balsamic aroma.

In several cases, especially when analyzing isomers or compounds with similar EI MS fingerprints, compounds are confirmed by using retention index (I) from different kinds of literature, e.g., different isomers having different polarity and boiling

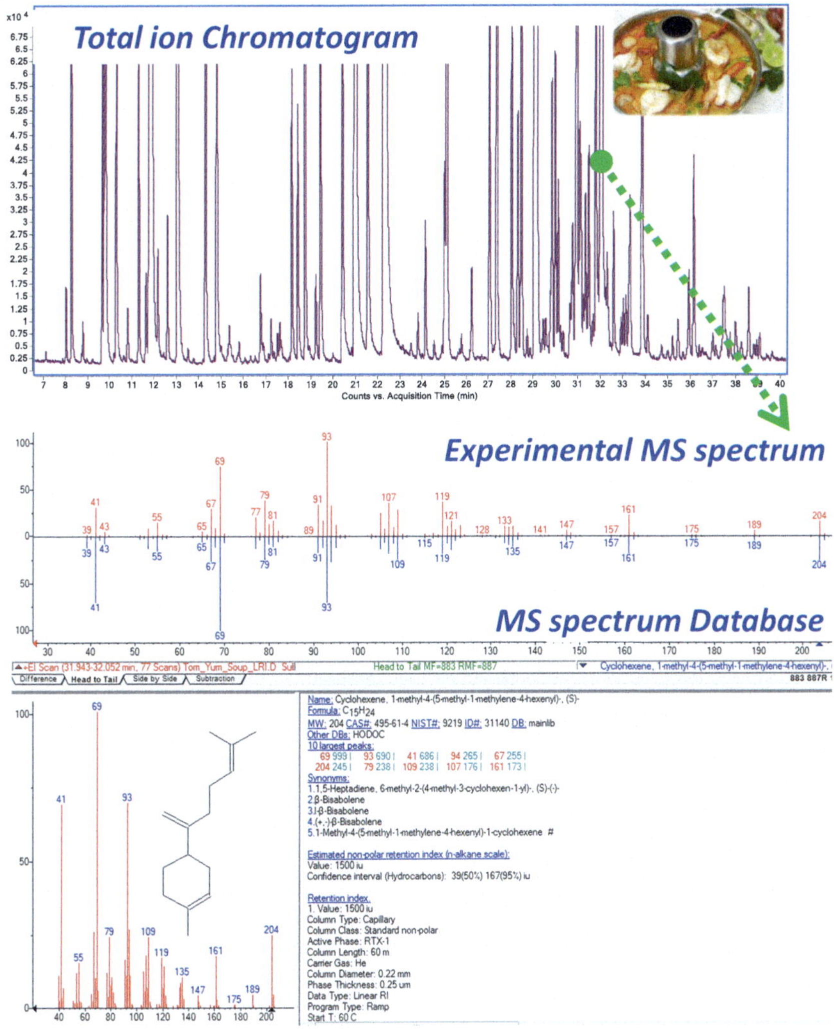

Fig. 2 Identification of a peak in GC-MS result with the possible compound identity proposed by comparison with the MS spectra in the National Institute of Standards and Technology (NIST) database

points. *I* value of a peak of interest was identified relative to a homologous series of alkanes, e.g., C6–C30, according to the van den Dool and Kratz formula [1], $I = 100n + 100\left(\dfrac{t_r(i) - t_r(n)}{t_r(n+1) - t_r(n)}\right)$, where $t_{r(i)}$, $t_{r(n)}$, and $t_{r(n+1)}$ are the retention times of the solute, alkane eluting immediately before the compound with the carbon number of n, and the alkane eluting immediately after the compound with the carbon number of $n + 1$, respectively.

GC-MS has been applied to provide volatile compound fingerprints of various food and beverage samples, such as ingredients, spices, meat, milk, vegetables, fruit, coffee, wine, or edible oil. The experimental process generally involves optimizing a stationary phase type, column dimension, carrier gas flow rate, and temperature program to maximize the number of separated peaks for each sample. However, the complexity of several samples could not be adequately analyzed with the conventional capability of one-dimensional (1D) GC. This has led to the development of more advanced techniques.

Multidimensional Gas Chromatography (MDGC) for Improved Fingerprinting of Food

Basic Concept of MDGC

Classical gas chromatography (GC) is one of the analytical separation techniques widely employed to separate volatile or semi-volatile compounds in many types of samples [2, 3]. Due to the use of only one column, the so-called one-dimensional gas chromatography (1DGC), complex sample (e.g., containing >100 compounds) analysis is still of concern with 1DGC separation and requires sample preparation, such as liquid–liquid extraction (LLE) or solid phase extraction (SPE), to reduce the interferences and improve analysis performance. It is thus time-consuming and still ineffective with the presence of unresolved peaks in 1D separation. To overcome the problems with 1DGC separation, two-dimensional gas chromatography (2DGC) has been applied for better separation performance [4]. Because of the selective combination of two different types of columns applied in 2DGC, a great many compounds in a complex sample can be effectively separated. 2DGC consists of an important key called the interface. It is located between the two columns and supports the transfer of pulses from the primary (^{1}D) column to the secondary (^{2}D) column. Hence, an effective interface is needed to perform 2D separation [4–6]. Moreover, the sampled portions in ^{1}D separation can be used to classify 2DGC into two classes, viz. comprehensive 2DGC (C2DGC or GC × GC) [7] and heartcut 2DGC (H/C 2DGC) [8]. The interface of C2DGC transfers all components from 1D to 2D separations; only single or multiple portions are transferred from 1D to 2D in H/C 2DGC.

Moreover, C2DGC has proved advantageous over H/C 2DGC owing to its complete multidimensional separation of the entire sample. In 1991, Liu and Philip [9] introduced comprehensive two-dimensional gas chromatography (C2DGC), reducing the long analysis time in classical 2DGC. The technique is conventionally performed within a single analytical run employing a short ^{2}D column with much faster 2D separation than 1D separation. All effluents from 1D separation were modulated and further separated on a 2D column in C2DGC. A modulator is an interface in C2DGC. This technique separated all sample components in both columns, resulting in the overall profile with fast analysis time. This technique is thus chosen to analyze complex samples in untargeted analysis. Nevertheless, the limited ^{2}D separation efficiency and lower ^{2}D resolution are the main drawbacks of C2DGC because of the use of shorter ^{2}D columns than H/C 2DGC. The use of a long ^{2}D column can solve this. A recent approach involves heart-cutting C2DGC (CH/C 2DGC) using sequential heartcuts with narrow windows to reduce the overall analysis time. Various publications are reporting the development of interfaces to allow comprehensive separation, such as valve-based modulators [10], thermal modulators [11], longitudinally modulated cryogenic system (LMCS) [12], and Deans switch (DS) [8]. A quality interface device ensures high sampling rates and effective transfer of the injected sample from ^{1}D to ^{2}D. A cryogenic modulator is the most commonly used modulator using liquid carbon dioxide or liquid nitrogen to trap or collect analytes from the 1D column. This approach has been reported for many applications of 2DGC. Conversely, it is still challenging to trap highly volatile compounds or solvents, and the ineffective release of low-volatile compounds requires additional devices. Therefore, valve- or flow-based modulation systems have been employed to overcome such problems.

Deans introduced the Deans switch (DS), an effluent-switching device, in 1965 [5, 13]. Many studies have reported using DS in H/C 2DGC because it offers heart-cut (H/C) operation, decreases fluctuations of retention times of eluted compounds, and improves analysis reproducibility and accuracy [5, 13–15]. Deactivated internal surface and pressure control were placed on both sides of DS, decreasing fluctuations of retention times of eluted compounds, improving reproducibility and accuracy of peak timing, and allowing multiple transferred components of different fractions during each run. There are many reports of the use of DS in various fields, such as natural products, petroleum, and foods. Our group also applied CH/C 2DGC to identify components in complex petrochemical and food samples [16–18].

Owing to the many advantages of DS, this device has been applied for enhanced CH/C 2D separation, albeit with a very long analysis time or requirement of a cryogenic trapping device, which is not environmentally friendly. In this work, we thus developed an additional concept and approach applying several Y split lines to improve the analysis time in CH/C 2DGC without using a cryogenic trap. In addition, this novel design will be applied to the analysis of highly volatile solvents and complex samples.

Modulators

One-dimensional gas chromatography (1DGC) offers high-impact separation despite the limitation in separating complex samples, such as foods, petrochemical samples, fragrance, forensic, and environmental samples, usually producing chromatograms with overlapping or co-eluted peaks. Therefore, this technique requires more steps in sample preparation for multi-component sample analysis, such as solid phase microextraction (SPME) [19], to reduce the interferences in complex samples and improve analysis performance. Analysis of complicated samples can usually be categorized into targeted and untargeted samples [20–22]. Targeted analysis is finding specific compounds or analytes in samples, such as heavy metals in environmental samples, organic acid preservatives in foods, aromatic compounds in fuels, or drugs in biological fluids by comparison with reference compounds. To decrease the complex matrix in targeted analysis, the sample preparation technique has been used before chromatography separation to reduce the interference compounds or matrix in the samples [23, 24]. In contrast, the untargeted analysis identifies unknown peaks in samples and studies the complete profile of samples, such as the study of the compound profile in fuels and components in biological samples. Furthermore, complex samples cause difficulty in analysis with 1DGC because of the overlapping peaks in the chromatogram results. To solve the 1DGC problems, 2DGC has been applied to separate the complex sample, which requires the second column inside the GC oven to increase the resolution of separations. An interface is placed between both columns to modulate sample compounds from ^{1}D to ^{2}D. The 2DGC instrument setup is illustrated in Fig. 3.

In 1984, the definition of multidimensional chromatography (MDC) was described and investigated by Giddings [4, 25], who proposed two multidimensional systems: continuous two-dimensional operation and coupled column assemblies. Simmons and Snyder first demonstrated the system and instrumental of MDGC in 1958 [26]. This separation technique has been developed parallel with 1DGC, e.g., involving column phase improvement, instrumentation, control programming, and data analysis. Nevertheless, MDGC is typically performed as a two-dimensional separation using two stationary phases of GC columns with different polarities. The other column properties allow compounds not separated in 1D separation and show the overlapping peaks to be separated in 2D separation. Thus, the results are displayed in a 2D contour plot. This contour plot combines ^{1}D and ^{2}D separations as illustrated in Fig. 4. The overlapping peaks from ^{1}D separation are separated in the contour plot. Furthermore, 2DGC involves using a special interface device to control the transfer of sample regions from ^{1}D to ^{2}D columns. There have been numerous reviews that reported many types of interfaces for both 2DGC types: C2DGC (or GC × GC) [27–29] and heartcut 2DGC (H/C 2DGC) [6, 14, 30, 31].

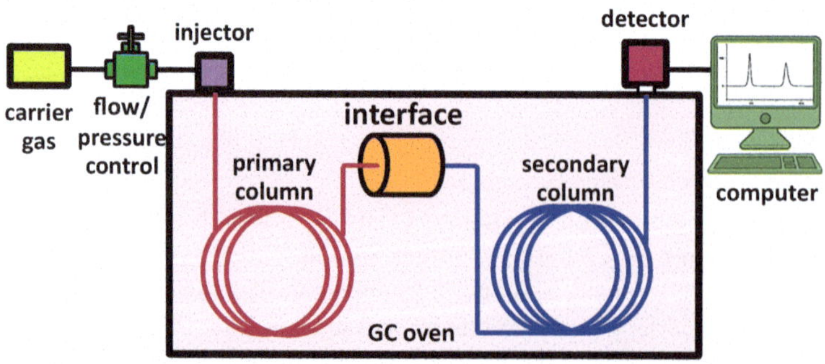

Fig. 3 The instrumental setup of 2DGC

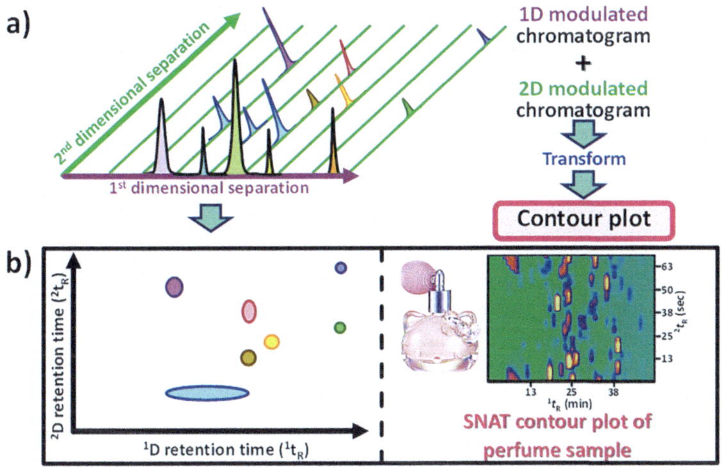

Fig. 4 Contour plot. (**a**) Aligned 1D modulated chromatograms. (**b**) Contour plot and SNAT contour plot of perfume sample

C2DGC interface is generally called a modulator, while H/C 2DGC employs a DS or switching valve as the interface.

The standard interfaces of C2DGC can be classified into three types: thermal modulator, valve-based modulator, and flow modulator. Liu and Phillip introduced the first GC × GC using a thermal modulator in 1991 [9]. The later development of the thermal modulator depends on high or low temperature compared with the oven temperature. A hot thermal modulator is a solid-state modulator containing a thick-film capillary column. The two related designs are an on-column thermal desorption modulator (using electrical current to release compounds from the thick-film capillary) and a rotating thermal modulator (with a moving modulator to trap and release compounds), as shown in Fig. 5a. However, these modulators have problems with the degradation of the stationary phase and contain moving parts. A cryogenic modulator was developed to reduce the moving parts, using low temperatures to trap analytes. The two related designs are (1) a longitudinally modulated cryogenic system (LMCS, with a moving cryogenic part to trap/release compounds) and (2) a jet-based modulator (applying one or two hot/cold points as well as the loop-based design for compound trap/release) as shown in Fig. 5b–d, respectively. Owing to the application of moving parts, causing column degradation is the main disadvantage of LMCS. The advantages of thermal modulators are peak or band focusing and modulation of only analytes, and they show difficulty in trapping highly volatile compounds and releasing low-volatile compounds as well as other limitations, such as cryogen consumption, system maintenance, and requirement of the optimization process.

A valve-based modulator is one of the GC × GC modulators introduced in 1998 that uses four ports of a heated six-port valve to switch the flow direction to either ^{2}D column or waste [32]. In contrast, it is not sensitive and effective because only

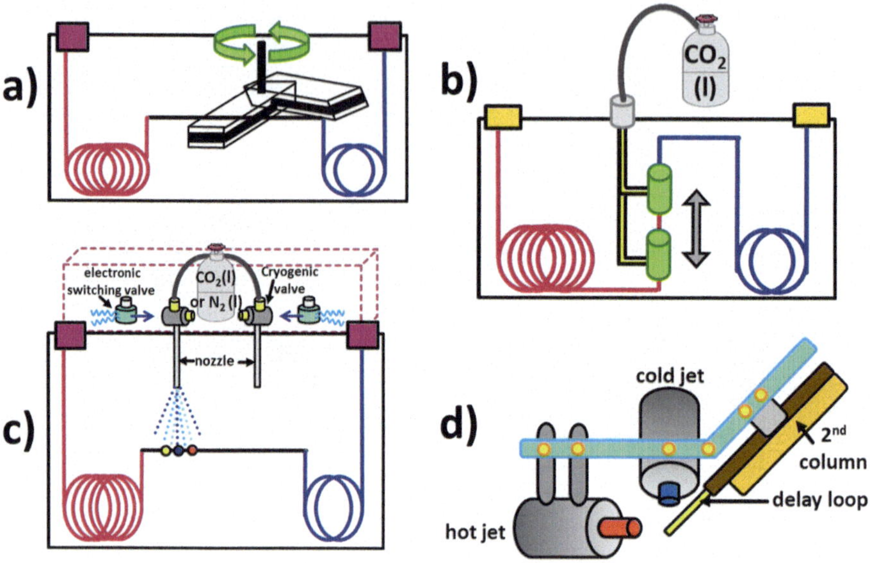

Fig. 5 Thermal modulators. (**a**) Rotating thermal modulator. (**b**) Longitudinally modulated cryogenic system (LMCS). (**c**) Jet-based modulator: two-point modulator. (**d**) Jet-based modulator: loop modulator

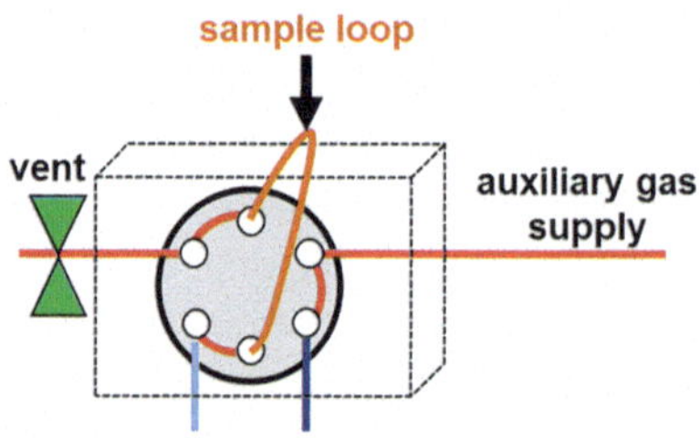

Fig. 6 Valve sampling loop (sample loop)

10–20% of analyte portions were transferred into the ^{2}D column. Therefore, the valve sampling loop was added into the valve system to avoid compound loss (Fig. 6). This type of valve modulator is classified as continuous flow. However, when flow rates in separation are changed, the sample loop size must also be changed to prevent compound loss, which is an inflexible system. To solve this problem, additional pneumatics were used to control the flow rate rather than change the loop size. Moreover, the continuous flow valve-based modulator is unsuitable with mass spectrometers (MS) due to limitations in vacuum pumping capacity. In 2004, Harynuk and Gorecki introduced another valve-based design called stopped flow, which overcomes the problems of continuous flow [33, 34]. Even this type uses a system similar to the continuous flow. The flow is stopped from the ^{1}D column before sampling to a ^{2}D column. This system allows a longer ^{2}D column with a narrow band injection, thus enhancing the second separation. Nevertheless, a ^{2}D

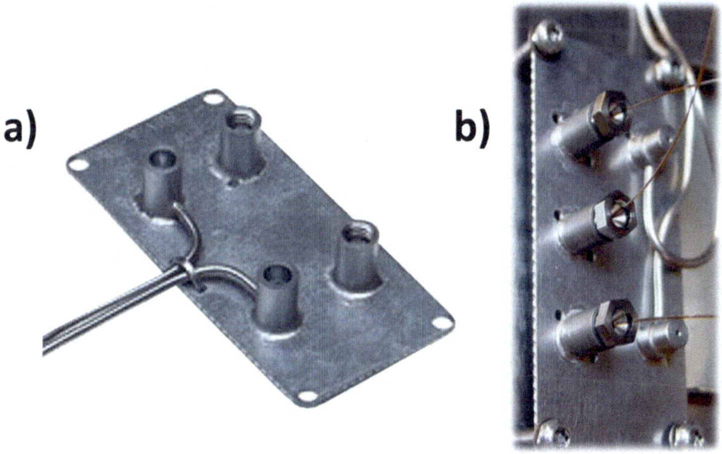

Fig. 7 Flow modulators. (**a**) Flow modulator and (**b**) Deans switch device

column cannot be applied for too long because of the accumulated amount of column bleeding and interruption of the temperature program in ^{1}D separation.

Alternatively, Seeley's group has developed the third type of modulator, a flow modulator enabling modulation of carrier gas [30, 35, 36]. The flow modulator uses the gas flow control in the capillary flow technology. A three-way solenoid valve is used to control the direction of flow. There are two types of flow modulators: flow modulator and H/C-based modulator using a Deans switch (DS) [5, 14] (Fig. 7). For conventional flow modulator, the three-way solenoid valve is periodically switched to enable the modulation process, which consists of two steps: collection and injection. In the collection step, the solenoid valve controls the flow to let the pulse (analyte portion) from the ^{1}D column fill the collection channel inside the modulator. The load time of the collection step must not be longer than the time to fill the collection channel. The injecting step involves injecting the pulse from the collection channel onto the ^{2}D column and toward the flame ionization detector (FID) or mass spectrometer (MS). This provides a stable and effective modulation process that modulates or traps highly volatile compounds. Nevertheless, besides the difficulty in optimizing, the flow modulator requires a high secondary flow rate (e.g., >20 mL/min), and the flow needs to be split for detection with a mass spectrometer.

In 2018, Cai and Stearns proposed a valve-based stopped flow modulator using hold-release of the ^{1}D column flow for long secondary separation time with 100% transfer of analytes [10]. The method precisely controlled the amount of ^{1}D effluent feeding into the ^{2}D column with a 6-port valve. The valve was placed inside a heating block to perform the compound movement. The head and end of the ^{1}D column were linked together via the valve, as shown in Fig. 8a. Afterward, the compounds were transferred onto the ^{2}D column, while the ^{1}D column flow slowed down and finally stopped (Fig. 8b). This system was applied to analyze a gasoline sample.

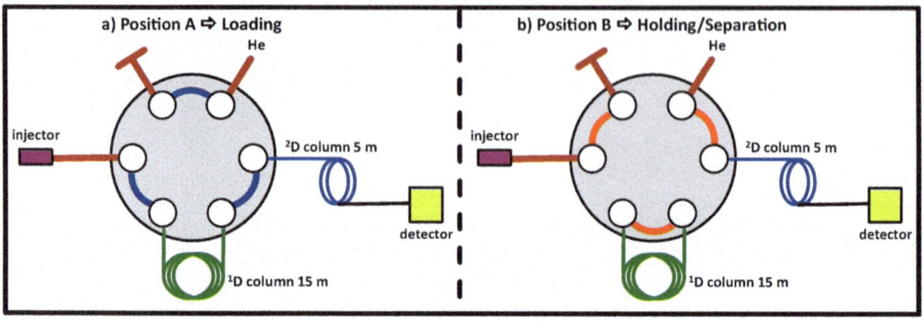

Fig. 8 A valve-based stopped flow modulator using hold-release. (**a**) Loading position and (**b**) holding/separation position. The image was reproduced with permission [10]

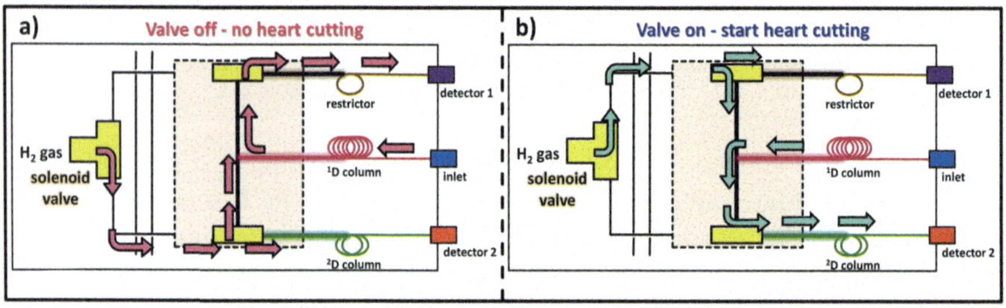

Fig. 9 (**a**) DS valve off step and (**b**) DS valve on step

Long analysis time is still required for separation with longer ^{2}D columns. Thus, this developed method modulates the eluent and does not focus on the analyte peaks.

Usually, CH/C is based on modulation using a DS. It is another type of flow modulator. It is also known as a microfluidic device and consists of two components, one inside and one outside the GC oven. The metallic part inside the oven contains no moving parts and is durable over various temperatures. For the part outside, it has a moving part and a solenoid valve. DS is often used to perform H/C of targeted compounds from a ^{1}D column to either a restrictor or a ^{2}D column. Thus, it is usually coupled with two detectors. DS has two main modes: valve off and valve on (Fig. 9). When the valve is in the "off" position, the flow is conventionally diverted to a short restrictor and the first detector. This provides the initial ^{1}D separation before the selection of the targeted region. Since the solenoid valve controls the flow, total flow transfers the sample from ^{1}D to the restrictor and on to the detector 1. With the "on" valve, the flow is directed from the ^{1}D to the ^{2}D column. Thus, the sample flows through ^{1}D and ^{2}D columns before reaching the second detector. The DS also applies a moderately high flow rate to the ^{2}D column (about 4 mL/min). Furthermore, there is no long analysis time. Moreover, it can be used to detect highly volatile compounds. A long second column can also be used, albeit with the main disadvantage of a long analysis time to complete the analysis.

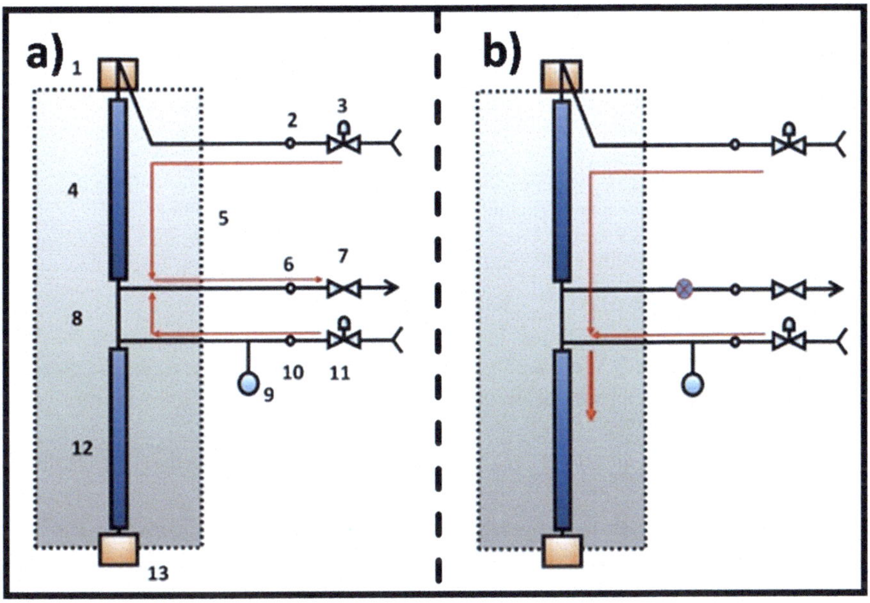

Fig. 10 The primary design of a Deans switch described by David R. Deans operated in two steps: (**a**) standby mode and (**b**) switching mode. This design consists of (1) an injector port, (2) the first tap, (3) the pressure controller, (4) ^{1}D column, (5) GC oven, (6) the second tap, (7) a needle valve, (8) a piece of capillary, (9) pressure gauge, (10) the third tap, (11) the second pressure controller, (12) ^{2}D column, and (13) detector (e.g., MS, FID, or nitrogen-phosphorus detector (NPD)). The image was reproduced with permission [14]

The first DS was described by David R. Deans in 1965 [5, 13, 37]. Figure 10 shows the primary design of DS. This approach has pressure controller before both ^{1}D and ^{2}D columns (the first and the second pressure controllers). The pressure controller provided the control of flow direction of effluent in this system. Hence, DS controls flow direction based on the pressure control.

The system consists of two modes: standby mode and switching mode. In the standby mode, it controlled the more significant flow of the second pressure than the first pressure. Thus, the effluent from the ^{1}D column was not transferred onto the ^{2}D column (no heart-cutting process) since the flow from the second controller forces effluent in the opposite direction. When switching mode, the effluent from the ^{1}D column was transferred from the ^{1}D column to the ^{2}D column (heart-cutting process), as the flow from the second controller is lower than the flow from the first controller. After introducing the first approach, several versions of the Deans switch were available from many companies, such as Shimadzu, Gerstel, Agilent, and Perkin-Elmer (Table 1).

DS from Agilent Technology will be used in this work. After various applications of Deans switch in CH/C 2DGC techniques [8, 14], numerous settings of DS will be performed on this work to achieve the analysis aim of complex samples. The main problem is no trapping, modulating, or focus on the analyte peaks. The standard

Table 1 The Deans switch is available from many companies which have different designs

Company	Deans switch design
Agilent	Capillary flow technology (CFT)
Gerstel	Multi-column switching (MCS)
Perkin-Elmer	Swafer microchannel flow technology
SGE	SilFlow microchannel device
Shimadzu	Multi-Deans switching system technology (MDSS)

way to focus or trap the peaks is to use a cryogenic modulator by cryo jet (jet-based modulator) with CO_2 or N_2 liquid. However, a cryogenic modulator is not the preferred interface for the 2DGC because of the high costs of the device, its environmentally unfriendly nature, difficulty in maintenance, and its inability and ineffectiveness to modulate highly volatile compounds.

There are many publications on the application of DS for the determination of complex samples, e.g., forensic, environmental, petrochemical, food, flavor, and biological samples [38–40], employing both classes of 2DGC, as shown in Table 2.

Petroleum samples (e.g., gasoline, diesel, or kerosene) are one of the complex samples containing many hydrocarbons and oxygenated compounds which can co-elute. The samples also show complicated chromatograms in classical 1DGC, which leads to difficulty in identification. Therefore, 2DGC has been used in petrochemical analysis to improve the separation. For high boiling fractions of petroleum samples, e.g., kerosene, diesel, and vacuum gas oils, both C2DGC and H/C 2DGC are used. C2DGC is more widely applied for group separations, whereas H/C 2DGC generally offers target analysis of selected analytes. For example, in 2014, Amer et al. [41] determined dichloromethane soluble products from high-sulfur Jordanian oil shale using C2DGC and H/C 2DGC combined with MS detection. Several groups of alkanes were separated from the other compounds in the sample. In addition, aromatics and sulfur compounds were identified with better MS match scores in a library than for the 1DGC-MS.

In 2020, Thongdorn-Ae et al. [18] reported the application of CH/C 2DGC to determine petrochemical products derived from palmitic acid oxidation. DS was applied as an interface to perform the cyclic multiple H/C strategy. A 0.2 min window was used to transfer a narrow band of compounds from a ^{1}D column to a ^{2}D column without cryogenic trapping devices, with the system shown in Fig. 11. The work results show an improvement of peak capacity 5840 with 235 compounds identified from library matching. Nevertheless, the total analysis time (15.3 h) is needed. However, using a wider H/C window can shorten the analysis time. This reduces the capacity to 390 with 1-h CH/C 2DGC analysis time. Thus, this is still time-consuming for high-efficiency analysis of complex samples.

Considering the limitations of all the modulation approaches described above, a new modulation approach or interface needs to be developed. In 2022, a splitter-based non-cryogenic artificial trapping (SNAT) modulator was designed [45]. Here, the performance of CH/C-2DGC was improved by developing a novel

Table 2 2DGC is applied for the analysis of various types of samples

Year	Class of 2DGC	Columns	Detector	Sample types	Analytes	Ref.
2014	H/C 2DGC	^{1}D column; Supelcowax (30 m × 0.05 mm × 0.25 μm) ^{2}D column; DB-17 ms (20 m × 0.18 mm × 0.18 μm)	MS (single quadrupole	High sulfur Jordanian oil shale	Dichloromethane soluble compounds	[41]
2015	H/C 2DGC	^{1}D column; Apolar column (50 m × 0.32 mm × 0.50 μm) ^{2}D column; sol-gel wax polar column (60 m × 0.32 mm × 0.25 μm)	MS (time of flight)	Strawberries	Key odorants	[42]
2015	H/C 2DGC	^{1}D column; HP-Innowax (30 m × 0.25 mm × 0.25 μm) ^{2}D column; DB-17 ms (30 m × 0.25 mm × 0.25 μm)	FID/MS (single quadrupole)	Cigarette	Nitroalkanes	[15]
2017	C2DGC	^{1}D column; weak-polar BPX-5 (30 m × 0.25 mm × 1.0 μm) ^{2}D column; a mid-polar BPX-50 (3 m × 0.10 mm × 0.10 μm)	FID/MS/TOFMS	Diesel fuel	Hydro-carbons	[43]
2018	H/C 2DGC	^{1}D column; SUPELCOWAX 10 phase (30 m × 0.25 mm × 0.25 μm) ^{2}D column; Stabilwax (30 m × 0.25 mm × 0.25 μm)	FPD/FID	Oxime samples	Oxime isomers	[8]
2019	C2DGC	^{1}D column; GsBP-624 (60 m × 0.32 mm × 1.8 μm) ^{2}D column; HP-INNOWax (2.5 m × 0.25 mm × 0.50 μm)	MS (single quadrupole)	Gases from an exhaust port of a household heating stove	Volatile organic compounds (VOCs)	[44]
2020	H/C 2DGC	^{1}D column; semi-standard nonpolar HP-5 (30 m × 0.25 mm × 0.25 μm) ^{2}D column; polar DB-wax column (60 m × 0.25 mm × 0.50 μm)	FID/MS (single quadrupole)	Petrochemical products	Volatile organic compounds (VOCs)	[18]

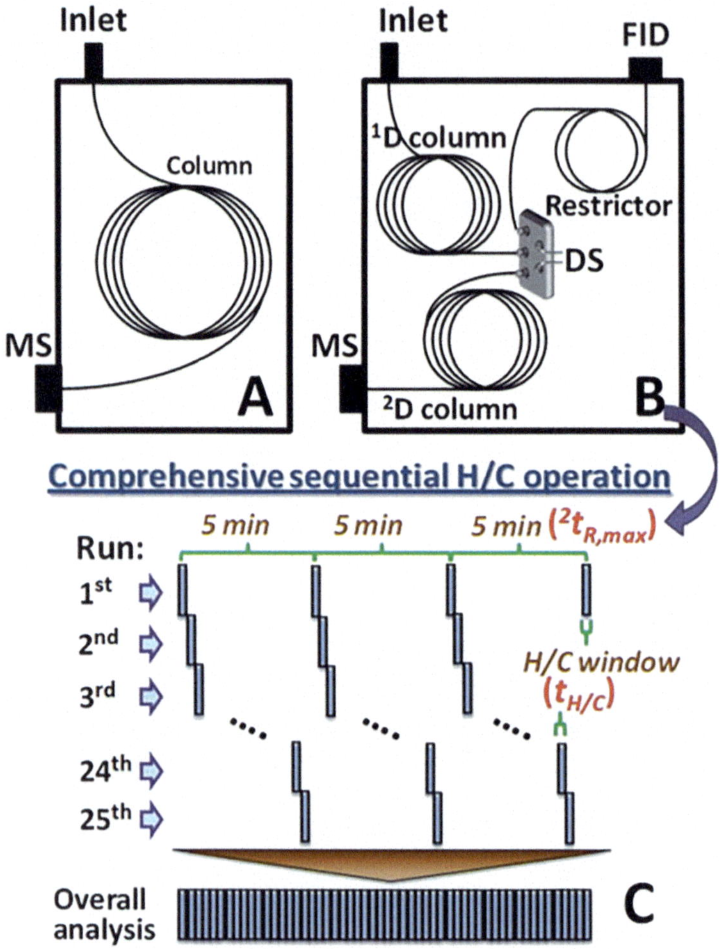

Fig. 11 Hyphenation systems: (**a**) 1DGC-MS, (**b**) CH/C MDGC-FID/MS, and (**c**) diagram illustrating the overall CH/C MDGC process. The image was reproduced with permission [18]

cryogenic-free system consisting of a DS and multiple splitters for analysis of highly volatile solvents and fuel samples. The design combined several split effluents, traveling for different periods, into a single effluent before an H/C event, which samples a wide H/C region in the ^{1}D separation into a narrow H/C window transferring onto the ^{2}D column. The system provides the trapping of compounds without the use of any cryogenic device or additional pressurized pulses. Such a design eliminates the long analysis time and is a cryogenic-free CH/C 2DGC.

The SNAT mechanism allows artificial modulation by mimicking such cryogenic phenomena. This is possible solely for a certain part of the sample zone after characteristic splitting, delaying (split pulses traveling with different void time, $t_{0,\text{split}}$) and combining the split pulses as illustrated for a modulation event by the steps

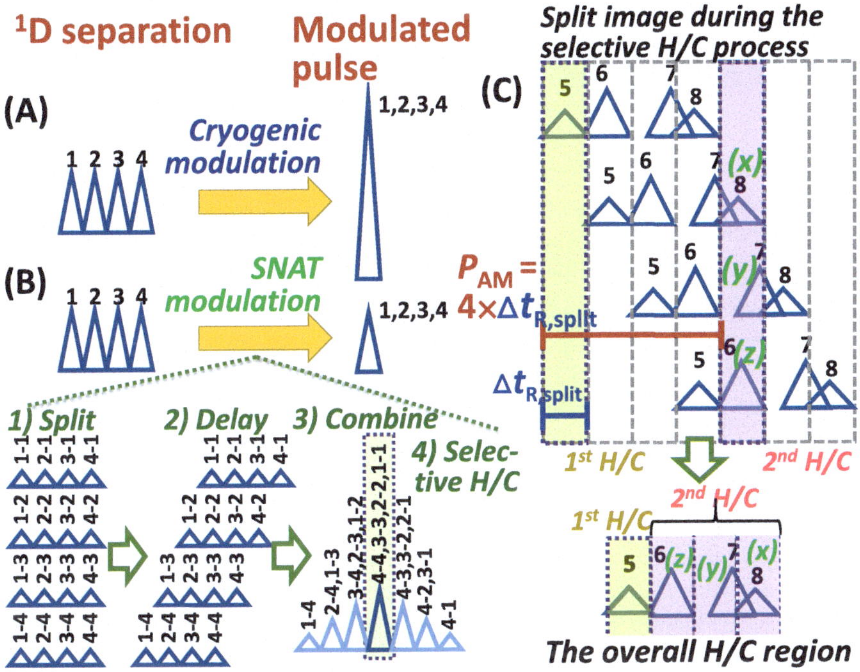

Fig. 12 Diagrams showing (**a**) conventional cryogenic trapping and (**b**) SNAT mechanisms with $n_{split} = 4$ for an artificial modulation of equally distributed peaks 1–4 with the same area and $\Delta t_{R,split}$ and (**c**) SNAT of randomly distributed peaks 5–8 with different areas. "–*x*" (where *x* = 1, 2, 3, or 4) in *B* indicates the *n*th split of each peak. (*x*), (*y*), and (*z*) in *C* represent the overlapped region within second H/C

1–3 in Fig. 12b. This can then be selectively sampled (or H/C) onto the ^{2}D column. After the selective H/C step, the other parts of split peaks are not sampled/transferred onto the ^{2}D column. This is equivalent to cryogenic focusing after the peak splitting (artificial split-and-trap process resulting in the modulated profile in Fig. 12b being the four-times split profile of that in Fig. 12a), providing that the H/C process in Fig. 12b covers the same zone as that of the trapped zone in Fig. 12a. The next H/C zone could be obtained within the next P_{AM}, defined as an artificial modulation period. This resulted in a similar modulated profile as that offered by a conventional cryogenic modulation period albeit with the periodic H/C mechanism. By heart-cutting all the artificially modulated zones to undergo ^{2}D separation, a comprehensive 2DGC analysis is obtained which samples the split overall ^{1}D peak profile to undergo ^{2}D separation. Note that the peak distribution with equal space is provided in Fig. 12a, b for the sake of simplicity. By viewing the split zones as pulses interval (see rectangular boxes in Fig. 12c), the concept is applicable for any kind of peak distribution. This is illustrated by the combined modulated regions from the first and second H/C (*x*, *y*, and *z*) in Fig. 12c containing split image of the whole peak profile obtained from ^{1}D separation. Further illustration of the split/

delay/combine mechanism and how peaks travel inside different parts of the system with the DS off (diverging the flow to FID) is provided as a clip in the supplementary information.

The successfulness of this mechanism relies on the splitting of a pulse of a sample from ^{1}D separation into a number of sub-pulses or peaks (n_{split}) with the same peak area and equal time difference ($\Delta t_{R,split}$ between any two adjacent split peaks of the same compound). $\Delta t_{R,split}$ can be experimentally measured for a given experimental condition. This leads to a specific region observed within every P_{AM} of $\Delta t_{R,split} \times n_{split}$ which contains the overlapped sub-pulse profile similar to that obtained from a cryogenic modulation within this P_{AM} (Fig. 12). All of these specific regions throughout ^{1}D separation time can then be selectively H/C to undergo ^{2}D separation, resulting in the comprehensive profile of the sub-pulses within a single injection. This is similar to the sample profile obtained from the comprehensive GC × GC with cryogenic modulation using the split ratio of n_{split} to 1. An ideal process for splitting of a pulse (or peak) should result in n_{split} split pulses each of which has the same amount (approximated herein with the same area) and $\Delta t_{R,split}$. This can be obtained using n_{split} splitters (restrictors) with different lengths and inner diameters. As illustrated in Fig. 13a, two-way splitting system may apply a 0.26 m × 0.10 mm (Splitter **1**), a 10 m × 0.25 mm restrictor (Splitter **2**) columns and two *y*-splitters connected at both ends of each splitter which resulted in $n_{split} = 2$ split pulses. These two columns have the same flow resistance and different void times ($t_{0,split}$) resulting in two split peaks with the same areas and $\Delta t_{R,split}$ of 5.00 min, Fig. 13b. By adjusting either ^{1}D column inlet pressure ($^1P_{in}$) or DS pressure (P_{DS}), $\Delta t_{R,split}$ can be finely tuned to result in a target $P_{AM} = \Delta t_{R,split} \times n_{split}$. This is experimentally illustrated by the relationship between P_{DS} and $\Delta t_{R,split}$ in Fig. 13c.

It should be noted at a fixed $\Delta t_{R,split}$ value that too long splitters **1** and **2** cause high back pressure for the system, while too short splitters lead to low ^{1}D column flow causing peak broadening. In order to increase separation performance, P_{AM}/t_{samp} is required to be higher [46, 47]. This could be straightforwardly obtained using more splitters increasing n_{split}. It is not practical to use several restrictors with different

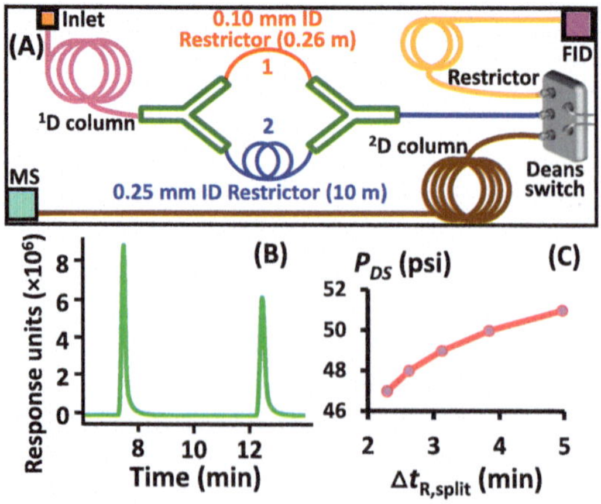

Fig. 13 SNAT analysis using Splitter **1**, Splitter **2**, and $n_{split} = 2$. (**a**) System configuration, (**b**) FID chromatogram of MeOH with the DS diverging the flow to FID, and (**c**) the experimental plot between P_{DS} and $\Delta t_{R,split}$ at 40 °C. The two peaks in B had the area of 9×10^7counts s

diameters due to the limited availability from the suppliers. A possible approach is to apply each splitter containing two restrictors with different ID (herein 0.25 and 0.10 mm) and different lengths in order to result in tunable flow resistance and $t_{0,\mathrm{split}}$ of each splitter. The example system of $n_{\mathrm{split}} = 4$ which applied splitters **1**, **2**, **3**, and **4** to result in four split peaks with about the same area and $\Delta t_{\mathrm{R,split}}$ is provided in Fig. 14.

In addition, the same group also successfully developed a new design called the multiloop splitter-based non-cryogenic artificial trapping (M-SNAT) which decreased the use of numerous splitters as well as eliminating leakage troubles in the primary system [48]. M-SNAT modulation technique was established which employed the ^{1}D semi-nonpolar, the ^{2}D polar, and deactivated fused silica (DFS) columns, splitters, and a DS. The DFS columns and splitters located between the ^{1}D column outlet and the DS were connected into multiple loops with progressive doubling of the length of the next loop. This enables duplex splitting mechanism, consisting of splitting of analyte pulses, the pulse delay and their combination, which leads to equal split peaks of the same analyte with the number of the split peaks (n_{split}) equal to 2^m (m = number of loops). This results in local profiles of artificially split-and-trapped analytes prior to their selective transfers onto the ^{2}D column by means of periodic multiple heartcuts (H/C). The developed SNAT approach is successful when the ratio of sampling time to modulation period ($t_{\mathrm{samp}}/P_{\mathrm{M}}$) is equal to n_{split}. The experimental setup of the M-SNAT system with $n_{\mathrm{split}} = 4$ is shown in Fig. 15. This contains two splitter loops containing two 0.18 μm ID and two 0.25 μm ID restrictors that were located after the ^{1}D column. The splitting process starts with a peak/pulse traveling from the ^{1}D column outlet to enter the first splitter loop (Fig. 15a). This splits the peak into two subpeaks (see *a* and *b* in the splitting diagram) traveling with different void time of $2\Delta t_{\mathrm{R,split}}$. *a* and *b* then enter the second splitter loop with the halved restrictor lengths and further split into four subpeaks (*a*1, *a*2, *b*1, and *b*2 with $\Delta t_{\mathrm{R,split}}$ between any two adjacent peaks, Fig. 15a). This design enables the split-delay-combination mechanism resulting in four equally split subpeaks of the solvents as shown by the experimental FID result in Fig. 15b. Periodic H/C (first to fourth) was then performed on the specific regions. In order to achieve a comprehensive 2DGC analysis, the H/C window should be equal to $\Delta t_{\mathrm{R,split}}$ which is also equal to the t_{R} difference between any two adjacent split peaks of the same compound in Fig. 15b. Furthermore, $P_{\mathrm{AM}} = \Delta t_{\mathrm{R,split}} \times n_{\mathrm{split}}$. The comprehensive

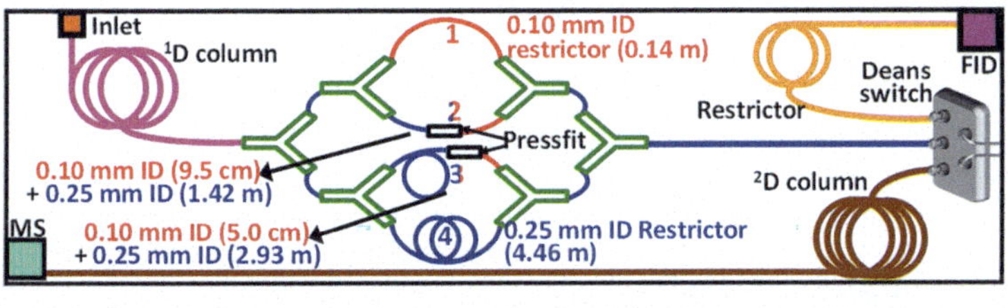

Fig. 14 System configuration; SNAT analysis using splitter **1**, **2**, **3**, and **4** consisting of 0.25 and/or 0.10 mm ID DFS columns and $n_{\mathrm{split}} = 4$

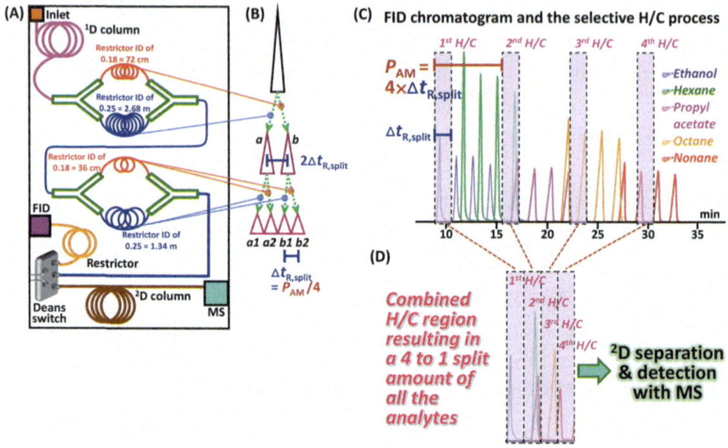

Fig. 15 M-SNAT analysis using $n_{split} = 4$ and 0.18 and 0.25 mm ID restrictors. (**a**) System configuration, (**b**) splitting diagram, (**c**) the overlaid split profiles of the solvent mixture obtained from FID with the DS diverging the flow to FID, and (**d**) the combined H/C region using $P_{AM} = 4.57$ min, $^{I}P_{DS}$, and isothermal separation at 40 °C

profile is confirmed in Fig. 15c where the whole profiles of all the solvents were H/C. Each H/C pulse then underwent ^{2}D separation.

The method was demonstrated for modulation of a wide range of compounds in petrochemical samples with $n_{split} = 8$ and 16 [48]. The application enables artificial split-and-trapped modulation mechanism without cryogen consumption and enhanced ^{2}D peak capacity ($^{2}n_{c}$) by use of a longer ^{2}D column.

M-SNAT was applied for analysis of volatile compounds in headspaces of axillary swab samples of PCR-proven COVID-19 positive and negative volunteers in Bangkok, Thailand [49]. The M-SNAT system configuration is shown in Fig. 16. Feature selection found 14 significant metabolites that differed between the COVID-19 positive and negative participants, including *p*-cymene, linalool, and 2,6,11-trimethyldodecane. The marker peak area thresholds were optimized by generating receiver operating characteristic (ROC) curves showing accuracy, sensitivity, and selectivity within the ranges of 94–98%, 93–97%, and 94–100%, respectively. The sweat-based COVID-19 screening method is also non-invasive and generates lower amounts of biohazard wastes (e.g., compared with the conventional rapid antigen test kits). While external validation of these tentative biomarkers is needed before implementing this approach in the clinic, this work demonstrates the ability of GC × GC-MS with the SNAT modulation system and chemometrics to explore the sweat metabolome of patients with SARS-CoV-2 infection.

MDGC for Food Analysis

The number of chemicals in a food sample is often larger than the analyte peak capacity in conventional GC using a single column, which could lead to several unresolved peaks that hinder the important signals in fingerprinting analyses and

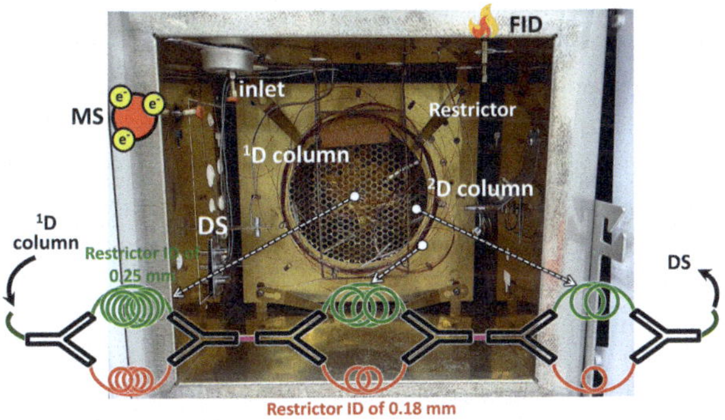

Fig. 16 M-SNAT system configuration for analysis of COVID-19 sweat samples. The image was reproduced with the permission [49]

adulteration tests. MDGC has thus been developed to improve the separation performance. The technique generally uses two columns with different selectivity toward a sample component. These two columns are connected sequentially via a device offering an effective heartcut (H/C) selectively transferring some analytes to undergo ^{2}D separation or modulation process transferring all the analytes to undergo ^{2}D separation. Each separated peak is now described by retention times and widths in the first- and second-dimensional separations, as well as the peak area and height, which could be used for the quantitative analysis. Compared with 1DGC, MDGC provides higher separation resolution and an improved number of separated peaks with reduced interferences. Due to the complex sample analysis capability, sample preparation steps (such as liquid phase extraction and solid phase extraction) can also be avoided, reducing sample loss during the preparation. This is crucial when the sample amount is limited. MDGC applications in food analysis have been reported with the example provided in Fig. 17.

There are two modes of separation in MDGC. A DS device can be applied to allow an effective H/C, selecting at least one specific pulse from a ^{1}D column to be further separated on a long ^{2}D column, e.g., 20–60 m, and this technique is called H/C MDGC. The other mode is recognized as GC × GC employing a modulator to offer a modulation process encompassing all pulses from a ^{1}D column to be separated onto a short ^{2}D column (e.g., <5 m). Such a modulation process occurs for each pulse within every modulation period (e.g., <12 s).

A universal instrumental approach to perform both H/C MDGC and GC × GC is shown in Fig. 17 which could contain one ^{1}D column and two ^{2}D columns with a microfluidic device called as Deans switch (DS, enabling flow switching to provide different modes of MDGC separation using a short ^{2}D column or a long ^{2}D column). H/C MDGC can be more suitable for high confidence analysis of targeted analytes that are unresolved from the matrix in ^{1}D separation, while GC × GC can be more suitable to obtain an overall profile or untargeted compound analysis within a short time. Both modes could be applicable with quantitative analysis via injections of a standard compound and construction of the calibration curve.

Characterization of chemical constituents in food (either qualitative or quantitative) by MDGC approaches is essential for the improved assessment of food safety and quality as well as authenticity and adulteration test. This enables the analysis

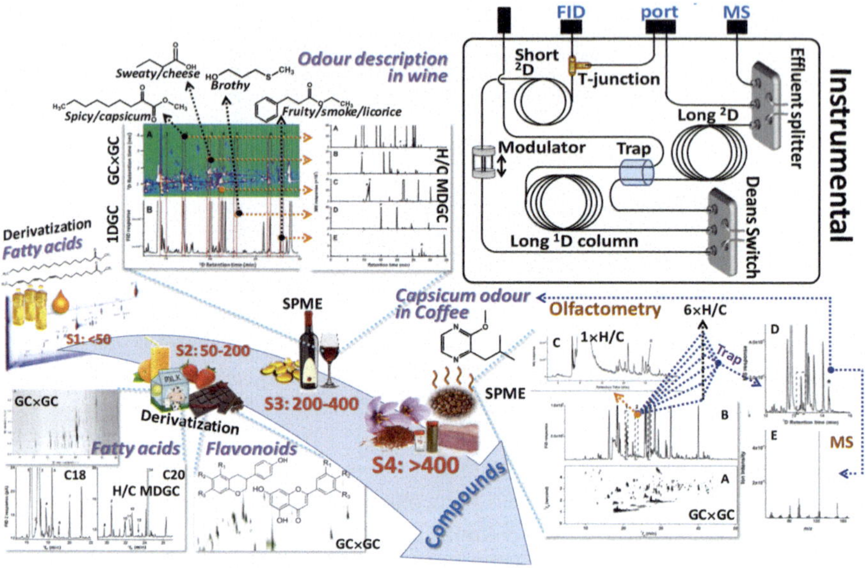

Fig. 17 Example MDGC instrumental configuration and the applications for analyses of food samples with different complexity: S1 = fatty acid methyl esters in derivatized vegetable oil, S2 = trimethylsilyl derivatives of flavonoids in derivatized chocolate, S3 = odor compounds in wine, and S4 = odor compounds in coffee. FID = flame ionization detector, port = olfactory port for sniffing analysis, and trap = cryogenic trap for selective trap and enrichment of target analytes. Adapted from [50]

and differentiation of food products with more descriptive and informative parameters for the evaluation of food compared to the conventional approach, which is based on external factors, e.g., smell, texture, flavor, or color. Figure 17 shows an example of MDGC application for analyzing food samples with different degrees of complexity according to number of volatile compounds. They could be classified into S1, S2, S3, or S4 group. S1 are relatively low-complexity samples such as FAME in derivatized vegetable oil containing ≤50 volatile compounds. The conventional 1DGC could analyze this sample group. S2 are medium-complexity samples consisting of 50–200 compounds, such as volatile compounds in essential oils. They could still be adequately analyzed with 1DGC. S3 are high-complexity samples containing 200–400 volatiles such as wine headspace or FAME in derivatized fish oil. S4 are very high-complexity samples (S4) consisting of >400 compounds, e.g., coffee headspace, spices, metabolomics, or industrially roasted food product samples. 1DGC analysis capability is generally insufficient to analyze the S3 and S4 groups. MDGC could thus be applied to separate and analyze these samples effectively. MDGC can also provide additional information of I in both first- and second-dimensional separations which can be useful for improved confidence in the identification of volatile compounds in a sample with any complexity.

It should be noted that good selection of the column set applied in MDGC requires orthogonality (degree of selectivity difference) between the two columns. This means each column should provide different selectivity toward different

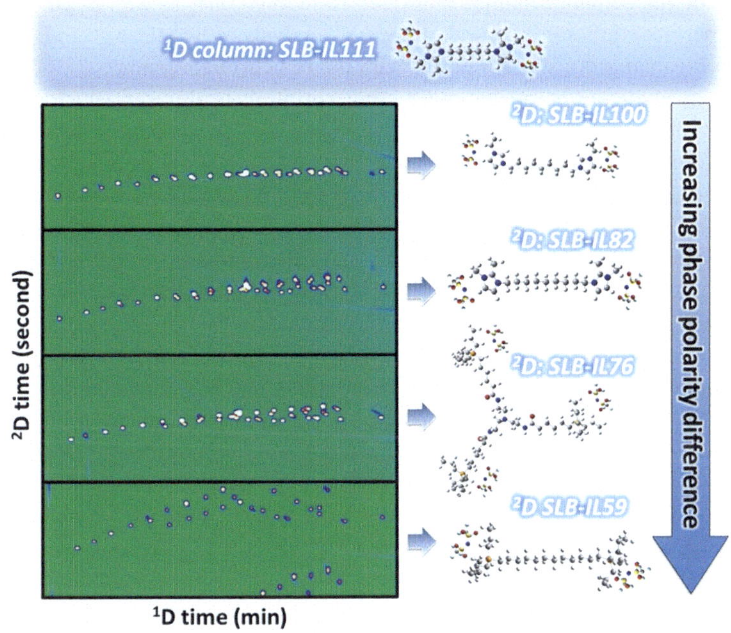

Fig. 18 GC × GC results for separations of fatty acid methyl esters using the column sets with increasing degrees of orthogonality from the top to the bottom. The image was reproduced with permission [51]

components of each sample. A better choice of column set is generally expected for the two columns with the higher degree of orthogonality. For example, by using SLB-IL111 with the highest polarity column as the ^{1}D column, use of a ^{2}D SLB-IL59 column with lower polarity (as indicated by the lower number after the code "SLB-IL") resulted in more effectively separated and distributed peaks of FAME in the 2D space. This is illustrated by the contour plots showing better separations from the top to the bottom (Fig. 18).

Adulteration Analysis with GC and GC × GC Fingerprinting

The role of GC fingerprinting in adulteration tests could be recognized by simulation of the results analyzing a model of authentic edible oil containing different fatty acid methyl esters (FAME) compared with the adulterated samples. Adulteration of the oil herein was performed by mixing various ratios of five different based oils containing the same set of FAME albeit with different amounts. The resulting chromatograms under isothermal separation using a poly(siloxane) column, which separated the adulterated samples with different relative amounts of the based oils, were simulated (blue chromatogram) compared with the authentic oil sample (red dotted chromatograms) in Fig. 19. This illustrates the technique's capability to differentiate the authentic sample from the adulterated ones as well as guide how to perform perfect adulteration based on the GC analysis of FAME, e.g., Fig. 19d with the blue chromatogram superimposed with the red chromatogram.

Example applications of GC × GC fingerprinting to differentiate food, petrochemical, and essential samples are shown in Fig. 20. These include differentiation of plum brandies from different sources according to the volatile compound analyses, good-quality and adulterated (poor-quality) fuel samples based on their different sulfur compound compositions, and agarwood oil samples with different grades based on different terpene compositions. With the enhanced separation power of GC × GC, their volatile compound fingerprints could be clearly separated even with eyes.

For more complicated cases, the fingerprinting results could be compared based on the similarity concepts which could be based on match/reverse match scores [55] (e.g., between the suspected and the authentic food samples), or based on statistical analysis such as principal component analysis (PCA). Using PCA, dimensionality of a huge dataset can be reduced by converting many variables into smaller ones that retain most of the original information set. Note that accuracy in sample classification is reduced with a smaller number of variables in a dataset. However, the data

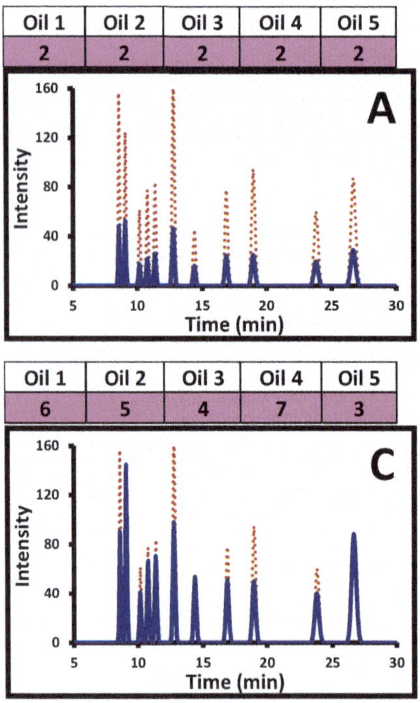

Fig. 19 Simulated GC-FAME chromatograms illustrating the ability of the method to distinguish the authentic oil (red chromatogram) from adulterated samples prepared with different blending ratios of oils 1–5 (blue chromatograms). Panels **A**–**C** show examples of unsuccessful matching where the adulterated oil cannot perfectly reproduce the authentic profile, while panel **D** demonstrates a "perfect adulteration" case, in which the corrected blend (blue) closely overlaps with the authentic target (red), indicating how the approach can also be used to guide the formulation of counterfeit mixtures

can be simply classified and visualized, e.g., each sample can be represented by a dot in a score plot as reported for analysis of different coffee samples [56].

M-SNAT modulation technique was developed, miniaturized, and applied in comprehensive two-dimensional gas chromatography (GC × GC) for analysis of four different types of cannabis samples traditionally processed and consumed in Thailand [57]. Selective H/C were then periodically performed to transfer the split-and-trap comprehensive profiles onto the ^{2}D column. These profiles after 2D separations were then detected by MS, which resulted in the intensity vs time profiles,

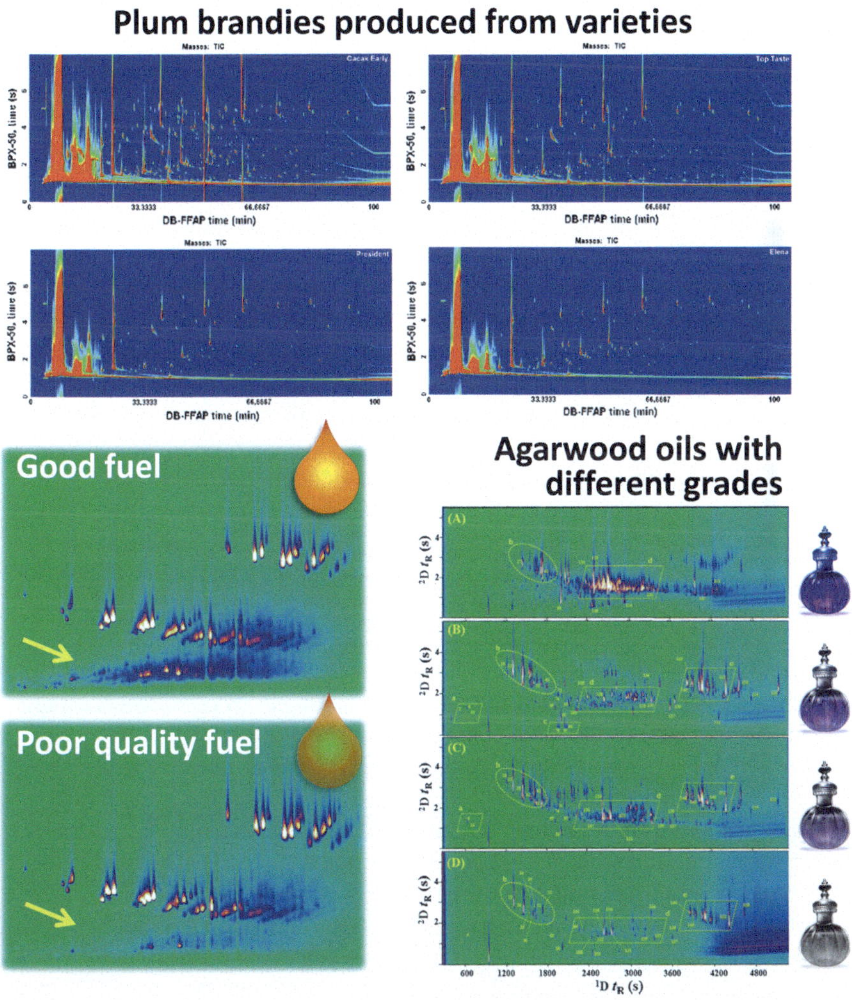

Fig. 20 Chemical fingerprint results of different plum brandies, fuel, and agarwood oil samples obtained using comprehensive two-dimensional gas chromatography (GC × GC). The image was reproduced with permission [52–54]

which can be considered as the artificial 1D modulated chromatograms as illustrated in Fig. 21b. This artificial modulation system was applied to analyze cannabis samples with enhanced ^{2}D peak capacity ($^{2}n_{c}$ ~ 15). These chromatograms were then transformed into the matrices before generating the contour plots shown in Fig. 21a. This showed effective modulation of a wide range of compounds with the carbon number lower than C6 and larger than C20. This can thus present the properties of this modulator that could trap highly volatile compounds and high molecular weight compounds. Note that the separation condition was selected with the significant degree of wraparound to effectively use the 2D separation space. This is required since the GC × GC system employed a single oven applying the same ^{1}D and ^{2}D temperature which limits the peak distribution in the 2D space.

The method was applied to analyze cannabis extracts using vegetable oils with or without frying process. This reveals 454 different peaks with 76, 92, 35, and 70 specific components specifically observed by using olive oil extraction (OE), fried OE, coconut oil extraction (CE), and fried CE, respectively [57]. The separated components can then be identified based on the conventional data analysis approach in GC × GC [47]. Under different procedures, each cannabis was extracted in olive or coconut oils at 110 °C (labeled as OE and CE, respectively) or under frying conditions (labeled as F-OE and F-CE, respectively). Volatile component distribution in different cannabis extracts is shown by the Venn diagram in Fig. 22. Twenty-six different peaks were observed in all the samples which were mainly identifiable as

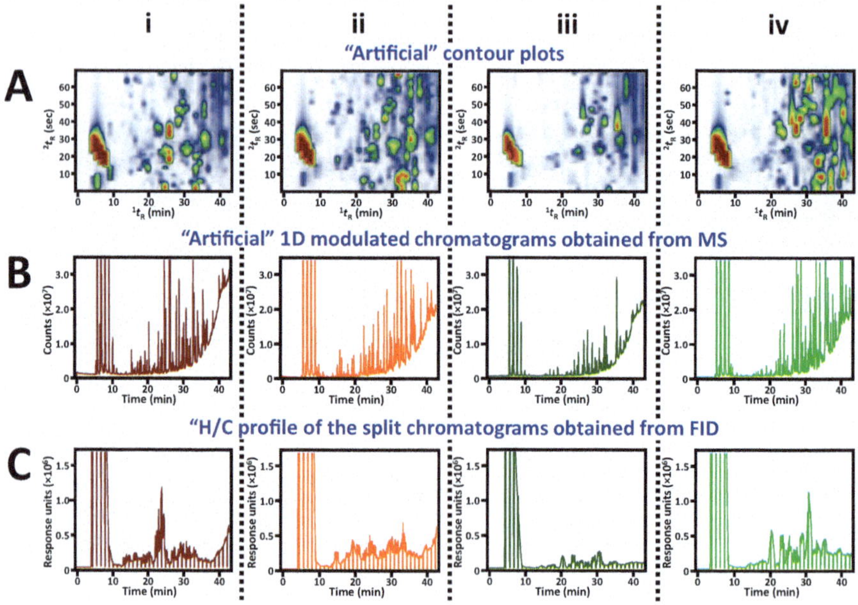

Fig. 21 M-SNAT analysis results for four cannabis samples: (i) OE, (ii) F-OE, (iii) CE, and (iv) F-CE. (**a**) Contour plots, (**b**) 1D modulated total ion chromatograms, and (**c**) FID profiles after H/C, using P_M = 1.143 min and n_{split} = 8

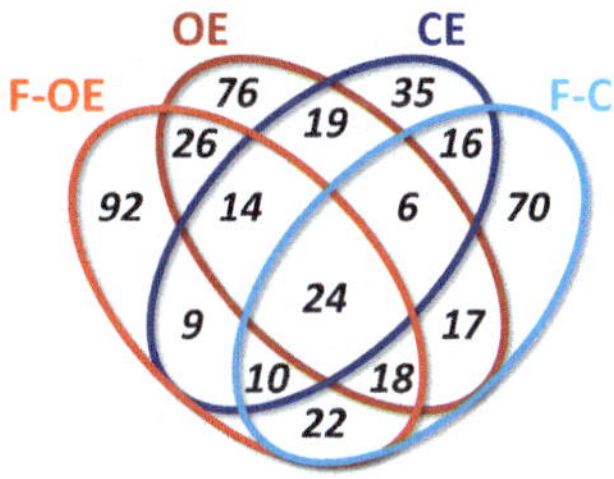

Fig. 22 Venn diagram showing numbers of volatile components observed in different cannabis extract samples by using M-SNAT GC × GC approach

alkanes (hexane, decane, dodecane, and 2,6,10-trimethyldodecane), ketones (camphenilone, α-cetone, and farnesyl acetone), aldehydes (nonanal and *trans*-2-decenal), and alcohols (*trans*-2-pinanol, α-terpineol, 2-methylisoborneol, widdrol, 2-hexadecanol). This indicates an efficient extraction of these hydrophobic components in both oils regardless of the frying process.

Conclusion

This chapter provides an overview of GC's capabilities and technologies in volatile compound fingerprinting and adulteration tests, focusing on food analysis. These commence from the conventional 1DGC analysis to the higher performance MDGC with different variants. With the expense of more complicated instrumental system configuration, MDGC offers excellent separation efficiency that serves advanced characterization of volatile compounds in food samples. The full capability of this technique is expected to be more widely applied in the future following the growing need for fingerprinting or monitoring of more complex food samples for adulteration, chemical information, toxicity, or the discovery of benefits of volatile and compounds.

References

1. Jiang M, Kulsing C, Nolvachai Y, Marriott PJ (2015) Two-dimensional retention indices improve component identification in comprehensive two-dimensional gas chromatography of saffron. Anal Chem 87(11):5753–5761. https://doi.org/10.1021/acs.analchem.5b00953
2. González M, Gallego M, Valcárcel M (1998) Simultaneous gas chromatographic determination of food preservatives following solid-phase extraction. J Chromatogr A 823(1):321–329. https://doi.org/10.1016/S0021-9673(98)00182-4
3. Gherghel S, Morgan RM, Arrebola-Liébanas J, Romero-González R, Blackman CS, Garrido-Frenich A, Parkin IP (2018) Development of a HS-SPME/GC–MS method for the analysis of volatile organic compounds from fabrics for forensic reconstruction applications. Forensic Sci Int 290:207–218. https://doi.org/10.1016/j.forsciint.2018.07.015
4. Giddings JC (1984) Two-dimensional separations: concept and promise. Anal Chem 56(12):1258A–1270A. https://doi.org/10.1021/ac00276a003
5. Deans DR (1965) An improved technique for back-flushing gas chromatographic columns. J Chromatogr A 18:477–481. https://doi.org/10.1016/S0021-9673(01)80403-9

6. Tranchida PQ, Purcaro G, Dugo P, Mondello L, Purcaro G (2011) Modulators for comprehensive two-dimensional gas chromatography. TrAC Trends Anal Chem 30(9):1437–1461. https://doi.org/10.1016/j.trac.2011.06.010
7. Paul D, Kurabayashi K (2016) First-principle modeling and characterization of thermal modulation in comprehensive two-dimensional gas chromatography using a microfabricated device. Sensors Actuators B Chem 231:135–146. https://doi.org/10.1016/j.snb.2016.02.132
8. Kulsing C, Nolvachai Y, Wong YF, Glouzman MI, Marriott PJ (2018) Observation and explanation of two-dimensional interconversion of oximes with multiple heart-cutting using comprehensive multidimensional gas chromatography. J Chromatogr A 1546:97–105. https://doi.org/10.1016/j.chroma.2018.02.035
9. Liu Z, Phillips JB (1991) Comprehensive two-dimensional gas chromatography using an on-column thermal modulator interface. J Chromatogr Sci 29(6):227–231. https://doi.org/10.1093/chromsci/29.6.227
10. Cai H, Stearns SD (2018) A comprehensive two-dimensional gas chromatography valve modulation method using hold-release primary column flow for long secondary separation time with 100% transfer. J Chromatogr A 1569:200–211. https://doi.org/10.1016/j.chroma.2018.07.064
11. Giri A, Khummueng W, Mercier F, Kondjoyan N, Tournayre P, Meurillon M, Ratel J, Engel E (2015) Relevance of two-dimensional gas chromatography and high resolution olfactometry for the parallel determination of heat-induced toxicants and odorants in cooked food. J Chromatogr A 1388:217–226. https://doi.org/10.1016/j.chroma.2015.01.045
12. Mitrevski B, Webster RL, Rawson P, Evans DJ, Choi H-K, Marriott PJ (2012) Multidimensional gas chromatography of oxidative degradation products in algae-derived fuel oil samples using narrow heartcuts and rapid cycle times. J Chromatogr A 1224:89–96. https://doi.org/10.1016/j.chroma.2011.12.051
13. Deans DR (1981) Use of heart cutting in gas chromatography: A review. J Chromatogr A 203:19–28. https://doi.org/10.1016/S0021-9673(00)80278-2
14. Sharif KM, Chin S-T, Kulsing C, Marriott PJ (2016) The microfluidic Deans switch: 50 years of progress, innovation and application. TrAC Trends Anal Chem 82:35–54. https://doi.org/10.1016/j.trac.2016.05.005
15. Wang X, Guo J, Shang J, Ding L, Zhao G, Xie F, Jia Y, Qin Y, Yu Y, Chen L, Zhang S (2015) Determination of nitroalkanes in mainstream cigarette smoke by heartcutting multidimensional gas chromatography system coupled with mass spectrometry detection. J Chromatogr A 1424:118–126. https://doi.org/10.1016/j.chroma.2015.10.078
16. Janta P, Pinyo D, Yodta Y, Vasasiri P, Weidenbach M, Pursch M, Yang X, Kulsing C (2021) Strategies towards simpler configuration and higher peak capacity with comprehensive multidimensional gas chromatography. RSC Adv 11:7946–7953. https://doi.org/10.1039/D0RA10495G
17. Janta P, Pinyo D, Yodta Y, Vasasiri P, Weidenbach M, Pursch M, Yang X, Kulsing C (2021) A multi-location peak parking approach for calculation of second dimensional retention indices for improved volatile compound identification with cryogen-free comprehensive heartcut two-dimensional gas chromatography. Anal Methods. https://doi.org/10.1039/D0AY01976C
18. Thongdorn-Ae N, Nhujak T, Janta P, Rueangthaweep A, Hinchiranan N, Kulsing C (2020) Cryogen-free comprehensive heartcut multidimensional gas chromatography using a Deans switch for improved analysis of petrochemical products derived from palmitic acid oxidation. Anal Methods 12:5160–5167. https://doi.org/10.1039/D0AY01527J
19. Tungkijanansin N, Alahmad W, Nhujak T, Varanusupakul P (2020) Simultaneous determination of benzoic acid, sorbic acid, and propionic acid in fermented food by headspace solid-phase microextraction followed by GC-FID. Food Chem 329:127161. https://doi.org/10.1016/j.foodchem.2020.127161
20. Magagna F, Valverde-Som L, Ruíz-Samblás C, Cuadros-Rodríguez L, Reichenbach SE, Bicchi C, Cordero C (2016) Combined untargeted and targeted fingerprinting with comprehensive two-dimensional chromatography for volatiles and ripening indicators in olive oil. Anal Chim Acta 936:245–258. https://doi.org/10.1016/j.aca.2016.07.005

21. Carvalho LM, Carvalho F, de Lourdes Bastos M, Baptista P, Moreira N, Monforte AR, da Silva Ferreira AC, de Pinho PG (2014) Non-targeted and targeted analysis of wild toxic and edible mushrooms using gas chromatography–ion trap mass spectrometry. Talanta 118:292–303. https://doi.org/10.1016/j.talanta.2013.09.038
22. Bignardi C, Cavazza A, Corradini C, Salvadeo P (2014) Targeted and untargeted data-dependent experiments for characterization of polycarbonate food-contact plastics by ultra high performance chromatography coupled to quadrupole Orbitrap tandem mass spectrometry. J Chromatogr A 1372:133–144. https://doi.org/10.1016/j.chroma.2014.10.104
23. Huang J, Gan N, Lv F, Cao Y, Ou C, Tang H (2016) Environmentally friendly solid-phase microextraction coupled with gas chromatography and mass spectrometry for the determination of biogenic amines in fish samples. J Sep Sci 39(22):4384–4390. https://doi.org/10.1002/jssc.201600893
24. Ding M, Liu W, Peng J, Liu X, Tang Y (2018) Simultaneous determination of seven preservatives in food by dispersive liquid-liquid microextraction coupled with gas chromatography-mass spectrometry. Food Chem 269:187–192. https://doi.org/10.1016/j.foodchem.2018.07.002
25. Giddings JC (1987) Concepts and comparisons in multidimensional separation. J High Resolut Chromatogr 10:319–323
26. Simmons MC, Snyder LR (1958) Two-stage gas-liquid chromatography. Anal Chem 30(1):32–35. https://doi.org/10.1021/ac60133a007
27. Amaral MSS, Nolvachai Y, Marriott PJ (2020) Comprehensive two-dimensional gas chromatography advances in technology and applications: biennial update. Anal Chem 92(1):85–104. https://doi.org/10.1021/acs.analchem.9b05412
28. Novaes FJM, Silva Junior AID, Kulsing C, Nolvachai Y, Bizzo HR, Neto FRDA, Rezende CM, Marriott PJ (2019) New approaches to monitor semi-volatile organic compounds released during coffee roasting using flow-through/active sampling and comprehensive two-dimensional gas chromatography. Food Res Int 119:349–358. https://doi.org/10.1016/j.foodres.2019.02.009
29. Nolvachai Y, Kulsing C, Sharif KM, Wong YF, Chin ST, Mitrevski B, Marriott PJ (2018) Multi-column trajectory to advanced methods in comprehensive two-dimensional gas chromatography. TrAC - Trends in Analytical Chemistry 106:11–20. https://doi.org/10.1016/j.trac.2018.06.012
30. Seeley JV (2012) Recent advances in flow-controlled multidimensional gas chromatography. J Chromatogr A 1255:24–37. https://doi.org/10.1016/j.chroma.2012.01.027
31. Górecki T, Harynuk J, Panić O (2004) The evolution of comprehensive two-dimensional gas chromatography (GC × GC). J Sep Sci 27(5–6):359–379. https://doi.org/10.1002/jssc.200301650
32. Bruckner CA, Prazen BJ, Synovec RE (1998) Comprehensive two-dimensional high-speed gas chromatography with chemometric analysis. Anal Chem 70(14):2796–2804. https://doi.org/10.1021/ac980164m
33. Oldridge N, Panic O, Górecki T (2008) Stop-flow comprehensive two-dimensional gas chromatography with pneumatic switching. J Sep Sci 31(19):3375–3384. https://doi.org/10.1002/jssc.200800265
34. Harynuk J, Górecki T (2004) Comprehensive two-dimensional gas chromatography in stop-flow mode. J Sep Sci 27(5–6):431–441. https://doi.org/10.1002/jssc.200301649
35. Seeley JV, Kramp F, Hicks CJ (2000) Comprehensive two-dimensional gas chromatography via differential flow modulation. Anal Chem 72(18):4346–4352. https://doi.org/10.1021/ac000249z
36. Seeley JV, Kramp FJ, Sharpe KS, Seeley SK (2002) Characterization of gaseous mixtures of organic compounds with dual-secondary column comprehensive two-dimensional gas chromatography (GC×2 GC). J Sep Sci 25(1–2):53–59. https://doi.org/10.1002/1615-9314(20020101)25:1/2<53::Aid-jssc53>3.0.Co;2-v
37. Deans DR (1968) A new technique for heart cutting in gas chromatography (1). Chromatographia 1(1):18–22. https://doi.org/10.1007/BF02259005

38. Prebihalo SE, Berrier KL, Freye CE, Bahaghighat HD, Moore NR, Pinkerton DK, Synovec RE (2018) Multidimensional gas chromatography: advances in instrumentation, chemometrics, and applications. Anal Chem 90(1):505–532. https://doi.org/10.1021/acs.analchem.7b04226
39. Seeley JV, Seeley SK (2013) Multidimensional gas chromatography: fundamental advances and new applications. Anal Chem 85(2):557–578. https://doi.org/10.1021/ac303195u
40. Amaral MSS, Marriott PJ (2019) The blossoming of technology for the analysis of complex aroma bouquets-a review on flavour and odorant multidimensional and comprehensive gas chromatography applications. Molecules 24(11):2080. https://doi.org/10.3390/molecules24112080
41. Amer MW, Mitrevski B, Roy Jackson W, Chaffee AL, Marriott PJ (2014) Multidimensional and comprehensive two-dimensional gas chromatography of dichloromethane soluble products from a high sulfur Jordanian oil shale. Talanta 120:55–63. https://doi.org/10.1016/j.talanta.2013.11.069
42. Cannon RJ, Agyemang D, Curto NL, Yusuf A, Chen MZ, Janczuk AJ (2015) In-depth analysis of Ciflorette strawberries (Fragaria × ananassa 'Ciflorette') by multidimensional gas chromatography and gas chromatography-olfactometry. Flavour Fragr J 30(4):302–319. https://doi.org/10.1002/ffj.3246
43. Wang FC-Y (2017) Comprehensive three-dimensional gas chromatography mass spectrometry separation of diesel. J Chromatogr A 1489:126–133. https://doi.org/10.1016/j.chroma.2017.01.091
44. Guan X, Zhao Z, Cai S, Wang S, Lu H (2019) Analysis of volatile organic compounds using cryogen-free thermal modulation based comprehensive two-dimensional gas chromatography coupled with quadrupole mass spectrometry. J Chromatogr A 1587:227–238. https://doi.org/10.1016/j.chroma.2018.12.025
45. Tungkijanansin N, Kulsing C (2022) Splitter-based non-cryogenic artificial trapping (SNAT): novel modulation mechanism in comprehensive two dimensional gas chromatography. J Chromatogr Open 2:100059. https://doi.org/10.1016/j.jcoa.2022.100059
46. Schöneich S, Gough DV, Trinklein TJ, Synovec RE (2020) Dynamic pressure gradient modulation for comprehensive two-dimensional gas chromatography with time-of-flight mass spectrometry detection. J Chromatogr A 1620:460982. https://doi.org/10.1016/j.chroma.2020.460982
47. Kulsing C, Nolvachai Y, Marriott PJ (2020) Concepts, selectivity options and experimental design approaches in multidimensional and comprehensive two-dimensional gas chromatography. TrAC Trends Anal Chem 130:115995. https://doi.org/10.1016/j.trac.2020.115995
48. Tungkijanansin N, Nolvachai Y, Varanusupakul P, Hinchiranan N, Kulsing C, Marriott PJ (2023) Development and application of a novel multiloop splitter-based non-cryogenic artificial trapping modulation system in comprehensive two-dimensional gas chromatography. Anal Chem 95(25):9437–9444. https://doi.org/10.1021/acs.analchem.2c04710
49. Tungkijanansin N, Giebelhaus RT, Schmidt SA, Nhujak T, Chatdarong K, Torvorapanit P, Harynuk JJ, Kulsing C (2024) Identification of coronavirus disease marker compounds in sweat with comprehensive two dimensional gas chromatography using multiloop splitter-based non-cryogenic artificial trapping modulation system. J Chromatogr Open 5:100113. https://doi.org/10.1016/j.jcoa.2023.100113
50. Nolvachai Y, Kulsing C, Marriott PJ (2017) Multidimensional gas chromatography in food analysis. TrAC Trends Anal Chem 96:124–137. https://doi.org/10.1016/j.trac.2017.05.001
51. Kulsing C, Hügel HM, Nolvachai Y, Marriott P (2015) Developments in gas chromatography using ionic liquid stationary phases. LCGC Europe 28:434–440
52. Vyviurska O, Matura F, Furdíková K, Špánik I (2017) Volatile fingerprinting of the plum brandies produced from different fruit varieties. J Food Sci Technol 54(13):4284–4301. https://doi.org/10.1007/s13197-017-2900-5
53. Kulsing C, Rawson P, Webster RL, Evans DJ, Marriott PJ (2017) Group-type analysis of hydrocarbons and sulfur compounds in thermally stressed merox jet fuel samples. Energy Fuel 31(9):8978–8984. https://doi.org/10.1021/acs.energyfuels.7b01119

54. Wong YF, Chin S-T, Perlmutter P, Marriott PJ (2015) Evaluation of comprehensive two-dimensional gas chromatography with accurate mass time-of-flight mass spectrometry for the metabolic profiling of plant–fungus interaction in Aquilaria malaccensis. J Chromatogr A 1387:104–115
55. Cordero C, Liberto E, Bicchi C, Rubiolo P, Schieberle P, Reichenbach SE, Tao Q (2010) Profiling food volatiles by comprehensive two-dimensional gas chromatography coupled with mass spectrometry: advanced fingerprinting approaches for comparative analysis of the volatile fraction of roasted hazelnuts (Corylus avellana L.) from different origins. J Chromatogr A 1217(37):5848–5858. https://doi.org/10.1016/j.chroma.2010.07.006
56. Thammarat P, Kulsing C, Wongravee K, Leepipatpiboon N, Nhujak T (2018) Identification of volatile compounds and selection of discriminant markers for elephant dung coffee using static headspace gas chromatography—mass spectrometry and chemometrics. Molecules 23(8):1910
57. Tungkijanansin N, Vimolmangkang S, Kulsing C, Varanusupakul P (2024) Comprehensive two-dimensional gas chromatography using a miniaturized multiloop splitter-based non-cryogenic artificial trapping (M-SNAT) modulation device for analysis of cannabis samples. J Pharm Biomed Anal 249:116395. https://doi.org/10.1016/j.jpba.2024.116395

Microfluidic Paper-Based Analytical Devices as Cost-Effective Tools for Food Adulteration

Hany Abd El-Raheem, Samah Abdelraheem, Rabiaa Helim, Heba Raheem, Ahmed F. A. Youssef, and Waleed Alahmad

Contents

H. A. El-Raheem
Environmental Engineering Program, Zewail City of Science and Technology, Giza, Egypt

Faculty of Postgraduate Studies for Advanced Sciences, Materials Science and Nanotechnology Department, Beni-Suef University, Beni-Suef, Egypt

S. Abdelraheem
Modern University for Technology and Information, Cairo, Egypt

R. Helim
Laboratory of Applied Energetics and Materials, University of Jijel, Jijel, Algeria

H. Raheem
Faculty of Pharmacy, Sohag University, Sohag, Egypt

A. F. A. Youssef
Environmental Engineering Program, Zewail City of Science and Technology, Giza, Egypt

Faculty of Science, Department of Chemistry, Cairo University, Giza, Egypt

W. Alahmad (✉)
Faculty of Science, Department of Chemistry, Chulalongkorn University, Bangkok, Thailand
e-mail: waleed.al@chula.ac.th

W. Alahmad, M. Ashfaq (eds.), *Advanced Techniques against Food Adulteration*,
https://doi.org/10.1007/978-3-031-99454-8_4

Abstract Microfluidic paper-based analytical devices (μPADs) have emerged as transformative tools for food adulteration detection, offering cost-effective, portable, and sustainable solutions. This chapter explores the advancements in μPAD technology, including fabrication methods, sample preparation techniques, and detection strategies. Various fabrication techniques such as photolithography, wax printing, laser cutting, and inkjet printing are discussed, each offering unique precision, resolution, and production efficiency advantages. This chapter emphasizes the importance of sample preparation, highlighting approaches like filtration, solvent extraction, and magnetic separation to improve compatibility with μPAD platforms. Detection strategies ranging from colorimetry to electrochemistry, chemiluminescence, and surface-enhanced Raman spectroscopy (SERS) demonstrate the versatility of μPADs in identifying harmful food contaminants, including pesticides, additives, pathogens, and allergens. Case studies illustrate their applications, such as melamine detection in milk, mercury in fish, and nitrite in meat. Concluding with future perspectives, this chapter underscores the need for enhancing fabrication processes, signal read-out technologies, and field-based applications to facilitate the widespread adoption of μPADs as reliable, accessible tools for ensuring food safety in resource-limited environments.

Keywords Microfluidic paper-based analytical devices · Food adulteration detection · Portable sensing technologies · Colorimetric and electrochemical sensors

Introduction

The global conversation about "food" now extends beyond food safety and quality to focus on the urgent need for reliable methods to assess authenticity, detect adulteration, identify toxins, and evaluate sustainability practices [1]. Efforts to strengthen the food supply chain and prevent intentional adulteration have intensified, with organizations like the U.S. Food and Drug Administration (FDA) taking the lead. The FDA emphasizes end-stage surveillance and monitors plantations in developing countries [2, 3]. For example, in 2018, the Bill & Melinda Gates

Foundation challenged innovators to develop crop monitoring technologies for low-income regions to ensure food safety and quality throughout the supply chain [4]. Detecting contaminants, assessing food freshness, and ensuring authenticity throughout the supply chain are crucial in the fight against malnutrition, which is responsible for nearly 50% of deaths in children under five. It also reduces waste and increases yield [5]. By developing simple, on-site, user-friendly electrochemical sensors, farmers can identify and address issues earlier, improving efficiency and productivity. These innovations are vital for enhancing food safety, combating malnutrition, and strengthening the global food supply chain [6, 7]. Although laboratories and specialized tests are essential in the meat, dairy, fish, and vegetable industries, they face limitations such as high costs, lengthy turnaround times, and difficulty recruiting skilled personnel, especially in remote areas. Current food analysis technologies are primarily restricted by the sample treatment step, which involves extraction, filtration, and dilution. Furthermore, these techniques rely on laboratory-based methods such as high-performance liquid chromatography, chemiluminescence, ultraviolet-visible spectrophotometry, and enzyme-linked immunosorbent tests [8–10]. There is a growing demand for rapid and cost-effective analytical tools that can be used at the point of need, require minimal expertise, and avoid bulky and expensive laboratory equipment [11]. The emergence of microfluidic technology enables the precise manipulation of fluids at the microscale, thereby facilitating the integration of sample preparation, reaction, separation, and detection processes within a limited spatial area of just a few square centimeters on a chip [12]. Microfluidic paper-based analytical devices (μPADs) offer significant advantages in terms of portability and enhanced capabilities for target separation, detection, and analysis [13]. These advantages are derived from the intrinsic properties of microfluidics, including a high surface area-to-volume ratio, increased surface tension, laminar flow, and capillary effects [14]. Paper has gained substantial attention as a sensing material in analytical chemistry due to its versatility, abundance, and sustainability [15, 16]. Its unique characteristics render it ideal for method development, managing complex matrices, and minimizing waste by-products. Moreover, the applications of paper-based platforms extend beyond the agri-food sector to encompass the detection of environmental pollutants, clinical biomarkers, and food additives [17, 18]. μPADs support a variety of detection techniques, such as optical, chemical, and hybrid systems, owing to their availability, chemical compatibility, and biodegradability [19]. These devices utilize capillary action, eliminating the need for external forces to transfer solutions and enhancing their simplicity and portability. Its surface is modified to prepare paper for critical applications using techniques such as screen printing, photolithography, etching, laser treatment, and plasma treatment. These modifications create hydrophobic barriers that confine fluids and samples to hydrophilic regions. This design enables μPADs to conduct multiple micro-bioreactions, sample preparation, and purification, rendering them highly versatile tools for various analytical and diagnostic applications [20]. Precise control of fluid flow is essential for the effective performance of μPADs. Key factors influencing flow rates include the capillary channel's radius and the fluid's surface tension. Omniphobic paper, recognized for its ability to repel most liquids, has emerged as a widely utilized material for managing fluid flow [21]. Various methods are employed to regulate flow, either by adjusting fluid velocity within porous channels or by functioning as valves to initiate or halt flow. For example, temperature can be manipulated to alter surfactant solubility in the paper. Additionally, measuring flow rates is critical to ensuring optimal μPAD functionality, underscoring the significance of flow manipulation in their design and application [22]. This chapter examines recent advancements in μPADs for detecting food adulteration. It provides a comprehensive overview of μPAD fabrication methods, sample preparation techniques, and their application in identifying food additives. This chapter concludes with a thorough evaluation of the performance and potential of μPADs in this domain.

Selection and Fabrication of μPADs

Selection of Papers

The type of paper utilized in μPADs is critical to their overall performance. Each paper type exhibits distinct properties that can be optimized to enhance sensitivity. Essential factors such as thickness, dimensions, adsorption characteristics, flow control, filtration efficiency, reagent capacity, paper grade, pore size, and wicking speed significantly influence the final analytical outcomes [23]. Since this study employed a base substrate for electrode fabrication, selecting an appropriate paper type is essential for achieving optimal analytical performance in μPADs [24]. Various paper types, including filter paper, waterproof paper, chromatography paper, and others, are employed in μPADs. Each possesses unique characteristics, indicating that the selection of paper is contingent upon the specific requirements of the assay [25–29]. Whatman No. 1 chromatography paper and filter paper are among the most frequently utilized substrates for μPADs, attributed to their uniform structure, robust wicking capacity, and high affinity for binding biomolecules, such as DNA, enzymes, and proteins [30, 31]. Nitrocellulose paper is also extensively used in bioassays due to its chemical groups that facilitate the covalent binding of biomolecules, including DNA, proteins, and enzymes [32]. Commonly available office and copy papers are also employed in fabricating μPADs. Although these papers contain lower cellulose content than others, incorporating substances such as calcium carbonate enhances their properties. However, these additives may modify the solution's pH and potentially affect the electrochemical measurements [33]. The selection of the paper substrate significantly influences the analytical performance of μPADs. Choosing the appropriate paper requires carefully considering the intended application, the electrode fabrication method, and the specific electrochemical technique [34]. A comparative study of carbon electrodes printed on office paper versus chromatography paper revealed notable differences in their performance. Chromatography paper, characterized by its porous structure, demonstrated superior electrochemical behavior owing to increased surface area, which improved solution penetration and led to enhanced electron transfer kinetics and well-defined redox peaks. Conversely, the hydrophobic nature of the office paper restricted the electrode surface area and solution contact, resulting in suboptimal electrochemical performance and absence of the faradaic current during the anodic stripping voltammetry (ASV) detection [35]. In our assessment, substrate selection should consider factors such as porosity, wicking properties, cost, commercial availability, and resistance to interference from impurities. Ultimately, the ideal paper substrate is contingent upon the method of electrode fabrication, the target application, and the selected electrochemical technique.

Fabrication Methods for Paper-Based Systems

The fabrication method plays a crucial role in the performance of μPADs. It is important to select a method compatible with the chosen paper type. This section provides a brief overview of common patterning techniques used in μPAD fabrication, considering factors such as paper properties and the desired functionality of the device.

Photolithography

Photolithography is a high-precision fabrication technique commonly used to create intricate patterns on paper, especially for paper-based analytical devices (PADs). The traditional process involves transferring geometric patterns from a photomask to a light-sensitive photoresist coated on paper. Although this method offers excellent resolution and precise control, it requires specialized equipment and a cleanroom environment. Recent advancements, such as adopting 3D-printed masks or direct laser writing, have streamlined the process, reducing costs and complexity. Photolithography is essential for creating well-defined hydrophobic and hydrophilic regions, crucial for the functionality of μPADs. Its main advantages include high resolution, design flexibility, and the ability to produce complex microchannels. However, the high cost and the need for specialized technical expertise continue to pose significant challenges [36, 37].

Wax Printing

Wax printing involves depositing wax patterns onto paper and melting the wax to form hydrophobic barriers that define fluidic channels. Since its introduction in 2009, the technique has seen substantial advancements, including developing more precise printing methodologies and improved heating sources, such as 3D wax printing and laser heating [38]. The resolution achievable with wax printing generally falls short of what photolithography provides, limiting its applicability in specific contexts. Additionally, the incompatibility of wax with organic solvents restricts its use in specific analytical procedures. The reliance on specialized wax printers and heating equipment can also pose challenges, particularly for laboratories with limited resources [39, 40]. Nonetheless, wax printing remains a promising approach for fabricating μPADs due to its simplicity, cost-effectiveness, and suitability for rapid prototyping. It is extensively employed in point-of-care diagnostics, environmental monitoring, and food safety assays. The selection of wax printing for a given application depends on factors, such as the required resolution, sample properties, and budgetary considerations [41].

Wax Dipping

While wax printing provides a straightforward method for fabricating μPADs, its dependence on wax printers, which are no longer in production, constitutes a significant limitation. Songjaroen et al. [41] introduced an alternative wax dipping technique to address this issue. This method employs a laser-cut iron mold to protect the desired hydrophilic channel regions on Whatman No. 1 paper. The paper mold assembly is subsequently submerged in molten wax at temperatures ranging from 120 to 130 °C for 1 h. The mold is removed after cooling, leaving behind the intended channel pattern. This technique negates the need for complex and expensive equipment or chemicals, rendering it a viable option for a resource-limited environment. However, it has certain drawbacks that are similar to those associated with traditional wax printing. The incompatibility of wax with organic solvents persists as a challenge, and variations in the dipping process can result in inconsistent outcomes. Furthermore, the resolution of the patterns produced is generally lower than that achieved with more advanced techniques [41].

Inkjet Printing

Ink offers considerable advantages for industrial production due to its compatibility with a diverse range of materials. Consequently, the development of μPADs via inkjet printing has gained increasing significance. This method requires only a standard desktop printer to produce patterns and sensing reagents, facilitating the fabrication of high-resolution designs. Commonly used inks for printing include alkyl ketene dimer (AKD), UV-curable ink, and hydrophobic sol-gel. However, replacing the ink within the cartridge may prove impractical. Additionally, the long-term storage of organic solutions in the ink cartridge is not recommended, as organic solvents can damage the printer. Further challenges arise from the need for distinct ink compositions for each printing cycle and the requirement for an extra heating step after ink deposition [42].

Stamping

In recent years, several patterned paper stamping technologies have been developed. The stamping techniques introduced by Curto et al. [43] utilize polydimethylsiloxane (PDMS) stamps that are impregnated with hydrophobic inks to create patterns on paper substrates. Although this method offers economic advantages for mass production, it is limited by resolution constraints and requires a separate stamp for each design. To address these challenges, Sun et al. [44] developed a suspended-droplet mode microfluidic paper-based device (SD-μPAD) that combines principles of stamping and wetting. This innovative approach uses a Teflon stamp to produce superhydrophobic patterns on paper, allowing for precise manipulation of liquid samples within the device. Although SD-μPADs can be fabricated quickly and cost-effectively, the need for a unique stamp for each design poses a significant limitation, potentially increasing both time and costs. Both stamping methods offer distinct benefits and drawbacks, highlighting the importance of selecting the appropriate technique based on the application's specific requirements [43].

Laser Cutting

Laser cutting is a precise and versatile method for fabricating paper-based devices. Intricate patterns, channels, and features can be cut or engraved into paper substrates using a focused laser beam, resulting in well-defined structures. The process starts with a digital design, which serves as the blueprint for the laser-cutting machine. Selecting the appropriate paper substrate is crucial to ensure accurate cuts and prevent excessive burning. Additionally, careful calibration of the laser cutter is essential to align the cutting process with the design specifications. Once the design is uploaded, the laser cutter follows the programmed path, using a high-energy laser beam to cut the substrate. Laser cutting offers numerous advantages, including high precision, minimal material waste, and the ability to create complex, customized designs. This makes it an excellent choice for applications, such as paper-based microfluidic devices and sensors [45].

Laser-Assisted Surface Modification

Laser-assisted surface modification is a precise and flexible technique for fabricating paper-based microfluidic devices. Utilizing CO_2 lasers or laser-assisted wax patterning creates intricate hydrophilic and hydrophobic patterns on paper substrates. CO_2 lasers selectively irradiate paper to form hydrophilic channels, while laser-assisted wax patterning involves melting wax to establish hydrophobic barriers. This method enables the creation of complex microfluidic networks with high precision. Its advantages, including design flexibility and intricate patterning capabilities, make it a promising approach for developing advanced point-of-care diagnostic tools and other applications in paper-based microfluidics [46].

Screen/Stencil Printing

Screen printing is an efficient, cost-effective method for fabricating paper-based microfluidic devices and is widely used in diagnostics and chemical analysis. The process begins with a digital design outlining microfluidic channels and test zones transferred to a mesh screen. Hydrophobic ink is applied, and a squeegee forces the ink through the mesh onto the paper substrate, forming precise patterns. Multiple layers or colors can be added by repeating the process. Proper drying ensures ink adhesion, while post-printing treatments, such as heat curing, enhance hydrophobicity and device functionality. This technique's flexibility and scalability make it ideal for producing durable, high-performance microfluidic devices for diverse applications [47, 48].

Electrode Drawing

Electrodes can be drawn on paper using pencils, carbon pencils, or conductive pens containing carbon or silver ink, offering a cost-effective and straightforward fabrication method. However, manual drawing often leads to low reproducibility due to inconsistent pressure. Advancements such as template masking and automated drawing systems have been introduced to address this. Template masking ensures consistent electrode placement and size, while computerized systems use software-controlled cutting printers to regulate writing speed and pressure, enhancing reproducibility. These techniques improve the precision and reliability of electrode fabrication, enabling the development of accurate and dependable paper-based microfluidic devices [49, 50].

Laser Scribing

Laser scribing is a precise and flexible method for creating conductive patterns on carbonaceous substrates like paper. A high-powered laser beam induces carbonization, transforming non-conductive materials into conductive carbon. This technique accommodates various substrates, including phenolic paper, resins, cellulosic materials, and textiles, offering high design flexibility and the ability to produce complex patterns. Although the fragile nature of paper poses challenges to consistent carbonization, laser scribing eliminates the need for reagents, masks, or complex processes. It provides a cost-effective and efficient approach for fabricating low-cost, flexible, and disposable paper-based electronics and sensors with strong potential for scalable production [51, 52].

Signal Read-Out

The choice of detection method for μPADs significantly influences the device's miniaturization, portability, ease of use, reliability, and cost-effectiveness [53]. A detection system typically involves three essential steps: specific recognition of the target molecule, conversion of this recognition event into a quantifiable signal, and subsequent signal processing to ensure accurate, reliable measurements suitable for the intended application [54]. Electrochemical and optical systems are commonly used as read-out devices.

Electrochemistry

Electrochemical detection provides several advantages, such as low cost, simplicity, portability, high sensitivity, miniaturization, and low power consumption. Electrochemical μPADs can employ microelectrodes, modified electrodes, or ion-selective electrodes, with blood glucose meters as a notable example [55]. These μPADs are not influenced by contaminants or lighting conditions, making them highly reliable. Screen printing is a common fabrication technique for these devices [56]. For example, Zhao and Liu [57] developed an electrochemical microfluidic paper-based immunosensor array for detecting HIV and HCV. They utilized a portable eight-channel potentiostat with Bluetooth connectivity to enable multiplexed detection. The device was fabricated using wax printing for the reaction zones and pen plotting for the electrodes, employing carbon and Ag/AgCl inks for the electrode materials. The electrochemical read-out could be triggered remotely via PC software or a smartphone app, facilitating real-time, portable, and cost-effective detection of both viruses, Fig. 1. In another example, Chen et al. [58] created a paper-based glucose detection system that uses electrochemiluminescence. This system employs a smartphone camera for signal detection, significantly lowering costs compared to traditional methods. Key components include a rechargeable battery, a carbon ink-printed μPAD, an electronic circuit, and a wax-printed microfluidic channel. The system achieved low detection limits of 0.017 mM in buffer and 0.030 mM in artificial urine, exceeding the sensitivity of other paper-based detection techniques.

Colorimetry

Colorimetry is a widely used and straightforward detection method for μPADs. Its advantages include easy reagent access, quick results, visual read-out, and a simple application process. The paper substrate provides a clear background for observing color changes resulting from chemical or enzymatic reactions [55]. Capillary action

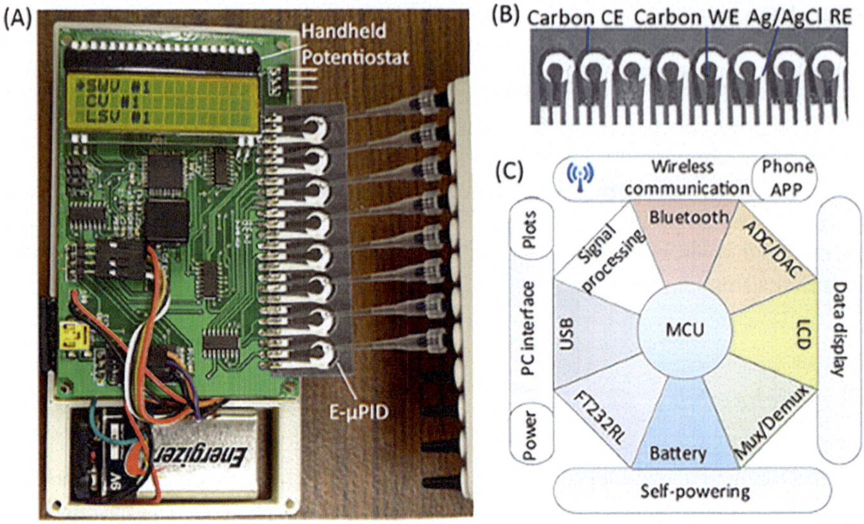

Fig. 1 A portable diagnostic platform combining an electrochemical immunosensor (E-IPIA) with a handheld potentiostat. (**a**) The handheld potentiostat with the inserted E-IPIA. (**b**) The E-IPIA itself. (**c**) The schematic diagram of the potentiostat circuit, built around a microcontroller unit (MCU). The image was reproduced with permission [57]

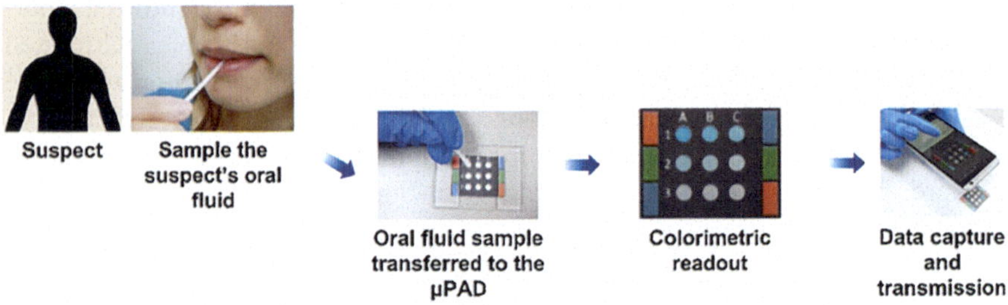

Fig. 2 A schematic representation of the proposed colorimetric paper-based ELISA system for drug abuse testing. The image was reproduced with permission [59]

facilitates the movement of analytes toward the test zone, enabling both qualitative and quantitative analysis. A common commercial example of colorimetric μPADs is pregnancy tests [54]. For instance, Chen et al. [59] developed a rapid, colorimetric μPAD-based ELISA for detecting ketamine, a drug often associated with abuse, Fig. 2. The results were analyzed using a smartphone app and ImageJ software, making the process portable, recordable, and shareable. Introducing 3,3′,5,5′-tetramethylbenzidine (TMB) to the paper-based device resulted in a color change for ketamine when using horseradish peroxidase in oral samples.

Fluorescence

Fluorescent detection relies on fluorophores and an appropriate light source to excite these molecules, with intense emitted photons serving as a measurable parameter. However, this technique is susceptible to self-fluorescence of paper materials, contributing to background noise. Notwithstanding this limitation, fluorescent detection demonstrates greater sensitivity compared to colorimetric methods, and the integration of quantum dots can further augment its sensitivity [60].

Chemiluminescence

Chemiluminescence is a simple yet highly sensitive detection method that non-specialists can use. However, it has some drawbacks, including high cost and the necessity for operation in dark conditions. Unlike fluorescence detection, which requires an external light source, chemiluminescence generates light through a chemical reaction, with intense light serving as the measurable parameter. In contrast, electrochemiluminescence produces luminescence through an electrochemical reaction, improving selectivity and reducing background signals. For example, Lu Chen et al. developed an innovative handheld paper-based bipolar electrode-electrochemiluminescence (P-BPE-ECL) system powered by a rechargeable battery and integrated with a smartphone for ECL signal detection. This design enhances portability and usability for point-of-care biochemical analysis. Their study introduces a straightforward, fast, and cost-effective screen-printing method for fabricating electrodes and microfluidic channels on paper, simplifying production steps to improve accessibility. The system demonstrated its analytical capability by detecting H_2O_2 and glucose, achieving detection limits of 1.75 μM for H_2O_2 and 0.017 mM for glucose in PBS, underscoring its potential for sensitive biochemical applications [58].

Surface-Enhanced Raman Spectroscopy (SERS)

SERS is a highly sensitive detection method that leverages electromagnetic phenomena. This technique is predicated on the principles of localized surface plasmons, which are excited by metallic nanoparticles, such as gold (Au) or silver (Ag). SERS enhances detection sensitivity by significantly amplifying the Raman scattering signals of molecules that adhere to metal surfaces via nanostructures. A paper substrate is often coated with metallic nanoparticles that bind to the analyte, augmenting the Raman signal. Given that the Raman spectrum indicates molecular polarizability, this method is particularly efficacious for detecting drugs containing non-polar groups [61, 62].

Preparation of Samples before Using PADs

Sample preparation remains a significant challenge in developing analytical methods for food analysis due to the complex composition of food matrices, which contain diverse organic and inorganic substances. Most μPAD applications focus on detecting small pesticides, organic molecules, and inorganic compounds. Common sample preparation techniques include dilution, filtration, and centrifugation. For instance, Silva et al. [63] diluted apple juice samples tenfold to detect *Salmonella typhimurium* using μPADs without additional preparation. However, precise separation methods are essential when analyzing macromolecules like proteins or fats in solid foods. For example, You's group used magnetic separation to detect *E. coli O157:H7* in sausage, involving homogenization in buffer, magnetic separation, and antibody labeling [64].

In contrast, many small molecules, such as additives or contaminants, are uniformly distributed in food and exhibit weak interactions with the food matrix. For these compounds, minimal sample preparation, such as solvent extraction, is often sufficient for adequate recovery. Several extraction techniques have been developed for this purpose, including dispersive solid-phase extraction, solid-phase extraction, and vortex-assisted liquid–liquid microextraction [65]. However, selecting a suitable solvent is critical in μPAD applications, as it affects the interaction with hydrophobic barriers [66]. For instance, Zhang et al. employed n-hexane to extract polychlorinated biphenyls detected on wax-printed μPADs using the buffer exchange principle [67]. Alternatively, fabrication methods must accommodate selective separation techniques necessary for complex food samples. Techniques such as laser cutting eliminate the need for organic solvents.

In contrast, solvent-free microwave extraction, imprinted polymer technology, aptamer-functionalized microfluidics, and metal-organic frameworks functionalized with antibodies and aptamers enable solvent-free sample preparation. Importantly, hydrophobic solvents and eluents are incompatible with μPADs, as they cannot flow through hydrophilic cellulose channels, underscoring the critical role of food sample preparation in μPAD applications [30, 68–71]. Another major challenge is selectivity, which significantly affects μPAD performance. Interferences in food samples can positively or negatively influence analytical results. Selective detection methods, such as 1,5-diphenylcarbazide for hexavalent chromium, help mitigate these issues [72]. However, combining selective sample preparation with appropriate detection methods becomes essential when analytes and interferences are too similar for differentiation. Imprinted polymer techniques, widely used in μPADs, offer high selectivity but may suffer from poor sensitivity or detection issues from inadequate sample preparation, ultimately limiting μPAD performance [69, 73].

The Application of PADs for Food Adulterant Analysis

Ensuring food safety is a critical global health priority, particularly in resource-limited regions. Prompt detection of food hazards is essential to prevent foodborne illnesses. Key food hazards include pathogens, toxins, allergens, antibiotic and hormone residues, preservatives, additives, chemical contaminants, pesticides, and herbicides, all of which present serious risks to human health, Fig. 3.

Chemical Toxicants and Adulterants

Food adulteration frequently involves the addition of substances such as sand, stones, or metallic compounds to increase weight or volume artificially. Another common practice is substituting high-quality ingredients with inferior or contaminated alternatives, such as rancid oils or mislabeled meat, posing significant health risks. To combat these issues, μPADs have been developed as effective tools for detecting harmful substances in food. For instance, melamine, a nitrogen-rich compound illegally added to products like milk and pet food to inflate apparent protein content, poses serious health risks, including kidney damage. A colorimetric sensor based on gold nanoparticles offers a reliable method for detecting melamine in milk. The nanoparticles interact with melamine, causing a visible color change from red to blue, which can be observed directly or quantified using a smartphone. This method detects melamine concentrations as low as 5.1 nM, exceeding regulatory safety thresholds [74]. Similarly, mercury contamination in salmon has been addressed using a μPAD functionalized with gold nanoparticles and HEDTO. Mercury presence induces nanoparticle aggregation, resulting in a distinct red-to-blue color change. This change can be captured with a smartphone and analyzed via image

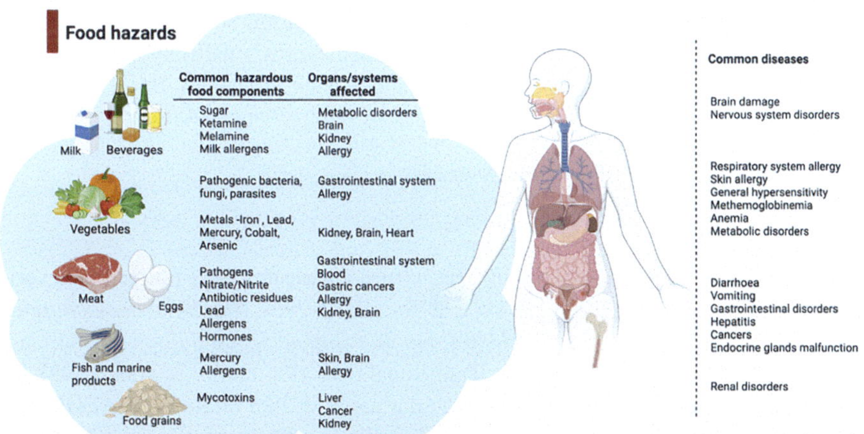

Fig. 3 Common food hazards, which can lead to both acute and chronic toxicities, affecting various organ systems

processing software, achieving a detection limit of 15 nM [75]. In a separate innovation, the EμPAD was developed to detect hazardous drugs like ketamine in beverages. This device employs a coating of zeolite nanoflakes combined with graphene oxide nanocrystals (Zeo-GO). When ketamine is present, the device generates an electrical signal proportional to its concentration. The method is highly efficient, providing results in just 2 s with a detection limit as low as 0.001 nM/mL and a range of 0.001–5 nM/mL [76].

Food Additives and Preservatives

Food additives and preservatives are chemicals used to enhance food's appearance, shelf life, flavor, and texture. While generally safe in small amounts, excessive consumption can pose health risks [77]. Researchers developed a cost-effective and sustainable biosensor to monitor sugar levels in beverages. The device features a hydrophilic cellulose paper disk impregnated with immobilized glucose oxidase mounted on a screen-printed carbon electrode. It accurately analyzes small sample volumes (5 μL) of commercial beverages, achieving detection limits comparable to high-performance liquid chromatography [77]. A microfluidic chip device using Whatman filter paper was developed using the Janovsky reaction to detect benzoic acid in food. The reaction zones on the μPAD were treated with sodium hydroxide, followed by conversion of benzoic acid to 3,5-dinitrobenzoic acid using KNO_3 and H_2SO_4. The μPAD was heated to 45 °C to facilitate the reaction, changing color from light to dark brownish-orange. The intensity, captured with a CMOS camera and analyzed via RGB software on a smartphone, correlates with the benzoic acid concentration [78]. Nitrite, commonly used as a meat preservative, poses health risks at elevated levels. The Griess reaction enables colorimetric detection of nitrite on μPADs. Nitrite reacts with sulfanilamide to form a diazonium salt, which couples with N-α-naphthyl-ethylenediamine to produce a magenta azo dye. The dye intensity, directly proportional to nitrite concentration, is measured using a smartphone camera and ImageJ software. This 15-min test achieves a sensitivity of 1.1 mg/kg, suitable for detecting nitrite in meat products and drinking water [79].

Foodborne Pathogens

Foodborne pathogens, including bacteria, viruses, parasites, and fungi, cause various illnesses. Common pathogens such as *Bacillus cereus*, *Staphylococcus aureus*, *E. coli* O157:H7, *Campylobacter*, *Listeria monocytogenes*, and *Salmonella* spp. are frequently linked to food poisoning. Traditional detection methods, such as culture techniques, require 5–7 days, while molecular methods like PCR and MALDI-TOF are labor-intensive and demand specialized facilities. μPADs provide a faster, more efficient alternative by using enzyme-substrate reactions to produce visible color changes in the paper matrix. For instance, a wax-printed filter paper assay detects *E. coli* O157:H7, *Salmonella enterica*, and *Listeria monocytogenes* in food samples using enzyme-specific reactions. β-galactosidase and chlorophenol red β-galactopyranoside (CPRG) identify *E. coli* by producing a red-violet color; bacterial esterase with 5-bromo-6-chloro-3-indolyl caprylate (MC) detects *Salmonella enterica* by producing purple; and phosphatidylinositol-specific phospholipase C with 5-bromo-4-chloro-3-indolyl-myo-inositol phosphate (X-InP) detects *Listeria monocytogenes* by producing blue color. This approach reduces bacterial enrichment time to 12 h or less and detects bacterial levels as low as 10^1 CFU/cm^2 in meat samples. A gold-based sensing platform with magnetic nanobead-peptide probes has been developed to detect *Staphylococcus aureus* in food samples. This method relies on detecting protease cleavage activity, which releases magnetic nanobeads from the sensor surface. It can identify *S. aureus* in foods such as ground beef, turkey sausage, lettuce, and milk, with detection limits ranging from 7 to 100 CFU/mL. The assay is rapid (completing in minutes), cost-effective, and label-free, and it remains stable for up to 6 months under cold storage [80]. For coliform detection in water, an open-channel μPAD was designed by printing omniphilic channels on omniphobic fluorinated paper. This device lyses coliform cells, releasing intracellular contents for mixing with detection reagents. By using a combination of high- and low-surface-tension liquids, the μPAD efficiently lyses *E. coli* and detects it at concentrations as low as 10^4 CFU/mL, utilizing the β-galactosidase enzyme specific to coliforms [81].

Food Allergens

Food allergies affect up to 10% of the population, with symptoms ranging from mild discomfort to severe, life-threatening anaphylaxis. While there is no cure, effective management relies on avoiding allergens, early detection, and emergency interventions. Common allergens include milk, eggs, tree nuts, wheat, peanuts, soy, shellfish, and fish [82]. One innovative μPAD, the xLFIA, employs a three-color test to detect milk, egg, and hazelnut proteins. Specific antibodies conjugated to gold and silver nanoparticles serve as colored markers on a nitrocellulose strip. The μPAD features three distinct lines, each corresponding to an allergen, and successfully detects allergen traces in commercial biscuits at concentrations as low as 0.1 mg/L. This immunoassay demonstrates excellent fluid control and simultaneous multi-allergen testing, making it highly efficient and versatile [82]. Another μPAD-based lateral flow immunoassay (LFI) was developed using modified cellulose material to detect allergenic proteins in food samples. The device delivers results within 15 min, including sample preparation. By refining geometric patterns, the sample flow rate was optimized, enabling high sensitivity and the detection of ovalbumin (a key egg allergen) at concentrations as low as 1 ppm in various foods [83]. Additionally, a sandwich immunoassay strip test using a nitrocellulose membrane was designed to detect the Bowman–Birk inhibitor, an antinutritional factor in soybeans. This test visually identified allergen concentrations up to 0.5 μg/mL, with enhanced sensitivity of 0.23 μg/mL when analyzed with the TSR3000 Membrane Strip Reader (BioDot, USA) [84].

Hormone and Antibiotic Residues in Food

17β-estradiol, a potent anabolic steroid hormone, is often used illegally to promote animal growth. Its residues in food and milk can accumulate in the food chain, causing endocrine disruption [85]. To address this, molecularly imprinted polymer (MIP)-grafted paper-based devices have been developed for detecting 17β-estradiol in milk. The optimal MIP synthesis involved acetonitrile, 17β-estradiol, (3-aminopropyl) triethoxysilane (APTES), and tetraethyl orthosilicate (TEOS) in a 1:12:12 ratio. This assay demonstrated high sensitivity, achieving a detection limit of 0.25 μg/L in spiked milk samples. These MIP-based paper devices provide a promising platform for sensitive, specific, and high-throughput food monitoring, with potential applications in multiplex assays [86]. To detect oxytetracycline and norfloxacin residues in pork, μPADs utilizing metal complexation techniques were employed, showcasing their versatility in analyzing diverse food contaminants [87]. μPAD technology represents a transformative advancement in food quality analysis, offering cost-effective, equipment-free detection. μPAD fabrication techniques, flow control strategies, and detection methodologies highlight the potential for broad applications in food monitoring.

Pesticides and Herbicides

Organophosphorus (OP) pesticides, including carbamates, are widely used in agriculture but pose significant chemical hazards in food. Carbamates are classified as endocrine-disrupting compounds capable of causing hormonal imbalances upon consumption [88]. Detecting these compounds in food is a critical step in ensuring food safety.

Microfluidic paper-based analytical devices (μPADs) have been developed for the colorimetric detection of carbamate pesticides by leveraging the inhibition of the acetylcholinesterase (AChE) enzyme. In this method, acetylthiocholine iodide is hydrolyzed by AChE into thiocholine, reacting with DTNB to produce a yellow compound detectable at 405 nm. The intensity of the yellow color decreases with increasing pesticide concentrations, allowing the degree of AChE inhibition to be quantified via a calibration curve. Results were validated against controls and various pesticide concentrations to ensure accuracy [89, 90]. A SERS-based technique employing gold nanoparticles (Au NPs) has also been developed to detect methyl parathion on fruit surfaces. Filter paper substrates soaked in Au NP solution were optimized using 4-mercaptobenzoic acid (4-MBA) and analyzed with a portable Raman spectrometer (785 nm laser). This method demonstrated a linear detection range of 0.018–0.354 μg/cm^2 with a detection limit of 0.011 μg/cm^2. Recovery rates from spiked fruit peels ranged from 94.09% to 98.72%, highlighting the method's reliability, reproducibility, and stability without requiring extensive sample preparation [89]. A μPAD-based colorimetric method was developed to detect glyphosate, a herbicide linked to organ toxicity and cancer. The method utilizes Mn–ZnS quantum dot (QD)-embedded MIPs. In glyphosate-free samples, ABTS is oxidized by H_2O_2, producing a dark green color. However, glyphosate binding to the Mn–ZnS QD-MIP inhibits •OH generation, resulting in a light green color. This color change enables the sensitive and selective detection of glyphosate in food grain samples [91].

Conclusion and Future Perspectives

Detecting hazardous materials in food remains a critical challenge for ensuring food safety, as insufficient quality control can lead to severe illnesses and fatalities. Conventional detection methods, while reliable, are often time-intensive and laborious. Unlike clinical samples such as blood or urine, food samples present unique complexities due to their diverse matrices, including components like insoluble fats, hydrophobic lipids, salts, and Maillard reaction products, which impede analyte detection. Addressing these challenges typically requires complex, multi-step sample preparation processes. Simplifying these processes to make food samples compatible with microfluidic paper-based analytical devices (μPADs) is vital for enhancing their effectiveness in food safety applications.

Future advancements in μPAD technology should focus on improving the detection systems to reduce their dependence on the traditional methods, such as scanners and grayscale image analysis, which compromise sensitivity and multiplexing accuracy. Alternative approaches, such as spectrometry and fluorescence-based assays, offer higher sensitivity but are hindered by the high costs of equipment and reagents. Creating cost-effective, equipment-free detection systems that maintain high sensitivity and multiplexing capabilities is crucial to address these challenges and facilitate the broader adoption of μPAD technology. Moreover, advancements in μPAD fabrication techniques that allow mass production at low cost, along with improved specificity and sensitivity, will further enhance their practical applications. When combined with innovations in the sample preparation methods, these improvements will establish μPADs as viable and accessible tools for extensive food safety monitoring. As μPAD applications expand into real-world contexts, increased market competition is anticipated to drive innovation and accelerate their acceptance and adoption. Such advancements will elevate food safety standards and play a significant role in the global initiative to effectively reduce foodborne risks.

Acknowledgments This work is supported by the Second Century Fund (C2F), Chulalongkorn University, Thailand. Waleed Alahmad gratefully acknowledges the Department of Chemistry, Faculty of Science, Chulalongkorn University, for the provision of research facilities, resources, and technical support.

References

1. Hamburg MA (2011) Advancing regulatory science. American Association for the Advancement of Science, Washington, pp 987–987
2. Strauss DM (2011) An analysis of the FDA food safety modernization act: protection for consumers and boon for business. Food Drug Law J 66(3):353–376
3. Anthes C, De Schutter O (2018) The food and agriculture Organization of the United Nations. In: Human rights in global health: rights-based governance for a globalizing world. Oxford Academic, Oxford, p 261
4. Li C, Li C, Yu H, Cheng Y, Xie Y, Yao W, Guo Y, Qian H (2021) Chemical food contaminants during food processing: sources and control. Crit Rev Food Sci Nutr 61(9):1545–1555
5. Racicot M, Zanabria R, Leroux A, Ng S, Cormier M, Tiwari A, Aklilu S, Currie R, Arsenault J, Griffiths M (2018) Quantifying the impact of food safety criteria included in the Canadian Food Inspection Agency risk assessment model for food establishments through expert elicitation. Food Control 92:450–463
6. Candel JJ (2014) Food security governance: a systematic literature review. Food Secur 6:585–601
7. Lan L, Yao Y, Ping J, Ying Y (2017) Recent progress in nanomaterial-based optical aptamer assay for the detection of food chemical contaminants. ACS Appl Mater Interfaces 9(28):23287–23301
8. Lin S, Lin X, Lou X-T, Yang F, Lin D-Y, Lu Z-W (2015) Rapid and sensitive SERS method for determination of Rhodamine B in chili powder with paper-based substrates. Anal Methods 7(12):5289–5294
9. Gao N, Huang P, Wu F (2018) Colorimetric detection of melamine in milk based on Triton X-100 modified gold nanoparticles and its paper-based application. Spectrochim Acta A Mol Biomol Spectrosc 192:174–180
10. Pérez-Ortega P, Lara-Ortega FJ, Gilbert-López B, Moreno-González D, García-Reyes JF, Molina-Díaz A (2017) Screening of over 600 pesticides, veterinary drugs, food-packaging contaminants, mycotoxins, and other chemicals in food by ultra-high performance liquid chromatography quadrupole time-of-flight mass spectrometry (UHPLC-QTOFMS). Food Anal Methods 10:1216–1244
11. Schilling UM (2014) Time is money—the economic impact of point of care on the emergency Department of a Tertiary Care University Hospital. Point Care 13(1):21–23
12. Noviana E, Ozer T, Carrell CS, Link JS, McMahon C, Jang I, Henry CS (2021) Microfluidic paper-based analytical devices: from design to applications. Chem Rev 121(19):11835–11885
13. Jiang W, Tang Q, Zhu Y, Gu X, Wu L, Qin Y (2024) Research progress of microfluidics-based food safety detection. Food Chem 441:138319
14. Weng X, Neethirajan S (2017) Ensuring food safety: quality monitoring using microfluidics. Trends Food Sci Technol 65:10–22
15. Cinti S, Minotti C, Moscone D, Palleschi G, Arduini F (2017) Fully integrated ready-to-use paper-based electrochemical biosensor to detect nerve agents. Biosens Bioelectron 93:46–51
16. Cinti S, Fiore L, Massoud R, Cortese C, Moscone D, Palleschi G, Arduini F (2018) Low-cost and reagent-free paper-based device to detect chloride ions in serum and sweat. Talanta 179:186–192

17. Cinti S, Proietti E, Casotto F, Moscone D, Arduini F (2018) Based strips for the electrochemical detection of single and double stranded DNA. Anal Chem 90(22):13680–13686
18. Wang C-C, Hennek JW, Ainla A, Kumar AA, Lan W-J, Im J, Smith BS, Zhao M, Whitesides GM (2016) A paper-based "pop-up" electrochemical device for analysis of beta-hydroxybutyrate. Anal Chem 88(12):6326–6333
19. Yehia AM, Farag MA, Tantawy MA (2020) A novel trimodal system on a paper-based microfluidic device for on-site detection of the date rape drug "ketamine". Anal Chim Acta 1104:95–104
20. Dou M, Sanjay ST, Benhabib M, Xu F, Li X (2015) Low-cost bioanalysis on paper-based and its hybrid microfluidic platforms. Talanta 145:43–54
21. Fu E, Downs C (2017) Progress in the development and integration of fluid flow control tools in paper microfluidics. Lab Chip 17(4):614–628
22. Noh H, Phillips ST (2010) Metering the capillary-driven flow of fluids in paper-based microfluidic devices. Anal Chem 82(10):4181–4187
23. Smith S, Madzivhandila P, Ntuli L, Bezuidenhout P, Zheng H, Land K (2019) Printed paper–based electrochemical sensors for low-cost point-of-need applications. Electrocatalysis 10:342–351
24. Camargo JR, Andreotti IA, Kalinke C, Henrique JM, Bonacin JA, Janegitz BC (2020) Waterproof paper as a new substrate to construct a disposable sensor for the electrochemical determination of paracetamol and melatonin. Talanta 208:120458
25. Qi J, Li B, Zhou N, Wang X, Deng D, Luo L, Chen L (2019) The strategy of antibody-free biomarker analysis by in-situ synthesized molecularly imprinted polymers on movable valve paper-based device. Biosens Bioelectron 142:111533
26. Nontawong N, Ngaosri P, Chunta S, Jarujamrus P, Nacapricha D, Lieberzeit PA, Amatatongchai M (2022) Smart sensor for assessment of oxidative/nitrative stress biomarkers using a dual-imprinted electrochemical paper-based analytical device. Anal Chim Acta 1191:339363
27. Lee SH, Lee JH, Tran V-K, Ko E, Park CH, Chung WS, Seong GH (2016) Determination of acetaminophen using functional paper-based electrochemical devices. Sensors Actuators B Chem 232:514–522
28. Cinti S, Basso M, Moscone D, Arduini F (2017) A paper-based nanomodified electrochemical biosensor for ethanol detection in beers. Anal Chim Acta 960:123–130
29. Boonyasit Y, Chailapakul O, Laiwattanapaisal W (2016) A multiplexed three-dimensional paper-based electrochemical impedance device for simultaneous label-free affinity sensing of total and glycated haemoglobin: the potential of using a specific single-frequency value for analysis. Anal Chim Acta 936:1–11
30. Alahmad W, Varanusupakul P, Varanusupakul P (2023) Recent developments and applications of microfluidic paper-based analytical devices for the detection of biological and chemical hazards in foods: a critical review. Crit Rev Anal Chem 53(2):233–252
31. Alahmad W, Sahragard A, Varanusupakul P (2021) Online and offline preconcentration techniques on paper-based analytical devices for ultrasensitive chemical and biochemical analysis: a review. Biosens Bioelectron 194:113574
32. Harpaz D, Eltzov E, Axelrod T, Marks RS, Tok AI (2020) Membrane type comparison and modification to modulate sample flow in paper diagnostics. Biochem Eng J 155:107483
33. Jiménez-Reyes M, Tenorio D, Rojas-Robles M, García-Rosales G (2018) Physicochemical behavior of several kinds of paper under gamma irradiation. Radiat Phys Chem 148:13–18
34. Pelton R (2009) Bioactive paper provides a low-cost platform for diagnostics. TrAC Trends Anal Chem 28(8):925–942
35. Martín-Yerga D, Álvarez-Martos I, Blanco-López MC, Henry CS, Fernández-Abedul MT (2017) Point-of-need simultaneous electrochemical detection of lead and cadmium using low-cost stencil-printed transparency electrodes. Anal Chim Acta 981:24–33
36. Asano H, Shiraishi Y (2015) Development of paper-based microfluidic analytical device for iron assay using photomask printed with 3D printer for fabrication of hydrophilic and hydrophobic zones on paper by photolithography. Anal Chim Acta 883:55–60

37. Kumar S, Kaushal JB, Lee HP (2024) Sustainable sensing with paper microfluidics: applications in health, environment, and food safety. Biosensors 14(6):300
38. Alahmad W, Uraisin K, Nacapricha D, Kaneta T (2016) A miniaturized chemiluminescence detection system for a microfluidic paper-based analytical device and its application to the determination of chromium (III). Anal Methods 8(27):5414–5420
39. Lu Y, Shi W, Jiang L, Qin J, Lin B (2009) Rapid prototyping of paper-based microfluidics with wax for low-cost, portable bioassay. Electrophoresis 30(9):1497–1500
40. Carrilho E, Martinez AW, Whitesides GM (2009) Understanding wax printing: a simple micropatterning process for paper-based microfluidics. Anal Chem 81(16):7091–7095
41. Alahmad W, Cetinkaya A, Kaya SI, Varanusupakul P, Ozkan SA (2023) Electrochemical paper-based analytical devices for environmental analysis: current trends and perspectives. Trends Environ Anal Chem 40:e00220
42. Yamada K, Henares TG, Suzuki K, Citterio D (2015) Paper-based inkjet-printed microfluidic analytical devices. Angew Chem Int Ed 54(18):5294–5310
43. Curto VF, Lopez-Ruiz N, Capitan-Vallvey LF, Palma AJ, Benito-Lopez F, Diamond D (2013) Fast prototyping of paper-based microfluidic devices by contact stamping using indelible ink. RSC Adv 3(41):18811–18816
44. Sun H, Li W, Dong Z-Z, Hu C, Leung C-H, Ma D-L, Ren K (2018) A suspending-droplet mode paper-based microfluidic platform for low-cost, rapid, and convenient detection of lead (II) ions in liquid solution. Biosens Bioelectron 99:361–367
45. Mahmud MA, Blondeel EJ, Kaddoura M, MacDonald BD (2018) Features in microfluidic paper-based devices made by laser cutting: how small can they be? Micromachines 9(5):220
46. Zhang Y, Liu J, Wang H, Fan Y (2019) Laser-induced selective wax reflow for paper-based microfluidics. RSC Adv 9(20):11460–11464
47. Oliveira AE, Pereira AC, de Resende MA (2023) Fabrication of low-cost screen-printed electrode in paper using conductive inks of graphite and silver/silver chloride. Electroanalysis 35(2):e202200093
48. Xiong C, Li M, Han Q, Zhao W, Dai L, Ni Y (2022) Screen printing fabricating patterned and customized full paper-based energy storage devices with excellent photothermal, self-healing, high energy density and good electromagnetic shielding performances. J Mater Sci Technol 97:190–200
49. Baharfar M, Rahbar M, Tajik M, Liu G (2020) Engineering strategies for enhancing the performance of electrochemical paper-based analytical devices. Biosens Bioelectron 167:112506
50. Prasertying P, Jantawong N, Sonsa-Ard T, Wongpakdee T, Khoonrueng N, Buking S, Nacapricha D (2021) Gold leaf electrochemical sensors: applications and nanostructure modification. Analyst 146(5):1579–1589
51. Arantes IV, Gongoni JL, Mendes LF, de Ataide VN, Ameku WA, Garcia PT, de Araujo WR, Paixão TR (2022) Electrochemical paper-based analytical devices. In: Paper-based analytical devices for chemical analysis and diagnostics. Elsevier, Amsterdam, pp 81–116
52. de Araujo WR, Frasson CM, Ameku WA, Silva JR, Angnes L, Paixão TR (2017) Single-step reagentless laser scribing fabrication of electrochemical paper-based analytical devices. Angew Chem 129(47):15309–15313
53. Kung C-T, Hou C-Y, Wang Y-N, Fu L-M (2019) Microfluidic paper-based analytical devices for environmental analysis of soil, air, ecology and river water. Sensors Actuators B Chem 301:126855
54. Silva-Neto HA, Arantes IV, Ferreira AL, do Nascimento GH, Meloni GN, de Araujo WR, Paixão TR, Coltro WK (2023) Recent advances on paper-based microfluidic devices for bioanalysis. TrAC Trends Anal Chem 158:116893
55. Kaneta T, Alahmad W, Varanusupakul P (2019) Microfluidic paper-based analytical devices with instrument-free detection and miniaturized portable detectors. Appl Spectrosc Rev 54(2):117–141
56. Fu L-M, Wang Y-N (2018) Detection methods and applications of microfluidic paper-based analytical devices. TrAC Trends Anal Chem 107:196–211

57. Zhao C, Liu X (2016) A portable paper-based microfluidic platform for multiplexed electrochemical detection of human immunodeficiency virus and hepatitis C virus antibodies in serum. Biomicrofluidics 10(2):024119
58. Chen L, Zhang C, Xing D (2016) Based bipolar electrode-electrochemiluminescence (BPE-ECL) device with battery energy supply and smartphone read-out: a handheld ECL system for biochemical analysis at the point-of-care level. Sensors Actuators B Chem 237:308–317
59. Chen C-A, Wang P-W, Yen Y-C, Lin H-L, Fan Y-C, Wu S-M, Chen C-F (2019) Fast analysis of ketamine using a colorimetric immunosorbent assay on a paper-based analytical device. Sensors Actuators B Chem 282:251–258
60. Hoang TX, Phan LMT, Vo TAT, Cho S (2021) Advanced signal-amplification strategies for paper-based analytical devices: a comprehensive review. Biomedicine 9(5):540
61. Noviana E, Carrão DB, Pratiwi R, Henry CS (2020) Emerging applications of paper-based analytical devices for drug analysis: a review. Anal Chim Acta 1116:70–90
62. Aham EC, Ravikumar A, Okoye CO, Ezeorba TPC, Arunjegan A, Tamilselvan G, Okagu IU, Zhang H, Zhang Z (2024) A comprehensive review of paper-based analytical devices for 17β-estradiol detection: current status and future perspectives. Microchem J 206:111486
63. Silva NF, Almeida CM, Magalhães JM, Gonçalves MP, Freire C, Delerue-Matos C (2019) Development of a disposable paper-based potentiometric immunosensor for real-time detection of a foodborne pathogen. Biosens Bioelectron 141:111317
64. You S-M, Jeong K-B, Luo K, Park J-S, Park J-W, Kim Y-R (2021) Based colorimetric detection of pathogenic bacteria in food through magnetic separation and enzyme-mediated signal amplification on paper disc. Anal Chim Acta 1151:338252
65. Madej K, Kalenik TK, Piekoszewski W (2018) Sample preparation and determination of pesticides in fat-containing foods. Food Chem 269:527–541
66. Liu Z, Hu J, Zhao Y, Qu Z, Xu F (2015) Experimental and numerical studies on liquid wicking into filter papers for paper-based diagnostics. Appl Therm Eng 88:280–287
67. Zhang D, Wang Y, Li C, Zhang X (2021) Polychlorinated biphenyl detection in organic solvents with paper-based analytical devices. Environ Technol 42(11):1766–1771
68. Chemat F, Fabiano-Tixier AS, Vian MA, Allaf T, Vorobiev E (2015) Solvent-free extraction of food and natural products. TrAC Trends Anal Chem 71:157–168
69. Ibarra I, Miranda J, Pérez-Silva I, Jardinez C, Islas G (2020) Sample treatment based on molecularly imprinted polymers for the analysis of veterinary drugs in food samples: a review. Anal Methods 12(23):2958–2977
70. Du F, Guo L, Qin Q, Zheng X, Ruan G, Li J, Li G (2015) Recent advances in aptamer-functionalized materials in sample preparation. TrAC Trends Anal Chem 67:134–146
71. Pérez-Cejuela HM, Herrero-Martínez JM, Simó-Alfonso EF (2020) Recent advances in affinity MOF-based sorbents with sample preparation purposes. Molecules 25(18):4216
72. Alahmad W, Tungkijanansin N, Kaneta T, Varanusupakul P (2018) A colorimetric paper-based analytical device coupled with hollow fiber membrane liquid phase microextraction (HF-LPME) for highly sensitive detection of hexavalent chromium in water samples. Talanta 190:78–84
73. Hua MZ, Li S, Wang S, Lu X (2018) Detecting chemical hazards in foods using microfluidic paper-based analytical devices (μPADs): the real-world application. Micromachines 9(1):32
74. Trapiella-Alfonso L, Costa-Fernandez JM, Pereiro R, Sanz-Medel A (2013) Synthesis and characterization of hapten-quantum dots bioconjugates: application to development of a melamine fluorescent immunoassay. Talanta 106:243–248
75. Shariati S, Khayatian G (2020) Microfluidic paper-based analytical device using gold nanoparticles modified with N, N′-bis (2-hydroxyethyl) dithiooxamide for detection of hg (ii) in air, fish and water samples. New J Chem 44(43):18662–18667
76. Narang J, Malhotra N, Singhal C, Mathur A, Chakraborty D, Anil A, Ingle A, Pundir CS (2017) Point of care with micro fluidic paper based device integrated with nano zeolite–graphene oxide nanoflakes for electrochemical sensing of ketamine. Biosens Bioelectron 88:249–257

77. Lawrence CSK, Tan SN, Floresca CZ (2014) A "green" cellulose paper based glucose amperometric biosensor. Sensors Actuators B Chem 193:536–541
78. Liu C-C, Wang Y-N, Fu L-M, Chen K-L (2018) Microfluidic paper-based chip platform for benzoic acid detection in food. Food Chem 249:162–167
79. Trofimchuk E, Hu Y, Nilghaz A, Hua MZ, Sun S, Lu X (2020) Development of paper-based microfluidic device for the determination of nitrite in meat. Food Chem 316:126396
80. Suaifan GA, Alhogail S, Zourob M (2017) Based magnetic nanoparticle-peptide probe for rapid and quantitative colorimetric detection of Escherichia coli O157: H7. Biosens Bioelectron 92:702–708
81. Snyder SA, Boban M, Li C, VanEpps JS, Mehta G, Tuteja A (2020) Lysis and direct detection of coliforms on printed paper-based microfluidic devices. Lab Chip 20(23):4413–4419
82. Anfossi L, Di Nardo F, Russo A, Cavalera S, Giovannoli C, Spano G, Baumgartner S, Lauter K, Baggiani C (2019) Silver and gold nanoparticles as multi-chromatic lateral flow assay probes for the detection of food allergens. Anal Bioanal Chem 411:1905–1913
83. Hua MZ, Lu X (2020) Development of a microfluidic paper-based immunoassay for rapid detection of allergic protein in foods. ACS Sens 5(12):4048–4056
84. Wang Y, Li Y, Wu J, Pei Y, Chen X, Sun Y, Hu M, Xing Y, Cao J, Li Z (2019) Development of an immunochromatographic strip test for the rapid detection of soybean Bowman-Birk inhibitor. Food Agric Immunol 30(1):1202–1211
85. Hirpessa BB, Ulusoy BH, Hecer C (2020) Hormones and hormonal anabolics: residues in animal source food, potential public health impacts, and methods of analysis. J Food Qual 2020(1):5065386
86. Xiao L, Zhang Z, Wu C, Han L, Zhang H (2017) Molecularly imprinted polymer grafted paper-based method for the detection of 17β-estradiol. Food Chem 221:82–86
87. Nilghaz A, Lu X (2019) Detection of antibiotic residues in pork using paper-based microfluidic device coupled with filtration and concentration. Anal Chim Acta 1046:163–169
88. Moreira S, Silva R, Carrageta DF, Alves MG, Seco-Rovira V, Oliveira PF, de Lourdes Pereira M (2022) Carbamate pesticides: shedding light on their impact on the male reproductive system. Int J Mol Sci 23(15):8206
89. Xie J, Li L, Khan IM, Wang Z, Ma X (2020) Flexible paper-based SERS substrate strategy for rapid detection of methyl parathion on the surface of fruit. Spectrochim Acta A Mol Biomol Spectrosc 231:118104
90. Beshana S, Hussen A, Leta S, Kaneta T (2022) Microfluidic paper based analytical devices for the detection of carbamate pesticides. Bull Environ Contam Toxicol 109(2):344–351
91. Sawetwong P, Chairam S, Jarujamrus P, Amatatongchai M (2021) Enhanced selectivity and sensitivity for colorimetric determination of glyphosate using Mn–ZnS quantum dot embedded molecularly imprinted polymers combined with a 3D-microfluidic paper-based analytical device. Talanta 225:122077

Next-Generation Solutions to Combat Food Adulteration: Integrating Advanced Detection Technologies and Risk Mitigation Strategies

Ahmed S. El-tahlawy and Waleed Alahmad

Contents

Abstract Food adulteration is a silent global threat, undermining consumer confidence, endangering public health, and disrupting fair trade. With increasingly sophisticated methods of fraud, conventional detection techniques often fall short in addressing these complexities. Advanced molecular technologies, including DNA barcoding, polymerase chain reaction (PCR), and next-generation sequencing (NGS), have emerged as transformative tools in the fight against food fraud. These methods provide exceptional accuracy, capable of identifying adulterants even in highly processed or intricate food matrices. Complementing these are advanced

A. S. El-tahlawy
Faculty of Science, Department of Chemistry, Chulalongkorn University, Bangkok, Thailand

Faculty of Veterinary Medicine, Food Hygiene, Safety, and Technology Department, Zagazig University, Zagazig, Egypt

W. Alahmad (✉)
Faculty of Science, Department of Chemistry, Chulalongkorn University, Bangkok, Thailand
e-mail: waleed.al@chula.ac.th

W. Alahmad, M. Ashfaq (eds.), *Advanced Techniques against Food Adulteration*,
https://doi.org/10.1007/978-3-031-99454-8_5

analytical tools — including electrochemical sensors, spectroscopy, and mass spectrometry — which offer rapid, sensitive, and often portable solutions for on-site detection and confirmation. Beyond detection, they enhance traceability, validate authenticity, and fortify the global food supply chain against contamination and fraud. The adoption of these cutting-edge approaches is fueled by their unmatched sensitivity to trace adulterants, adaptability to new challenges, and alignment with global demands for transparency and accountability. As supply chains grow more interconnected, these technologies are pivotal in meeting regulatory requirements and restoring public confidence. By integrating molecular biology and advanced analytical technologies, the food industry can establish robust systems that safeguard authenticity, ensure safety, and promote ethical trade practices. These innovations are not merely advancements; they represent the foundation of a future where food safety is uncompromising, and consumer trust is unwavering.

Keywords Food safety · Molecular technology · Advanced Analytical Tools · Food Authentication

Introduction

Food is more than sustenance; it symbolizes culture, a driver of economies, and a cornerstone of human health. In an era where global supply chains link producers to consumers across continents, food integrity has never been more vital or vulnerable [1, 2]. Food adulteration, the intentional or unintentional addition of inferior or harmful substances to food products, represents one of the greatest challenges to food safety today. It is a covert, multifaceted threat that spans local and global markets, impacting consumer trust, public health, and economic stability [3, 4]. Historically, food adulteration has ranged from simply diluting premium ingredients with cheaper substitutes to using hazardous chemicals to enhance appearance or prolong shelf life [5–7]. However, the problem has grown more complex in modern times. The globalization of food systems, coupled with the increasing demand for convenience and profitability, has created opportunities for sophisticated adulteration practices that are difficult to detect using traditional methods [8]. From mislabeled fish species and fake honey to dairy and meat products laced with harmful additives, the consequences of food adulteration are far-reaching, often resulting in severe health risks, regulatory breaches, and significant economic losses [9–11].

Conventional methods of detecting food adulteration, such as chemical tests and microscopic analysis, have been valuable tools for decades. However, these approaches are often limited by their specificity, sensitivity, and applicability to processed or multi-component foods. With the rise of advanced adulteration techniques, the limitations of these traditional methods have become increasingly apparent [12–14]. As such, there is an urgent need for innovative, precise, reliable solutions that address the challenges posed by evolving adulteration practices. In this context, molecular biology and DNA-based techniques have emerged as revolutionary tools in food safety and authenticity. These advanced methodologies leverage the unique properties of DNA to achieve unparalleled accuracy in identifying food components. DNA is remarkably stable under various conditions and can be extracted from trace amounts of biological material, making it an ideal target for food adulteration analysis [15]. Technologies such as DNA barcoding, polymerase chain reaction (PCR), and next-generation sequencing (NGS) have opened new frontiers in the detection and prevention of food fraud [16–18]. The importance of these molecular techniques lies in their ability to detect adulterants with exceptional sensitivity and their potential to enhance traceability and transparency within global food supply chains. For example, DNA barcoding allows for precisely identifying species in mislabeled seafood [17, 19], while PCR techniques can pinpoint specific adulterants in processed food products [20, 21]. Moreover, the advent of NGS has made it possible to analyze complex food matrices, unraveling even the most concealed instances of adulteration [22]. These tools are not merely supplementary to traditional methods; they represent a transformative leap forward, offering a comprehensive solution to one of the most persistent challenges in food safety. As regulatory bodies and consumers demand higher standards of authenticity and transparency, integrating molecular biology and DNA-based techniques into food safety frameworks has become essential. These methods align with global efforts to promote ethical trade practices, protect public health, and foster consumer confidence. Beyond detection, they play a critical role in risk mitigation, enabling proactive measures to prevent adulteration and ensure the integrity of food systems. This chapter explores the profound impact of molecular and DNA-based techniques in combating food adulteration, delving into their scientific principles, practical applications, and future potential. By addressing this critical issue with cutting-edge solutions, we can build a safer, more transparent, and sustainable food landscape that meets the needs of both producers and consumers in an increasingly interconnected world.

Modern Trends in Food Adulteration

Food adulteration, defined as the intentional contamination or substitution of products with substandard, harmful, or cheaper substances, has evolved dramatically over the years [3, 4]. It is not a new phenomenon; food adulteration has been practiced since ancient times to increase profits or deceive consumers [23, 24]. However, as food systems have globalized, and consumer awareness has heightened, the nature of food adulteration has become more sophisticated. Today, adulteration is driven not only by economic motives but also by technological advancements, supply chain complexities, and geopolitical considerations. In this section, we explore the modern trends in food adulteration, focusing on the evolution of adulteration techniques, the high-risk commodities most commonly targeted, and the regulatory challenges that continue to hinder effective food safety enforcement.

Evolution of Adulteration Techniques

The history of food adulteration reflects changing technological, economic, and regulatory landscapes. In the early stages, adulteration was relatively crude and straightforward. Common examples included diluting milk with water [25], mixing cheaper oils with more expensive ones [26], or adding coloring agents to fruit juices to make them appear more appealing [27]. These types of adulteration were often easy to detect through fundamental sensory analysis or simple chemical tests. However, as technology advanced, so did the sophistication of food fraud techniques. The advent of chemical additives, synthetic substances, and even novel methods like genetic manipulation made food adulteration much more challenging to detect. These modern adulteration techniques are designed to mimic authentic food's characteristics closely, making it challenging for consumers and even regulatory authorities to distinguish between real and fake products without advanced testing methods [28, 29]. A significant turning point in the evolution of food adulteration came with the global expansion of the food trade. As food products moved across borders, the complexity of food adulteration techniques also increased. Food items began to be adulterated with more complex substances to increase their volume or shelf life and to reduce production costs. The globalization of supply chains has resulted in adulteration becoming a cross-border issue, where the regulatory standards of one country may differ from those of another, allowing the illicit practices to spread unchecked [30].

Another significant evolution in food adulteration was the development of molecular manipulation and synthetic biology. With biotechnology, fraudsters can genetically modify food products, making it even more challenging to detect adulteration without highly sophisticated techniques [31]. For example, genetically modified crops, synthetic meats, or cultured seafood may be sold as natural or organic products without proper labeling, misleading consumers and raising ethical concerns [32]. Today, modern adulteration techniques include mixing cheaper ingredients into premium products and using food-grade chemicals, dyes, and additives to alter the appearance, texture, and nutritional profile of foods. For instance, melamine has been used to falsely increase the apparent protein content of milk [33, 34]. At the same time, synthetic dyes are added to mimic the natural color of high-value spices, such as saffron or turmeric [35]. Such methods are designed to evade basic tests and exploit gaps in food regulation systems.

High-Risk Commodities and Common Adulterants

Certain food products are more susceptible to adulteration due to their high market value, widespread use, and complex production or distribution processes. Fraudsters typically target these high-risk commodities because even a tiny degree of adulteration can lead to significant financial gain. Common adulterants used in food fraud vary by product, but they generally include cheaper, harmful, or synthetic ingredients that are added to mimic the characteristics of more expensive, authentic ingredients [36, 37]. Table 1 provides a detailed overview of some high-risk commodities targeted by food adulteration and lists the common adulterants used in these products.

Regulatory Gaps and Enforcement Challenges

Despite food safety regulations in many countries, significant gaps remain in their enforcement and implementation. While some countries have developed stringent food safety laws and inspection systems, others struggle with underfunded, poorly

Table 1 High-risk commodities and common adulterants in the food industry

Commodity	Common adulterants	Details	References
Spices and herbs	Starches, synthetic dyes, crushed seeds, synthetic colorants, flour, marigold petals	High-value spices like saffron and turmeric are often adulterated with cheaper substances, such as rice flour, lead chromate, or marigold petals to enhance color and weight	[36, 38]
Olive oil	Soybean oil, sunflower oil, palm oil, other vegetable oils	Extra virgin olive oil is frequently diluted with cheaper vegetable oils like soybean or sunflower oil to increase volume and reduce cost	[39, 40]
Honey	Sugar syrup, high-fructose corn syrup, glucose, maltose	Honey is often adulterated with cheaper sugar syrups or artificial sweeteners to mimic the viscosity and sweetness of real honey, compromising its quality and authenticity	[41, 42]
Meat and seafood	Water, fat, starches, non-authentic animal parts (e.g., pork or non-halal meat), fish substitution	Meat products such as beef or lamb are adulterated by adding water starch, using lower-quality animal parts, or mislabeling cheaper fish species	[7, 43]
Milk and dairy	Water, starch, detergents, synthetic proteins (e.g., melamine), non-dairy fats	Milk is often diluted with water, adulterated with starches or synthetic chemicals like melamine, and dairy products can be substituted with cheaper fats or non-dairy ingredients	[22, 44]
Cheese	Vegetable fats, starches, non-dairy fat, water, milk protein substitutes	Cheese is sometimes adulterated with cheaper vegetable oils, starches, or milk protein substitutes to reduce costs	[45, 46]
Wheat flour	Chalk powder, starches, soapstone, starch, synthetic whitening agents	Flour can be adulterated with cheaper substances like chalk powder or starches to increase its bulk and improve its texture or whiteness	[47]
Tea and coffee	Used tea leaves, colored leaves, starches, sawdust, cheap coffee species	High-grade tea is frequently adulterated with colored leaves or powdered starch, while coffee can be diluted with cheaper coffee species or other fillers	[48]
Juices and beverages	Artificial sweeteners, water, cheaper fruit juices, synthetic colors, preservatives	Fruit juices like orange or apple juice are often diluted with water, cheaper fruit juices, or artificial colors to mimic the original product's color and flavor	[27, 49]
Chocolate and cocoa	Starch, cheap fats, vegetable oils, non-cocoa substitutes, synthetic flavoring agents	Cocoa products, especially chocolate, are adulterated with vegetable oils, starch, or non-cocoa fats, such as palm oil, to enhance texture and reduce costs	[50]

equipped regulatory bodies that are ill-equipped to tackle the growing problem of food adulteration. This disparity in regulation, enforcement, and food safety practices creates opportunities for food fraud to flourish in areas with less stringent oversight [51]. One major challenge is the lack of standardized, accessible testing methods for detecting a wide range of adulterants. Food adulteration techniques have become increasingly sophisticated, with fraudsters using cutting-edge technologies to circumvent traditional detection methods. Regulatory authorities in many countries still rely on outdated or inadequate testing methods that cannot identify modern adulterants, often present in trace amounts or disguised by advanced manufacturing processes [28]. Moreover, the global nature of food supply chains complicates enforcement efforts. Food products move across borders at an unprecedented rate, and adulteration can easily occur in one country and be detected only after the product reaches a foreign market. The absence of international food safety standards and protocols contributes to this issue. Some countries have robust regulations and advanced detection systems, while others do not, making it challenging to ensure the safety of imported foods. This regulatory fragmentation creates opportunities for unscrupulous actors to exploit weaker regulatory systems. Another enforcement challenge is the informal sector. In many countries, food production and distribution occur in informal markets where regulations are challenging to enforce. Food vendors, small-scale producers, and street food vendors often operate outside the formal regulatory framework, making it harder for authorities to monitor and ensure food safety [52].

In some cases, food fraud is prevalent in these informal markets because of the lack of oversight and the ease of introducing adulterated products into the supply chain. Lastly, there is a growing need for more public–private partnerships in food safety regulation. Many food companies, particularly in the retail and processing sectors, have the resources to implement their stringent safety standards. Still, these efforts must be aligned with governmental regulations to ensure that food safety is consistently maintained across the supply chain. The engagement of the private sector, in collaboration with regulatory authorities, can help close the gaps in enforcement and strengthen the overall food safety framework.

Advanced Analytical Techniques for Food Adulteration Detection

Overview of Molecular Techniques

Molecular biology techniques have revolutionized the field of food safety, offering unprecedented accuracy and reliability in detecting food adulteration. These techniques utilize the unique properties of DNA, a stable, specific, and universally present biomolecule, to identify and authenticate food components. Unlike traditional methods, molecular approaches can detect adulteration even in processed or multi-ingredient foods, providing a robust solution to modern challenges in food fraud detection [53]. Key molecular biology techniques used in food adulteration analysis include DNA barcoding, polymerase chain reaction (PCR), real-time PCR (qPCR), next-generation sequencing (NGS), and loop-mediated isothermal amplification (LAMP). Each method has distinct advantages, ranging from high specificity and sensitivity to rapid analysis and adaptability to various food matrices.

DNA Barcoding

DNA barcoding is a cutting-edge molecular technique that has revolutionized how we identify species, offering a precise and reliable method for detecting food fraud and mislabeling. The technique hinges on amplifying and sequencing short, standardized DNA regions that serve as "barcodes" for species identification (Fig. 1a). These genetic markers are typically conserved across species, making them highly

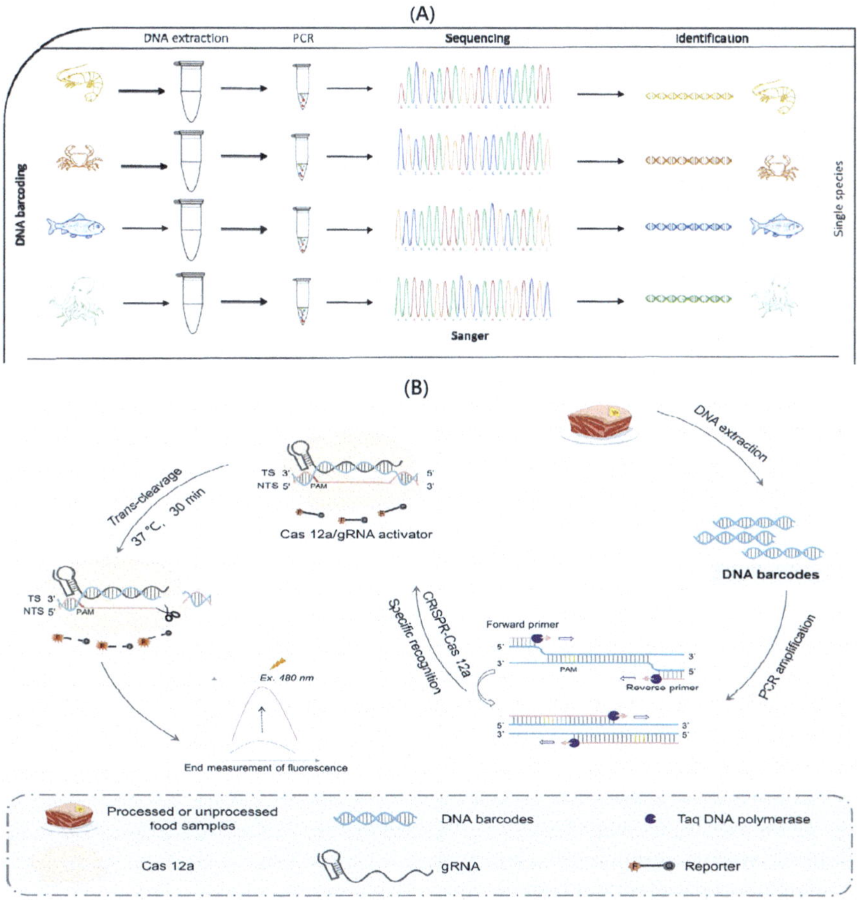

Fig. 1 (**A**) Schematic illustration outlines the key steps in DNA barcoding for identifying seafood species [17]. (**B**) Workflow of DNA barcoding and CRISPR-Cas12a system for food adulteration detection. The image was reproduced with permission [54]

effective for distinguishing even closely related organisms [55–57]. For animals, the most widely used barcode is the mitochondrial gene cytochrome c oxidase I (*COI*), a component of the mitochondrial respiratory chain [56]. This gene is favored because it has a relatively high mutation rate, which provides enough variation to differentiate between species while remaining stable within a species [58]. The *COI* gene is especially effective in identifying fish and other seafood, often subject to mislabeling in the marketplace [59]. COI has also been successfully used to uncover fraudulent practices, such as substituting cheaper fish species for high-value ones like tuna or grouper [60]. In plants, ribulose-1,5-bisphosphate carboxylase (*rbcL)* and maturase K (*matK*) genes are commonly employed for DNA barcoding [61]. The *rbcL* gene is a chloroplast DNA marker that plays a key role in photosynthesis. It is highly conserved within plant species, which allows for reliable identification across a wide range of plant taxa [62]. The *matK* gene, another chloroplast marker, offers a higher degree of variation than *rbcL* and is often used in combination to enhance identification accuracy, particularly for more closely related plant species [63]. These genetic markers have proven essential in combating fraud in herbal products, where adulteration with cheaper or non-authentic species is a significant concern.

The application of DNA barcoding extends beyond food safety into broader fields, such as conservation, ecology, and biodiversity studies. By creating a comprehensive genetic database of species, DNA barcoding also enables rapid species identification in cases of illegal wildlife trade, invasive species monitoring, and the protection of endangered species [56, 64]. In food safety, the technique has been indispensable in detecting species substitution and ensuring that products labeled as "organic" or "premium" are accurately represented. For instance, a study by Neo et al. [65] found that up to 30% of seafood in global markets could be mislabeled, with cheaper fish such as tilapia sold as snapper or cod. Similarly, DNA barcoding has exposed the fraud of exotic meats, where venison or kangaroo meat is often substituted with more common and less expensive meats like beef or pork [66, 67]. This mislabeling not only deceives consumers but also poses potential health risks, particularly if allergens or toxins exist in the substituted species. By utilizing DNA barcoding, authorities and industries can ensure better food traceability, combat fraud, and enhance consumer confidence. The growing availability of DNA barcoding databases and the decreasing cost of sequencing technologies further improve the accessibility and practicality of this method for routine food safety monitoring and regulatory enforcement.

Integrating DNA barcoding with CRISPR-based systems presents a promising approach for enhancing food adulteration detection (Fig. 1b). While DNA barcoding effectively identifies species based on genetic markers like cytochrome c oxidase I (*COI*) for animals and *rbcL* and *matK* for plants. CRISPR technology can amplify the sensitivity and specificity of this detection by enabling targeted genetic modifications. By combining the high precision of CRISPR-Cas12 systems with the extensive genetic information obtained from DNA barcoding, we can develop more sophisticated tools to identify mislabeling and species substitution and detect the presence of adulterants at the molecular level. CRISPR could be used to create biosensors that rapidly recognize specific gene sequences associated with fraudulent species, significantly improving the speed and accuracy of food testing processes. This synergy of technologies can also offer better scalability, allowing for on-site, real-time detection of food fraud in diverse markets, ensuring food safety while protecting both consumers and the integrity of the food industry.

CRISPR-Based Detection Systems

The CRISPR-Cas system, widely known for its revolutionary impact on genetic editing, has recently emerged as a cutting-edge tool for food adulteration detection. This powerful molecular technology, which leverages the ability of bacteria to "remember" viral DNA sequences and use them to defend against future infections, has been ingeniously adapted for detecting genetic material in complex food samples [68]. The simplicity, precision, and versatility of CRISPR-based systems make them an exciting and highly effective approach for detecting food adulteration at the molecular level.

Genetic Precision in Food Adulteration Detection

At the heart of CRISPR-based detection systems are the CRISPR/Cas9 or, more commonly, CRISPR/Cas12/Cas13 enzymes, which can recognize specific DNA or RNA sequences within a sample [69]. These enzymes are guided by short RNA molecules known as guide RNAs (gRNAs), which are designed to bind to complementary DNA or RNA sequences from target adulterants [70]. When the CRISPR enzyme recognizes and binds to its target sequence, it activates a detectable signal, which could be fluorescence, color change, or other outputs, making it possible to identify the presence of the adulterant with remarkable specificity [71].

The CRISPR/Cas system's genetic precision allows it to differentiate between similar sequences, ensuring that only the adulterant of interest is detected. For instance, if a food sample is suspected of being adulterated with a certain species' DNA, the CRISPR/Cas system can be engineered to recognize the specific genetic markers associated with that species. This is particularly beneficial for detecting food fraud, such as species substitution, where DNA from one species is substituted for another, whether in meat products like beef or fish or high-value commodities like honey or olive oil [68, 72].

Targeting Genetic Sequences with High Sensitivity

The CRISPR-based detection system's ability to target specific genetic sequences allows for the detection of adulterants at very low concentrations, offering an extraordinary level of sensitivity [73]. This is crucial in food adulteration, where the adulterant is often added in small amounts to avoid detection. For example, if a cheaper meat species is introduced into a higher-end product, the amount of adulterant may be as little as 1–2%. However, the CRISPR/Cas system can detect even these minute amounts by amplifying and highlighting the presence of specific genetic markers associated with the adulterant [74]. The high sensitivity of CRISPR-based systems, coupled with their ability to work in complex food matrices, makes them invaluable for detecting both known and unknown adulterants. Unlike traditional PCR or NGS, which require more extensive sample preparation and sophisticated equipment, CRISPR-based detection can be performed on-site, providing fast and reliable results [75]. This attribute is particularly advantageous in the food industry, where speed and accuracy in detecting adulteration are critical to maintaining food safety and consumer trust.

Real-Time and Multiplex Detection Capabilities

One of the most appealing features of CRISPR-based systems is their potential for real-time detection. When CRISPR/Cas enzymes are used with fluorescence-based reporters, the detection process can be monitored as it happens, providing immediate feedback on the presence of adulterants [76]. This real-time capability accelerates the detection process and enhances the precision of results, enabling rapid decision-making in food safety inspections. Furthermore, CRISPR technology offers the possibility of multiplexing, where multiple targets can be detected simultaneously within a single reaction. By using different gRNAs for different adulterants, CRISPR systems can be adapted to detect a wide array of potential food frauds in one test [77]. This is particularly important when testing for multiple forms of adulteration in complex food matrices, such as meat products that may contain multiple species or vegetable oils mixed with cheaper alternatives. Multiplex CRISPR-based detection systems offer an efficient, comprehensive solution to food authenticity testing, making it possible to test for a broad range of adulterants in a single step.

Field-Deployable CRISPR Diagnostic Tools

In addition to laboratory-based detection, CRISPR-based systems are increasingly designed for field deployment, offering a portable, user-friendly solution for on-site adulteration detection. Using paper-based diagnostic devices/microfluidic chips, CRISPR assays can be made more accessible and cost-effective, enabling non-experts to perform food authenticity tests in real-time, at the point of sale, or even during transportation. These portable CRISPR diagnostic tools can revolutionize food safety surveillance, allowing food inspectors, producers, and even consumers to rapidly screen for adulterants without the need for specialized laboratory equipment or technical expertise [78, 79]. The simplicity of CRISPR-based detection also reduces the costs associated with food safety testing, making it a viable option for widespread use across the food supply chain.

Broader Applications in Food Safety and Transparency

Beyond detecting adulterants in meat and fish products, CRISPR-based detection systems hold promise for a wide range of applications in food safety, including the detection of genetically modified organisms (GMOs), allergens, and pathogens. For example, CRISPR/Cas13, which targets RNA instead of DNA, can identify specific pathogens or viruses in food products, adding another layer of protection against foodborne illnesses [69, 80]. Additionally, CRISPR-based methods can assist in ensuring the transparency of food labeling. They can be used to verify the authenticity of organic foods, trace the origin of raw materials, and confirm whether products labeled as "GMO-free" are genuinely free of genetic modification [81]. This capability can help build consumer confidence by ensuring that food products meet their standards, fostering greater transparency and accountability in the food industry.

Polymerase Chain Reaction (PCR)

Polymerase chain reaction (PCR) is one of the most powerful tools in molecular biology, offering unparalleled sensitivity and specificity in detecting food adulteration [82]. By amplifying specific DNA sequences, PCR allows the identification of even trace amounts of adulterants in food products, which are often undetectable by traditional methods. The process involves a series of thermal cycles: denaturation, annealing, and extension, leading to the exponential amplification of target DNA sequences, analyzed to confirm the presence of adulterants [83].

The ability of PCR to selectively target unique genetic markers makes it an invaluable tool in the detection of adulteration across various food categories, including meat, dairy, and plant-based products. For example, in the meat industry, PCR can identify species-specific DNA, making it possible to detect the presence of undeclared meat species or mislabeling of meat products [84–86]. In dairy products, PCR can distinguish between different milk species, such as cow, goat, or sheep, ensuring that the product complies with labeling standards [87, 88].

One of the advanced variants of PCR, real-time PCR (qPCR), offers additional advantages by providing quantitative data. Unlike traditional PCR, which only confirms the presence or absence of a target DNA sequence, qPCR can measure the amount of DNA present in a sample, enabling a more detailed understanding of the degree of adulteration. This allows for real-time adulteration level monitoring, offering qualitative and quantitative insights into food composition [77]. By tracking the amplification process in real time, qPCR provides more precise and reproducible results, which are crucial for regulatory purposes and quality control in the food industry.

Multiplex PCR takes PCR's capabilities a step further by allowing the simultaneous amplification of multiple DNA targets within a single reaction. This technique is beneficial when dealing with complex food matrices, where a single adulterant may not change the product's characteristics. Multiplex PCR can efficiently detect multiple adulterants or contaminants, such as foreign species or genetically modified organisms (GMOs), in a single test, reducing the time and cost associated with individual tests for each adulterant [72, 85]. By focusing on specific genetic markers unique to particular species or varieties of ingredients, PCR enhances our ability to monitor the integrity of food products at a molecular level. This makes it a highly reliable method for identifying subtle adulterations that could go unnoticed. Furthermore, the high sensitivity and versatility of PCR-based methods ensure that even low-level adulterations in complex food products are detected with high precision, giving regulators and manufacturers the tools they need to protect consumers and maintain food safety standards. This sophisticated molecular technology not only provides a level of assurance regarding food authenticity but also helps to safeguard public health by identifying potential allergens, contaminants, or other harmful substances that may be introduced during food processing or distribution. As such, PCR continues to revolutionize how we approach food testing, offering insight and reliability that were previously unattainable with conventional testing methods.

Next-Generation Sequencing (NGS)

Next-generation sequencing (NGS) has revolutionized the field of molecular biology and is rapidly emerging as a powerful tool for detecting food adulteration. Unlike traditional methods such as PCR, NGS provides a comprehensive and detailed view of the genetic makeup of food, allowing for the identification of adulterants with unparalleled accuracy and depth [89]. At its core, NGS involves the parallel sequencing of millions of DNA fragments, enabling high-throughput, accurate, and large-scale analysis of genetic material. This technology allows for the simultaneous identification of multiple species in a food sample, making it particularly useful for detecting both known and unknown adulterants. By sequencing the entire genome or targeted regions of the DNA, NGS can uncover subtle genetic markers indicative of adulteration that might be missed by other methods [90, 91].

One of the key advantages of NGS over conventional PCR methods lies in its ability to provide a holistic view of the food's genetic content. While PCR targets specific DNA sequences for amplification, NGS reads entire genomic regions in a single run [77, 92]. This means that NGS can detect not only the adulterants added to the food but also any potential microbial contaminants or genetic modifications that may be present. For instance, in the case of meat adulteration, NGS can detect traces of DNA from species not declared on the label, whether from plant-based fillers or other animal sources [93, 94]. The genetic basis for this advanced detection technique lies in the ability to map out the entire genomic signature of a food sample. By comparing the sequences obtained from the sample against known genetic databases, NGS can pinpoint exactly which organisms or adulterants are present [95]. This comparative genomics approach enables the detection of adulteration at the species, strain, or even variant level, providing detailed insights into the composition of food products. For example, when used in the analysis of dairy products, NGS can identify DNA sequences from non-dairy plant-based additives or identify animal species that are not typically included in certain types of dairy (such as the presence of beef or pork in milk products) [96, 97]. Similarly, in the case of honey adulteration, NGS can differentiate between authentic floral sources and sugar syrups that may be used to adulterate the product [98, 99], offering clear genetic signatures that other methods may fail to detect.

Furthermore, the versatility of NGS allows for its integration with other genomic technologies, such as metagenomics, to provide a more comprehensive analysis of food composition [16]. Through metagenomic sequencing, NGS can identify the genetic material of the food itself and the microbiota present, which could provide insights into the food's processing history or potential contamination risks [100, 101]. The growing use of NGS in food authenticity and safety testing promises a new frontier in food quality assurance. As genomic databases expand, the potential for identifying food fraud through NGS will become even more precise and reliable. Moreover, the capacity of NGS to detect adulterants across diverse food matrices, from processed foods to fresh produce, makes it an indispensable tool for food regulators, manufacturers, and consumers alike.

Loop-Mediated Isothermal Amplification (LAMP)

In recent years, advancements in molecular biology techniques have revolutionized the detection of food adulteration. Among these, loop-mediated isothermal amplification (LAMP) has emerged as a powerful tool for identifying adulterants in various food products. LAMP is a highly sensitive, rapid, and cost-effective diagnostic method that amplifies DNA at a constant temperature, typically between 60 °C and 65 °C, eliminating the need for thermal cycling [102]. This innovative approach is advantageous for on-site detection due to its simplicity, portability, and reduced reliance on sophisticated laboratory equipment [103]. Its ability to detect adulterants across diverse food matrices, including dairy, meat, and processed products, highlights its potential as a game-changing solution for ensuring food authenticity and safety [104]. The underlying principle of LAMP involves using a set of specially designed primers that target multiple regions of the DNA sequence, creating a loop structure that enables rapid amplification. This unique amplification mechanism makes LAMP highly specific, with results typically visible within 30–60 min [105].

Genetically, LAMP takes advantage of the specificity and robustness of DNA sequences to detect even minute adulterant traces. The primers used in LAMP are designed to recognize distinct genomic regions, ensuring high specificity for the target DNA, such as adulterant species or unauthorized genetic material [106]. These primers can be tailored to target specific genes, making LAMP an incredibly versatile tool for detecting various adulterants, from plant species to animal DNA and even pathogenic microorganisms [107]. For example, LAMP can distinguish between species-specific DNA sequences in meat products, ensuring the product's authenticity [108]. Moreover, LAMP offers a high level of sensitivity, capable of detecting very low concentrations of target DNA, which is particularly beneficial in food fraud, where adulterants are often added in minute amounts. This heightened sensitivity is due to the unique amplification process that generates many DNA copies in a short period [109]. Furthermore, the LAMP method can be coupled with various detection formats, including colorimetric assays, turbidity measurements, or fluorescence, making it adaptable to different testing environments and enhancing its practical applicability [110, 111]. In addition to its speed, simplicity, and cost-effectiveness, LAMP's compatibility with multiplex assays is another significant advantage. By incorporating multiple primer sets into a single reaction, multiplex LAMP can detect several food adulterants simultaneously, providing a comprehensive analysis in one step [112]. This feature is especially valuable when testing for complex adulteration, where multiple adulterants may be present in the food sample.

Digital PCR (dPCR)

Digital PCR (dPCR) represents a significant advancement in molecular diagnostics, offering exceptional precision and sensitivity for detecting food adulteration. Unlike conventional PCR, which provides relative quantification of DNA through amplification cycles, dPCR achieves absolute quantification by partitioning the sample into thousands of individual micro-reactions. Each partition undergoes amplification, and the presence or absence of amplification is used to count DNA molecules directly [85, 113]. This partitioning, performed in microfluidic chips or droplets, enables dPCR to detect low-abundance targets in complex food matrices, even when adulterants are present in trace amounts. By providing a highly accurate measurement of DNA concentration without calibration curves or standard controls, dPCR is particularly effective for detecting minute levels of adulterants, such as species-specific DNA in meat products or trace contaminants in food ingredients [114, 115]. From a genetic perspective, dPCR offers several advantages over traditional PCR methods. While conventional PCR relies on relative quantification by comparing amplification signals to a threshold, dPCR counts the actual number of target DNA molecules in each partition, eliminating issues, such as PCR inhibition, amplification bias, and non-specific binding. This allows for a much more precise and reproducible detection of adulterants [116]. Additionally, dPCR can quantify both target and non-target DNA simultaneously, making it highly versatile [117]. For example, it can detect both the adulterant species and the authentic food DNA in the same reaction, providing comprehensive information on the food's composition [118]. dPCR's ability to provide absolute quantification without needing standard curves or reference materials further enhances its applicability in detecting food adulteration. It can be particularly beneficial in situations where adulterants are unknown or where the exact concentration of adulterants is critical for regulatory compliance [119]. One of the key aspects of digital PCR that enhances its appeal for food safety applications is its robustness in complex food matrices [120]. Food samples are often heterogeneous, and traditional PCR methods may struggle with inhibition from other components in the sample. The high resolution and partitioning of dPCR mitigate these issues, enabling more reliable results even in complex food matrices, such as processed foods, dairy products, and beverages [72, 121, 122].

Mass Spectrometry (MS)

Mass spectrometry (MS) has emerged as a powerful and sophisticated tool for detecting food adulteration, providing a level of sensitivity, specificity, and accuracy that surpasses traditional methods [123]. Unlike techniques that rely on the amplification of genetic material or the identification of specific biomarkers, mass spectrometry offers a comprehensive approach by analyzing the mass-to-charge ratio of ions, revealing detailed molecular compositions of complex food matrices [124]. This ability to measure and identify minute amounts of adulterants has made MS an invaluable tool in ensuring food authenticity and safety (Fig. 2). The principle of MS involves ionizing food samples and measuring the resulting ions' masses, which provides a molecular fingerprint of the sample. This fingerprint can then be compared against known spectra to identify the presence of adulterants [126]. The versatility of MS lies in its ability to detect a wide range of substances, from synthetic chemicals to natural compounds [127], making it an ideal method for uncovering adulteration in food products that may not be detectable using conventional techniques, such as PCR or LAMP. One of the major advantages of MS is its ability to detect both organic and inorganic adulterants with a high degree of sensitivity and precision. For example, in the case of food oils, MS can identify adulterants, such as lower-cost vegetable oils mixed with olive oil or other premium oils, which may be undetectable using less advanced techniques [128]. Similarly, MS can detect trace levels of contaminants, such as pesticides, heavy metals, and synthetic food additives, which are commonly used in food fraud [129]. This detection level is crucial, as adulterants are often added in small amounts, and MS can provide the resolution necessary to uncover these subtle discrepancies in food composition.

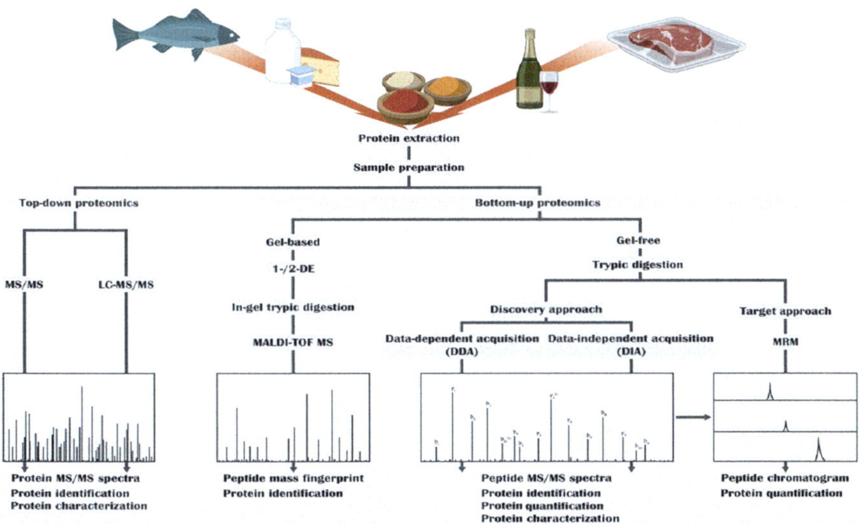

Fig. 2 A schematic diagram illustrating the typical proteomics workflows used for food authentication and quality assessment [125]

MS can be used in conjunction with DNA barcoding or next-generation sequencing (NGS) to detect adulteration in animal-based products when applied to genetic-based adulteration. For instance, in meat products, MS can differentiate between species-specific protein markers, allowing the detection of unauthorized animal species mixed into beef, chicken, or fish products [130]. The ability to profile proteins and peptides associated with specific animal tissues enhances the specificity of MS, enabling the detection of understated adulterations, such as meat substitution or contamination with low-value meat species.

Furthermore, combining MS with advanced data analysis tools, such as chemometrics and machine learning algorithms, has enhanced its capacity for identifying and quantifying adulterants in food products [131]. Chemometrics involves using statistical methods to analyze complex datasets obtained from MS, allowing for identifying patterns or trends that may indicate adulteration [132]. Conversely, machine learning algorithms enable MS to improve its accuracy over time by learning from past analyses, making it even more reliable for detecting new or emerging adulterants [123]. One noteworthy advancement is the use of high-resolution mass spectrometry (HRMS), which offers greater precision and sensitivity than conventional MS methods. HRMS can detect even the smallest molecular differences, making it particularly useful for identifying food fraud involving complex chemical additives or subtle alterations in food composition [133].

Isotope ratio mass spectrometry (IRMS), a subset of MS, has also gained prominence for food authenticity testing, especially for determining food products' geographical origin or authenticity. IRMS analyzes the ratio of stable isotopes, such as carbon and nitrogen isotopes, in food samples to trace their origin [134]. This technique has been widely used to authenticate products like honey and olive oil, where adulteration may involve mixing products from different regions or countries [125, 126]. The isotopic signature provides a unique "food fingerprint," making it possible to verify whether the product matches its claimed origin. The ability of MS to detect food adulterants at the molecular level opens the door for not only identifying traditional forms of food fraud but also discovering novel adulterants that may not have been anticipated by regulatory bodies or existing testing protocols. This capability makes MS a forward-thinking solution in the ever-evolving landscape of food safety and authenticity. With the increasing complexity of food production systems and the growing concerns about food fraud globally, MS provides an indispensable tool for ensuring that consumers receive the products they expect, free from harmful or misleading adulterants.

Moreover, integrating MS into real-time, point-of-consumption testing platforms is becoming more feasible, with developments in miniaturized mass spectrometers and portable analytical devices. These advancements pave the way for on-site food analysis, allowing for quicker detection of adulteration, which is essential for rapid-response interventions in food safety crises.

Raman Spectroscopy

Raman spectroscopy has rapidly emerged as a powerful and versatile technique for detecting food adulteration, offering unique advantages in non-destructive, rapid, and detailed analysis [135]. This optical technique relies on the inelastic scattering of monochromatic light, where the frequency of scattered light shifts concerning the vibrational modes of the molecules in the sample. These vibrational shifts provide a molecular fingerprint, which can be used to identify and quantify the chemical composition of the food, enabling the detection of adulterants and contaminants at molecular levels [136].

The appeal of Raman spectroscopy in food safety lies in its ability to analyze complex food matrices with minimal sample preparation. Unlike traditional chemical methods that may require lengthy procedures or hazardous reagents, Raman spectroscopy is non-invasive and can be performed directly on the food sample without altering its integrity. This makes it particularly useful for on-site detection of food adulteration, offering real-time analysis in field settings [137]. The use of Raman spectroscopy has already been explored in numerous applications, including the identification of fraudulent substances such as synthetic additives [138], preservatives [139], and even toxic contaminants like pesticides or heavy metals [140, 141].

At the molecular level, Raman spectroscopy excels in detecting subtle differences in the molecular vibrations between adulterants and authentic food ingredients. For example, in olive oil, Raman spectroscopy can distinguish between olive oil and cheaper oils (e.g., sunflower or palm oil) based on characteristic peaks that correspond to the molecular vibrations of different fatty acids present in each oil type [142]. Similarly, in milk, Raman spectra can reveal the presence of starch or other foreign substances that may have been added to dilute the milk [143]. This specificity is rooted in the fact that the Raman spectra of each substance are unique due to the distinct molecular vibrations of its chemical bonds, allowing Raman to act as a molecular fingerprint for the identification of food ingredients.

One of the unique features of Raman Spectroscopy is its ability to detect food adulteration without the need for any genetic manipulation or DNA analysis, unlike PCR or LAMP [144]. While DNA-based methods are highly effective at identifying biological adulterants, Raman spectroscopy offers a complementary advantage by identifying a wider range of chemical adulterants, including synthetic and inorganic substances. It can also be used to detect adulterants in highly processed foods where DNA-based detection may not be as effective, such as in beverages [145], oils [146], or powdered products [147]. This ability to analyze natural and synthetic components makes Raman spectroscopy indispensable for comprehensive food safety monitoring.

Recent advancements in Raman spectroscopy, particularly surface-enhanced Raman scattering (SERS), have significantly enhanced its potential for detecting food adulteration. SERS, an advanced variation of traditional Raman spectroscopy, utilizes the interaction between incident light and molecular vibrations to create a unique spectral fingerprint of a material. Its unique advantage lies in using specially engineered metallic nanostructures, such as gold or silver nanoparticles, which amplify Raman scattering signals by several orders of magnitude [148]. This amplification enables the detection of extremely low concentrations of target analytes, often in the range of parts per billion (ppb) or even parts per trillion (ppt), making SERS ideal for identifying adulterants in complex food matrices [149]. The sensitivity and specificity of SERS are particularly beneficial for detecting trace amounts of contaminants in high-end food products like honey [150] and olive oil [151], as well as unapproved additives [152], harmful chemicals [153], and species-specific DNA [154]. By leveraging the enhanced Raman signal, SERS allows for precisely identifying adulterants based on their unique molecular vibrational modes, ensuring robust food safety testing and fraud prevention. The application of chemometrics in conjunction with Raman spectroscopy has also improved its utility for complex food adulteration scenarios. Chemometric techniques, such as principal component analysis (PCA) or partial least squares (PLS), allow for the analysis of complex spectral data by identifying patterns or trends in the Raman spectra [155]. This approach enables the classification of food samples based on their authenticity and the simultaneous detection of multiple adulterants.

Nuclear Magnetic Resonance

Nuclear magnetic resonance (NMR) spectroscopy has rapidly emerged as a powerful and sophisticated analytical tool in food adulteration detection. Unlike many traditional techniques that focus on specific chemical markers, NMR provides a holistic and non-destructive approach by characterizing food's molecular structure and dynamics at the atomic level. This enables identifying known and unknown adulterants, offering significant advantages in food authenticity testing [156]. At its core, NMR spectroscopy measures the interaction between atomic nuclei and a magnetic field, providing unique information about the chemical environment of atoms within a molecule [157]. NMR is beneficial in food adulteration because it allows for the analysis of complex food matrices, such as oils, beverages, dairy, and meat, without extensive sample preparation. This non-invasive nature makes NMR ideal for routine food quality control and safety monitoring [158].

Applications of NMR in Food Adulteration Detection

NMR technology is a powerful tool in food analysis, excelling in detecting adulteration by providing a detailed "fingerprint" of a food's molecular composition. This unique fingerprint can be compared to reference spectra to identify the presence of adulterants, ensuring the authenticity and quality of food products [159]. Recent advancements in NMR research have been extensively reviewed, with studies focusing on specific applications such as the identification of food authenticity [160], the assessment of quality parameters in fruits [161], and correlations between water distribution, mobility, water-holding capacity, and meat quality [162]. Additionally, NMR applications have been explored for particular foods, including dairy products [163], Salmon [164], and fermented foods [165], highlighting its versatility and significance in food science. Moreover, NMR can identify isotopic signatures in food that may indicate contamination or the addition of foreign substances. For instance, NMR can detect isotopic shifts in the chemical structure of meat that could suggest the presence of non-native animal DNA or the use of growth hormones and antibiotics [166]. This makes NMR a powerful tool for ensuring the integrity of meat and seafood products.

High Sensitivity and Specificity

One of the most significant advantages of NMR is its high sensitivity and reproducibility. It can detect even trace amounts of adulterants in food samples, making it highly effective in identifying fraudulent practices where adulterants are added in small quantities [159]. The ability to detect subtle molecular changes, such as slight variations in the fatty acid profile [167] or the presence of synthetic chemicals [156], ensures that NMR is a reliable tool for high-precision food analysis. Unlike some molecular techniques like PCR, which target specific DNA sequences, NMR provides a more comprehensive analysis by observing the total molecular structure. This allows for the detection of a wider range of adulterants, even those that are not genetically based, such as chemical contaminants [168], preservatives [169], or artificial flavorings [170]. The method's ability to detect multiple adulterants simultaneously is another compelling reason for its growing use in the food industry.

Genomic Insights and DNA-Based Adulteration Detection

While NMR does not directly target DNA sequences like PCR, it can still provide valuable genomic insights when coupled with other techniques. For example, NMR can be used in combination with DNA barcoding or PCR-based methods to confirm the presence of specific species in food products [171, 172]. NMR can offer a comprehensive and multi-dimensional approach to food authenticity testing by analyzing the molecular composition and combining this data with genetic markers. In particular, NMR can assist in confirming species-specific markers identified through DNA-based techniques. In cases where genetic material is difficult to amplify due to degradation or processing, NMR can provide complementary data to strengthen the evidence of adulteration [173]. Furthermore, NMR can be used to cross-check the results of genetic analysis, improving the accuracy and reliability of food safety assessments [174].

Metabolomics

Metabolomics, the comprehensive study of small molecules (metabolites) produced by living organisms, is rapidly emerging as a revolutionary approach to food adulteration detection [175]. Unlike traditional methods that focus on detecting specific DNA sequences or chemical additives, metabolomics offers a holistic view of the biochemical composition of food, providing a powerful tool for identifying adulteration patterns that may otherwise go undetected. This technique analyzes the metabolic profile of food products, capturing a wide range of metabolites, including amino acids, lipids, sugars, organic acids, and volatile compounds, all of which contribute to the unique "fingerprint" of a given food [176]. The key advantage of metabolomics in food adulteration detection lies in its ability to detect not only known adulterants but also unexpected contaminants or fraudulent ingredients that may not be identified through targeted analysis. By examining the full spectrum of metabolites, metabolomics can reveal subtle changes in food composition caused by the addition of cheaper substitutes or harmful adulterants [177]. This is particularly valuable in cases of food fraud where the adulterants are not easily identifiable through traditional methods, such as PCR or DNA barcoding.

From a genetic perspective, the metabolic pathways in food are influenced by the product's plant, animal, or microbial origin. Genetic expression, environmental factors, and post-harvest processing conditions shape these metabolic profiles. For example, in plant-based foods, metabolomic analysis can identify variations in the levels of secondary metabolites (e.g., flavonoids, alkaloids, or terpenes), which may differ depending on the plant species or the presence of adulterants like cheaper or non-native ingredients [178]. Similarly, metabolomics can detect alterations in fat and protein content in animal-derived products like milk or meat, indicating adulteration with non-authentic or low-quality substitutes [179]. The process of metabolomic analysis typically involves the extraction of metabolites from food samples, followed by advanced techniques, such as mass spectrometry (MS) or nuclear magnetic resonance (NMR) spectroscopy [180]. MS provides high sensitivity and resolution for detecting and quantifying metabolites. At the same time, NMR offers detailed structural information, allowing researchers to identify metabolites based on their molecular fingerprint (Fig. 3). Together, these methods enable the detection of minute changes in the biochemical composition of food, making it possible to identify even trace amounts of adulterants [182]. Furthermore, using sophisticated multivariate statistical analyses, metabolomic data can be interpreted to distinguish between authentic and adulterated food products, providing highly accurate results.

One of the most compelling applications of metabolomics in food adulteration detection is its ability to perform untargeted screening, a key feature that sets it apart from traditional methods. In untargeted metabolomics, the entire metabolite profile of a sample is analyzed without preselecting specific compounds for detection. This approach is beneficial for detecting complex adulteration strategies, such as when multiple ingredients are added to food products or when adulterants are introduced in small quantities [181]. For instance, in honey adulteration, metabolomics can

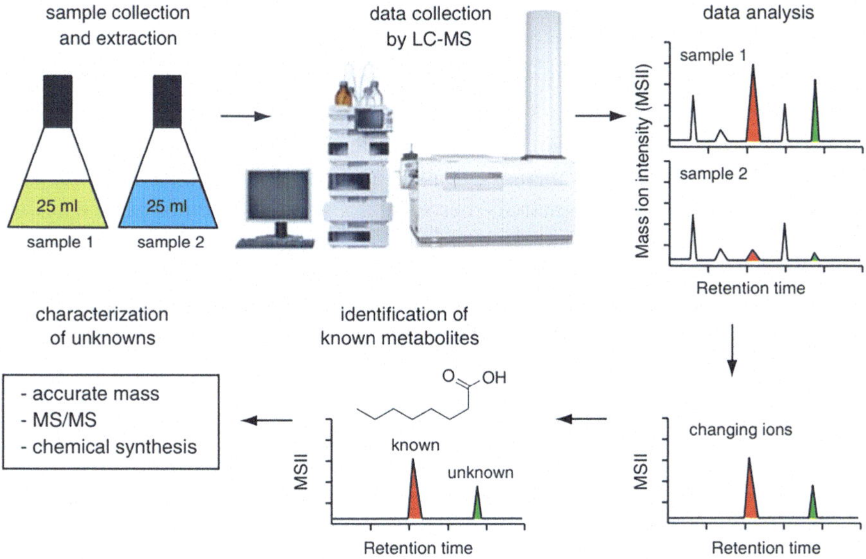

Fig. 3 The general process for conducting an untargeted metabolomics analysis. The image was reproduced with permission [181]

identify the presence of sugar syrups or cheaper sweeteners, even when they are present in concentrations too low to be detected by conventional methods [41].

Moreover, metabolomics can also be applied to detecting food authenticity beyond adulteration. By comparing the metabolic profiles of authentic food samples with those of suspected adulterated samples, researchers can determine the geographic origin, quality, and even the farming practices behind the product [183]. This is particularly valuable in products like olive oil and spices, where authenticity and origin verification are critical for consumers. For example, the metabolic signature of olive oil can be used to differentiate between extra virgin olive oil and adulterated olive oil mixed with cheaper oils based on the distinct levels of fatty acids, sterols, and phenolic compounds present in the oil [184].

The genetic underpinning of metabolomics lies in the interaction between genes and metabolites. Genes encode enzymes that drive metabolic processes, and alterations in gene expression due to environmental factors or genetic modifications can result in changes to the metabolic profile [185]. For example, adding genetically modified organisms (GMOs) or using specific agricultural practices can alter the metabolic pathways in plants, leading to detectable differences in the food's biochemical composition. Metabolomics provides a dynamic snapshot of how genetic factors influence food quality and how food fraud may alter these profiles [186]. Another strength of metabolomics in food adulteration detection is its non-targeted nature, which allows it to identify new markers of adulteration that have not been previously recognized. This characteristic is especially valuable in addressing emerging food fraud practices, where traditional detection methods may fail to identify novel adulterants [187].

Electrochemical Biosensors

In recent years, electrochemical biosensors have emerged as a powerful and innovative tool for detecting food adulteration. These sensors offer a unique combination of specificity, sensitivity, portability, and cost-effectiveness, making them an attractive alternative to traditional methods of food analysis [188]. Electrochemical biosensors detect specific interactions between a biological element (such as an enzyme, antibody, or nucleic acid) and the target analyte in the food sample. This interaction results in a measurable change in an electrical signal, such as current, voltage, or impedance, which can be used to identify and quantify the presence of adulterants [189].

Principle of Electrochemical Biosensors

Electrochemical biosensors typically have three main components: the biological recognition element, the transducer, and the detection system [190]. The biological recognition element, such as an antibody or nucleic acid probe, is designed to specifically bind to the adulterant or target molecule in the food sample. When the target interacts with the recognition element, it triggers a biochemical response that the transducer converts into an electrical signal. The transducer, which may include materials like carbon nanotubes, gold electrodes, or graphene, amplifies the signal, allowing for accurate detection of even low concentrations of adulterants. Finally, the detection system processes the electrical signal, providing real-time results for analysis [191].

Genetic Applications in Electrochemical Biosensors for Food Adulteration Detection

The role of genetic components in electrochemical biosensors has significantly enhanced the specificity and sensitivity of food adulteration detection. Genetic-based biosensors utilize DNA or RNA sequences as recognition elements to identify specific genetic markers associated with food adulterants, such as unauthorized species or genetically modified organisms (GMOs). For example, DNA probes designed to target species-specific genetic sequences allow for the detection of animal-derived DNA in meat products, ensuring that the product is authentic and not substituted with cheaper or non-declared species [192]. In food authenticity, electrochemical biosensors can also detect gene sequences that indicate contamination with genetically modified (GM) crops or allergens [193]. For instance, a biosensor could be designed to recognize specific genetic sequences associated with GM soybeans or maize, allowing for rapidly detecting GM material in food products [194]. Similarly, biosensors can detect the genetic markers of allergens in food, offering a sensitive method for safeguarding consumer health [195].

Moreover, electrochemical biosensors have been developed to detect microbial contamination in food, where genetic analysis plays a critical role [196]. By targeting specific pathogenic DNA sequences, these sensors can provide real-time identification of foodborne pathogens, such as *Escherichia coli* [197] or *Salmonella* species [198], before they pose a risk to consumers. Using genetic primers or probes that bind specifically to pathogenic DNA ensures that the biosensors offer high sensitivity and accuracy, even at low pathogen concentrations.

Advantages of Electrochemical Biosensors in Food Adulteration Detection

1. High sensitivity and specificity: Electrochemical biosensors can achieve detection limits as low as picomolar or even femtomolar concentrations, making them highly sensitive for detecting minute quantities of adulterants. The use of genetic recognition elements, such as DNA or RNA probes, further enhances specificity by allowing for the detection of adulterants based on their unique genetic signatures [199].
2. Portability and on-site testing: One of the key advantages of electrochemical biosensors is their portability. These sensors can be miniaturized and integrated into handheld devices, enabling on-site, real-time testing without laboratory equipment. This is particularly beneficial for field testing, allowing for rapid detection of food adulterants at points of sale, during food processing, or in supply chains [200].
3. Cost-effectiveness: Electrochemical biosensors are relatively inexpensive to produce and operate compared to traditional analytical methods, such as PCR or mass spectrometry. This makes them an attractive option for widespread adoption in food safety applications, especially in resource-limited settings or areas with limited access to sophisticated laboratory facilities [201].
4. Rapid detection: Electrochemical biosensors provide results in minutes, offering significant advantages over more time-consuming methods like PCR, which can take several hours. The quick turnaround time of electrochemical biosensors enables faster decision-making in food safety monitoring, minimizing the risk of adulterated products reaching consumers [202].
5. Multiplexing capability: Electrochemical biosensors can be adapted to detect adulterants simultaneously using multiplexed assays. By incorporating multiple recognition elements into a single sensor, it is possible to detect several types of adulterants in one analysis, such as different species of animal DNA in meat products or various types of pesticides in plant-based foods [203].

Recent Advances and Applications

Recent developments in nanotechnology and materials science have further improved the performance of electrochemical biosensors for food adulteration detection. Nanomaterials such as gold nanoparticles, carbon nanotubes, and graphene oxide have been integrated into sensor designs to enhance sensitivity, improve electron transfer, and increase the surface area for binding interactions. These innovations have led to the development of highly efficient biosensors that can detect even trace levels of adulterants in complex food matrices [204]. For instance, researchers have designed electrochemical biosensors for detecting milk adulteration with starch, detergent, or urea [44, 205, 206]. These sensors work by detecting changes in the electrochemical properties of the sample in response to the presence of adulterants, offering a rapid, reliable, and cost-effective means of quality control. Similarly, electrochemical biosensors have been developed to detect honey adulteration with sugar syrups or cheaper, lower-quality honey, ensuring that consumers receive authentic, unadulterated products [207]. Another exciting application of electrochemical biosensors is in detecting foodborne pathogens and toxins. By using genetic probes that target specific pathogenic DNA sequences, electrochemical biosensors can rapidly identify contamination in food products, allowing for immediate intervention and prevention of foodborne illness outbreaks [208, 209]. These sensors can monitor microbial contamination in various food products, including meat, dairy, seafood, and fresh produce. In addition to food adulteration and pathogen detection, electrochemical biosensor systems are increasingly used for allergen detection in food products [210]. These biosensors utilize specific antibodies or aptamers that bind to allergens, triggering an electrochemical response that can be measured and quantified. By integrating nanomaterials like gold nanoparticles and carbon nanotubes, these biosensors offer enhanced sensitivity and faster detection times, even at low concentrations of allergens (Fig. 4). This is

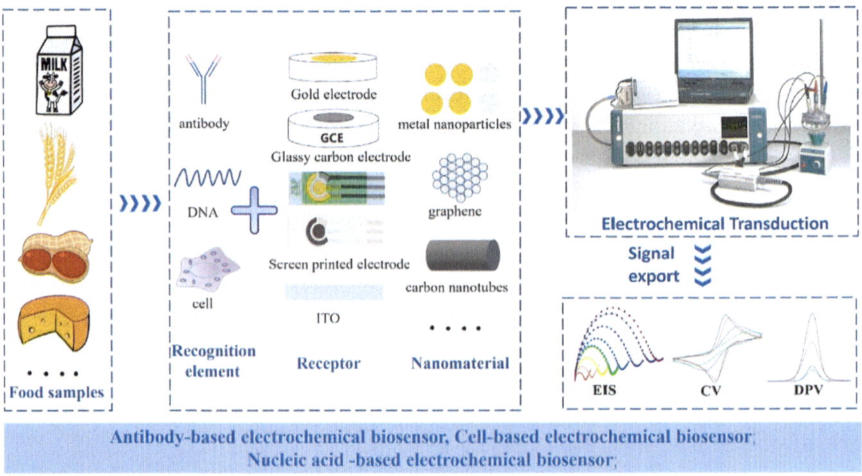

Fig. 4 An electrochemical biosensor platform for detecting allergens in food products. The image was reproduced with permission [210]

particularly valuable for individuals with food allergies, as it allows for rapid, on-site testing of food products to ensure they are free from harmful allergens like peanuts, gluten, or dairy.

Conclusion and Future Directions

In the face of growing concerns over food safety, authenticity, and adulteration, advanced detection technologies are revolutionizing how we protect consumers and ensure the integrity of our food supply. From DNA-based techniques like PCR, DNA barcoding, and next-generation sequencing (NGS) to a range of molecular-level tools such as nuclear magnetic resonance (NMR) spectroscopy, loop-mediated isothermal amplification (LAMP), and others, these methods provide unmatched sensitivity, accuracy, and versatility. Their ability to identify even trace amounts of adulterants, quantify fraud, and uncover species-level mislabeling makes them indispensable in modern food authentication and regulatory frameworks. Integrating these technologies into food safety systems ensures consumer trust and upholds ethical trade practices globally. However, as food supply chains evolve and become more complex, the future of food adulteration detection demands even greater innovation. Future directions should focus on the miniaturization and portability of detection devices to enable real-time, on-site monitoring. The development of multi-omics approaches, combining genomics, proteomics, and metabolomics, holds the potential to provide a more comprehensive analysis of food authenticity. In addition, artificial intelligence (AI) and machine learning are expected to play a transformative role in handling large datasets generated by these techniques, enhancing food adulteration detection systems' speed, precision, and predictive capabilities. Moreover, integrating these advanced methods with blockchain technology could ensure transparent traceability throughout the food supply chain, creating an unbroken record of food origin and quality. This would empower regulatory authorities, producers, and consumers alike with real-time, verifiable information. In conclusion, the combination of cutting-edge technologies, interdisciplinary approaches, and digital innovations marks the future of food adulteration detection. By bridging science, technology, and policy, we can achieve a robust, secure, and transparent food system that safeguards public health and promotes global food security. The road ahead holds exciting prospects, where food safety evolves beyond a regulatory requirement to become a shared responsibility for a healthier and more sustainable future.

Acknowledgments This work is supported by the Second Century Fund (C2F), Chulalongkorn University, Thailand. The authors gratefully acknowledge the Department of Chemistry, Faculty of Science, Chulalongkorn University, for the provision of research facilities, resources, and technical support.

References

1. Hariram N, Mekha K, Suganthan V, Sudhakar K (2023) Sustainalism: an integrated socio-economic-environmental model to address sustainable development and sustainability. Sustainability 15(13):10682
2. Manik S, Ramani A, Majumder R, Hossain S, Hazra T (2024) Traditional foods for festivity: linking food diversity with socio-cultural aspects, traditional foods: the reinvented super-foods. Springer, Cham, pp 69–88
3. Banti M (2020) Food adulteration and some methods of detection, review. Int J Nutr Food Sci 9(3):86–94
4. Haji A, Desalegn K, Hassen H (2023) Selected food items adulteration, their impacts on public health, and detection methods: a review. Food Sci Nutr 11(12):7534–7545
5. Jha SN, Jaiswal P, Grewal MK, Gupta M, Bhardwaj R (2016) Detection of adulterants and contaminants in liquid foods—a review. Crit Rev Food Sci Nutr 56(10):1662–1684
6. Lohumi S, Lee S, Lee H, Cho B-K (2015) A review of vibrational spectroscopic techniques for the detection of food authenticity and adulteration. Trends Food Sci Technol 46(1):85–98
7. Jurica K, Brčić Karačonji I, Lasić D, Bursać Kovačević D, Putnik P (2021) Unauthorized food manipulation as a criminal offense: food authenticity, legal frameworks, analytical tools and cases. Foods 10(11):2570
8. Esteki M, Regueiro J, Simal-Gándara J (2019) Tackling fraudsters with global strategies to expose fraud in the food chain. Compr Rev Food Sci Food Saf 18(2):425–440
9. Smaoui S, Tarapoulouzi M, Agriopoulou S, D'Amore T, Varzakas T (2023) Current state of milk, dairy products, meat and meat products, eggs, fish and fishery products authentication and chemometrics. Foods 12(23):4254
10. Khan MA, Hossain ME, Islam MS, Rahman MS, Sudhakaran PO, Dey MM (2023) A systematic review of fish adulteration and contamination in Bangladesh: a way forward to food safety. Rev Aquac 15(4):1574–1589
11. Brar DS, Pant K, Krishnan R, Kaur S, Rasane P, Nanda V, Saxena S, Gautam S (2023) A comprehensive review on unethical honey: validation by emerging techniques. Food Control 145:109482
12. He Y, Bai X, Xiao Q, Liu F, Zhou L, Zhang C (2021) Detection of adulteration in food based on nondestructive analysis techniques: a review. Crit Rev Food Sci Nutr 61(14):2351–2371
13. Goyal R, Singha P, Singh SK (2024) Spectroscopic food adulteration detection using machine learning: current challenges and future prospects. Trends Food Sci Technol 146:104377
14. Ndlovu PF, Magwaza LS, Tesfay SZ, Mphahlele RR (2022) Destructive and rapid non-invasive methods used to detect adulteration of dried powdered horticultural products: a review. Food Res Int 157:111198
15. Baptista M, Cunha JT, Domingues L (2021) DNA-based approaches for dairy products authentication: a review and perspectives. Trends Food Sci Technol 109:386–397
16. Haynes E, Jimenez E, Pardo MA, Helyar SJ (2019) The future of NGS (next generation sequencing) analysis in testing food authenticity. Food Control 101:134–143
17. Fernandes TJ, Amaral JS, Mafra I (2021) DNA barcode markers applied to seafood authentication: an updated review. Crit Rev Food Sci Nutr 61(22):3904–3935
18. Feng J, Lan H, Wu Z, Pan D (2024) Novel integration of lateral flow strip and point-of-care isothermal amplification techniques for meat adulteration detection: a comprehensive review. Trends Food Sci Technol 152:104698
19. Di Pinto A, Marchetti P, Mottola A, Bozzo G, Bonerba E, Ceci E, Bottaro M, Tantillo G (2015) Species identification in fish fillet products using DNA barcoding. Fish Res 170:9–13
20. Behr M, Garlant L, Pietretti D, Pellegrin C, Lievens A, Sanfeliu AB, Maquet A, Alvarellos L (2024) A robust set of qPCR methods to evaluate adulteration in major spices and herbs. Food Control 165:110623

21. Hazra T, Sharma V, Sharma R, Arora S (2017) A species specific simplex polymerase chain reaction-based approach for detection of goat tallow in heat clarified milk fat (ghee). Int J Food Prop 20(sup1):S69–S75
22. Shawky E, Nahar L, Nassief SM, Sarker SD, Ibrahim RS (2024) Dairy products authentication with biomarkers: a comprehensive critical review. Trends Food Sci Technol 147:104445
23. Kowalska A (2018) The study of the intersection between food fraud/adulteration and authenticity. Acta Univ Agric Silvic Mendelianae Brun 66(5):1275–1286
24. Charlebois S, Schwab A, Henn R, Huck CW (2016) Food fraud: an exploratory study for measuring consumer perception towards mislabeled food products and influence on self-authentication intentions. Trends Food Sci Technol 50:211–218
25. Nascimento CF, Santos PM, Pereira-Filho ER, Rocha FR (2017) Recent advances on determination of milk adulterants. Food Chem 221:1232–1244
26. Huq A, Uddin I, Ahmed E, Siddique M, Zaher M, Nigar S (2022) Fats and oils adulteration: present scenario and rapid detection techniques. Food Res 6(1):5–11
27. Xu L, Xu Z, Liao X (2022) A review of fruit juice authenticity assessments: targeted and untargeted analyses. Crit Rev Food Sci Nutr 62(22):6081–6102
28. Choudhary A, Gupta N, Hameed F, Choton S (2020) An overview of food adulteration: concept, sources, impact, challenges and detection. Int J Chem Stud 8(1):2564–2573
29. Sen M (2021) Food chemistry: role of additives, preservatives, and adulteration. In: Food chemistry: the role of additives, preservatives and adulteration. Wiley, Hoboken, pp 1–42
30. Gayialis SP, Kechagias EP, Papadopoulos GA, Masouras D (2022) A review and classification framework of traceability approaches for identifying product supply chain counterfeiting. Sustainability 14(11):6666
31. Haider A, Iqbal SZ, Bhatti IA, Alim MB, Waseem M, Iqbal M, Mousavi Khaneghah A (2024) Food authentication, current issues, analytical techniques, and future challenges: a comprehensive review. Compr Rev Food Sci Food Saf 23(3):e13360
32. Bawa A, Anilakumar K (2013) Genetically modified foods: safety, risks and public concerns—a review. J Food Sci Technol 50(6):1035–1046
33. Ritota M, Manzi P (2018) Melamine detection in milk and dairy products: traditional analytical methods and recent developments. Food Anal Methods 11:128–147
34. Rajpoot M, Bhattacharya R, Sharma S, Gupta S, Sharma V, Sharma AK (2021) Melamine contamination and associated health risks: gut microbiota does make a difference. Biotechnol Appl Biochem 68(6):1271–1280
35. Luzardo-Ocampo I, Ramírez-Jiménez AK, Yañez J, Mojica L, Luna-Vital DA (2021) Technological applications of natural colorants in food systems: a review. Foods 10(3):634
36. Momtaz M, Bubli SY, Khan MS (2023) Mechanisms and health aspects of food adulteration: a comprehensive review. Foods 12(1):199
37. Aslam N, Fatima R, Altemimi AB, Ahmad T, Khalid S, Hassan SA, Aadil RM (2024) Overview of industrial food fraud and authentication through chromatography technique and its impact on public health. Food Chem 460:140542
38. Castillejos-Mijangos LA, Acosta-Caudillo A, Gallardo-Velázquez T, Osorio-Revilla G, Jiménez-Martínez C (2022) Uses of FT-MIR spectroscopy and multivariate analysis in quality control of coffee, cocoa, and commercially important spices. Foods 11(4):579
39. Lozano-Castellón J, López-Yerena A, Domínguez-López I, Siscart-Serra A, Fraga N, Sámano S, López-Sabater C, Lamuela-Raventós RM, Vallverdú-Queralt A, Pérez M (2022) Extra virgin olive oil: a comprehensive review of efforts to ensure its authenticity, traceability, and safety. Compr Rev Food Sci Food Saf 21(3):2639–2664
40. Tan CH, Kong I, Irfan U, Solihin MI, Pui LP (2021) Edible oils adulteration: a review on regulatory compliance and its detection technologies. J Oleo Sci 70(10):1343–1356
41. Bose D, Padmavati M (2024) Honey authentication: a review of the issues and challenges associated with honey adulteration. Food Biosci 61:105004
42. Biswas A, Naresh K, Jaygadkar SS, Chaudhari SR (2023) Enabling honey quality and authenticity with NMR and LC-IRMS based platform. Food Chem 416:135825

43. Spychaj A, Goderska K, Fornal E, Montowska M (2021) A practical approach to identifying processed white meat of Guinea fowl, rabbit, and selected fish species using end-point PCR. Int J Food Sci 2021(1):7710462
44. Patil GB, Wani SP, Bafna PS, Bagul VS, Kalaskar MG, Mutha RE (2024) Milk adulteration: from detection to health impact. Food Humanity 3:100339
45. Abedini A, Salimi M, Mazaheri Y, Sadighara P, Sani MA, Assadpour E, Jafari SM (2023) Assessment of cheese frauds, and relevant detection methods: a systematic review. Food Chem X 19:100825
46. Silva LK, Santos LS, Ferrão SP (2022) Application of infrared spectroscopic techniques to cheese authentication: a review. Int J Dairy Technol 75(3):490–512
47. Pandiselvam R, Sruthi N, Kumar A, Kothakota A, Thirumdas R, Ramesh S, Cozzolino D (2023) Recent applications of vibrational spectroscopic techniques in the grain industry. Food Rev Int 39(1):209–239
48. Kennedy SP, Gonzales P, Roungchun J (2021) Coffee and tea fraud, food fraud. Elsevier, Amsterdam, pp 139–150
49. Mondal D, Equbal MZ, Jain A, Mishra V (2024) Disposable paper-based sensor array for detection of orange juice adulteration. J Food Meas Charact 18(7):1–12
50. Amin S, Mir SR (2018) Application of authentication and traceability in chocolate, fingerprinting techniques in food authentication and traceability. CRC Press, Boca Raton, pp 363–382
51. Nayak R, Jespersen L (2022) Development of a framework to capture the maturity of food safety regulatory and enforcement agencies: insights from a Delphi study. Food Control 142:109220
52. Hoffmann V, Moser C, Saak A (2019) Food safety in low and middle-income countries: the evidence through an economic lens. World Dev 123:104611
53. Ibrahim M, Detroja A, Sheth BP, Bhadja P, Sanghvi G, Bishoyi AK (2024) Existing status and future advancements of adulteration detection techniques in herbal products. Mol Biol Rep 51(1):151
54. Wu Y, Liu J, Li H-T, Zhang T, Dong Y, Deng S, Lv Y, He Q, Deng R (2022) CRISPR-Cas system meets DNA barcoding: development of a universal nucleic acid test for food authentication. Sensors Actuators B Chem 353:131138
55. DeSalle R, Goldstein P (2019) Review and interpretation of trends in DNA barcoding. Front Ecol Evol 7:302
56. Antil S, Abraham JS, Sripoorna S, Maurya S, Dagar J, Makhija S, Bhagat P, Gupta R, Sood U, Lal R (2023) DNA barcoding, an effective tool for species identification: a review. Mol Biol Rep 50(1):761–775
57. Nehal N, Choudhary B, Nagpure A, Gupta RK (2021) DNA barcoding: a modern age tool for detection of adulteration in food. Crit Rev Biotechnol 41(5):767–791
58. Skibinski DO, Ghiselli F, Diz AP, Milani L, Mullins JG (2017) Structure-related differences between cytochrome oxidase I proteins in a stable heteroplasmic mitochondrial system. Genome Biol Evol 9(12):3265–3281
59. Filonzi L, Ardenghi A, Rontani PM, Voccia A, Ferrari C, Papa R, Bellin N, Nonnis Marzano F (2023) Molecular barcoding: a tool to guarantee correct seafood labelling and quality and preserve the conservation of endangered species. Foods 12(12):2420
60. Kotsanopoulos KV, Exadactylos A, Gkafas GA, Martsikalis PV, Parlapani FF, Boziaris IS, Arvanitoyannis IS (2021) The use of molecular markers in the verification of fish and seafood authenticity and the detection of adulteration. Compr Rev Food Sci Food Saf 20(2):1584–1654
61. Jamdade R, Upadhyay M, Al Shaer K, Al Harthi E, Al Sallani M, Al Jasmi M, Al Ketbi A (2021) Evaluation of Arabian vascular plant barcodes (rbcL and matK): precision of unsupervised and supervised learning methods towards accurate identification. Plants 10(12):2741
62. Zhu J, Huang Y, Chai W, Xia P (2024) Decoding the chloroplast genome of Tetrastigma (Vitaceae): variations and phylogenetic selection insights. Int J Mol Sci 25(15):8290

63. Jiang S, Chen F, Qin P, Xie H, Peng G, Li Y, Guo X (2022) The specific DNA barcodes based on chloroplast genes for species identification of Theaceae plants. Physiol Mol Biol Plants 28(4):837–848
64. Smart U, Cihlar JC, Budowle B (2021) International wildlife trafficking: a perspective on the challenges and potential forensic genetics solutions. Forensic Sci Int Genet 54:102551
65. Neo S, Kibat C, Wainwright BJ (2022) Seafood mislabelling in Singapore. Food Control 135:108821
66. Adenuga BM, Montowska M (2023) A systematic review of DNA-based methods in authentication of game and less common meat species. Compr Rev Food Sci Food Saf 22(3):2112–2160
67. Quinto CA, Tinoco R, Hellberg RS (2016) DNA barcoding reveals mislabeling of game meat species on the US commercial market. Food Control 59:386–392
68. Suliman Maashi M (2024) CRISPR/Cas-based aptasensor as an innovative sensing approaches for food safety analysis: recent progresses and new horizons. Crit Rev Anal Chem 54(7):2599–2617
69. Singh M, Bindal G, Misra CS, Rath D (2022) The era of Cas12 and Cas13 CRISPR-based disease diagnosis. Crit Rev Microbiol 48(6):714–729
70. Qin J, Wang W, Gao L, Yao SQ (2022) Emerging biosensing and transducing techniques for potential applications in point-of-care diagnostics. Chem Sci 13(10):2857–2876
71. Li Y, Man S, Ye S, Liu G, Ma L (2022) CRISPR-Cas-based detection for food safety problems: current status, challenges, and opportunities. Compr Rev Food Sci Food Saf 21(4):3770–3798
72. Kumar P, Rani A, Singh S, Kumar A (2022) Recent advances on DNA and omics-based technology in food testing and authentication: a review. J Food Saf 42(4):e12986
73. Hajikhani M, Zhang Y, Gao X, Lin M (2023) Advances in CRISPR-based SERS detection of food contaminants: a review. Trends Food Sci Technol 138:615–627
74. Qiu M, Zhang J, Pang L, Zhang Y, Zhao Q, Jiang Y, Yang X, Man C (2022) Recent advances on CRISPR/Cas system-enabled portable detection devices for on-site agri-food safety assay. Trends Food Sci Technol 129:364–387
75. Lu B, Guo Z, Zhong K, Osire T, Sun Y, Jiang L (2023) State of the art in CRISPR/Cas system-based signal conversion and amplification applied in the field of food analysis. Trends Food Sci Technol 135:174–189
76. Mao Z, Lei H, Chen R, Ren S, Liu B, Gao Z (2023) CRISPR molecular detection techniques: advances from single to multiple detection methods. TrAC Trends Anal Chem 166:117198
77. Lanubile A, Stagnati L, Marocco A, Busconi M (2024) DNA-based techniques to check quality and authenticity of food, feed and medicinal products of plant origin: a review. Trends Food Sci Technol 149:104568
78. Kapoor A, Ramamoorthy S, Sundaramurthy A, Vaishampayan V, Sridhar A, Balasubramanian S, Ponnuchamy M (2024) Based lab-on-a-chip devices for detection of agri-food contamination. Trends Food Sci Technol 147:104476
79. Boodhoo N, Shoja Doost J, Sharif S (2024) Biosensors for monitoring, detecting, and tracking dissemination of poultry-borne bacterial pathogens along the poultry value chain: a review. Animals 14(21):3138
80. Lu Y, Yang H, Bai J, He Q, Deng R (2024) CRISPR-Cas based molecular diagnostics for foodborne pathogens. Crit Rev Food Sci Nutr 64(16):5269–5289
81. Khaliq EA, Perveen N, Hamid D, Hameed M, Asif M, Arif M (2024) Novel CRISPR-based genome editing systems for crop improvement. In: OMICs-based techniques for global food security. Wiley, Hoboken, pp 245–268
82. Denyingyhot A, Srinulgray T, Mahamad P, Ruangprach A, Sa S, Saerae T, Vesaratchavest M, Dahlan W, Keeratipibul S (2022) Modern on-site tool for monitoring contamination of halal meat with products from five non-halal animals using multiplex polymerase chain reaction coupled with DNA strip. Food Control 132:108540

83. Mualim M, Latif H, Pisestyani H, Rahayu P (2024) Analysis of species adulteration in beef sausage using real-time polymerase chain reaction in Makassar. Indonesia Vet World 17(10):2355
84. Elbarbary NK, Darwish WS, Fotouh A, Dandrawy MK (2024) Unveiling the mix-up: investigating species and unauthorized tissues in beef-based meat products. BMC Vet Res 20(1):380
85. Vishnuraj M, Kumar NA, Vaithiyanathan S, Barbuddhe S (2023) Authentication issues in foods of animal origin and advanced molecular techniques for identification and vulnerability assessment. Trends Food Sci Technol 138:164–177
86. Uddin SMK, Hossain MM, Chowdhury ZZ, Johan MRB (2021) Short targeting multiplex PCR assay to detect and discriminate beef, buffalo, chicken, duck, goat, sheep and pork DNA in food products. Food Add Contamin Part A 38(8):1273–1288
87. Kourkouli A, Thomaidis N, Dasenaki M, Markou A (2024) Novel and sensitive touchdown polymerase chain reaction assays for the detection of goat and sheep milk adulteration with cow milk. Molecules 29(8):1820
88. Su Y, Meng L, Wang J, Zhao Y, Zheng N (2024) Simultaneous detection of eight dairy-derived components using double-tube multiplex qPCR based taqman probe. Foods 13(20):3213
89. Yang H, He S, Feng Q, Xia S, Zhou Q, Wu Z, Zhang Y (2024) Navigating the depths of seafood authentication: technologies. Regul Future Prosp Measure Food 14:100165
90. Katara A, Chand S, Chaudhary H, Chaudhry V, Chandra H, Dubey RC (2024) Evolution and applications of next generation sequencing and its intricate relations with chromatographic and spectrometric techniques in modern day sciences. J Chromatogr Open 5:100121
91. Domrazek K, Jurka P (2024) Application of next-generation sequencing (NGS) techniques for selected companion animals. Animals 14(11):1578
92. Kumar R, Das SP, Choudhury BU, Kumar A, Prakash NR, Verma R, Chakraborti M, Devi AG, Bhattacharjee B, Das R (2024) Advances in genomic tools for plant breeding: harnessing DNA molecular markers, genomic selection, and genome editing. Biol Res 57(1):80
93. Alam A, Usman M, Parveen B, Ahmad S (2024) Future directions and emerging trends in medicinal plant DNA barcoding. In: Ethnopharmacology and OMICS advances in medicinal plants volume 1: uncovering diversity and ethnopharmacological aspects. Springer, Cham, pp 219–237
94. Negi A, Lakshmi P, Praba K, Meenatchi R, Pare A (2021) Detection of food adulterants in different foodstuff. In: Food chemistry: the role of additives, preservatives and adulteration. Wiley, Hoboken, pp 117–164
95. Gan Y, Qi G, Hao L, Xin T, Lou Q, Xu W, Song J (2024) Analysis of whole-genome as a novel strategy for animal species identification. Int J Mol Sci 25(5):2955
96. Hong E, Lee SY, Jeong JY, Park JM, Kim BH, Kwon K, Chun HS (2017) Modern analytical methods for the detection of food fraud and adulteration by food category. J Sci Food Agric 97(12):3877–3896
97. Zhang S, Chen J, Gao F, Su W, Li T, Wang Y (2024) Foodomics as a tool for evaluating food authenticity and safety from field to table: a review. Foods 14(1):15
98. Danieli PP, Lazzari F (2022) Honey traceability and authenticity, Review of current methods most used to face this problem. J Agric Sci 66(2):101–119
99. Smith C, Cokcetin N, Truong T, Harry E, Hutvagner G, Bajan S (2021) Cataloguing the small RNA content of honey using next generation sequencing. Food Chem Mol Sci 2:100014
100. Sadurski J, Polak-Berecka M, Staniszewski A, Waśko A (2024) Step-by-step metagenomics for food microbiome analysis: a detailed review. Foods 13(14):2216
101. Okoye CO, Jiang H, Nazar M, Tan X-L, Jiang J (2024) Redefining modern food analysis: significance of omics analytical techniques integration, chemometrics and bioinformatics. TrAC Trends Anal Chem 175:117706
102. Sharma AS, Lee NY (2024) Advancements in visualizing loop-mediated isothermal amplification (LAMP) reactions: a comprehensive review of colorimetric and fluorometric detection strategies for precise diagnosis of infectious diseases. Coord Chem Rev 509:215769

103. Soroka M, Wasowicz B, Rymaszewska A (2021) Loop-mediated isothermal amplification (LAMP): the better sibling of PCR? Cells 10(8):1931
104. Eş I, Khaneghah AM (2024) Advancing food quality assurance: integrating microneedle technology with advanced analytical methods. Nano Today 54:102115
105. Becherer L, Borst N, Bakheit M, Frischmann S, Zengerle R, von Stetten F (2020) Loop-mediated isothermal amplification (LAMP)–review and classification of methods for sequence-specific detection. Anal Methods 12(6):717–746
106. Wang Y, Feng J, Tian X (2019) Application of loop-mediated isothermal amplification (LAMP) for rapid detection of Atlantic cod (Gadus morhua), Pacific cod (Gadus macrocephalus) and haddock (Melanogrammus aeglefinus). Mol Cell Probes 47:101420
107. Ran G, Ren L, Han X, Liu X, Li Z, Pang D, Ouyang H, Tang X (2016) Development of a rapid method for the visible detection of pork DNA in halal products by loop-mediated isothermal amplification. Food Anal Methods 9:565–570
108. Wang S, Song H, Wang T, Xue H, Fei Y, Xiong X (2024) Recent advancements with loop-mediated isothermal amplification (LAMP) in assessment of the species authenticity with meat and seafood products. Crit Rev Food Sci Nutr 65:2214–2235
109. Wang N, Zhang J, Xiao B, Li H, Chen J, Sun X, Huang F, Li H, Chen A (2023) Integration of in-cassette lysis, purification, and lateral flow strips-based sensor for rapid and on-site detection of yak milk adulteration. Sensors Actuators B Chem 394:134309
110. Garg N, Ahmad FJ, Kar S (2022) Recent advances in loop-mediated isothermal amplification (LAMP) for rapid and efficient detection of pathogens. Curr Res Microb Sci 3:100120
111. Rasheed S, Ikram M, Ahmad D, Abbas MN, Shafique M (2024) Advancements in colorimetric and fluorescent-based sensing approaches for point-of-care testing in forensic sample analysis. Microchem J 206:111438
112. Das D, Masetty M, Priye A (2023) Based loop mediated isothermal amplification (Lamp) platforms: integrating the versatility of paper microfluidics with accuracy of nucleic acid amplification tests. Chem 11(3):163
113. Hossain MM, Abidin SASZ, Bujang A, Taib MN, Sagadevan S, Johan MR, Nizar NNA (2023) TaqMan multiplex qPCR for detecting animal species in meat and meat products: development, recent advances and future prospects. Food Control 150:109761
114. Cao L, Cui X, Hu J, Li Z, Choi JR, Yang Q, Lin M, Hui LY, Xu F (2017) Advances in digital polymerase chain reaction (dPCR) and its emerging biomedical applications. Biosens Bioelectron 90:459–474
115. Chen J, Zhang Y, Chen C, Zhang Y, Zhou W, Sang Y (2020) Identification and quantification of cassava starch adulteration in different food starches by droplet digital PCR. PLoS One 15(2):e0228624
116. Zhang H, Cao L, Brodsky J, Gablech I, Xu F, Li Z, Korabecna M, Neuzil P (2024) Quantitative or digital PCR? A comparative analysis for choosing the optimal one for biosensing applications. TrAC Trends Anal Chem 174:117676
117. Zhang H, Yan Z, Wang X, Gaňová M, Korabečná M, Zahradník P, Chang H, Neuzil P (2022) Digital PCR system development accelerator—a methodology to emulate dPCR results. Sensors Actuators B Chem 358:131527
118. Floren C, Wiedemann I, Brenig B, Schuetz E, Beck J (2015) Species identification and quantification in meat and meat products using droplet digital PCR (ddPCR). Food Chem 173:1054–1058
119. Li B, Yu M, Xu W, Chen L, Han J (2023) Comparison of PCR techniques in adulteration identification of dairy products. Agriculture 13(7):1450
120. Lubis HN, Mohd-Naim NF, Alizul NN, Ahmed MU (2016) From market to food plate: current trusted technology and innovations in halal food analysis. Trends Food Sci Technol 58:55–68
121. Agrimonti C, Bottari B, Sardaro MLS, Marmiroli N (2019) Application of real-time PCR (qPCR) for characterization of microbial populations and type of milk in dairy food products. Crit Rev Food Sci Nutr 59(3):423–442

122. Pierboni E, Rondini C, Torricelli M, Ciccone L, Tovo GR, Mercuri ML, Altissimi S, Haouet N (2018) Digital PCR for analysis of peanut and soybean allergens in foods. Food Control 92:128–136
123. Augusti R, Fulgêncio ACC, Nogueira HM, Gomes JCL, dos Santos LB, de Macedo AN, Porto BLS, Sena MM, Almeida MR (2024) Enhancing food authentication screening through the integration of chemometrics and ambient ionization mass spectrometry: a comprehensive review. Trends Food Sci Technol 147:104480
124. Nwachukwu SC, Edo GI, Jikah AN, Emakpor OL, Akpoghelie PO, Agbo JJ (2024) Recent advances in the role of mass spectrometry in the analysis of food: a review. J Food Meas Charact 18:1–16
125. Chien HJ, Zheng YF, Wang WC, Kuo CY, Hsu YM, Lai CC (2023) Determination of adulteration, geographical origins, and species of food by mass spectrometry. Mass Spectrom Rev 42(6):2273–2323
126. Dou X, Zhang L, Yang R, Wang X, Yu L, Yue X, Ma F, Mao J, Wang X, Zhang W (2023) Mass spectrometry in food authentication and origin traceability. Mass Spectrom Rev 42(5):1772–1807
127. Shi L, Habib A, Bi L, Hong H, Begum R, Wen L (2024) Ambient ionization mass spectrometry: application and prospective. Crit Rev Anal Chem 54(6):1584–1633
128. Sudhakar A, Chakraborty SK, Mahanti NK, Varghese C (2023) Advanced techniques in edible oil authentication: a systematic review and critical analysis. Crit Rev Food Sci Nutr 63(7):873–901
129. Mukherjee AG, Renu K, Gopalakrishnan AV, Veeraraghavan VP, Vinayagam S, Paz-Montelongo S, Dey A, Vellingiri B, George A, Madhyastha H (2023) Heavy metal and metalloid contamination in food and emerging technologies for its detection. Sustainability 15(2):1195
130. Varunjikar MS, Lie KK, Lundebye A-K, Belghit I, Ørnsrud R, Berntssen M, Lecrenier M, Oveland E, Palmblad N, Rasinger J (2024) Proteomics for food and feed authentication in the circular food chain. Trends Food Sci Technol 153:104710
131. Taheri S, Andrade JCD, Conte-Junior CA (2024) Emerging perspectives on analytical techniques and machine learning for food metabolomics in the era of industry 4.0: a systematic review. Crit Rev Food Sci Nutr. https://doi.org/10.1080/10408398.2024.2435597
132. Liu C, Zuo Z, Xu F, Wang Y (2023) Authentication of herbal medicines based on modern analytical technology combined with chemometrics approach: a review. Crit Rev Anal Chem 53(7):1393–1418
133. Shen Q, Wang S, Wang H, Liang J, Zhao Q, Cheng K, Imran M, Xue J, Mao Z (2024) Revolutionizing food science with mass spectrometry imaging: a comprehensive review of applications and challenges. Compr Rev Food Sci Food Saf 23(4):e13398
134. Ye H, Yang J, Xiao G, Zhao Y, Li Z, Bai W, Zeng X, Dong H (2023) A comprehensive overview of emerging techniques and chemometrics for authenticity and traceability of animal-derived food. Food Chem 402:134216
135. Wang K, Li Z, Li J, Lin H (2021) Raman spectroscopic techniques for nondestructive analysis of Agri-foods: a state-of-the-art review. Trends Food Sci Technol 118:490–504
136. Allakhverdiev ES, Kossalbayev BD, Sadvakasova AK, Bauenova MO, Belkozhayev AM, Rodnenkov OV, Martynyuk TV, Maksimov GV, Allakhverdiev SI (2024) Spectral insights: navigating the frontiers of biomedical and microbiological exploration with Raman spectroscopy. J Photochem Photobiol B Biol 252:112870
137. Arroyo-Cerezo A, Jiménez-Carvelo AM, González-Casado A, Ruisánchez I, Cuadros-Rodríguez L (2023) The potential of the spatially offset Raman spectroscopy (SORS) for implementing rapid and non-invasive in-situ authentication methods of plastic-packaged commodity foods–application to sliced cheeses. Food Control 146:109522
138. Passoni DDJ, Pacheco MTT, Silveira L (2020) Raman spectroscopy for the identification of differences in the composition of automobile lubricant oils related to SAE specifications and additives. Instrum Sci Technol 49(2):164–181

139. Hassoun A, Carpena M, Prieto MA, Simal-Gandara J, Özogul F, Özogul Y, Çoban ÖE, Barba FJ, Marti-Quijal FJ (2020) Use of spectroscopic techniques to monitor changes in food quality during application of natural preservatives: a review. Antioxidants 9(9):882
140. Guo Z, Chen P, Yosri N, Chen Q, Elseedi HR, Zou X, Yang H (2023) Detection of heavy metals in food and agricultural products by surface-enhanced Raman spectroscopy. Food Rev Intl 39(3):1440–1461
141. Xu M-L, Gao Y, Han XX, Zhao B (2017) Detection of pesticide residues in food using surface-enhanced Raman spectroscopy: a review. J Agric Food Chem 65(32):6719–6726
142. Covaciu F-D, Berghian-Grosan C, Feher I, Magdas DA (2020) Edible oils differentiation based on the determination of fatty acids profile and Raman spectroscopy—a case study. Appl Sci 10(23):8347
143. He H, Sun D-W, Pu H, Chen L, Lin L (2019) Applications of Raman spectroscopic techniques for quality and safety evaluation of milk: a review of recent developments. Crit Rev Food Sci Nutr 59(5):770–793
144. Hassoun A, Måge I, Schmidt WF, Temiz HT, Li L, Kim H-Y, Nilsen H, Biancolillo A, Aït-Kaddour A, Sikorski M (2020) Fraud in animal origin food products: advances in emerging spectroscopic detection methods over the past five years. Foods 9(8):1069
145. Li L, Cao X, Zhang T, Wu Q, Xiang P, Shen C, Zou L, Li Q (2022) Recent developments in surface-enhanced Raman spectroscopy and its application in food analysis: alcoholic beverages as an example. Foods 11(14):2165
146. Zhao H, Zhan Y, Xu Z, Nduwamungu JJ, Zhou Y, Powers R, Xu C (2022) The application of machine-learning and Raman spectroscopy for the rapid detection of edible oils type and adulteration. Food Chem 373:131471
147. Arendse E, Nieuwoudt H, Magwaza LS, Nturambirwe JFI, Fawole OA, Opara UL (2021) Recent advancements on vibrational spectroscopic techniques for the detection of authenticity and adulteration in horticultural products with a specific focus on oils, juices and powders. Food Bioprocess Technol 14:1–22
148. López-Lorente ÁI (2021) Recent developments on gold nanostructures for surface enhanced Raman spectroscopy: particle shape, substrates and analytical applications. A review. Anal Chim Acta 1168:338474
149. Neng J, Wang J, Wang Y, Zhang Y, Chen P (2023) Trace analysis of food by surface-enhanced Raman spectroscopy combined with molecular imprinting technology: principle, application, challenges, and prospects. Food Chem 429:136883
150. Yan S, Li Y, Peng Y, Ma S, Han D (2022) Detection of nitrofurans residues in honey using surface-enhanced Raman spectroscopy. J Food Sci 87(7):3318–3328
151. Camerlingo C, Portaccio M, Delfino I, Lepore M (2019) Surface-enhanced Raman spectroscopy for monitoring Extravirgin olive oil bioactive components. J Chem 2019(1):9537419
152. Zhang Y, Wang H, Ni C, Wang Q, Lin T (2024) Three-dimensional nanoporous gold/gold nanoparticles substrate for surface-enhanced Raman scattering detection of illegal additives in food. Spectrochim Acta A Mol Biomol Spectrosc 323:124879
153. Kim S, Joo J-H, Kim W, Bang A, Choi HW, Moon SW, Choi S (2021) A facile, portable surface-enhanced Raman spectroscopy sensing platform for on-site chemometrics of toxic chemicals. Sensors Actuators B Chem 343:130102
154. Khalil I, Yehye WA, Julkapli NM, Sina AAI, Chowdhury FI, Khandaker MU, Hsiao VK, Basirun WJ (2021) Simultaneous detection of dual food adulterants using graphene oxide and gold nanoparticle based surface enhanced Raman scattering duplex DNA biosensor. Vib Spectrosc 116:103293
155. Wang H-P, Chen P, Dai J-W, Liu D, Li J-Y, Xu Y-P, Chu X-L (2022) Recent advances of chemometric calibration methods in modern spectroscopy: algorithms, strategy, and related issues. TrAC Trends Anal Chem 153:116648
156. Balthazar CF, Guimarães JT, Rocha RS, Pimentel TC, Neto RP, Tavares MIB, Graça JS, Alves Filho EG, Freitas MQ, Esmerino EA (2021) Nuclear magnetic resonance as an analytical tool for monitoring the quality and authenticity of dairy foods. Trends Food Sci Technol 108:84–91

157. Bisht B, Kumar V, Gururani P, Tomar MS, Nanda M, Vlaskin MS, Kumar S, Kurbatova A (2021) The potential of nuclear magnetic resonance (NMR) in metabolomics and lipidomics of microalgae-a review. Arch Biochem Biophys 710:108987
158. Cao R, Liu X, Liu Y, Zhai X, Cao T, Wang A, Qiu J (2021) Applications of nuclear magnetic resonance spectroscopy to the evaluation of complex food constituents. Food Chem 342:128258
159. Hatzakis E (2019) Nuclear magnetic resonance (NMR) spectroscopy in food science: a comprehensive review. Compr Rev Food Sci Food Saf 18(1):189–220
160. Mannina L, Segre A (2002) High resolution nuclear magnetic resonance: from chemical structure to food authenticity. Grasas Aceites 53(1):22–33
161. Butz P, Hofmann C, Tauscher B (2005) Recent developments in noninvasive techniques for fresh fruit and vegetable internal quality analysis. J Food Sci 70(9):R131–R141
162. Pearce KL, Rosenvold K, Andersen HJ, Hopkins DL (2011) Water distribution and mobility in meat during the conversion of muscle to meat and ageing and the impacts on fresh meat quality attributes—a review. Meat Sci 89(2):111–124
163. Mariette F (2009) Investigations of food colloids by NMR and MRI. Curr Opin Colloid Interface Sci 14(3):203–211
164. Marcone MF, Wang S, Albabish W, Nie S, Somnarain D, Hill A (2013) Diverse food-based applications of nuclear magnetic resonance (NMR) technology. Food Res Int 51(2):729–747
165. Brigante F, Solovyev P, Bontempo L (2024) Nuclear magnetic resonance applications in fermented foods and plant-based beverages: challenges and opportunities. Food Rev Int 40:1–28
166. Kanokruangrong S, Kebede B, Carne A, Stewart I, Bekhit AE-DA (2025) Metabolomic investigation of fresh beef, lamb and venison using nuclear magnetic resonance spectroscopy in relation to colour stability. Food Chem 463:141447
167. Li J, Vosegaard T, Guo Z (2017) Applications of nuclear magnetic resonance in lipid analyses: an emerging powerful tool for lipidomics studies. Prog Lipid Res 68:37–56
168. Shomaji S, Masna NVR, Ariando D, Deb Paul S, Horace-Herron K, Forte D, Mandal S, Bhunia S (2021) Detecting dye-contaminated vegetables using low-field NMR relaxometry. Foods 10(9):2232
169. Ohtsuki T, Sato K, Sugimoto N, Akiyama H, Kawamura Y (2012) Absolute quantitative analysis for sorbic acid in processed foods using proton nuclear magnetic resonance spectroscopy. Anal Chim Acta 734:54–61
170. Fan K, Zhang M (2019) Recent developments in the food quality detected by non-invasive nuclear magnetic resonance technology. Crit Rev Food Sci Nutr 59(14):2202–2213
171. Ragupathy S, Thirugnanasambandam A, Vinayagam V, Newmaster SG (2024) Nuclear magnetic resonance fingerprints and mini DNA markers for the authentication of cinnamon species ingredients used in food and natural health products. Plants 13(6):841
172. Kanwal N, Musharraf SG (2024) Analytical approaches for the determination of adulterated animal fats and vegetable oils in food and non-food samples. Food Chem 460:140786
173. Cermakova E, Lencova S, Mukherjee S, Horka P, Vobruba S, Demnerova K, Zdenkova K (2023) Identification of fish species and targeted genetic modifications based on DNA analysis: state of the art. Foods 12(1):228
174. Krishnan P, Kruger N, Ratcliffe R (2005) Metabolite fingerprinting and profiling in plants using NMR. J Exp Bot 56(410):255–265
175. Li S, Tian Y, Jiang P, Lin Y, Liu X, Yang H (2021) Recent advances in the application of metabolomics for food safety control and food quality analyses. Crit Rev Food Sci Nutr 61(9):1448–1469
176. Selamat J, Rozani NAA, Murugesu S (2021) Application of the metabolomics approach in food authentication. Molecules 26(24):7565
177. Mialon N, Roig B, Capodanno E, Cadiere A (2023) Untargeted metabolomic approaches in food authenticity: a review that showcases biomarkers. Food Chem 398:133856
178. Pavagadhi S, Swarup S (2020) Metabolomics for evaluating flavor-associated metabolites in plant-based products. Meta 10(5):197

179. Thiruvengadam M, Kim J-T, Kim W-R, Kim J-Y, Jung B-S, Choi H-J, Chi H-Y, Govindasamy R, Kim S-H (2024) Safeguarding public health: advanced detection of food adulteration using nanoparticle-based sensors. Crit Rev Anal Chem 13:1–21
180. Salem MA, Perez de Souza L, Serag A, Fernie AR, Farag MA, Ezzat SM, Alseekh S (2020) Metabolomics in the context of plant natural products research: from sample preparation to metabolite analysis. Meta 10(1):37
181. Vinayavekhin N, Saghatelian A (2010) Untargeted metabolomics. Curr Protoc Mol Biol 90(1):1–24
182. Fraga-Corral M, Carpena M, Garcia-Oliveira P, Pereira AG, Prieto M, Simal-Gandara J (2022) Analytical metabolomics and applications in health, environmental and food science. Crit Rev Anal Chem 52(4):712–734
183. Wadood SA, Nie J, Song Y, Li C, Rogers KM, Khan WA, Khan A, Xiao J, Liu H, Yuan Y (2024) Authentication of edible herbal materials and food products using mass spectrometry based metabolites and inorganic constituents. Food Chem 463:141424
184. Nasr EG, Epova EN, Sebilo M, Larivière D, Hammami M, Souissi R, Abderrazak H, Donard OF (2022) Olive oil traceability studies using inorganic and isotopic signatures: a review. Molecules 27(6):2014
185. Troisi J, Cavallo P, Colucci A, Pierri L, Scala G, Symes S, Jones C, Richards S (2020) Metabolomics in genetic testing. Adv Clin Chem 94:85–153
186. Benevenuto RF, Venter HJ, Zanatta CB, Nodari RO, Agapito-Tenfen SZ (2022) Alterations in genetically modified crops assessed by omics studies: systematic review and meta-analysis. Trends Food Sci Technol 120:325–337
187. Liu J, Zhao H, Yin Z, Dong H, Chu X, Meng X, Li Y, Ding X (2023) Application and prospect of metabolomics-related technologies in food inspection. Food Res Int 171:113071
188. Curulli A (2021) Electrochemical biosensors in food safety: challenges and perspectives. Molecules 26(10):2940
189. Van Dorst B, Mehta J, Bekaert K, Rouah-Martin E, De Coen W, Dubruel P, Blust R, Robbens J (2010) Recent advances in recognition elements of food and environmental biosensors: a review. Biosens Bioelectron 26(4):1178–1194
190. Grieshaber D, MacKenzie R, Vörös J, Reimhult E (2008) Electrochemical biosensors-sensor principles and architectures. Sensors 8(3):1400–1458
191. Huang X, Zhu Y, Kianfar E (2021) Nano biosensors: properties, applications and electrochemical techniques. J Mater Res Technol 12:1649–1672
192. Barbosa C, Nogueira S, Saraiva R, Chaves S (2017) Next generation sequencing in food authenticity and safety. In: Food traceability and authenticity. CRC Press, Boca Raton, pp 164–179
193. Zheng Y, Karimi-Maleh H, Fu L (2022) Advances in electrochemical techniques for the detection and analysis of genetically modified organisms: an analysis based on bibliometrics. Chem 10(5):194
194. Feriotto G, Borgatti M, Mischiati C, Bianchi N, Gambari R (2002) Biosensor technology and surface plasmon resonance for real-time detection of genetically modified roundup ready soybean gene sequences. J Agric Food Chem 50(5):955–962
195. Curulli A (2022) Recent advances in electrochemical sensing strategies for food allergen detection. Biosensors 12(7):503
196. Wang K, Lin X, Zhang M, Li Y, Luo C, Wu J (2022) Review of electrochemical biosensors for food safety detection. Biosensors 12(11):959
197. Qazi RA, Aman N, Ullah N, Jamila N, Bibi N (2024) Recent advancement for enhanced *E coli* detection in electrochemical biosensors. Microchem J 196:109673
198. Mahari S, Gandhi S (2022) Recent advances in electrochemical biosensors for the detection of salmonellosis: current prospective and challenges. Biosensors 12(6):365

199. Kalambate RP, Kalambate PK, Laiwattanapaisal W (2024) Revolutionizing melamine detection: cutting-edge advances from traditional analyses to state-of-the-art electrochemical sensors. Next Mater 3:100085
200. Rasheed S, Kanwal T, Ahmad N, Fatima B, Najam-ul-Haq M, Hussain D (2024) Advances and challenges in portable optical biosensors for onsite detection and point-of-care diagnostics. TrAC Trends Anal Chem 173:117640
201. Wu J, Liu H, Chen W, Ma B, Ju H (2023) Device integration of electrochemical biosensors. Nat Rev Bioeng 1(5):346–360
202. Singh A, Sharma A, Ahmed A, Sundramoorthy AK, Furukawa H, Arya S, Khosla A (2021) Recent advances in electrochemical biosensors: applications, challenges, and future scope. Biosensors 11(9):336
203. Sultana S, Azlan A, Desa MNM, Mahyudin NA (2023) Multiplex platforms in biosensor based analytical approaches: opportunities and challenges for the speciation of animal species in food chain. Food Control 149:109727
204. Eivazzadeh-Keihan R, Noruzi EB, Chidar E, Jafari M, Davoodi F, Kashtiaray A, Gorab MG, Hashemi SM, Javanshir S, Cohan RA (2022) Applications of carbon-based conductive nanomaterials in biosensors. Chem Eng J 442:136183
205. Shalileh F, Sabahi H, Dadmehr M, Hosseini M (2023) Sensing approaches toward detection of urea adulteration in milk. Microchem J 193:108990
206. Singh M (2023) Nanosensor platforms for detection of milk adulterants. Sens Actuators Rep 5:100159
207. Morais SL, Castanheira M, Santos M, Domingues VF, Delerue-Matos C, Barroso MF (2024) Electrochemical Genosensors as a new approach to plant DNA detection and quantification for honey authentication. Eng Proc 82(1):79
208. Gupta R, Raza N, Bhardwaj SK, Vikrant K, Kim K-H, Bhardwaj N (2021) Advances in nanomaterial-based electrochemical biosensors for the detection of microbial toxins, pathogenic bacteria in food matrices. J Hazard Mater 401:123379
209. Bobrinetskiy I, Radovic M, Rizzotto F, Vizzini P, Jaric S, Pavlovic Z, Radonic V, Nikolic MV, Vidic J (2021) Advances in nanomaterials-based electrochemical biosensors for foodborne pathogen detection. Nano 11(10):2700
210. Sheng K, Jiang H, Fang Y, Wang L, Jiang D (2022) Emerging electrochemical biosensing approaches for detection of allergen in food samples: a review. Trends Food Sci Technol 121:93–104

Optical Sensor Arrays for Food Adulteration Detection

Bahram Hemmateenejad and Farshad Ardakani

Contents

Abstract Food adulteration significantly threatens public health, consumer trust, and economic stability, necessitating advanced detection methods to ensure food safety and authenticity. Optical sensor arrays have emerged as a promising solution due to their simplicity, cost-effectiveness, and potential for on-site analysis. This

B. Hemmateenejad (✉) · F. Ardakani
Chemistry Department, Shiraz University, Shiraz, Iran
e-mail: hemmatb@shirazu.ac.ir

W. Alahmad, M. Ashfaq (eds.), *Advanced Techniques against Food Adulteration*,
https://doi.org/10.1007/978-3-031-99454-8_6

chapter provides a comprehensive overview of optical sensor arrays, focusing on their principles, architectures, and applications in detecting food adulteration. Unlike traditional selective sensors, optical sensor arrays employ semi-selective or cross-reactive elements to generate unique response patterns, or "fingerprints," enabling the identification and quantification of adulterants in complex food matrices. This chapter explores the fundamental principles of optical sensor arrays, including their ability to detect changes in optical properties, such as color, fluorescence, and absorbance. It also discusses the classification of sensor elements, including substrates, and the role of digital imaging and multivariate statistical analysis in signal processing. Applications of optical sensor arrays are detailed across various food products, including coffee, soft drinks, milk, water, sugars, meat, honey, oils, and saffron, demonstrating their versatility and effectiveness in combating food adulteration. By integrating optical sensing with advanced computational techniques, these arrays offer rapid, reliable, and accessible solutions for food safety and quality assurance, paving the way for a future where food authenticity is more easily verifiable.

Keywords Sensor · Optical properties · Signal processing · Computational techniques · Food adulteration

Introduction

Food adulteration remains a significant global challenge, threatening public health, undermining consumer trust, and causing economic losses. The ever-growing complexity of food supply chains and the sophisticated techniques employed in adulteration demand the development of advanced detection methods [1]. Among various emerging technologies, optical sensor arrays have garnered substantial attention due to their simplicity, cost-effectiveness, and high potential for on-site analysis. This chapter explores the principles, architectures, and applications of optical sensor arrays for detecting food adulteration, focusing on leveraging their ability to provide rapid and reliable results.

Optical sensor arrays represent a paradigm shift from traditional selective sensors to versatile platforms capable of analyzing complex mixtures. Unlike selective sensors designed to target specific analytes, sensor arrays mimic biological sensing systems, relying on a combination of semi-selective or cross-reactive elements. When exposed to different samples, these arrays produce unique response patterns, or "fingerprints," enabling them to identify and quantify adulterants in diverse food matrices [2].

The fundamental principles underlying optical sensor arrays are rooted in their capacity to detect and interpret changes in optical properties, such as color, fluorescence, and absorbance, upon interaction with analytes. These changes are captured as digital signals, which are subsequently processed using multivariate statistical tools. Such integration of optical sensing with advanced computational techniques allows for comprehensive analysis, enhancing the sensitivity and specificity of detection. This chapter delves into the critical components of optical sensor arrays, starting with their fundamental principles and extending to the classification of sensor elements, including reversible and cumulative types. The stabilization of sensor elements and the role of substrates in ensuring robust and reproducible responses are also discussed. Additionally, this chapter highlights the importance of digital image processing in signal acquisition and the interpretation of complex datasets through multivariate statistical methods, such as principal component analysis (PCA), partial least squares-discriminant analysis (PLS-DA), linear discriminant analysis (LDA), and support vector machines (SVMs). Applications of optical sensor arrays in detecting food adulteration span a wide range of products, including beverages (e.g., coffee, soft drinks, milk, and water), sugars and sweeteners, meats, honey, oils, and saffron. Each application is explored in detail, demonstrating the versatility and effectiveness of these sensors in identifying common adulterants and ensuring food authenticity. By providing a comprehensive overview of optical sensor arrays, this chapter aims to equip researchers, practitioners, and stakeholders with the knowledge required to harness this cutting-edge technology for combating food adulteration. The advancements in this field promise a future where food safety and quality assurance are more accessible and reliable than ever before.

Selective Sensor Versus Sensor Array

A chemical sensor is a device that selectively responds to a specific analyte through physical or chemical interactions. It consists of two main components: a receptor, which interacts with the analyte and alters its properties, and a transducer, which converts these changes into a signal output that can be either qualitative or quantitative [3–6]. The conventional sensors are typically designed using a "lock and key" methodology as a selective sensor, also known as "1 vs. 1" (a single molecular probe is employed for a single analyte) or host–guest interactions between specific receptors and specific analytes (Fig. 1a) [8, 9]. This renders them capable of detecting a single analyte with high sensitivity, operating simply and selectively, exhibiting good sensitivity, and providing high space-time resolution. Indeed, this traditional sensor frequently encounters interference due to its structural or chemical similarities with the target analyte. Furthermore, the synthesis of a highly selective receptor for the complex analyte, or an unknown structure or each analyte in a complex biological sample, is not a practical proposition, and differentiating substances, such as proteins, in complex systems like human metabolic processes, foods, urine, serum, etc., can be challenging. Sensor arrays have been developed to differentiate and detect target analytes in complex systems that can identify slight differences [10]. Such arrays are designated "N vs. N" sensor arrays, and multiple molecular probes are employed for numerous analytes (Fig. 1b) [8]. The design of these sensors is rational and inspired by the mammalian olfactory and taste system [11]. The sense of taste enables the detection of fundamental tastes, including sweet,

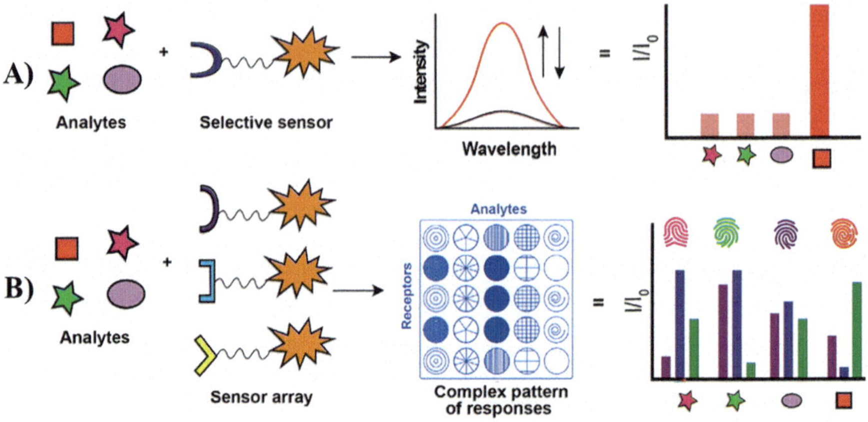

Fig. 1 Schematic representation of (**a**) "selective sensor" with selectivity for a particular analyte and (**b**) "sensor array" that produces a fingerprint response for each analyte. Adapted with permission from [7]

sour, salty, bitter, and umami. In contrast, the olfactory sense identifies the aromas and flavors associated with food and other substances [12, 13]. Therefore, a sensor array comprising numerous sensing units displays a diverse spectrum of responses to a range of target analytes through physical or chemical processes, and in comparison to a conventional single sensor, a sensor array can distinguish a diverse range of targets without the need for prior identification [14]. Furthermore, these sensors can discern slight variations in chemical composition between analytes. This enables the qualitative or quantitative detection of the target analytes [11]. This specificity is derived from the pattern recognition of combined sensing units, which produce a fingerprint response for each analyte.

Basic Principle of Optical Sensor Array

An optical sensor array comprises an array of chemical sensors that utilize electromagnetic radiation to produce an analytical signal within a transduction element. The impact of radiation on the sample is gauged by assessing a specific optical parameter, including absorbance, reflectance, luminescence, and fluorescence, within the sensing units embedded within the array [15]. This enables identifying and detecting specific target analytes through analyzing output signals (the most commonly employed techniques are based on the principles of light absorption or light emission). Moreover, the sensors above can measure not only the intensity of light but also many additional related properties, including lifetime, refractive index, and scattering [16, 17]. Additionally, they can be designed to cover different regions of the spectrum, including infrared (IR), near-infrared (NIR), ultraviolet (UV), and visible (VIS) [18, 19]. Optical sensors must satisfy two fundamental requirements: firstly, they must be capable of interacting with analytes selectively, and secondly, they must be able to report or provide feedback on such interactions. This feedback can be conveyed through chromophores (a molecule that absorbs light at a particular wavelength and reflects color as a result) or fluorophores (a chemical compound that emits light when excited by a light source). The preference is for those with very high extinction coefficients or high quantum yields, respectively [19]. These sensors demonstrate a range of selectivity, from semi-selective, cross-reactive behavior with multiple analytes to highly selective responses to a specific analyte or even a class of analytes [20].

Optical sensors based on chemical processes offer numerous benefits compared to their conventional electricity-based counterparts. These instruments are characterized by several key features, the foremost of which are their selectivity, immunity to electromagnetic interference, and safety when used in the presence of flammable and explosive substances. They also exhibit sensitivity, affordability, and non-destructive capabilities and offer diverse functionalities. Furthermore, they can be easily miniaturized, enabling multiple analyses to be conducted using a single control instrument at a central location [15, 17]. The history of colorimetry as an analytical technique is robust, with colorimetric sensors facilitating straightforward quantification through visual observation. The sensors in question rely on chemo-responsive dyes that interact with the analytes in question, resulting in a change of color depending on the chemical environment, either in reflected or absorbed light [21]. Traditional sensors typically utilize three visible channels red, green, and blue (RGB). In contrast, modern designs employ hyperspectral imaging, which uses a wide array of wavelengths, including non-visible light, thereby facilitating comprehensive analysis. Colorimetric sensor arrays are preferred in laboratory and industrial contexts due to their specificity and affordability. In contrast, molecular emission techniques, such as fluorescence and phosphorescence, offer exceptional sensitivity and specificity, achieving detection levels that are markedly higher than those of conventional methods. Although intensity measurement in these luminescence techniques is attractive due to its cost-effectiveness, it is subject to challenges from variables such as fluctuations in light intensity and sample turbidity [2, 22–24]. In this chapter, the discourse frequently centers on applying colorimetric sensor arrays (CSAs) and fluorometric sensor arrays (FSAs) as methodologies for analytical purposes.

Types of Sensor Array

The evolution of human sensory perception is inextricably linked to nutritional intake and daily activities, particularly regarding the gustatory and olfactory senses, which enable the recognition of smell and taste. Electronic devices, known as electronic tongue (E-tongue) and electronic nose (E-nose), can mimic these sensory functions [12, 25]. The operational principles of the sensor array for detecting and identifying analytes are illustrated in Fig. 2, which also depicts the concept of an

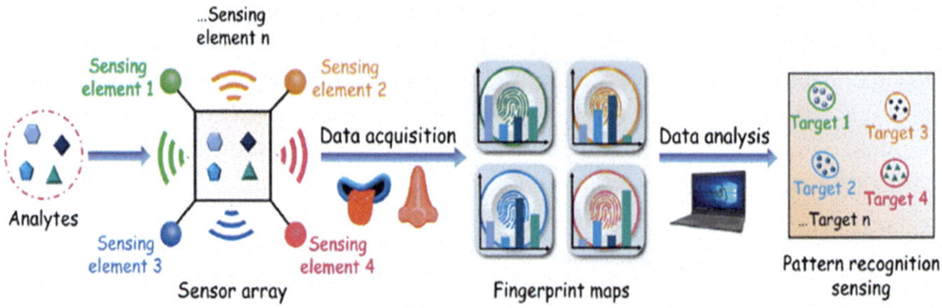

Fig. 2 Mechanisms of sensor arrays for classifying and identifying analytes. Adapted with permission from [26]

electric tongue and nose [26–28]. Thanks to sophisticated brain signal processing, humans have a complex sense of smell, capable of identifying millions of odors despite having only 400 olfactory receptors. An E-nose is an instrument that emulates the human sense of smell to detect odors and volatile compounds in food. It functions via a series of stages involving sensors that interact with the chemical substances in the foodstuff. Each sensor exhibits sensitivity to a specific range of volatile compounds, thus enabling a comprehensive range of odor identification [29, 30].

An E-tongue is a sophisticated analytical device that emulates a sense of taste and quantifies a range of chemical compounds dissolved in liquids. It employs an array of sensors with distinct mechanisms to analyze intricate liquid mixtures through sophisticated data processing techniques. The E-tongue can discern various tastes and identify specific chemical compounds by modifying the input signals. Incorporating sensory capabilities, particularly those about taste and olfaction, enhances the overall experience of food and beverage consumption. Deploying sensors that emulate these human senses enables the food production industry to enhance process automation and quality control. These sensors provide essential data for analysis in situations where human expert panels may not be feasible, thus making them valuable for automated industrial processes, assessing harmful substances, and cost management [31, 32]. By employing advanced technology, including sensor arrays and machine learning algorithms, these devices can analyze complex chemical and biological compounds found in liquids and gases [31]. The E-nose olfactory system is used in various fields, especially for assessing food quality by monitoring processes, determining shelf life, evaluating damage and toxicity, and conducting quality control. However, it has a limitation in that it cannot identify specific compounds in samples, a feature provided by gas chromatography/mass spectrometry (GC/MS) analysis. The E-nose technology utilizes a method called "electronic volatile fingerprinting" [33–35].

Classes of Optical Sensor Elements

Optical chemical sensing employs a variety of sensor elements as indicators, with colorimetric sensors relying on light absorption and luminescent sensors depending on light emission being the most prevalent. These chemo-responsive optical sensors can be classified according to their chemical interactions into two categories: reversible sensor elements and cumulative sensor elements [2, 15, 19, 26, 36].

Reversible Sensor Elements

Reversible sensors employ interactions between analytes and indicators at equilibrium to generate discernible color alterations. Such dyes include pH indicators, host–guest indicators, and vapochromic or solvatochromic indicators (colorants with large permanent dipoles), all serving various detection purposes. The initial and most pervasive category of sensor elements are those that function as pH indicators. These substances are sensitive to donate and accept hydrogen (Brønsted acids or bases), which can be observed by analyzing their UV-Vis absorption spectra. Litmus, a historic pH indicator derived from lichens, exemplifies this concept. A wide range of pH indicators is sourced from nature, particularly anthocyanin dyes found in fruits and vegetables, with pKa values spanning below 0–14, indicating their diverse applicability. Furthermore, these pH indicators can be combined with various other compounds, including porphyrins, phenylboronic acid, and others, for diverse applications [37]. The subsequent classification of sensor elements is that of host–guest indicators known as Lewis acids and bases, respectively [38, 39]. These are defined by their electron exchange capabilities, with Lewis acids accepting electron pairs and Lewis bases donating them. Metalloporphyrins are the preferred choice of Lewis acidic dyes for identifying metal-ligating volatiles due to the significant colorimetric changes that occur upon ligand binding and their established role as a known host [40]. The evolution of supramolecular chemistry has led to the development of crown ethers, cryptands, cyclodextrins, and calixarenes, which bind metal ions dependent on size. Significant progress in CSAs using metalloporphyrins has improved ligand recognition through Lewis interactions, allowing for the sensitive detection of metal-ligating compounds; these sensors can be enhanced by modifying the metal centers [39]. In contrast, many strongly smelling compounds, including sulfides, amines, carboxylic acids, and phosphines, act as Lewis bases and assume the role of guests. Lewis basic compounds linked to microbial metabolites are also vital for microbial defense. Lewis acidic dyes are commonly employed for anion detection [41], whereas Lewis basic dyes are used to detect cations [42–44]. In other categories, dyes undergo color changes in response to variations in their surrounding polarity, affecting their states' stabilization. This phenomenon is observed in solvatochromic and vapochromic indicators [45–47]. Dyes with significant dipole moment differences exhibit positive solvatochromism in polar environments. Examples of such dyes include thiazines and merocyanines, which typically have a "push–pull" structure. The quantification of interactions between dyes and solvents is achieved through the utilization of solvent polarity scales, such as Reichardt's ET(30) and Taft's π^* scale. Vapochromic materials also undergo color changes due to the intercalation of solvent molecules within their crystal structures. These changes are driven by weak interactions, including coordination to metal sites, π–π stacking, and hydrogen bonding. However, these materials face challenges, including slow response times and weak interactions. The introduction of nanoscale coordination complexes may improve their sensitivity and selectivity by facilitating faster diffusion and increasing the interaction surface area [48].

Cumulative Sensor Elements

The utilization of cumulative sensors gives rise to the involvement of irreversible chemical reactions (there is no equilibrium), including those employing (1) "'reduction–oxidation" indicators and (2) chromogenic precipitating or aggregative indicators (e.g., plasmonic nanoparticles and nanoscale transition metal sulfides) [20, 49–52]. The term "reduction–oxidation" (redox) indicator describes colorimetric reagents that undergo a distinctive color transformation when oxidized or reduced. In most media, the involvement of protons in electron transfer processes has led to classifying these indicators into two principal categories: pH-dependent and pH-independent [52]. Some organic dyes, including methylene blue, BODIPY, Nile blue, o-dianisidine, diphenylamine, and 2,2′-bipyridine, have been observed to undergo reversible redox reactions.

The final category comprises chromogenic indicators that change color in response to particle aggregation, such as metal salt precipitation. With the advent of diverse nanomaterials, including plasmonic nanostructures (such as gold nanoparticles and silver nanoparticles), noble-metal nanoclusters, quantum dots, carbon dots (graphene and carbon nanotubes), upconversion nanostructures, and metal-organic framework, the number of sensing elements in this category is increasing [20, 53]. These contemporary nanomaterial methodologies are based on mechanisms, including absorption, quenching, and scattering, which result in significant alterations in color or fluorescence. It is essential to emphasize that the functionality of these sensors is contingent upon the ability to control the nanostructure of the sensing materials. To illustrate, gold, silver, or copper nanoparticle agglomerates can be utilized in "off–on" fluorescent probes, wherein fluorescence quenching occurs when attached fluorophores are present and recovery is induced by analyte binding [54, 55]. The appropriate functionalization of particles and choice of fluorophore create a versatile sensing platform. In applications involving ultra-trace detection, particularly at concentrations below parts per billion, there is significant potential for developing cumulative or dosimetric measurement sensors.

Substrates for Optical Sensors

E-tongues and E-noses are constructed in two principal forms: liquid-supported array (LSA) and solid-supported array (SSA). The former is created through the direct addition of the solutions of the sensing elements into a series of vials (e.g., 96 or 384-well ELISA kit). This configuration of sensor arrays is the most prevalent and traditional due to the ease of adjusting the concentration of receptors and indicators in each unit, thereby allowing for a high degree of flexibility concerning experimental configuration. Conversely, the SSAs may provide a superior option for specific reagents that are unstable in solution and require in situ use. Also, handling the devices based on SSAs is more convenient, and fewer reagents are usually necessary for fabricating SSAs. The preparation of solid arrays typically involves the incorporation of sensor elements onto solid supports, including materials such as cellulose acetate, filter paper, nylon, nitrocellulose, ethylcellulose, silicone polystyrene, poly(tetrafluoroethylene) (PTFE), poly(methyl methacrylate) (PMMA), polyethylene, polystyrene (PS), and polyvinylidene difluoride (PVDF) (direct deposition, which can be achieved with the use of a pin or micropipette), or begins with creating hydrophobic barriers on paper substrates using techniques like wax printing or photolithography. Sensor elements are then deposited onto the surface of the solid support using methods, such as capillary spotting, pipetting, inkjet printing, or dipping with stainless steel pins. To prevent issues caused by the free diffusion of these reagents, which can lead to inaccurate results when interacting with analyte solutions, immobilization and enrichment techniques are crucial for enhancing the sensitivity and accuracy of the CSA. There are two primary methods of attaching cumulative sensor elements (nanoparticles) to paper: "wet-end addition" and "surface treatment" [56, 57]. Wet-end addition involves incorporating nanoparticles into individual fibers before forming the paper sheet. Conversely, the surface treatment commonly used for optical applications consists of impregnating or coating the paper with nanoparticles, the efficiency of which is influenced by the water-repellency and porosity of the paper. Direct assembly [58], layer-by-layer (LbL) deposition [59], and in situ assembly [60] are the three main techniques for coating or impregnating nanoparticles. Direct assembly involves the separate preparation of nanoparticles before paper application and requires careful management of the nanoparticles to prevent aggregation. LbL deposition involves the adsorption of oppositely charged polyelectrolytes onto a substrate. For the first time, specific polyelectrolyte modifications are used to create a cationic charge on cellulose. In situ assembly, as demonstrated with gold nanoparticles (AuNPs) on cellulose substrates, involves the attachment of metal complex ions to modified cellulose followed by reduction to produce nanoparticles [61]. To circumvent the potential for human error and manual intervention in the addition of dye to the substrate, as well as the simultaneous addition of an array of dyes to the substrate, and ultimately reduce the time required for this process, the "printed arrays" method was introduced. However, it should be noted that this method has only been successfully applied to a limited number of arrays [62]. An additional technique designed explicitly for fluorescent sensors is the utilization of fiber-optic arrays. This technique entails the chemical etching of bundles of minute silica fiber-optic cables, forming a 2D array of microwells. The microwells distribute fluorescent probes randomly within individual beads [63].

Chemical sensors face a significant challenge due to their sensitivity to ambient humidity, which can affect their ability to accurately detect low concentrations of volatile organic compounds (VOCs). Variations in moisture can alter water vapor levels, compromising sensor performance. However, using hydrophobic materials like PVDF membranes in colorimetric or fluorometric sensor arrays has been shown to mitigate this humidity sensitivity across a wide range of relative humidity levels, from 10% to over 95%. This enhancement enables the effective use of these sensor arrays in aqueous solutions [64]. Selecting an appropriate substrate and dye for a solid array is paramount. Molecular dyes are deposited as a viscous layer or on a solid array, thereby facilitating rapid access of the analytes to the dye. In contrast, non-molecular dyes do not penetrate the solid array (paper), resulting in a significant reduction in the sensitivity of the analytical response, particularly in the case of colorimetric or fluorometric measurements [65, 66].

SSAs offer several advantages over liquid sensor arrays, including biodegradability, user- and field-friendly characteristics, and cost-effectiveness, and the necessity for manual liquid handling and pipetting of sensors and analytes is eliminated. Nevertheless, solid sensor arrays typically demonstrate reduced sensitivity, rendering them less effective at differentiating analytes with analogous chemical compositions relative to their liquid sensor array counterparts. Despite their widespread deployment, solid sensor arrays are often preferred for gaseous analytes, given their capacity to detect a broader range of gases. However, these arrays have inherent limitations, including the necessity for solvents and liquids in conducting tests, which restrict their suitability for field deployment due to logistical constraints and user-friendliness concerns.

In light of the expanding field of sensor arrays, it is imperative to investigate the interactions between the sensor elements and the analyte, even the substrate (cellulose in paper), to ascertain the underlying chemical reactions. Unfortunately, this issue has not been addressed well in this context, and its absence is keenly felt. An in-depth examination of these interactions can yield invaluable insights that will advance our understanding in this field.

A diverse range of intermolecular and intramolecular interactions exists, which generate a continuum in the strength of the interactions (i.e., change in enthalpy). The weakest interaction is dispersion forces or van der Waals between nonpolar and non-charged molecules and atoms (i.e., alkanes, noble gases, and halogens). Other weakest interactions are dipole–dipole interactions, hydrogen bonding, charge-transfer π–π molecular complexation, and ion–dipole interactions, and the strongest interactions are covalent or ionic bonds and electrostatic ion–ion interactions. The interactions mentioned above are not isolated occurrences; they occur continuously, with one transitioning smoothly into another. The advantage of incorporating stronger interactions in sensor arrays is twofold: firstly, it allows the sensor array to achieve greater intrinsic sensitivity, and secondly, it will enable the array to achieve more excellent chemical selectivity. Furthermore, stronger interactions facilitate a considerably broader spectrum of chemical interactions, extending beyond the limitations of mere physical absorption. This enables the exploration of significantly higher dimensionality and the enhancement of the ability to distinguish between compounds with subtle differences or complex mixtures [2, 19, 67].

Digital Imaging, Color Spaces, and Signal Recording

Colorimetric readouts in sensor assays are divided into two primary categories: equipment-free and equipment-dependent methods [68]. Equipment-free methods rely on naked-eye detection for qualitative and semi-quantitative assessments (see Fig. 3a), where distinct color changes indicate the presence of an analyte, yielding binary results. Conversely, in equipment-dependent methods, digital images can be captured in some ways, including by scanners, digital cameras, and smartphones (particularly when analyte concentrations are low or visual signals are minimal), as shown in Fig. 3a. In contrast, most of the equipment for fluorometric readouts is a UV lamp and a smartphone or digital camera to take an image [71]. Different imaging equipment yield varying image qualities, making it essential to utilize the same equipment throughout the analysis to minimize errors. While scanners and digital cameras require a computer for processing and storing images, smartphones offer

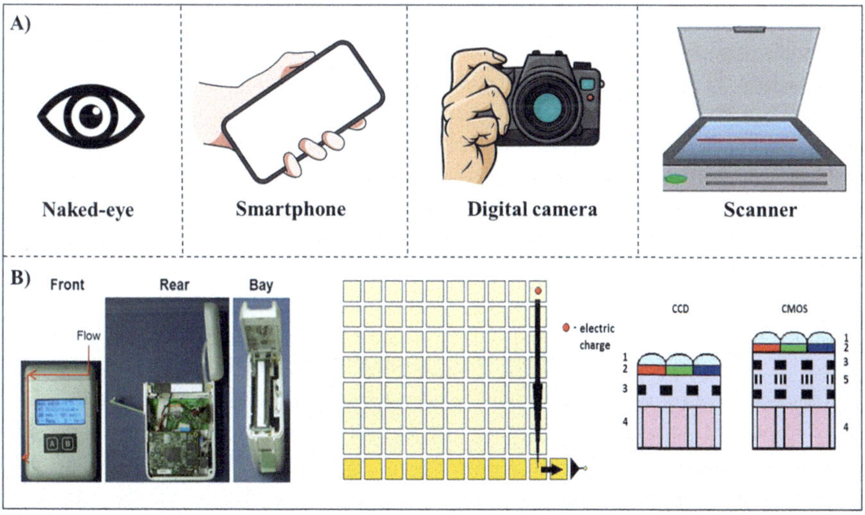

Fig. 3 (**a**) Schematic of naked-eye, smartphone, digital camera, and scanner as instrumentation and (**b**) photographs of the hand-held reader including front, rear, and cartridge bay views. Adapted with permission from [69], schematic representation of a CCD array and comparison of the CMOS and CCD arrays: (1) microlenses, (2) color filter, (3) signal transmission path, (4) photodiodes, (5) transistors and amplifiers. Adapted with permission from [70]

the advantage of using applications to analyze high-quality images instantly. Due to their imaging capabilities and high-speed processors, smartphones are emerging as preferable tools for image acquisition. Reliance on smartphones necessitates consideration of specific issues, such as the relationship between pixel count and RGB separation, sensitivity to scattered light, color noise influenced by sensor size and temperature, and stability of automatic focusing and exposure settings.

It is critical to ensure stable lighting conditions to enhance the quality of images obtained, as the accuracy of results is directly tied to effectively controlling ambient illumination and the camera's proximity to the analyte. Colorimetric and fluorometric evaluations depend on light reflection from materials, which change colors due to reactions between analytes and reagents. In smartphones, certain filters assign specific colors to each pixel when recording sample signals. Conversely, flatbed scanners are ubiquitous instruments in laboratory settings, employed for imaging sensor arrays in colorimetric array sensing due to their high resolution, accurate positioning, and consistent illumination. However, in applications beyond the laboratory, there is a necessity for the creation of portable, low-noise, and cost-effective smartphone and digital cameras that are capable of performing onboard analysis.

Image sensors in these devices capture color by quantifying red, green, and blue optical lines and converting optical signals to electrical signals. There are three main types of image sensors: charge-coupled devices (CCD), contact image sensors (CIS), and complementary metal-oxide semiconductors (CMOS). The innovative hand-held optoelectronic reader (Fig. 3b) is an efficient and compact device designed for the on-site analysis of colorimetric data from a variety of gaseous and liquid analytes, including biomarkers, explosives, and food products [69, 70]. The device employs a color CIS with a linear array of CMOS detectors, which requires less power and occupies less space than traditional CCDs (Fig. 3b). The device employs timed pulses of red, green, and blue light, utilizing a disposable colorimetric array to enhance the detection of gaseous analytes. Additionally, it is equipped with a diaphragm micropump for controlled analyte flow. The integrated electronics facilitate real-time measurements of reflectivity in primary colors and incorporate modules for statistical analysis of the data collected. This reader's practical applications include rapid, low-noise assessments and the capability for real-time statistical evaluations, thus making it a valuable tool for on-site analytical tasks [70].

The assignment of filters to specific color spaces is carried out through various methods. RGB is the most commonly utilized color model. Other color models include CMYK (cyan, magenta, yellow, and black) and HSB (hue, saturation, and brightness). $L^*a^*b^*$ is another model, where L^* represents the luminance or lightness component, which ranges from 0 to 100, and the parameters a and b represent the chromaticity. The chromatic components, * (green to red) and b^* (blue to yellow), range from −120 to 120, whereas YUV (where Y represents the luma component, or brightness, which ranges from 0 to 100% in most applications) comprises the chrominance components, U and V, which represent the blue-luminance and red-luminance, respectively [72].

The RGB color system, utilized for capturing images, has been designed to emulate the functionality of the human visual system. In this system, the three matrices

(R, G, and B) are employed to store information regarding the intensities of the components mentioned above. This permits the formation of a relationship between analyte concentration and the color components. The selection of these colors follows the specifications set forth by the International Color Consortium (ICC), which has defined the range from zero to 255 (8-bit format) or 0 to 1 (fractional format), referring to these as channel intensities. For instance, red is denoted as [$R = 255$, $G = 0$, and $B = 0$], white as [$R = 255$, $G = 255$, and $B = 255$], black as [$R = 0$, $G = 0$, and $B = 0$], and violet as [$R = 238$, $G = 130$, and $B = 238$] [71, 72].

To determine the colorimetric and fluorometric properties of an array-based sensor, measuring the color variations of each sensor within the array during exposure to the target analyte is essential. This involves capturing images of the sensor array both before and during exposure, allowing for creating a real-time color difference profile by subtracting pre-exposure images from those taken during exposure [73]. These profiles are then compared against a library of previously documented color change patterns from known analytes and concentrations. Effective visualization is achieved through maps that display absolute color differences, overcoming any discrepancies caused by variations in printing techniques or minor lighting changes. Digital subtraction of red, green, and blue values creates a color difference maps (CDMs), while averaging sensor spot centers minimizes edge artifacts. The resulting patterns can be compressed into $3N$ vectors, where N represents the number of sensor spots, making storage efficient [71, 74].

$$\Delta R = R_a - R_b.$$

$$\Delta G = G_a - G_b.$$

$$\Delta B = B_a - B_b.$$

Here, a represents after, and b represents before. ΔR, ΔG, and ΔB are the color differences.

Multivariate Statistical Analysis

The analysis of cross-reactive arrays utilizes two primary statistical methods: descriptive (clustering) and classification techniques. Clustering analysis employs unsupervised techniques and focuses on identifying trends and relationships within data without prior knowledge of analyte classes. Although useful for qualitative evaluations and reducing redundant elements, it lacks predictive accuracy for unknown interactions. Conversely, classification analysis employs supervised techniques that utilize known analyte class information to categorize samples accurately. The classification accuracy reflects the algorithm's efficacy developed from a training set, which is crucial for identifying unknown samples. This process necessitates a significant number of samples for effective algorithm training.

Descriptive analysis is often conducted before classification to enhance the sensor array's configuration [75].

The CSAs were first introduced for qualitative (classification) analysis; they produce an eye-catching colorimetric scheme that can be used to identify the analyzed samples from each other. However, the color values correlate with the concentration; therefore, CSAs have also been employed for quantitative analysis.

For qualitative analysis, the data collected from CSAs can be analyzed by either unsupervised clustering methods, such as principal component analysis (PCA) and hierarchical cluster analysis (HCA), or supervised methods, such as linear discriminant analysis (LDA) and partial least squares-discriminant analysis (PLS-DA). The unsupervised methods search for clusters among the samples without having any information about class labels, whereas the supervised methods make classification models based on class labels of the objects. For example, in PCA, the data are projected into new orthogonal lower-dimensional space, such that the new dimensions preserve the highest variances of the original data. LDA also projects the data into a lower-dimensional space. However, the new dimensions are calculated to preserve the maximum class separation [76].

All the above-mentioned methods are linear models and assume data normality. When these assumptions are invalid, nonlinear and nonparametric methods can be employed. Different machine learning methods, such as random forest (RF) and decision trees (DTs), artificial neural networks (ANNs), and support vector machines (SVMs) can be employed [76].

Applications to Food Adulteration

The quality and safety of food are fundamentally connected to human health, which emphasizes the importance of monitoring the food industry. The complexity of food's chemical composition presents challenges in detecting contaminants and assessing quality. Traditional analytical methods, particularly GC/MS, can be unwieldy when dealing with similar complex mixtures. A promising alternative lies in using chemical sensor arrays, which not only facilitate the detection of food adulteration but also predict optimal storage conditions and assess quality. Advancements in chemical sensing and chemometrics have led to the adoption of electronic nose and tongue technologies across various food sectors, including meat [77], herbal medicine [78], fruit pickle samples [79], wine [80], seafood [81], and so on.

Coffee

Coffee is a major global beverage, driving a multi-billion-dollar industry. For producers and traders, effective techniques for analyzing the quality of coffee beans are critical, especially given the limitations of human sensory panels, which are subjective and subject to variability caused by several factors. These panels cannot detect counterfeit coffee blends; premium brands can be adulterated with lower-quality substitutes. While coffee's aroma, due to numerous volatile compounds, is a unique identifier, chemically analyzing and differentiating different types of coffee remains challenging. This complexity is due to the more than 1000 chemical compounds in roasted coffee beans and the influence of the roasting process on flavor and aroma development [82, 83].

In 2010, Suslick and colleagues presented a CSA (E-nose) for analyzing the aroma of various coffees, including whole bean and ground varieties. This method demonstrated the ability to differentiate between ten distinct commercial coffee brands and distinguish between beans roasted at varying temperatures and durations. The sensor elements employed were metalloporphyrins, pH indicators, vapochromic dyes, and metal salts, which were applied on a polyethylene terephthalate (PET) film. The digital images of the array were obtained using a flatbed scanner, and the statistical methods employed were PCA and HCA [84]. Different roasting degrees in coffee present distinct aromatic profiles. Research conducted by Kim et al. in 2018 introduced a CSA (E-nose) capable of detecting volatile compounds linked to the coffee aroma, enabling differentiation among various roast levels. Medium roast coffee exhibited significantly higher concentrations of Strecker aldehydes and α-diketones than other roasts. Additionally, specific sulfur compounds were more prevalent in medium-dark and dark roasts, while dimethyl sulfide was uniquely identified in light roast samples. The detection methods employed utilized various pH indicators alongside polyethylene glycol (PEG)-400 and reagents such as 2,4-dinitrophenyl hydrazine (DNPH) and Ellman's reagent, applied to PVDF membranes, with PCA and HCA used for analytical evaluation coffee [85]. In 2024, Mo developed an FSA (E-nose) to discriminate 40 coffee samples from 12 geographical origins. The sensor elements were two types of polymer-templated silver nanoclusters (polyacrylic acid-AgNCs and polymethacrylic acid-AgNCs) on polyethersulfone membrane for the detection of organic acids and coffee (see Fig. 4a). The sensor components displayed distinct fluorescence response characteristics when engaged with organic acids through PCA and RF algorithms [86].

Soft Drinks

The history of liquors is extensive and remains significant in contemporary society, enhancing enjoyment and relationships. Liquors are categorized into three main types: fermented liquors, including wine and beer; distilled liquors, such as whiskey and vodka; and prepared beverages, such as cocktails and liqueurs. The wide array of flavors results from over 1000 distinct flavor substances, shaped by factors such as origin, age, brand, and blending techniques. However, the global alcoholic beverage market faces challenges regarding product authenticity and safety, with risks of adulteration leading to potential hazards like methanol poisoning. Therefore, there

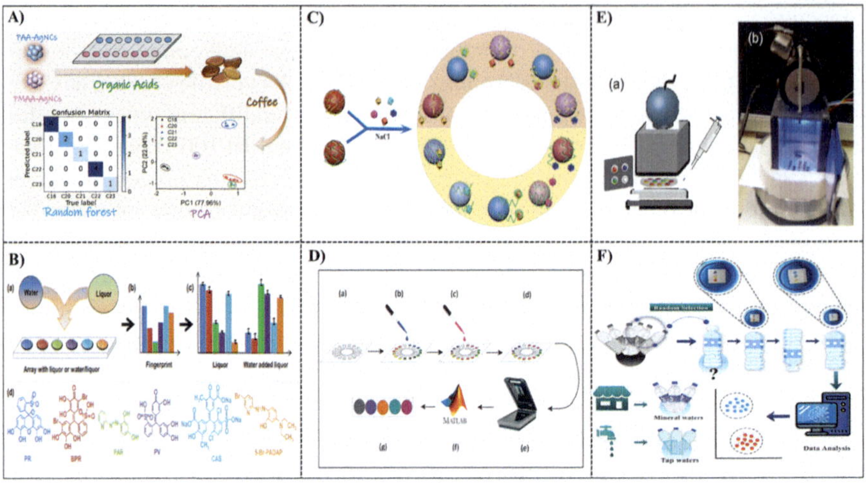

Fig. 4 (**a**) Discrimination of complex mixtures of coffee aromas by FSA. Adapted with permission from [86]. (**b**) Six commercial solvent discoloration dyes detecting Chinese liquors. Adapted with permission from [87]. (**c**) Schematic of CSA to monitor milk adulteration. Adapted with permission from [88]. (**d**) Schematic of the steps for fabricating of and measurement by the μPAD to discriminate eight antibiotics in milk. Adapted with permission from [89]. (**e**) Combinatorial map of the chemo-sensitive spots, sketch of the experimental setup to identify toxic compounds in water. Adapted with permission from [90]. (**f**) Image of the dip-type sensor array and schematic procedure of its application in discriminating water samples. Adapted with permission from [91]

is an urgent need for reliable quality control measures to assess various types of liquor, especially among different grades of the same brand [92].

Compounds exhibiting intramolecular charge transfer (ICT) are highly regarded for their ability to distinguish between various types of liquor due to their sensitivity to the environmental conditions of the liquors. In 2017, Han synthesized a novel ICT fluorescent molecule, (E)-5-((4-(diethylamino)-2-hydroxybenzylidene)amino) isophthalic acid (DHAIA), which demonstrated a significant response to changes in solvent microenvironment. This response manifested as variations in UV-Vis absorption and fluorescence. The molecule's interactions with different liquors allowed for the development of a liquor fingerprint, facilitating the identification of various brands and their authenticity [93]. In 2018, Li et al. developed a hand-held CSA (E-nose) that effectively differentiates among 14 liquors, including scotch, bourbon, rye whiskies, brandy, and vodka. This innovative sensor comprises 36 elements (pH indicators, metal–dye salts, redox indicators, chromogenic dyes, and solvatochromic dyes) featuring various chemically responsive inks that utilize eight distinct responsive mechanisms, such as aldehyde/ketone sensitivity and pH responsiveness. The sensor elements, embedded in hydrophobic matrices (polypropylene), demonstrate notable color changes when exposed to different liquor vapors. Additionally, the array employs HCA, PCA, and SVM for its discrimination capabilities [94]. Sun et al. introduced a CSA designed to differentiate Chinese liquors in liquid form (in 2020). The array, which includes 6 solvatochromic dyes (E-tongue) in a 96-well plate as substrate, operates on the principle of these dyes' responses to local polarity. Utilizing a pattern recognition technique, the sensor array achieved 100% accuracy in distinguishing 21 varieties of Chinese liquors. Additionally, it demonstrated the ability to detect diluted liquors, which is linked to changes in solvent polarity resulting from dilution. This suggests significant potential for the array's use in assessing the quality of Chinese liquors (see Fig. 4b). The statistical methods employed included PCA and HCA [87].

Milk

Milk is a vital human food source, providing a rich source of proteins, calcium, and vitamins B2 and B12. Several factors, including the cow's diet, genetics, and the environment, influence milk's nutritional quality. The deterioration of milk, primarily due to microbial growth, affects its safety, taste, and texture. While pasteurization effectively eliminates harmful pathogens, it does not entirely eradicate heat-resistant spores, which continue to present persistent challenges in the dairy industry. To guarantee quality, commercial milk products are standardized through pooling and thermal processing to enhance shelf stability and diminish contamination. Mastitis is a prevalent health issue in dairy cows, underscoring the necessity for more efficacious diagnostic methods. Moreover, over 100 chemical compounds have been identified in fermented milk, with pivotal flavor contributors including acetaldehyde, ethanol, diacetyl, and acetoin [95, 96].

Aminoglycoside antibiotics in animal husbandry can lead to antibiotic residues in animal-derived food products, posing potential risks to human health. Therefore, it is crucial to detect these residues in items like milk. In 2018, Yan developed a CSA (E-tongue) using two single-stranded DNA oligonucleotides as nonspecific receptors, utilizing gold nanoparticles (AuNPs) for the detection and quantification of amino acids (AAs). The process involved DNA detachment from AuNPs upon adding different AAs, causing varying degree of AuNPs aggregation due to salt. The method successfully identified five AAs in aqueous solutions and complex milk matrices with nearly 100% accuracy, employing HCA (see Fig. 4c) [88]. In 2022, Taghizadeh-Behbahani et al. introduced a microfluidic paper-based optical tongue (CSA) to discriminate 8 antibiotics in food (milk and egg samples). They used 5 metal-indicator complexes as sensor elements on the Whatman grade 2 filter paper, and discrimination and clustering were based on the differential interaction of the antibiotics with these elements, done by HCA and LDA (see Fig. 4d) [89]. Qin et al. (2024) introduced a cost-effective and efficient CSA (E-tongue) for rapidly detecting and differentiating antibiotic-resistant psychrophilic bacteria in raw milk. This array utilizes gold nanoparticles modified with D-amino acid as colorimetric probes. The interaction between the nanoparticles and psychrophilic bacteria leads to a noticeable color change within 6 h. Four primary psychrophilic bacteria were distinguished based on their metabolic responses to various D-amino acids by analyzing LDA response patterns. This method effectively identifies psychrophilic bacteria and assesses antibiotic susceptibility, thus mitigating the risk of bacterial infections and minimizing the adverse effects of antibiotic misuse [97]. In the same year, Ghohestani et al. developed an FSA (E-tongue) for the detection and quantification of sources of milk, including goat, camel, sheep, and cow milk. Furthermore, the method was employed to detect and quantify three adulterants, namely sodium hypochlorite, hydrogen peroxide, and formaldehyde, in milk. Eight nanoclusters (NCs) based on AuNCs, CuNCs, CdNCs, and AuAgNCs with different capping agents were employed as sensor elements on the Whatman grade 2 filter membrane. The adulteration of milk derived from disparate species gave rise to discrepancies in the fluorescence of the distinctively functionalized NCs. The ensuing CDMs afforded a unique pattern for the diverse species of milk, which is advantageous for the detection of adulterants. LDA, PCA, PCA-LDA, and PLS were used for detection and quantification [98].

Water

Water's flavor is heavily influenced by its chemical composition, particularly its salt content, where cations and anions contribute through various interactions. Volatile organic compounds also play a role, highlighting that water's flavor perception extends beyond mere taste. Water consumption is vital for human health as it provides essential minerals, with mineral water sourced from underground aquifers offering a pure and consistent composition related to local geology. In contrast, municipal water may contain microorganisms and comes from surface or groundwater sources, often requiring chlorination for safety, which can alter its chemical properties, such as pH and hardness. It is crucial to distinguish between mineral water and tap water, as bottled mineral water may sometimes include tap water [99, 100].

Identifying toxic compounds with varying chemical characteristics in water is a significant concern. In 2015, Goncalves et al. developed optical sensor arrays (E-tongue) for the threats above, namely dimethylmethylphosphonate (DMMP), cyclohexanone, piperazine, 5-fluorouracil (5-FU), imidacloprid, and paraoxon. The researchers employed metalloporphyrins, pH indicators, and their combinations, which were combined with polyvinyl chloride (PVC) and di(2-ethylhexyl) sebacate (DOS) as sensor elements on Thermanox plastic. PCA was then used to identify the toxic compounds (see Fig. 4e) [90]. Heavy metal ions (e.g., Cu^{2+}, Cd^{2+}, Pb^{2+}, etc.) represent an increasingly significant environmental concern. These non-biodegradable, highly toxic, persistent, and hazardous elements are especially problematic at low concentrations, particularly in water samples. In 2020, Sharifi and colleagues published a paper reporting a colorimetric Cu(II) assay in rainwater and tap water samples using a 3D origami paper-based analytical device. The authors doped pyrocatechol violet and chrome azurol S into PVC to fix the elements on the Whatman grade 2 filter paper [101]. One year later, Sharifi and colleagues developed a CSA (E-tongue) incorporating 6 silver and gold nanoparticles (nanozymes) and different organic capping agents as sensor elements in a 96-well ELISA kit to discriminate 29 anions in water. The catalytic activity of the nanoparticles, serving as enzyme mimics, was modulated by the anions, which influenced the oxidation of the 3,3',5,5'-Tetramethylbenzidine (TMB) substrate. Standard statistical methods, including PCA and HCA, were employed to analyze the results [102]. Low molecular weight compounds, such as catechols, exhibit high solubility in water. Even at very low concentrations, they have the potential to be harmful to human health. In 2016, Wang et al. introduced a mobile laboratory CSA (E-tongue) that combined pH indicators with phenylboronic acid. This was used to differentiate between 7 water-soluble catechols at varying concentrations. The researchers employed PCA, LDA, and HCA for classifications and clustering. The formation of complexes between phenylboronic acid and these catechols reduces their pKa values, increasing acidity in the sensing units. This causes color shifts in the pH indicators, which can be captured using a smartphone camera and processed remotely [103]. Very recently, Hatamifar et al. employed dip-type colorimetric paper-based sensors (E-tongue) to differentiate between bottled mineral water and tap water. Three organic dyes (bromothymol blue, bromophenol blue, and methyl red), immobilized on Whatman filter paper using a deep eutectic solvent, were employed as the sensor elements. The sensor array was utilized to analyze 40 brands of mineral water and 26 tap water samples from disparate regions of Shiraz and other Iranian cities. PCA and LDA were employed to cluster and classify water samples based on their source (see Fig. 4f) [91].

Sugars and Sweeteners

Sweeteners are essential food additives, categorized into natural sugars and artificial sweeteners. The increasing popularity of artificial sweeteners is attributed to their enhanced sweetness and potential benefits for weight and insulin management while providing minimal or no calories. However, their safety is debated, leading to usage restrictions for certain sweeteners like aspartame, particularly in foods for infants, and the establishment of regulated concentration limits. These artificial sweeteners represent 62% of the sweetener market and are widely used in beverages and pharmaceuticals. Since 1977, the FDA has approved several non-nutritive sweeteners, including saccharin, aspartame, acesulfame-K, and sucralose, while cyclamate remains banned in the USA despite its approval in over 100 countries due to potential carcinogenic concerns in animal studies. Evaluating sweeteners requires intricate analytical techniques such as ion chromatography and high-performance liquid chromatography (HPLC), demanding highly sensitive equipment [104, 105]. This underscores the necessity for developing more practical and efficient methodologies for the routine quality control of foodstuffs (sensor array).

In 2008, Lim et al. developed a rapid liquid sensing array method (E-tongue) to detect and identify a variety of sugars and related compounds. This method utilizes a nanoporous sol-gel matrix to immobilize indicators, successfully differentiating 15 types of monosaccharides, disaccharides, and artificial sweeteners through HCA, PCA, and LDA. The solution was prepared using a mixture of tetramethyl orthosilicate (TMOS), methyltrimethoxysilane (MTMS), methanol, and nano-pure water in a specific molar ratio, followed by the incorporation of pH indicators on a hydrophilic membrane [106]. One year later, Musto et al. introduced a disposable CSA (E-tongue) capable of detecting and distinguishing various natural and artificial sweeteners, including "*N*" Low, Stevita, Equal, Splenda, Domino sugar, and Sugar in the Raw (see Fig. 5a). This sensor array, which retained most elements from earlier work by the Suslick group, utilized turmoil-encapsulated pigments fixed on a hydrophilic porous membrane. This entrapment method facilitated swift interactions between aqueous analytes and hydrophilic indicators while preventing leaching, ensuring rapid and consistent responses. The application of LDA, PCA, and HCA for data clustering and classification was also discussed [107]. In 2011, Musto et al. introduced innovative differential sensing platforms for detecting sugars, saccharides, and artificial sweeteners through indicator displacement assays (IDAs) for colorimetric sensing (E-tongue). This method relies on the competitive affinity between two guests and a host, allowing the analyte to displace the indicator from the host, thus altering the indicator's spectroscopic characteristics. Indicators featuring cis-diols, such as alizarin red and bromopyrogallol red, to enhance the detection of saccharides, were optimized to develop an IDA sugar sensor. LDA, HCA,

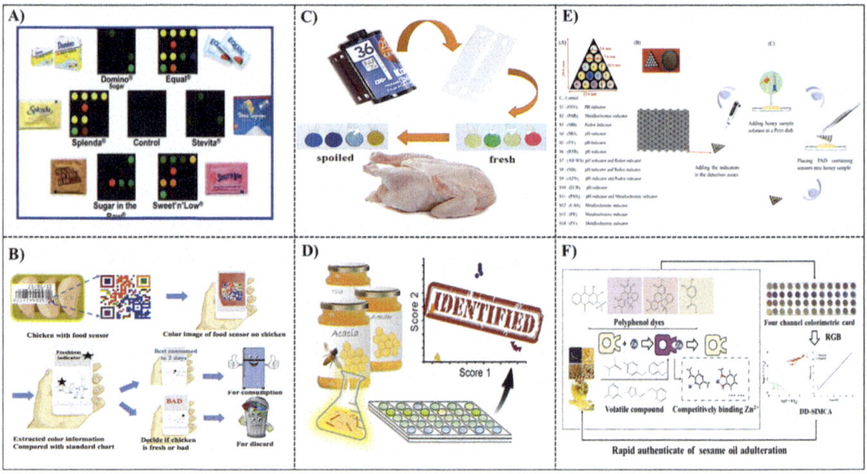

Fig. 5 (**a**) CDMs for six commonly used natural and artificial sweetener packets. Adapted with permission from [107], (**b**) The processing and monitoring of devices to detect the emanation of gases from chicken meat. Adapted with permission from [108]. (**c**) Procedure for preparation of CSA as naked-eye detection of chicken meat spoilage. Adapted with permission from [109]. (**d**) Schematic of device as optical sensor array for discriminating syrups and honey. Adapted with permission from [110]. (**e**) Schematic of sensor fabrication and analytical procedure for honey assay. Adapted with permission from [111]. (**f**) Schematics of the mechanism of oil adulteration. Adapted with permission from [112]

and PCA were also utilized for clustering and classifying sugars, saccharides, and artificial sweeteners [113].

Meat

The rising demand for meat, fish, and chicken is not matched by improvements in production quality control, leading to bacterial contamination due to inadequate cleaning and poor environmental conditions. Ineffective storage practices further compromise product freshness, while certain feeding methods and long storage periods can intensify lipid oxidation. Spoilage, characterized by changes in appearance and odor, results from a complex interplay of biological processes, storage duration, packaging, and microbial activity, with microbial by-products being the main spoilage agents in chicken during refrigeration. This underlines the urgent need for efficient spoilage detection methods, with sulfurous compounds and biogenic amines being key freshness indicators. Current analytical approaches for monitoring quality are often complicated and labor-intensive, emphasizing the necessity for simplified, rapid identification techniques. Traditional classification methods based on color, texture, and aroma are frequently insufficient. While advanced techniques like GC-MS and Fourier-transform infrared (FT-IR) spectroscopy are used to analyze volatile compounds in meat, they demand significant time and specialized expertise for effective application [114].

The first application of CSA (E-nose) for monitoring meat spoilage was reported by Salinas, who used an array of 13 unique dyes, including pH indicators, Lewis acids, and natural dyes, embedded in three inorganic materials: UVM-7, silica, and alumina. Notably, the sensor could distinguish between samples stored just 2 days apart. The chromogenic data collected from the array underwent analysis using qualitative PCA and quantitative PLS techniques. The PCA analysis demonstrated the system's effectiveness in categorizing chicken samples based on their freshness and providing a 3D visual representation of sample clustering [115]. Later, Chen et al. created an economical CSA (E-nose) to detect volatile compounds, featuring a distinctive colorimetric fingerprint. This array utilized 3 pH indicators and 9 metalloporphyrin materials to print chemically responsive dyes on a C2 reverse silica gel flat plate. To evaluate chicken freshness, they implemented a classification algorithm, specifically orthogonal linear discriminant analysis (OLDA) combined with the AdaBoost algorithm, resulting in a method referred to as AdaBoost OLDA [116]. In 2016, Li et al. developed a disposable CSA (E-nose) paired with a handheld device to assess the freshness of 5 meat products: beef, chicken, fish, pork, and shrimp. This sensor array is highly sensitive to gaseous analytes, specifically amines and sulfides, even at low concentrations. Comprising 20 sensor elements, each element was based on a 2-methoxyethanol solution integrated with organically modified silica sol-gel (ormosil) as the isolating matrix. The ormosils include a blend of tetraethoxysilane, triethoxyoctylsilane, and triethoxyphenethylsilane, which undergo polymerization during and post-printing, supported by stainless steel rectangular pins on polypropylene membranes. The hand-held device allowed for accurate differentiation of the headspace of types of meat over storage periods ranging from newly purchased to 4 days, utilizing PCA, HCA, and SVM for clustering and distinguishing meat volatiles by type and storage duration [117]. Chen et al. then employed a combination of 3 sensor elements, namely Nile red, zinc tetraphenylporphyrin (Zn-TPP), and methyl red, on paper with used PDMS, plastic, and tape to detect the emanation of gases from chicken meat (E-nose). The sensor was affixed to the surface of the meat sample or situated within the interior packaging structure. The condition of the meat product can be monitored remotely via a smartphone camera by photographing the sensor. Color information is then extracted from the photograph and compared with the standard chart, determining the meat's quality. The accuracy of the sensor platform was validated by monitoring the aging and eventual spoilage of chicken over 14 days under different temperature conditions (between −5 and 20 °C), with the final step involving the application of PCA as a clustering method (see Fig. 5b) [108].

In 2020, Magnaghi et al. devised a colorimetric sensor system (E-nose) to monitor deterioration processes in various meat products, including chicken, beef, pork, and cod fillets. The sensor elements were a panel of pH indicators and a colorimetric dye selective for thiols. These dyes were embedded into an anion exchange cellulose sheet. The array was placed over a tray containing the sample, either meat or fish (and not enriched to any extent with spoilage products). A progressive color change was observed in the 6 spots. The authors then used PCA to compare the spoilage processes for each kind of food, followed by an analysis of the evolution of the array color [118]. The primary objective in fabricating a robust sensor for detecting meat spoilage is the sensor's resistance to humidity and the acceptability of the color change that occurs at the initial stage of the spoilage process. In 2024, Banan-Khorshid and Hemmateenejad developed a CSA as a smart label for the naked-eye detection of chicken meat spoilage. They utilized pH indicators as sensor elements immobilized on a triacetyl cellulose (TAC) membrane (see Fig. 5c) [109]. An excess of amine in meat poses a substantial health risk. Alimohammadi et al. developed a CSA (E-nose) designed to identify hazardous amines and evaluate meat freshness in various types, including fish, lamb, and chicken. Utilizing ninhydrin-mediated transition metal-doped carbon dots (TMCDs) deposited on a paper substrate, the sensor exhibits distinct color changes upon interaction with different metal cations. It could detect 29 different amines, accurately distinguish between them, and recognize binary and ternary mixtures of amine solvents. The sensor array has been applied in meat quality and freshness assessments, employing PCA, HCA, and LDA for data analysis [119].

Honey

Honey is a natural sweet substance created by bees from plant nectar, featuring a complex chemical composition that varies based on its botanical and geographic origins. It contains over 600 volatile compounds, such as aldehydes, organic acids, and terpenes, contributing to its nutritional value and appeal in the food industry. However, its high market price raises concerns about fraud, including adulteration with sugars or using sugar-fed bees. The composition of honey, including glucose and fructose, influences its flavor and aroma. At the same time, both monofloral and polyfloral varieties offer distinct flavors and potential medicinal benefits, extending their use in pharmaceuticals and cosmetics. Quality assessment of honey typically involves analytical techniques like HPLC and conductivity measurement. However, there is a need for simpler and more accurate methods due to the intricacies of these assessments [120, 121]. In 2016, Tahir et al. employed CSA (E-nose) to differentiate between 6 honey varieties of varying botanical origins. A total of 58 aroma compounds were identified. Twenty abundant or active compounds were selected for characterizing honey aroma. The sensor elements included 6 metalloporphyrins and 8 hydrophobic nanoporous film membrane pH indicators. Subsequently, the honeys were discriminated against using HCA and PCA. Additionally, partial least squares regression (PLSR) was employed to indicate the relationship between sensory descriptors and aroma compounds [122]. In 2018, Bojanowski et al. published a report presenting a cross-reactive sensor array. This comprised a two-component probe system based on three viologen substituted boronic acids, a poly(aryleneethynylene) (PAE), and a further number of simple PAEs. Its purpose was the discrimination of syrups and 27 different honeys. This is illustrated in Fig. 5d. The fluorescence turn-on array can detect saccharide composition, whereas the PAEs can discriminate against trace amounts of colored components in the honey samples. The LDA was employed to classify syrups and honey [110].

In 2021, Chaharlangi et al. presented a paper-based CSA (E-nose) for the discrimination of monofloral European honey. Functionalized gold (AuNPs) and silver (AgNPs) nanoparticles using 6 different reducing and/or capping agents were used as sensor elements and deposited on the PVDF membrane to measure honey aroma. The color intensity of the AuNPs and AgNPs exhibited differential changes when they interacted with volatile organic compounds present in the headspace of the honey samples. The authors employed PLS, SVM, and LDA to discriminate between the honey [123]. In 2024, Masoomi, S. et al. deployed a CSA (E-nose) based on an electronic tongue for the differentiation of 27 Iranian honey samples originating from diverse geographical and botanical sources, as well as the detection of 14 instances of adulteration (the utilization of disparate types of syrup in honey). The sensor elements were 14 pH, metal, and redox indicators immobilized by PVC on the Whatman grade 2 filter paper (see Fig. 5e). Given the importance of the differentiation between heated and raw honey, given that the former is associated with increased levels of 5-hydroxymethylfurfuraldehyde (HMF), four honey samples and one mixed sample were subjected to heating procedures. The CSA proved highly effective in discriminating between heated and raw honey samples by applying PCA on digitized colorimetric data [111].

Oil

Oils are vital to a healthy diet, containing beneficial compounds, such as essential fatty acids and fat-soluble vitamins. They are extensively utilized in various fields, including cooking, medicine, and cosmetics. However, oil adulteration, defined as the addition of inferior or inexpensive oils and synthetic additives (such as artificial colors), negatively impacts these products' quality and nutritional value. In response, research has focused on developing analytical techniques, such as electronic noses and tongues, to detect adulteration practices [124].

In 2020, Liu and colleagues developed a CSA (E-tongue) to detect sesame oil adulteration. The sensor array employed a set of polyphenol dyes, including catechol violet, Alizarin Red S, bromocatechol red, and pyrogallol red, as sensor elements. The sensor elements were arranged in an array to allow for the competitive coordination of Zn^{2+} at their interface with volatile authenticity markers (VAMs) in the sesame oil. The coordination of Zn^{2+} with the VAMs resulted in a color change (see Fig. 5f). For this purpose, the researchers employed the use PLS. The PLSR model was also utilized to validate the linear correlation between polyphenol dyes' RGB value and sesame oil's adulteration concentration [112]. In 2021, Huang et al. introduced a CSA (E-nose) to detect adulterated extra virgin olive oil (EVOO) mixed with soybean oil and corn oil. The fatty acid composition of the vegetable oil in question determines the composition of its VOCs. The sensor elements were pH indicators and metalloporphyrins added to hydrophobic PVDF membranes. PCA and LDA were used for this purpose [125].

Cherry seed oil is a cosmetic oil of significant economic value often adulterated with more affordable oils, such as sesame and sunflower. Consequently, adulteration detection is of paramount importance. In 2025, Chaharlangi et al. developed a paper-based method, known as the CSA (E-nose), for adulteration detection in cherry seed oils. The method involved using a CAS printed with 20 pH and redox indicators on a reverse-phase silica gel plate. The CSA was then exposed to the VOCs present in the sample, and PCA and PLS-DA were employed for analysis and finally compared to lab-based headspace gas-chromatography ion-mobility spectrometry (GC-IMS) [126].

Saffron

Saffron, known scientifically as Crocus sativa, thrives in various climatic conditions, especially at altitudes between 600 and 1700 m [127]. The intricate harvesting process demands around 150,000 flowers to yield 1 kg of saffron, contributing to its high market value. The flower's stigmas are highly regarded for their aroma, color, taste, and medicinal benefits. Despite efforts to boost production and explore new cultivation areas, challenges such as temperature and humidity optimization persist. Care is vital during harvesting and drying to maintain quality, which is determined

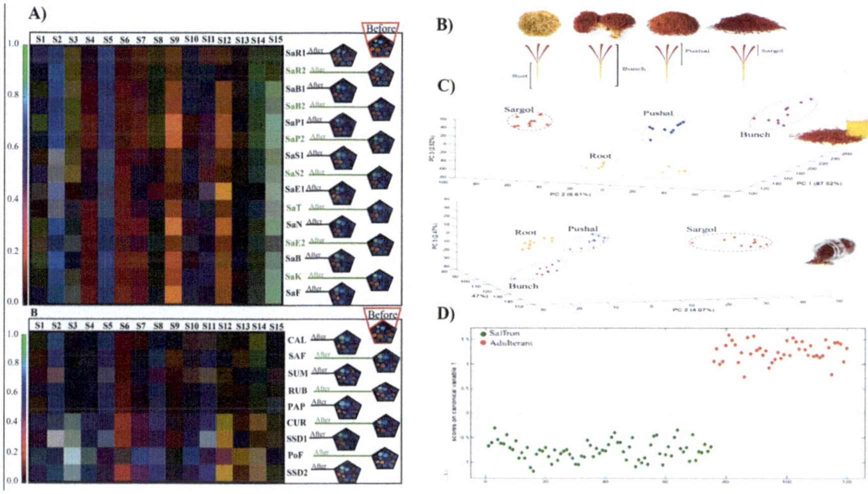

Fig. 6 (A-D) CDMs and categories of saffron, category clustering by PCA using extracted samples and untreated solid samples, and LDA plot for classifying saffron from possible adulterant herbs. Adapted with permission from [129]

by key components, such as crocins, safranal, and picrocrocin. Saffron's quality can be jeopardized through adulteration, which diminishes the concentration of these essential compounds [128]. In 2024, Masoomi, et al. developed an optical nose FSA to authenticate saffron. The sensor elements were metallic nanoclusters and carbon dots with varying functional groups. The device could differentiate saffron from other herbs and pure saffron from its mixtures with other herbs based on the discrepancy in their chemical compositions. The sensor array could also distinguish between four distinct saffron organs exhibiting varying qualities. For classification, the authors employed PLS, LDA, and PCA (see Fig. 6) [129].

Conclusion and Future Trends

Optical sensor arrays represent a transformative approach for addressing the global challenge of food adulteration. By leveraging the principles of cross-reactive sensing and advanced computational analysis, these arrays offer a versatile, cost-effective, and rapid solution for detecting adulterants in a wide range of food products. The integration of optical sensing with multivariate statistical tools, such as PCA, LDA, and machine learning algorithms, has significantly enhanced these systems' sensitivity, specificity, and reliability. Applications in various food sectors, including coffee, soft drinks, milk, water, sugars, meat, honey, oils, and saffron, have demonstrated the potential of optical sensor arrays to ensure food authenticity and safety.

Looking ahead, several trends and advancements are expected to shape the future of optical sensor arrays in food adulteration detection:

1. *Miniaturization and portability*: Developing compact, portable, and user-friendly optical sensor devices will enable on-site analysis, making these technologies more accessible for field applications, including remote or resource-limited settings.
2. *Integration with IoT and smart devices*: Incorporating Internet of Things (IoT) technologies and smartphone-based platforms will facilitate real-time monitoring and data sharing, enhancing the efficiency of food safety inspections and quality control processes.
3. *Enhanced sensitivity and selectivity*: Ongoing research into novel sensor materials, such as advanced nanomaterials and functionalized polymers, will improve the sensitivity and selectivity of optical sensor arrays, allowing for the detection of trace-level adulterants and complex mixtures.
4. *Machine learning and AI integration*: Using machine learning and artificial intelligence (AI) algorithms will further refine the data analysis capabilities of optical sensor arrays, enabling more accurate pattern recognition, classification, and prediction of food adulteration.
5. *Multimodal sensing*: Combining optical sensor arrays with other sensing modalities, such as electronic noses and tongues, will provide a more comprehensive food quality analysis, capturing both volatile and non-volatile compounds.
6. *Sustainability and biodegradability*: Developing eco-friendly and biodegradable sensor materials will align with global sustainability goals, reducing the environmental impact of sensor production and disposal.
7. *Standardization and regulatory compliance*: Establishing standardized protocols and regulatory frameworks for using optical sensor arrays in food safety will ensure consistency, reliability, and widespread adoption across the food industry.

In conclusion, optical sensor arrays are promising to revolutionize food safety and quality assurance. As research and technological advancements evolve, these systems will play an increasingly critical role in safeguarding public health, enhancing consumer trust, and mitigating economic losses associated with food adulteration. The future of optical sensor arrays lies in their ability to adapt to emerging challenges, integrate with cutting-edge technologies, and provide robust, scalable solutions for ensuring the authenticity and safety of our food supply.

References

1. Sawyer WE, Izah SC (2024) Unmasking food adulteration: public health challenges, impacts and mitigation strategies. ES General 4:1091
2. Askim JR, Mahmoudi M, Suslick KS (2013) Optical sensor arrays for chemical sensing: the optoelectronic nose. Chem Soc Rev 42(22):8649–8682

3. Albert KJ et al (2000) Cross-reactive chemical sensor arrays. Chem Rev 100(7):2595–2626
4. Atashbar M, Krishnamurthy S, Korotcenkov G (2010) Basic principles of chemical sensor operation. Chem Sens Fundamentals Sens Mater 1:1–62
5. Gründler P (2017) Chem Sens ChemTexts 3(4):16
6. Anzenbacher P Jr et al (2010) A practical approach to optical cross-reactive sensor arrays. Chem Soc Rev 39(10):3954–3979
7. Mitchell L, New EJ, Mahon CS (2021) Macromolecular optical sensor arrays. ACS Appl Polym Mater 3(2):506–530
8. Ma Y et al (2018) Optical colorimetric sensor arrays for chemical and biological analysis. Science China Chem 61:643–655
9. Röck F, Barsan N, Weimar U (2008) Electronic nose: current status and future trends. Chem Rev 108(2):705–725
10. Collins BE, Wright AT, Anslyn EV (2007) Combining molecular recognition, optical detection, and chemometric analysis. In: Creative chemical sensor systems. Springer, Cham, pp 181–218
11. Lee JH et al (2020) A behavior-learned cross-reactive sensor matrix for intelligent skin perception. Adv Mater 32(22):2000969
12. Tibaduiza D et al (2024) Electronic tongues and noses: a general overview. Biosensors 14(4):190
13. Mahanti NK et al (2024) Enhancing food authentication through E-nose and E-tongue technologies: current trends and future directions. Trends Food Sci Technol 150:104574
14. Banica F-G (2012) Chemical sensors and biosensors: fundamentals and applications. Wiley, Hoboken
15. Lobnik A, Turel M, Urek SK (2012) Optical chemical sensors: design and applications. Adv Chem Sens 3:5
16. Gründler P (2007) Chemical sensors: an introduction for scientists and engineers. Springer, Cham
17. McDonagh C, Burke CS, MacCraith BD (2008) Optical chemical sensors. Chem Rev 108(2):400–422
18. Zaera F (2012) Probing liquid/solid interfaces at the molecular level. Chem Rev 112(5):2920–2986
19. Li Z, Askim JR, Suslick KS (2018) The optoelectronic nose: colorimetric and fluorometric sensor arrays. Chem Rev 119(1):231–292
20. Bigdeli A et al (2017) Nanoparticle-based optical sensor arrays. Nanoscale 9(43):16546–16563
21. Nassau K (2003) The physics and chemistry of color: the 15 mechanisms. The science of color. Wiley, Hoboken, pp 247–280
22. Lakowicz J (2006) Principles of fluorescence spectroscopy. University of Maryland School of Medicine, Baltimore, p 132
23. Podbielsk H et al (2006) Optical chemical sensors. Springer, Cham, pp 353–385
24. Qazi HH, Mohammad AB, Akram M (2012) Recent progress in optical chemical sensors. Sensors 12(12):16522–16556
25. Suslick KS (2004) An optoelectronic nose: "seeing" smells by means of colorimetric sensor arrays. MRS Bull 29(10):720–725
26. Li T et al (2023) Recent progress in sensor arrays: from construction principles of sensing elements to applications. ACS Sens 8(3):994–1016
27. Del Valle M (2012) Sensor arrays and electronic tongue systems. Int J Electrochem 2012(1):986025
28. Podrażka M et al (2017) Electronic tongue—a tool for all tastes? Biosensors 8(1):3
29. Bushdid C et al (2014) Humans can discriminate more than 1 trillion olfactory stimuli. Science 343(6177):1370–1372
30. Buck L, Axel R (1991) A novel multigene family may encode odorant receptors: a molecular basis for odor recognition. Cell 65(1):175–187

31. Peveler WJ (2024) Food for thought: optical sensor arrays and machine learning for the food and beverage industry. ACS Sens 9(4):1656–1665
32. Kumar A, Castro M, Feller J-F (2023) Review on sensor array-based analytical technologies for quality control of food and beverages. Sensors 23(8):4017
33. Rabehi A et al (2024) Advancements and prospects of electronic nose in various applications: a comprehensive review. Appl Sci 14(11):4506
34. Modesti M et al (2021) E-nose and olfactory assessment: teamwork or a challenge to the last data? The case of virgin olive oil stability and shelf life. Appl Sci 11(18):8453
35. Bonah E et al (2020) Application of electronic nose as a non-invasive technique for odor fingerprinting and detection of bacterial foodborne pathogens: a review. J Food Sci Technol 57:1977–1990
36. Li Z, Suslick KS (2020) The optoelectronic nose. Acc Chem Res 54(4):950–960
37. Kolthoff IM, Rosenblum C (1937) Acid-base indicators. Macmillan, London
38. Rakow NA, Suslick KS (2000) A colorimetric sensor array for odour visualization. Nature 406(6797):710–713
39. Paolesse R et al (2017) Porphyrinoids for chemical sensor applications. Chem Rev 117(4):2517–2583
40. Jensen WB (1978) The Lewis acid-base definitions: a status report. Chem Rev 78(1):1–22
41. Deka B (2019) Anion receptor chemistry and its importance. Recent Trends Pharm Sci 6:106
42. Citterio D et al (2007) pH-independent fluorescent chemosensor for highly selective lithium ion sensing. Anal Chem 79(3):1237–1242
43. Bharadwaj PK (2017) Metal ion binding by laterally non-symmetric macrobicyclic oxa-aza cryptands. Dalton Trans 46(18):5742–5775
44. Xu M et al (2010) Cyclodextrin supramolecular complex as a water-soluble ratiometric sensor for ferric ion sensing. Langmuir 26(6):4529–4534
45. Reichardt C, Welton T (2011) Solvents and solvent effects in organic chemistry. Wiley, Hoboken
46. Reichardt C (1994) Solvatochromic dyes as solvent polarity indicators. Chem Rev 94(8):2319–2358
47. Marini A et al (2010) What is solvatochromism? J Phys Chem B 114(51):17128–17135
48. Cich MJ et al (2010) Enantiomerically selective vapochromic sensing. Sensors Actuators B Chem 149(1):199–204
49. Lin H, Suslick KS (2010) A colorimetric sensor array for detection of triacetone triperoxide vapor. J Am Chem Soc 132(44):15519–15521
50. Gao M, Tang BZ (2017) Fluorescent sensors based on aggregation-induced emission: recent advances and perspectives. ACS Sens 2(10):1382–1399
51. Mahmoudi M et al (2016) Identification of nanoparticles with a colorimetric sensor array. ACS Sens 1(1):17–21
52. Hulanicki A, Głąb S (1978) Redox indicators: characteristics and applications. Pergamon, Oxford
53. Feigl F, Anger V (2012) Spot tests in inorganic analysis. Elsevier, Amsterdam
54. Bhopate DP et al (2015) A highly selective and sensitive single click novel fluorescent off–on sensor for copper and sulfide ions detection directly in aqueous solution using curcumin nanoparticles. New J Chem 39(9):7086–7096
55. Ng SM, Koneswaran M, Narayanaswamy R (2016) A review on fluorescent inorganic nanoparticles for optical sensing applications. RSC Adv 6(26):21624–21661
56. Ngo YH et al (2011) Paper surfaces functionalized by nanoparticles. Adv Colloid Interf Sci 163(1):23–38
57. Fukahori S et al (2007) Effect of void structure of photocatalyst paper on VOC decomposition. Chemosphere 66(11):2136–2141
58. Dong BH, Hinestroza JP (2009) Metal nanoparticles on natural cellulose fibers: electrostatic assembly and in situ synthesis. ACS Appl Mater Interfaces 1(4):797–803

59. Pinto RJ et al (2007) Electrostatic assembly and growth of gold nanoparticles in cellulosic fibres. J Colloid Interface Sci 312(2):506–512
60. Morales-Narváez E et al (2015) Nanopaper as an optical sensing platform. ACS Nano 9(7):7296–7305
61. Ghasemi F et al (2023) Based optical nanosensors–a review. Anal Chim Acta 1238:340640
62. Long J et al (2011) A colorimetric array of metalloporphyrin derivatives for the detection of volatile organic compounds. Mater Sci Eng B 176(16):1271–1276
63. LaFratta CN, Walt DR (2008) Very high density sensing arrays. Chem Rev 108(2):614–637
64. Janzen MC et al (2006) Colorimetric sensor arrays for volatile organic compounds. Anal Chem 78(11):3591–3600
65. McGaughey O (2006) Development of a generic multi-analyte optical sensor platform for fluorescence-based sensing. Dublin City University, Dublin
66. Kemling JW, Suslick KS (2011) Nanoscale porosity in pigments for chemical sensing. Nanoscale 3(5):1971–1973
67. Mezenov YA et al (2019) Metal–organic frameworks in modern physics: highlights and perspectives. Adv Sci 6(17):1900506
68. Morbioli GG et al (2017) Technical aspects and challenges of colorimetric detection with microfluidic paper-based analytical devices (μPADs)-A review. Anal Chim Acta 970:1–22
69. Askim JR, Suslick KS (2015) Hand-held reader for colorimetric sensor arrays. Anal Chem 87(15):7810–7816
70. Sliwinska M et al (2014) Food analysis using artificial senses. J Agric Food Chem 62(7):1423–1448
71. Fernandes GM et al (2020) Novel approaches for colorimetric measurements in analytical chemistry–a review. Anal Chim Acta 1135:187–203
72. Fan Y et al (2021) Digital image colorimetry on smartphone for chemical analysis: a review. Measurement 171:108829
73. Stern A, Javidi B (2006) Three-dimensional image sensing, visualization, and processing using integral imaging. Proc IEEE 94(3):591–607
74. Chen R, Li X, Li J (2018) Object-based features for house detection from RGB high-resolution images. Remote Sens 10(3):451
75. Lavine BK, Workman J Jr (2013) Chemometrics. Anal Chem 85(2):705–714
76. Smith R, Chemometrics (2015) Forensic chemistry: fundamentals and applications. Wiley, Hoboken, pp 469–503
77. Lin Y et al (2024) Fast real-time monitoring of meat freshness based on fluorescent sensing array and deep learning: from development to deployment. Food Chem 448:139078
78. Hemmateenejad B et al (2017) Development of colorimetric sensor array for discrimination of herbal medicine. J Iran Chem Soc 14:595–604
79. Bordbar MM, Tashkhourian J, Hemmateenejad B (2018) Qualitative and quantitative analysis of toxic materials in adulterated fruit pickle samples by a colorimetric sensor array. Sensors Actuators B Chem 257:783–791
80. Zhang C, Bailey DP, Suslick KS (2006) Colorimetric sensor arrays for the analysis of beers: a feasibility study. J Agric Food Chem 54(14):4925–4931
81. Gonzalez-Martin A et al (2010) An array-based sensor for seafood freshness assessment. Bull Korean Chem Soc 31(11):3084–3092
82. Oliveira BP et al (2024) Unveiling sensory quality in coffee: a data-driven exploration. Res Sq. https://doi.org/10.21203/rs.3.rs-4824307/v1
83. Poisson L et al (2017) The chemistry of roasting—Decoding flavor formation. In: The craft and science of coffee. Elsevier, Amsterdam, pp 273–309
84. Suslick BA, Feng L, Suslick KS (2010) Discrimination of complex mixtures by a colorimetric sensor array: coffee aromas. Anal Chem 82(5):2067–2073
85. Kim S-Y, Kang B-S (2018) A colorimetric sensor array-based classification of coffees. Sensors Actuators B Chem 275:277–283

86. Mo Y et al (2024) A machine learning-assisted fluorescent sensor array utilizing silver nanoclusters for coffee discrimination. Spectrochim Acta A Mol Biomol Spectrosc 322:124760
87. Sun S et al (2020) A colorimetric sensor array for the discrimination of Chinese liquors. Analyst 145(21):6968–6973
88. Yan S et al (2018) Identification of aminoglycoside antibiotics in milk matrix with a colorimetric sensor array and pattern recognition methods. Anal Chim Acta 1034:153–160
89. Taghizadeh-Behbahani M, Shamsipur M, Hemmateenejad B (2022) Detection and discrimination of antibiotics in food samples using a microfluidic paper-based optical tongue. Talanta 241:123242
90. Goncalves CG et al (2016) Detection of diverse potential threats in water with an array of optical sensors. Sensors Actuators B Chem 236:997–1004
91. Hatamifar Y, Shojaeifard Z, Hemmateenejad B (2024) Discrimination of bottled mineral water from tap water using a dip-type colorimetric paper-based sensor array and chemometrics. Spectrochim Acta A Mol Biomol Spectrosc 321:124719
92. Yu Y et al (2024) Optical sensor array for the discrimination of liquors. J Future Foods 4(1):48–60
93. Han J et al (2017) A chemosensor for fingerprinting liquors: implication for differentiation and identification. Sensors Actuators B Chem 248:101–107
94. Li Z, Suslick KS (2018) A hand-held optoelectronic nose for the identification of liquors. ACS Sens 3(1):121–127
95. Magan JB et al (2021) Compositional and functional properties of milk and dairy products derived from cows fed pasture or concentrate-based diets. Compr Rev Food Sci Food Saf 20(3):2769–2800
96. Cimmino F et al (2023) Role of milk micronutrients in human health. Front Biosci Landmark 28(2):41
97. Qin Y et al (2024) Colorimetric sensor array for the rapid distinction and detection of various antibiotic-resistant psychrophilic bacteria in raw milk based-on machine learning. Food Chem X 22:101281
98. Ghohestani E, Tashkhourian J, Hemmateenejad B (2024) Rapid detection and quantification of milk adulterants using a nanoclusters-based fluorescent optical tongue. Food Chem 456:139973
99. Platikanov S et al (2013) Influence of minerals on the taste of bottled and tap water: a chemometric approach. Water Res 47(2):693–704
100. Sheibani E, Mohammadi A (2018) The impacts of water compositions on sensory properties of foods and beverages cannot be underestimated. Food Res Int 108:101–110
101. Sharifi H, Tashkhourian J, Hemmateenejad B (2020) A 3D origami paper-based analytical device combined with PVC membrane for colorimetric assay of heavy metal ions: application to determination of cu (II) in water samples. Anal Chim Acta 1126:114–123
102. Sharifi H, Tashkhourian J, Hemmateenejad B (2021) An array of metallic nanozymes can discriminate and detect a large number of anions. Sensors Actuators B Chem 339:129911
103. Wang Y et al (2016) A mobile laboratory for rapid on-site analysis of catechols from water samples with real-time results production. RSC Adv 6(84):80885–80895
104. Mattes RD, Popkin BM (2009) Non-nutritive sweetener consumption in humans: effects on appetite and food intake and their putative mechanisms. Am J Clin Nutr 89(1):1–14
105. Castro-Muñoz R et al (2022) Natural sweeteners: sources, extraction and current uses in foods and food industries. Food Chem 370:130991
106. Lim SH et al (2008) A colorimetric sensor array for detection and identification of sugars. Org Lett 10(20):4405
107. Musto CJ, Lim SH, Suslick KS (2009) Colorimetric detection and identification of natural and artificial sweeteners. Anal Chem 81(15):6526–6533
108. Chen Y et al (2017) Low cost smart phone diagnostics for food using paper-based colorimetric sensor arrays. Food Control 82:227–232

109. Banan-khorshid Z, Hemmateenejad B (2024) A colorimetric sensor array based on transparent photographic film (triacetyl cellulose polymer) as a smart label for naked eye detection of chicken meat spoilage. Microchem J 196:109671
110. Bojanowski NM et al (2018) An optical sensor array discriminates syrups and honeys. Chem Eur J 24(17):4255–4258
111. Masoomi S, Sharifi H, Hemmateenejad B (2024) A paper-based optical tongue for characterization of Iranian honey: identification of geographical/botanical origins and adulteration detection. Food Control 155:110052
112. Liu R et al (2022) Visual classification for sesame oil adulteration detection and quantification of compounds used as adulterants using flavor compounds targeted array sensor in combination with DD-SIMCA and PLS. Sensors Actuators B Chem 357:131335
113. Musto CJ, Suslick KS (2010) Differential sensing of sugars by colorimetric arrays. Curr Opin Chem Biol 14(6):758–766
114. Andersen HJ, Oksbjerg N, Therkildsen M (2005) Potential quality control tools in the production of fresh pork, beef and lamb demanded by the European Society. Livest Prod Sci 94(1–2):105–124
115. Salinas Y et al (2012) Monitoring of chicken meat freshness by means of a colorimetric sensor array. Analyst 137(16):3635–3643
116. Chen Q et al (2014) Evaluation of chicken freshness using a low-cost colorimetric sensor array with AdaBoost–OLDA classification algorithm. LWT - Food Sci Technol 57(2):502–507
117. Li Z, Suslick KS (2016) Portable optoelectronic nose for monitoring meat freshness. ACS Sens 1(11):1330–1335
118. Magnaghi LR et al (2020) Colorimetric sensor array for monitoring, modelling and comparing spoilage processes of different meat and fish foods. Foods 9(5):684
119. Alimohammadi M et al (2023) An optical nose based on array of metal-doped carbon dots for identification of hazardous amines and assessing meat freshness. Sensors Actuators B Chem 393:134274
120. Machado AM et al (2020) Honey volatiles as a fingerprint for botanical origin—a review on their occurrence on monofloral honeys. Molecules 25(2):374
121. Soares S et al (2017) A comprehensive review on the main honey authentication issues: production and origin. Compr Rev Food Sci Food Saf 16(5):1072–1100
122. Tahir HE et al (2016) Discrimination of honeys using colorimetric sensor arrays, sensory analysis and gas chromatography techniques. Food Chem 206:37–43
123. Chaharlangi M et al (2021) A paper-based colorimetric sensor array for discrimination of monofloral European honeys based on gold nanoparticles and chemometrics data analysis. Spectrochim Acta A Mol Biomol Spectrosc 247:119076
124. Srivastava S et al (2024) Advances in detection technology for authentication of vegetable oils: a comprehensive review. Heliyon 10:e34759
125. Huang L, Wang M, Liu H (2022) Identification of adulterated extra virgin olive oil by colorimetric sensor array. Food Anal Methods. https://doi.org/10.1007/s12161-021-02141-x
126. Chaharlangi M et al (2024) Based optical nose arrays and untargeted GC-IMS for the adulteration detection of cherry seed oils. Microchem J 209:112610
127. Dhanapal AR et al (2024) Medicinal spices and herbs from India: therapeutic properties and molecular mechanism of saffron (Crocus sativus L.). In: Medicinal spices and herbs from India. Springer, Cham, pp 315–342
128. Lachguer K et al (2024) Saffron (Crocus sativus L.) cultivation and properties: a review. Int J Hortic Sci Technol 13:365–384
129. Masoomi S, Sharifi H, Hemmateenejad B (2024) An optical-nose device based on fluorescent nanomaterials sensor array for authentication of saffron. Sensors Actuators B Chem 405:135365

Electrochemical Sensors in Food Adulteration Detection

Simge Er Zeybekler

Contents

Abstract Food adulteration poses serious risks to public health and consumer safety. Traditional methods for detecting adulteration are often time-consuming, costly, and require laboratory equipment. Electrochemical sensors present a promising alternative due to their high sensitivity, selectivity, portability, and low-cost operation. This chapter summarises the use of electrochemical sensors in detecting various types of food fraud, including hormones, dyes, and artificial sweeteners. Different types of electrochemical sensors, such as amperometric, potentiometric, and impedimetric sensors, along with their operating principles, recent advances in sensor design and fabrication, and their potential future role in food safety monitoring are briefly discussed.

S. Er Zeybekler (✉)
Department of Engineering Sciences, Faculty of Engineering and Architecture, İzmir Kâtip Çelebi University, İzmir, Turkey

Department of Biochemistry, Faculty of Science, Ege University, İzmir, Turkey

W. Alahmad, M. Ashfaq (eds.), *Advanced Techniques against Food Adulteration*,
https://doi.org/10.1007/978-3-031-99454-8_7

Keywords Food adulteration · Electrochemical sensors · Nanobiotechnology · Food safety

Introduction

Food adulteration refers to intentionally including chemical substances not supposed to be present in specific food or beverage products. This unethical practice often reduces production costs or artificially enhances the product's apparent quantity [1]. The prevalence of food adulteration is rising, posing significant threats to public health and impacting the stability of both economic and social progress. Regrettably, millions of individuals lose their lives each year due to illnesses linked to unhealthy and contaminated food [2]. Adherence to food regulations is maintained through stringent controls implemented throughout the production process. However, food adulteration remains widespread across many products, including dairy, meat, honey, fruit juices, vegetable oils, snacks, jams, cereals, pre-packaged foods, and alcoholic beverages. However, the use of these artificial chemical agents poses significant risks to human health, potentially leading to acute and chronic toxicity and associated adverse health outcomes, including cancer and organ failure. Consequently, applying these chemicals in food production must be conducted with the utmost care, adhering to stringent safety protocols and remaining within the reference values established by food regulations. To mitigate spoilage of fruits during harvesting, processing, and transportation or to facilitate the sale of fruits out of season, vendors frequently harvest fruits before achieving optimal maturity and employ chemical agents for artificial ripening. Chemicals such as ethylene, ethanol, methanol, propylene, methyl jasmonate, ethylene glycol, ethephon, and calcium carbide are commonly utilized for this purpose [3]. Conversely, gibberellic acid, alpha-naphthyl acetic acid, and oxytocin are used as growth hormones to cultivate fruits and vegetables. While oxytocin is a mammalian hormone and a veterinary pharmaceutical, its application on vegetable crops poses significant risks. Nonetheless, it is extensively employed to enhance the size and coloration of various crops, including bottle gourds, bitter melons, pumpkins, and cucumbers. Artificial sweeteners are administered into one side of the fruit using syringe pumps to substitute for the natural sweetness. Introducing sugar or sugar solutions externally is a widespread method of juice adulteration. Additionally, saccharin, high fructose corn syrup, partially hydrolyzed cane syrup, and beet mid-invert sugar elevate the Brix value, creating a composition that resembles authentic fruit juice and honey [4]. Natural pigments such as chlorophyll, annatto, and caramel are commonly found in fruits and vegetables. However, retailers often prefer synthetic dyes because of their advantages over natural alternatives. The cost-effectiveness and availability of artificial dyes play a significant role in this preference. Some colors used in processed fruits and vegetables are permitted, while others are banned, including Rhodamine B, auramine, methane yellow, Congo red, Orange II, and malachite green [5]. Detergents improve the consistency of dairy products, fats, and oils. In addition to detergents, other substances, including salt, glucose, starch, neutralizers, powdered soap, surfactants, and color additives, are incorporated to modify aspects, such as whiteness, viscosity, thickness, and solid-fat content. Moreover, the incorporation of gelatin, stabilizers, enzymes, and additional external agents has demonstrated significant enhancement in the consistency of food products [6].

Consuming synthetic fruits treated with ripening agents can be carcinogenic and may lead to health issues, such as dizziness, nausea, and kidney failure [7]. Likewise, artificial coloring substances have demonstrated carcinogenic and genotoxic effects [8]. The use of growth hormones is linked to an increased risk of uterine cancer [9]. Consumption of foods preserved with formaldehyde has been associated with health complications affecting the kidneys, liver, and lungs [10]. The substitution of certain food items can also result in poisoning, allergic reactions, cardiovascular problems, and diabetes. Table 1 outlines several significant food additives and their potential health impacts.

The methodologies employed for detecting food adulteration encompass various analytical techniques, such as high-performance liquid chromatography (HPLC), recognized as the gold standard, as well as gas chromatography-mass spectrometry (GC-MS), Fourier transform infrared spectroscopy (FT-IR), and nuclear magnetic resonance spectroscopy (NMR). Nevertheless, these methods incur significant costs and entail intricate processes requiring specialized personnel [19]. The integration of advanced sensors with portable devices is essential for the on-site detection of counterfeiting at the point of sale. This approach would eliminate the need for laboratory-based analyses, reducing costs and expediting product recall.

Within this framework, electrochemical sensors have emerged as an up-and-coming category of analytical devices, characterized by their high sensitivity, specificity, portability, and capability for real-time assessment. These electrochemical sensing platforms operate on the principle that a specific analyte in a sample induces discernible changes in electrical properties, such as current, potential, or

Table 1 Adulterants in food and their potential health impacts

Adulterants	Health impacts	References
Artificial ripening agents	Gastrointestinal irritation with nausea, mental disorientation, lasting skin damage, carcinogenic, headaches, feelings of dizziness, changes in mood, drowsiness, diarrhea, and renal failure	[11, 12]
Artificial coloring agents	Cancers, diabetes, genotoxic effects, metabolic diseases, hyperactivity, allergic reactions, and microbial infection, brain injury	[4, 5]
Artificial sweetening agents	Migraine attacks, bladder cancer, clastogenic, genotoxic	[4, 8]
Substituted meat products	Cardiovascular diseases, allergic reactions, and diabetes	[13–15]
Illicit alcohol	Blindness, nausea, vomiting, organ failure, dizziness, confusion, and even death	[16]
Oil substitution	Heart failure, cancer, loss of vision, epidemic dropsy, glaucoma	[17, 18]

impedance. Such alterations facilitate the precise measurement of analyte concentrations within food matrices, which are pertinent to the issue of food adulteration. Various sensing methodologies support electrochemical sensors, including amperometric, potentiometric, conductometric, and impedimetric techniques. Moreover, recent advancements in sensor materials, particularly the utilization of nanomaterials such as carbon-based nanocomposites, nanoparticles, and conductive polymers, have significantly enhanced the sensitivity of these devices. Integrating these nanomaterials with microfluidic and miniaturized systems has further augmented the capabilities of electrochemical sensors. These innovations significantly improve the performance of detecting critical molecules associated with food adulteration across complex food matrices.

The Fundamentals of Electrochemical Detection

Electrochemical sensing is predicated on the redox reactions occurring at the surface of the working electrode in the presence of the analyte. The resultant current generated during these reactions is converted into an electrical signal, which can be quantified and interpreted as directly proportional to the concentration of the analyte. These sensors operate through a framework that includes a working electrode to oxidize or reduce the analyte, a reference electrode providing a stable potential for the working electrode, and a counter electrode that allows electrons to flow through the circuit, thus completing the electrical loop. This setup is placed in an electrolyte solution that facilitates the effective transport of ions vital for redox reactions [20, 21]. Electrochemical detection techniques encompass several methodologies, including amperometry, voltammetry, impedance spectroscopy, and potentiometry (Fig. 1). Amperometry measures current at a fixed potential, reflecting the rate of redox reactions and the concentration of the analyte. In contrast, voltammetry involves changing the potential at the working electrode to assess the corresponding current. This method allows for the identification and quantification of analytes by leveraging their unique redox potentials through techniques, such as cyclic voltammetry (CV), differential pulse voltammetry (DPV), square-wave voltammetry (SWV), and linear sweep voltammetry (LSV). Impedance spectroscopy examines the impedance across a spectrum of frequencies, yielding valuable insights into the kinetics of redox processes and the dynamics at the electrode–electrolyte interface. Potentiometry assesses the potential difference between electrodes while minimizing current flow, correlating the measured potential to ion concentration by applying the Nernst equation [23, 24]. Each of these methods provides essential information for comprehensive electrochemical analysis.

The efficacy of electrochemical sensors is fundamentally dependent on nanomaterials such as graphene oxide, reduced graphene oxide (rGO), carbon nanotubes (CNT), and nanoparticles employed in modifying the working electrode. These

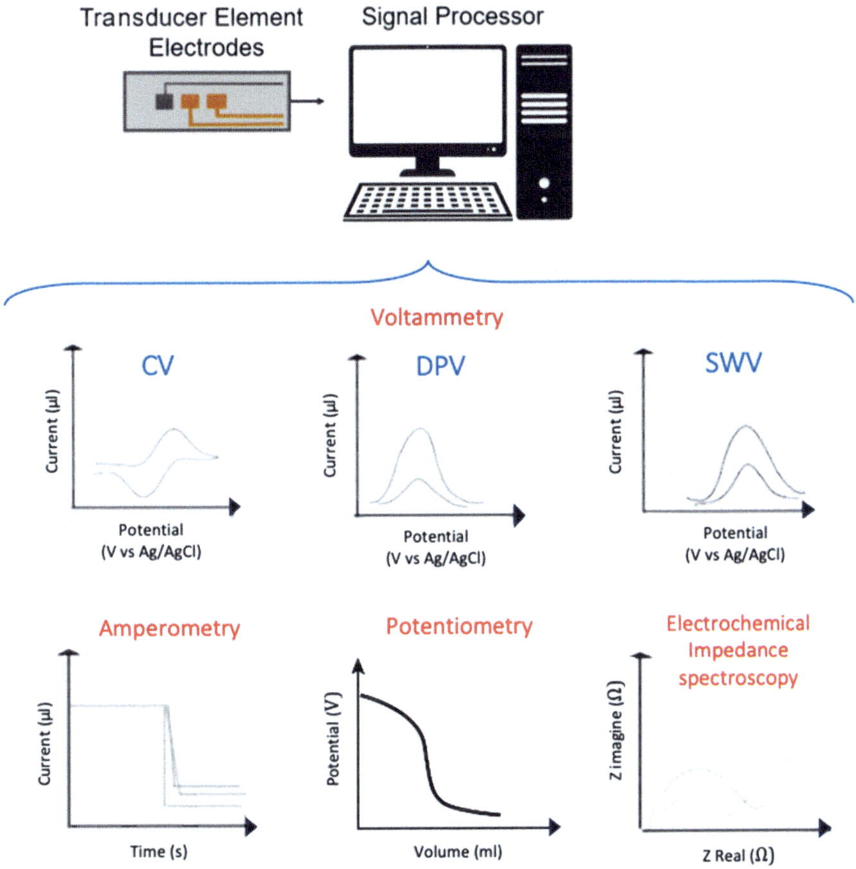

Fig. 1 Various electrochemical techniques. (Adapted with permission [22])

modifications enhance the active surface area of the electrode, facilitating a more excellent binding of analytes to the electrode surface (Fig. 2). Consequently, this leads to the advancement of electrochemical sensors characterized by low detection limits and high sensitivity. Furthermore, surface modifications that utilize nanomaterials with superior electrical conductivity, such as graphene, nanoparticles, and conducting polymers, are designed to reduce response times by increasing the electron transfer rate on the electrode surface [25]. These advancements are crucial for identifying low levels of hazardous substances, such as pathogens, toxins, pesticides, and heavy metals, which pose serious threats to food safety and security. These innovations give electrochemical sensors a strong real-time, on-site food product monitoring solution. They provide a cost-effective alternative that achieves rapid and sensitive measurements, outperforming traditional analytical methods [26].

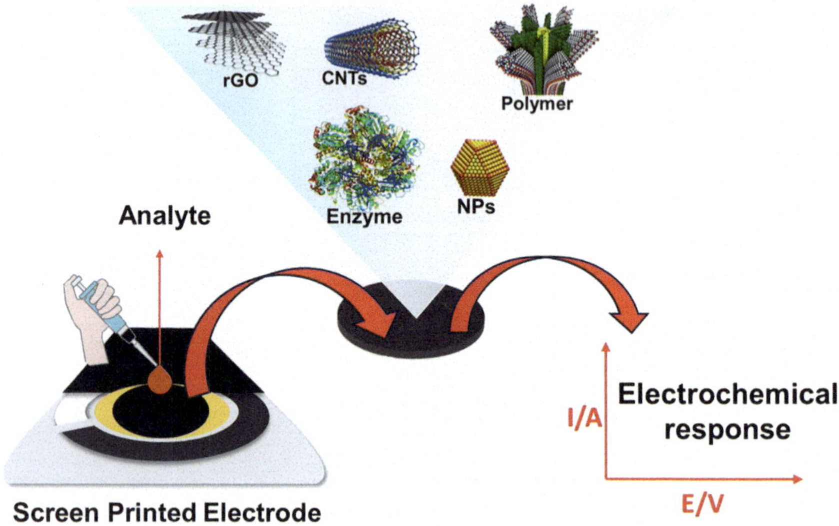

Fig. 2 A simplified illustration depicting electrochemical sensors that utilize different nanomaterials

Electrochemical Detection of Artificial Ripening Agents

Ethylene (C_2H_4) is a multifunctional phytohormone that contributes significantly to plant growth. It triggers the ripening of fruits, is involved in the aging of leaves and flowers, and facilitates seed sprouting and blooming, among other functions [27]. Because of its analogous molecular structure, acetylene (C_2H_2) shares fruit-ripening characteristics with ethylene. Both compounds are unsaturated and possess similar molecular sizes, making them capable of ripening fruits [3]. Calcium carbide (CaC_2), a carcinogen, is usually prohibited due to some toxic chemicals generated during fruit ripening. When CaC_2 is applied to fruits, it interacts with moisture and releases acetylene, further stimulating the ripening process [7]. Molecularly imprinted polymers (MIPs) feature synthetic binding sites created explicitly for target molecules. They are typically produced through copolymerization methods that involve template molecules and functional monomers. Once the template molecules are extracted, recognition cavities are generated within the polymer tailored to interact with the template molecule [28]. Integrating MIPs into electrochemical sensors facilitates the development of analytical devices suitable for food analysis. These sensors are characterized by their selectivity and chemical, thermal, and mechanical stability. Additionally, they offer advantages regarding reusability, ease of preparation, and cost-effectiveness [29]. Ghatak et al. developed an electrochemical sensor based on dysprosium particles-polyacrylamide cross-linked-ambroxol (AMB) imprinted polymer to detect CaC_2 treatment in mango. AMB, known for its mucolytic properties, was recognized as a marker for artificial ripening in this research, as the level of AMB decreased after applying calcium carbide (CaC_2) to mangoes.

The linear detection range was 10–500 μM, and the limit of detection (LOD) was 6.48 μM/L [30]. Similarly, Ghatak et al. developed an electrochemical sensor based on the MIP to detect the presence of acetylene by measuring its biomarker, furaneol (FU). The concentration of FU decreases significantly when exposed to CaC_2, unlike naturally ripped mangoes. FU is a flavoring molecule found in mangoes and other citrus fruits. FU-imprinted electrochemical sensor was developed in the presence of FU (template), methacrylic acid (MAA, functional monomer), ethylene glycol dimethacrylate (EGDMA, cross-linker), and graphene powder (G). The linear detection range was 0.5–500 μM, and the LOD was 11 nM [9]. Similarly, Kim et al. produced an MIP-based electrochemical sensor for detecting FU in strawberries. This innovative electrochemical sensor utilized poly(methacrylic acid) (PMAA) as the sensing material, while polyaniline (PANI) served as the redox-active component. The linear detection range was 0.1–5.5 mg/mL, and the LOD was 0.0814 mg/mL [31]. Because of enzymes' significant catalytic efficiency and specific interactions, enzyme-based electrochemical biosensors rank among the most sophisticated and commercially viable bioanalytical instruments. These biosensors identify electroactive compounds following enzymatic catalysis, either directly on the electrode or through a mediator [32]. Enzymatic electrochemical biosensors provide rapid, real-time, and accurate detection of analytes [33]. Ramachandra et al. developed an enzymatic electrochemical biosensor utilizing a ceria (CeO_2) nano-interface in conjunction with the enzyme acetylcholinesterase (AChE). The biosensor employed a platinum (Pt) working electrode to detect CaC_2 in mangoes. The performance of the Pt/CeO_2/AChE biosensor exhibited a detection range of 1–20 nM, with an LOD determined to be 0.6 nM. Electrochemical gas sensors function based on redox reactions. They typically feature a gas-permeable membrane that links the electrochemical cell to the gas-exposed environment. Based on their operating principle, electrochemical gas sensors can be categorized as conductometric, amperometric, or potentiometric [34]. Wang et al. developed an affordable electrochemical gas sensor employing ZnO nanosheets for the detection of ethylene in bananas, achieving response and recovery times of 8 s and 20 s, respectively. The minimum concentration that could be detected was 10 ppm [35]. Zevenbergen et al. developed an electrochemical ethylene gas sensor that uses a thin ionic liquid layer as an electrolyte. The LOD of 760 ppb was reached, with a linear response observed up to 10 ppm. Sekhar et al. produced an electrochemical sensor using a reduced graphene oxide (rGO)-copper (Cu) nanocomposite to detect ethene. It was reported that this electrochemical sensor could detect ethene concentrations as low as 10 ppb, which is particularly valuable for monitoring purposes after harvest [36].

Ethephon is a plant growth regulator commonly utilized to encourage the ripening of various vegetable products before harvest, aid in fruit gathering, and hasten ripening processes after harvest [37]. Research indicates that ethephon significantly inhibits acetylcholinesterase activity in long-term blood plasma and erythrocytes [38]. A phosphonic acid group within the compound's molecular structure imparts high polarity and low volatility [39]. As a result, the direct quantification of ethephon through liquid or gas chromatography presents considerable challenges.

Zhang et al. developed an electrochemical sensor that utilizes a polyaniline (PANI)-modified stannic oxide (SnO_2) nanocomposite as a modification agent for the electrode surface. The researchers demonstrated that the SnO_2@PANI nanocomposite-modified electrode surfaces exhibited superior performance to both pristine SnO_2 and PANI-modified electrode surfaces. This enhanced performance was attributed to the synergistic effects between SnO_2 and PANI. The SnO_2@PANI electrochemical sensor demonstrated a detection range of 0.01 to 5 ng mL^{-1}, with an LOD of 4.76 pg mL^{-1} [38]. Li et al. developed an MIP-based electrochemical sensor using gold nanoparticles and o-phenylenediamine to detect apple ethephon. The performance of the ethephon electrochemical sensor exhibited a detection range of 1.0×10^{-13} to 1.0×10^{-9} mol/L, with an LOD determined to be 2.78×10^{-14} mol/L [40]. Nanozymes are complex nanomaterials that exhibit enzymatic properties, effectively mimicking the characteristics of natural enzymes. Based on their reaction mechanisms, they are primarily classified into two categories: oxidoreductases and hydrolases. The oxidoreductase category encompasses various enzymes, including peroxidases, oxidases, and superoxide dismutases [41]. Gold nanoparticles (GNPs) exhibit notable peroxidase-like activity, enabling the conversion of colorless 3,3′,5,5′-tetramethylbenzidine (TMB) into blue TMBox in the presence of H_2O_2 [42]. Natural enzymes are susceptible to environmental conditions and typically exhibit peak activity only under optimal circumstances [43]. In contrast, nanozymes are considerably more stable and are not easily influenced by external environmental factors. Furthermore, the larger specific surface area of nanozymes facilitates the presence of extensive reactive regions, resulting in enhanced enzymatic activity compared to their natural counterparts [44]. Owing to their robust structure, nanozymes are ideal nanomaterials for rapidly and effectively monitoring food adulteration in real time. Ji et al. developed an electrochemical sensor based on iron oxide@gold-platinum (Fe_3O_4@Au–Pt) nanozymes to detect ethephon in mangoes. This study demonstrated that ethephon inhibited acetylcholinesterase activity, resulting in reduced formation of H_2O_2. Consequently, Fe_3O_4@Au–Pt could not effectively catalyze the transformation of TMB into ox-TMB due to the lack of enough H_2O_2, which caused a significant increase in the TMB current signal. Therefore, in this study, electrochemical measurements were conducted with the understanding that the concentration of ethephon was positively related to the TMB current signal. The Fe_3O_4@Au–Pt nanozyme-based electrochemical sensor displayed a detection range of 0.05–6.0 mmol/L, with an LOD determined to be 2.01 nmol/mL [45]. Photoelectrochemical (PEC) analysis involves converting optical energy into electrical energy. When a photoactive material is exposed to light, it becomes excited and generates electron–hole pairs. The photogenerated electrons subsequently migrate to the electrode surface, resulting in a measurable photocurrent. PEC analysis exhibits greater sensitivity and reduced background noise than traditional electrochemical methods, owing to the spatial separation between the excitation light source and the electrical output signal [46]. Additionally, the electronic readouts employed in the PEC methodology contribute to its cost-effectiveness, simplicity of equipment requirements, and potential for miniaturization [47]. Yan et al. developed a photochemical sensor using a gold nanoparticle/graphite phase carbon nitride/non-integral bismuth oxychloride/bismuth oxybromide (AuNPs/g-C_3N_4/$BiOCl_{0.5}Br_{0.5}$) ITO electrode to detect the ethephon. The surface plasmon resonance (SPR) effects from AuNPs and the g-C_3N_4/$BiOCl_{0.5}Br_{0.5}$ heterojunction facilitate the generation and separation of photogenerated carriers, significantly improving the photocurrent response. When ethephon is present, the composite of AuNPs/g-C_3N_4/$BiOCl_{0.5}Br_{0.5}$ reacts with ethephon to trigger an electrocatalytic redox reaction, leading to a reduction in photocurrent intensity, thus enabling a specific response to ethephon. The PEC sensor demonstrated a detection range of 20.0–63.0 nM, with an LOD of 6.9 nM [48].

Electrochemical Detection of Growth Hormones

Gibberellic acid is a crucial growth hormone utilized to enhance the growth of fruits and vegetables and stands as one of the key hormones within the gibberellin category. However, recent studies have indicated that elevated levels of gibberellic acid may lead to cancer, chronic toxicity, and various other health issues [49]. Therefore, it is essential to accurately measure the concentration of gibberellic acid and analyze it using a sensitive and selective method. Ghanbary et al. created an electrochemical sensor for the detection of gibberellic acid (GA3) utilizing 5-ethyl-5-phenyl barbituric acid (EPBA) on a pencil graphite (PG) electrode modified with polyvinylpyrrolidone (PVP). Integrating PVP as an electron-transport matrix and EPBA as an electron transfer mediator accomplished a flawless synergistic effect in the electrochemical oxidation of GA3. The EPBA/PVP-based electrochemical sensor showed a detection range of 15–225 nmol/L, with an LOD determined to be 4.97 nmol/L [50]. Aliahmadi et al. developed a Ca/Mg-layered double hydroxide (LDH) based electrochemical sensor using a glassy carbon electrode (GCE) to detect GA3. LDH represents a category of materials characterized by their ability to exchange anions, which consists of mixed metal hydroxides. These compounds display significant anion-exchanging capabilities. Hydrotalcite is recognized as the most prominent and essential ore in this category; therefore, these materials are called hydrotalcite-like structures [51]. The Ca/Mg-LDH/GCE sensors displayed a detection range of 0.029–0.297 mM, with an LOD of 0.009 mM for GA3. Han-Zhao et al. developed an electrochemical luminescence (ECL) sensor based on MIP to detect GA3. The electrochemiluminescence (ECL) method, which utilizes MIPs, merges ECL's exceptional sensitivity with the superior selectivity offered by molecular imprinting technology and is becoming more prevalent for analysis and detection. Despite photons' high energy, their quantity is relatively low, leading to lower sensitivity. ECL resonance energy transfer (ECL-RET) involves the donor transferring power to the acceptor following an electrochemical reaction at the surface of the electrode, enabling the acceptor to absorb energy and emit light. Thus, improved sensitivity can be achieved [52, 53]. They utilized the tris(bipyridine)ruthenium(II) ($Ru(bpy)_3^{2+}$) as the probe and gold-deposited semiconducting graphitic carbon nitride (g-C_3N_4/Au) as the energy donor. GA3 imprinted polymer surface was obtained with o-phenylenediamine via electropolymerizing on the GCE. The emission peak of g-C_3N_4/Au aligns with the absorption peak of $Ru(bpy)_3^{2+}$, facilitating resonance RET between $Ru(bpy)_3^{2+}$and g-C_3N_4/Au. This interaction significantly boosts the ECL intensity of $Ru(bpy)_3^{2+}$. When $Ru(bpy)_3^{2+}$ was introduced to the electrode surface via the imprinted cavities, the energy from the excited state of g-C_3N_4/Au was efficiently conveyed to $Ru(bpy)_3^{2+}$, leading to a strong ECL signal. This mechanism significantly enhances the system's detection sensitivity. A linear relationship was observed between ECL intensities and GA3 concentrations, ranging from 4.00×10^{-14} to 7.00×10^{-11} mol/L, with an LOD of 1.64×10^{-14} mol/L [54].

Similarly, Li et al. developed an MIP-based ECL sensor to detect GA3 through competitive binding between the GA3 and the Rhodamine B (RhB)-labeled GA3 (RhB-GA3). After the competitive binding process, the residual RhB-GA3 on the MIP underwent electro-oxidation, forming RhB oxide. This transformation markedly improved luminol's previously weak electrochemiluminescence (ECL) signal. A notable reduction in the ECL signal was observed when RhB-GA3 was replaced with GA3 molecules in the experimental samples. ECL intensities showed a linear correlation with GA3 concentrations from 1×10^{-11} to 3×10^{-9} mol/L, achieving a detection limit of 3.45×10^{-12} mol/L [10]. Graphene, a nanomaterial composed of a layer of carbon atoms organized in a hexagonal structure and bonded with sp^2 bonds, remains highly appealing in scientific and technological fields. The main feature of this material is its exceptional electrical conductivity [55]. Sheikh-Mohseni et al. developed graphene nanosheet-modified carbon paste electrode (CPE) to detect GA3. The modified electrode demonstrated effective electrocatalytic activity for the oxidation of gallic acid due to its excellent conductivity, minimal electron transfer resistance, and catalytic properties. The performance of the modified electrode showed a response to GA within the concentration range of 3.0×10^{-5} to 1.5×10^{-4} M, with an LOD determined to be 1.1×10^{-7} M [56]. Magnetite nanoparticles (Fe_3O_4 NPs), characterized as a semimetallic metal oxide with a cubic inverse spinel structure, demonstrate exceptional adsorption capabilities, impressive electrocatalytic characteristics, and natural electrical conductivity [57, 58]. Incorporating Fe_3O_4 NPs onto the electrode surface improves electrode surface area, mass, electron transfer rate, selectivity, sensitivity, and, most importantly, a better signal-to-noise ratio. Electrochemical sensors utilizing Fe_3O_4 exhibit significant benefits, such as affordability, high sensitivity, and quick response times [59]. Lian-Ming et al. developed an MIP-based electrochemical sensor based on Fe_3O_4@Au and L-cysteine-GA3 (L-Cys-GA3). MIP/Fe_3O_4@Au was produced via electropolymerization of L-Cys in the presence of GA3. The findings indicated that the concentration of GA3 varying from 1.0×10^{-11} to 1.0×10^{-8} M exhibited a linear correlation, which had a detection limit of 2.6×10^{-12} M. Cong et al. developed an MIP-based electrochemical sensor that effectively incorporates multiple labeled dendrimers. Dendrimers are hyper-branched nanostructured macromolecules with numerous functional groups located on a spherical surface [60]. In the context of this research, ferrocenecarboxylic acid (FcA) modified polydiethylenetriaminepentaacetic acid-glycol ester (PDTPA) was employed to label gibberellic acid (GA3), leading to the synthesis of the compound GA3/PDTPA/(FcA)n. They mentioned that the template molecule was subject to multiple labeling, potentially allowing the cavity to correspond with various FcA molecules. Compared to the mono-labeling approach, the performance of the MIP-based electrochemical sensor was significantly enhanced. The experimental findings showed that the MIP-based electrochemical sensor was responsive to GA3 detection within a concentration range of 1.0×10^{7} M to 2.0×10^{9} M, having a detection limit of 9.3×10^{10} M [61]. You et al. developed a probe-free electrochemical immunosensor to detect GA3 in soybean. Probe-free electrochemical sensors offer redox probe molecules that enable electrochemical activity without requiring external redox probe solutions [62]. Given these attributes, developing and implementing this type of sensor are attracting increased attention and significance in contemporary research and industry. Conversely, dopamine (DA) is an important neurotransmitter. It features catechol and amine functional groups that can polymerize under alkaline conditions (pH 8.5). Polydopamine (PDA) can be generated on a solid surface with reactive catechol and quinone groups [63]. The catechol groups present in PDA have been shown to enhance the charge transfer rate [64]. Furthermore, the functional groups of PDA can interact with the carboxyl (-COOH) groups of antibodies through either a Schiff base reaction or a nucleophilic reaction [65]. The electrode surface was modified with ferrocene carboxaldehyde (Fc), graphene oxide (GO-COOH), and multi-walled carbon nanotubes (MWNT-COOH). Fc was employed to create electrochemical signals. The developed immunosensor demonstrated the broadest detection range (100 aM to 1 mM) and the lowest detection limit (17.4 aM) for GA3 reported to date [66]. Li et al. created an electrochemical immunosensor using a copper ion-labeled antigen to detect GA3 through a competitive immunoreaction in a coupled immunoassay. The findings indicated that the linear detection range was 1–150 μg/mL [67].

1-Naphthalacetic acid, a naphthalene derivative, is a chemical that promotes plant growth [68]. Depending on the plant species, the application dose is between 20 mg/L and 100 mg/L. It shows acute toxicity for human health. Thus, the maximum residue limit (MRL) for this chemical by the European Union for a wide range of fruits, vegetables, seeds, legumes, and other food products ranges from 0.6 mg kg^{-1} to 0.15 mg kg^{-1}. Additionally, the maximum residue level for 1-naphthalacetic acid in nuts and citrus fruits has been determined to be 0.05 mg/kg [69]. Carbon-based materials are frequently used to modify electrode surfaces [29, 30]. Carbon nanotubes (CNTs), carbon-based materials, are commonly used to enhance electrochemical signals with their high conductivity and surface area [70]. Multi-walled carbon nanotubes (MWCNTs), a type of carbon nanotube, have several walls that can have a unique carbon atom, a large surface area, and a very high adsorption capacity. Conversely, metal nanoparticles such as magnetic, zinc, or silver nanoparticles have outstanding conductivity, stability, and a high surface-to-volume ratio, which enhance the electron transfer rate between the target molecule and the electrode surface. Metal nanoparticles have been widely used to produce novel electrochemical sensing platforms with all these features [71]. Prabhu et al. developed a zinc oxide nanoparticles-multi-walled carbon nanotube-based electrochemical sensor to detect 1-naphthalacetic acid. The findings indicated that the concentration of 1-naphthalacetic acid varying from 0.2×10^{-8} to 7.0×10^{-6} M exhibited a linear correlation, which had a detection limit of 1.1 nM [68]. Lü et al. created an electrochemical sensor using MWNT film-coated GCE to detect 1-naphthalacetic acid. The Ca/Mg-LDH/GCE sensors demonstrated a detection range of 1×10^{-8} to 2×10^{-6} mol/L, with an LOD of 2.5×10^{-9} mol/L [72]. Acetylene black (AB) is a carbon black that can be used to develop electrochemical sensors with its porous structure and excellent electrical conductivity [73]. Huang et al. compared dihexadecyl hydrogen phosphate (DHB)-modified electrode and AB/DHB-modified electrode performance to detect 1-naphthalacetic acid. They indicated that AB/DHB-modified electrode exhibited increased oxidation peak current compared to DHB-modified electrode. This means the AB/DHB-modified electrode showed more sensitivity in detecting 1-naphthalacetic acid. The AB/DHB-modified sensor exhibited a detection range 4×10^{-8} to 5×10^{-6} mol/L, with an LOD of 1×10^{-8} mol/L [74]. Label-free electrochemical immunosensors are designed to directly ascertain the immune complex (antigen–antibody complex) by quantifying the physical changes complex formation brings. Consequently, these label-free electrochemical immunosensors present an extensive linear detection range, a low LOD, and swift measurements at a reduced cost due to their label-free detection and capacity for quantitative measurement [75]. Liang et al. created an electrochemical immunosensor using a modified 3-thiopropionic acid (MPA) gold electrode. The anti-1-naphthalacetic acid polyclonal antibody was immobilized on the MPA-modified gold electrode using carbodiimide chemistry. This developed immunosensor displayed a detection range of 0.05–2 g/mL, with an LOD of 0.035 g/mL [76]. Electrochemical pretreatment of boron-doped diamond (BDD) electrodes can enhance their effectiveness in electroanalytical applications. A cathodically treated electrode has a hydrogen-terminated surface, whereas an anodically treated BDD electrode possesses an oxygen-terminated surface. The main advantage of BDDE is the broad, usable potential range, which facilitates measurement at highly positive potential values while minimizing the risk of electrode surface benefit of BDDE [77]. Janíková et al. developed BDDE to detect 1-naphthalacetic acid. This developed immunosensor exhibited an LOD of 0.09 μg/mL [78]. A subsequent study presents an MIP-based electrochemical sensor for detecting 1-naphthyl acetic acid.

Oxytocin, a neurosecretory peptide hormone composed of nine distinct amino acid residues, is vital in biological systems as a neurotransmitter. Oxytocin is given to milk-producing animals by injection into plant parts, such as the stem. In this way, the milk yield of the animal increases. Two synthetic variants of oxytocin, namely Pitocin and Syntocinon, are employed to augment milk production in cattle. The absence of international regulations about the permissible use of exogenous oxytocin further complicates this issue. Consuming food products contaminated with oxytocin may adversely affect human growth, potentially resulting in the premature onset of puberty, hormonal imbalances, nervous breakdown, uterine contractions, and increased water retention. The Food and Drug Adulteration Prevention Act of 1960 bans the use of oxytocin in vegetables. The Drug and Cosmetic Act of 1940 also designates it as a Schedule H1 medication requiring a physician's prescription for supply [79]. Utilizing bioelectrodes with a reduced carbon footprint offers an innovative strategy for tackling the increasing challenge of electronic waste. Biodegradable polymers have emerged as sustainable and renewable alternatives to synthetic materials. Carbon nanofibers (CNFs) are carbon structures with a cylindrical shape, typically ranging from about 100 to 300 nm in diameter and lengths extending from several micrometers up to 200 μm [80]. They display high porosity, reinforcing strength, and numerous reactive carbon edges [81]. All these features make them promising nanomaterials for electrochemical sensor applications. Mehrotra et al. developed a chitosan-carbon nanofiber (CNF) membrane and optimized it to apply electrochemical sensors. Surface analysis of the membrane indicated a crystalline structure characterized by uniform particle distribution. The measured surface area was 25.52 m^2/g, with a pore volume of 0.0233 cm^3/g. Modifications to the membrane facilitated the creation of a bioelectrode capable of detecting exogenous oxytocin in milk. Electrochemical impedance spectroscopy was employed for quantitative oxytocin levels, resulting in a linear detection range of 10–105 ng/mL. The LOD was found to be 24.98 ± 11.37 pg/mL with a sensitivity of 2.77×10^{-10} Ω/log ng mL^{-1}/mm^2 for the analysis of oxytocin within milk samples. The recovery rates were calculated to be between 90.85% and 113.34%. The application of the chitosan-CNF membrane underscores its environmental compatibility while advancing the development of eco-friendly, disposable materials for sensing applications [82]. Numerous studies have demonstrated that doping graphene nanosheets with heteroatoms, such as nitrogen (N), phosphorus (P), and boron (B), can significantly enhance their electrochemical properties. The charge polarization between carbon atoms and these dopants improves electrical conductivity by facilitating increased electron transport [83]. While both nitrogen-doped and phosphorus-doped graphene exhibit enhanced performance in electrochemical systems when utilized individually, graphene that is dual-doped with nitrogen and phosphorus (NPG) has been shown to surpass the capabilities of single-doped graphene. This improvement can be attributed to the synergistic effects resulting from the combination of the two dopants. Thomas et al. conducted an electrochemical oxytocin assessment by employing a modified nitrogen and phosphorus co-doped graphene (NPG) electrode. The study focused on the electro-oxidation behavior of oxytocin utilizing the square-wave pulse voltammetry (SWPV) technique. The researchers established the linear determination range for oxytocin in two distinct segments: 0.1–10 nM and 15–95 nM, while the LOD was determined to be 40 pM. Furthermore, the developed NPG sensor was successfully utilized to quantitatively assess oxytocin in various matrices, including milk, fruit juice, and biological fluids [84]. Metal-organic frameworks (MOFs) are crystalline porous nanomaterials characterized by their superior properties, including large surface areas and adjustable pore sizes, which render them suitable for various electrochemical applications [85]. Existing literature indicates that monometallic MOFs generally exhibit inferior electron transfer capabilities and chemical stability compared to bimetallic MOFs. The enhanced properties of bimetallic MOFs can be attributed to the synergistic effects arising from increased active sites and optimized pore sizes [86]. Consequently, bimetallic MOFs are regarded as promising candidates for use as electrocatalysts. However, MOFs display lower electrical conductivity than graphene. This limitation is primarily due to weak orbital overlap and constrained electron mobility resulting from the interaction between metal ions and organic linkers [87, 88]. Therefore, an effective strategy may involve integrating bimetallic MOFs with highly conductive materials like graphene. This approach leverages synergistic catalytic activity to develop advanced electrochemical sensors with the desired performance characteristics. Abedeen et al. produced an electrochemical sensor for detecting oxytocin in milk samples based on Zn-Cu metal-organic framework (Zn-Cu-MOF) modified reduced graphene oxide (rGO). They indicated that the synergistic interactions between the components significantly enhanced the electrocatalytic activity of the Zn–Cu MOF/rGO nanocomposite. This developed MOF-based electrochemical sensor exhibited a 40–400 nM detection range, with an LOD of 1.1 nM [89]. Al-Taee et al. explored the electrochemical characteristics of oxytocin using MWCNT-polyfurfurylamine-modified glassy carbon electrodes through the DPV technique. Their study centered on the reduction peak of the cysteine component, optimizing the experimental conditions to understand the electrochemical behavior of oxytocin better. They indicated that a plot of current versus concentration gives two straight lines: $15.9–79.6 \times 10^{-10}$ M and $95.4–298.3 \times 10^{-10}$ M [90]. Rai et al. created a catalyst composed of copper, alginate, carboxymethylcellulose, dextran, and reduced graphene oxide (Cu/Alg./CMC/Dex/r-GO) to convert o-nitroaniline into 1,2-benzenediamine for the detection of oxytocin. The findings indicate that the LOD for oxytocin in phosphate buffer was established at 0.230 ng/mL [91].

Electrochemical Detection of Artificial Sweeteners

Sugar refers to saccharides, categorized as sweet carbohydrates found in food. These come in simple forms known as simple sugars (monosaccharides like glucose, fructose, and galactose) and in compound forms as disaccharides, such as sucrose, lactose, and maltose [92]. Saccharin, aspartame, acesulfame potassium, neotame, sucralose, advantame, steviol glycosides, and thaumatin are sugar substitutes as food additives approved by the Food and Drug Administration (FDA).

At the same time, sugar alcohols that are less sweet than sugar, such as sorbitol, xylitol, lactitol, mannitol, erythritol, and maltitol, can also be used as sugar substitutes [93]. Sugar alcohols are substances that do not fall into the categories of either alcohols or sugars. Instead, they possess characteristics of both types of compounds [92]. The FDA prohibits using salts such as calcium cyclamate, sodium cyclamate, magnesium cyclamate, and potassium cyclamate as sweeteners in the United States [94]. Sugar has been identified as the third most frequently counterfeited product globally. The incidence of sugar counterfeiting in the food industry is increasing. A primary motivation for this trend is reducing costs while enhancing food products' visual appeal. Sugar counterfeiting in food products has risen markedly, raising significant concerns within the industry and among consumers. This alarming trend is primarily driven by two key factors: the desire for cost reduction in food production and the quest to enhance the esthetic appeal of food items. While artificial sweeteners are legally permitted within certain limits, the unauthorized use of these substances in higher concentrations can lead to various adverse health effects. The health implications of excessive artificial sweetener consumption are particularly concerning. Research suggests that such practices may contribute to a heightened risk of multiple cancers, provoke gastrointestinal disturbances, and induce neurological issues [95]. Additionally, consumers may experience changes in taste perception, leading to a cycle of overconsumption and dependency on artificial sweeteners. Allergic reactions, metabolic disruptions, and cardiovascular complications are further notable risks associated with overuse [96]. Given these potential dangers, regulatory bodies must implement strict guidelines on artificial sweeteners in food products. Plastics, chalk, urea, and washing soda are the most common substances misused as sugar counterfeits [92]. Chalk, primarily made of calcium carbonate, which gives it its white hue, is a common adulterant in sugar. It is composed of fine, soft particles that are easily ground into powder. Chalk absorbs moisture, enhances sugar's longevity, lightens sugar's color, and usually increases its sugariness. Using large quantities of chalk can result in decreased appetite, lead poisoning, constipation, kidney stones, and dental issues. Urea appears as a white crystalline or granular solid, typically odorless, although it may sometimes emit a faint ammonia scent. It is used to lighten sugar's color. However, excessive urea can lead to skin infections and irritation [97]. Plastics are lightweight synthetic organic polymers with low thermal conductivity and chemical stability. Their moldability can resemble sugar crystals. However, plastics pose a global issue because they create numerous health risks [98]. Washing soda is used to lighten the color of sugar [92]. Each presents significant health risks to consumers, making stringent regulations essential to public safety. Furthermore, there is a pressing need to advance compact devices that facilitate immediate analysis of these substances on-site. This would allow for timely and accurate assessments of their safety and potential health impacts.

Honey has gained a notorious reputation as one of the most frequently counterfeited food products globally. This is primarily due to its composition, which is overwhelmingly made up of various sugars. Unscrupulous producers often resort to deceptive practices, seeking to cut production costs by adulterating pure honey with inexpensive artificial sweeteners. The methods of adulteration can be classified into two main categories for honey: direct and indirect. Direct adulteration is the more straightforward approach, where substances are deliberately mixed into pure honey. Conversely, indirect adulteration poses a more significant challenge to detect. This method occurs when bees inadvertently consume honey mixed with chemicals, industrial sugars, and other contaminants, which then find their way into the honey they produce. The complexity of indirect adulteration makes it significantly more challenging to identify than direct contamination, highlighting the importance of vigilance in preserving honey's purity [99, 100]. This practice misleads consumers and diminishes genuine honey's quality and unique flavor profile. Honey adulteration can harm consumers' health by raising blood sugar levels, triggering insulin release, and leading to type II diabetes, weight gain, obesity, elevated blood lipid levels, and high blood pressure [101].

Flexible sensors are remarkable devices that excel at detecting the subtle deflections of soft materials. This ability significantly boosts their efficiency in identifying critical elements, such as biomarkers, food samples, electrolytes, and various biomolecules. These innovative platforms are designed for in situ recognition, allowing for the real-time monitoring of specific analytes directly at the point of interest [102]. Advanced materials and cutting-edge structural designs can enhance the performance of these sensors. Their unique adaptability allows them to conform to dynamic and irregular surfaces, significantly reducing the risks of degradation from external factors. This characteristic is vital for ensuring accurate and reliable data collection over time. At the core of these flexible sensors is a foundational substrate that offers robust support for various associated components, including conductive materials and integrated circuits. This sturdy backbone is crucial for maintaining the integrity and effectiveness of the sensory systems, enabling them to deliver precise readings in various applications [103]. The materials such as polyethylene terephthalate (PET) or polyimide (PI) are suitable substrates for producing flexible sensor systems. Chugh et al. introduced an innovative, flexible enzymatic sensor to detect glucose adulteration in honey. This sensor utilizes polyethylene terephthalate (PET) as its substrate, enhancing its versatility and adaptability. On the working electrode, a conductive polymer made of graphene oxide (GO)-modified poly(3,4-ethylene dioxythiophene): poly(styrene sulfonate) (PEDOT: PSS) is employed, providing excellent electrical conductivity and sensitivity. The sensor demonstrated an impressive linear detection range for glucose concentrations from 0.29 to 26.06 mM/L. It achieved a remarkably low LOD at 0.08 mM, making it a valuable tool for ensuring the purity of honey [104].

The electronic tongue is an innovative device designed to replicate the complex functions of human taste perception. Due to its robust design and user-friendly operation, its development has gained momentum for various industrial applications [105]. The efficiency of its sensors hinges on certain materials' ability to absorb substances and facilitate their transformation into ions. As a low-selectivity device, the electronic tongue excels at identifying different analytes within complex solution mixtures [100]. It employs diverse scientific principles, including electrochemical techniques such as amperometry, potentiometry, and cyclic voltammetry. In voltammetric E-tongues, generating a vector comprising K voltammetric measurements for each working electrode is possible. This can be executed across various techniques, including cyclic and square-wave voltammetry. As depicted in Fig. 3, the intricacy of employing a multi-working electrode (multi-WE) system is quite significant. This complexity underscores the importance of implementing variable selection algorithms, which are crucial in extracting the most valuable insights from the rich data generated by the electrochemical device. Such algorithms help distill information, ensuring researchers and practitioners can focus on the most relevant and informative measurements to drive their analyses and conclusions. Additionally, it utilizes mass change detection and various optical methods, like luminescence, absorbance, and reflectance, alongside enzymatic approaches to enhance its functionality [107, 108]. In electrochemical techniques, specific sensors constructed from simple metals or electrodes combined with advanced materials—such as metal

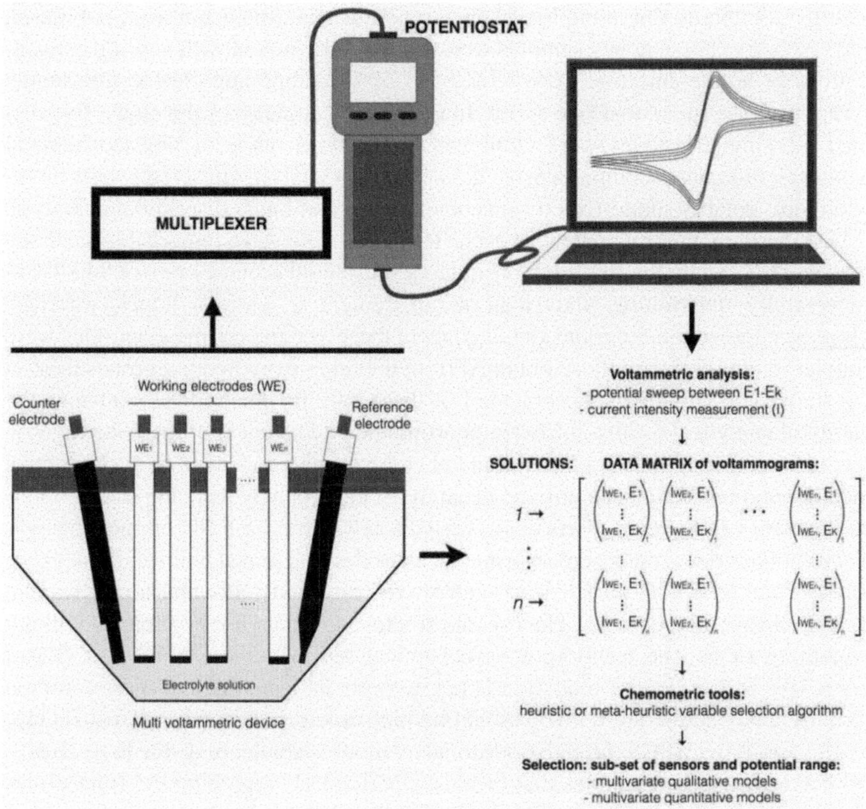

Fig. 3 Database containing signal profiles generated by a voltammetric E-tongue device with sensor array (K working electrodes), based on the analysis of each sample (M samples). Adapted with permission [106]

oxides, carbon paste, screen-printed layouts, or graphite-epoxy formulations—are particularly noteworthy. These sensors are typically submerged in a liquid solution and generate voltage readings relative to a stable potentiometric reference electrode. While numerous sensor designs have emerged, only a few have successfully integrated essential characteristics such as high selectivity, long-lasting durability, and impressive sensitivity. These attributes are critical for the electronic tongue's adequate performance across various applications, bridging the gap between conventional taste testing and modern analytical methods [109, 110].

Ciursa et al. delved into the innovative application of chemometric analysis, paired with a voltammetric tongue, to effectively distinguish between honey samples based on their geographical and botanical origins. The research specifically focuses on identifying potential adulterants in various types of honey, showcasing its significance in ensuring product purity. A sophisticated cyclic voltammetric electronic tongue was developed to achieve this, utilizing multiple materials such as silver, gold, glass, and platinum. This sophisticated configuration was created to identify adulteration in monofloral honey types: acacia, tilia, sunflower, polyfloral, and raspberry. These honey samples were intentionally mixed with varying concentrations of adulterants 5%, 10%, 20%, and 50%, comprising corn syrup, maple syrup, inverted sugar, and rice syrup. In the analytical phase of the study, five distinct working electrodes were employed alongside a range of physicochemical parameters to gather comprehensive data. The data analysis involved robust methodologies, notably support vector machine (SVM) and linear discriminant analysis (LDA), built on a multivariate approach. The SVM achieved a remarkable accuracy rate of 100%, while the LDA demonstrated a substantial accuracy rate of 94.87% in successfully determining the sugar syrup content in various honey varieties. Moreover, the voltammograms obtained from the silver and copper electrodes were simpler to interpret than those obtained from the other experimental probes used in the study. This highlights the potential of these specific electrodes to enhance the clarity of analytical results, further supporting the effectiveness of the voltammetric tongue in detecting honey adulteration [111]. Electrodes enhanced with conductive metallopolymers have been utilized in catalytic, electroanalytical, and sensor applications [16, 17]. Revenga-Parra et al. created a new Schiff base Ni complex that was electropolymerized onto screen-printed electrodes integrated with carbon nanotubes. Combined with partial least squares regression (PLSR) and a multivariate model, this method detected glucose and fructose levels in honey samples without separating steps. The resulting electrochemical sensor demonstrated an LOD of 1.2×10^{-5} M for glucose oxidation [112]. Reports indicate that the most common form of honey adulteration involves adding high-fructose corn syrup (HFCS) [113, 114]. Identifying HFCS through traditional methods is challenging due to its chemical makeup, which resembles that of honey, particularly regarding the fructose-to-glucose ratio [115]. Huang et al. developed an innovative electrochemical chamber explicitly designed to detect adulteration in high-fructose corn syrup. This advanced technique utilizes two identical gold electrodes to measure the variations in impedance between pure and adulterated syrup samples. The analysis focuses on logarithmic impedance (Log Z) over a frequency range of 1–105 Hz and evaluates charge transfer resistance (RCT). The researchers demonstrated that pure high-fructose corn syrup could be distinguished from its adulterated counterparts with remarkable clarity and repeatability, achieving an impressive variation of less than 3% across triplicate tests. Due to its efficiency, this method was portrayed as a rapid screening solution suited for routine analysis. Remarkably, the data acquisition time for recording Log Z and RCT was approximately 4 min, highlighting the technique's practicality for quick assessments in quality control processes [116]. Singh et al. developed an innovative sensor based on MIPs tailored for the electrochemical detection of acesulfame-K, a widely used artificial sweetener. This sensor was created by electropolymerizing a delicate ortho-phenylenediamine film, which forms the foundation for its sensing capabilities. The rigorous validation of the methodology demonstrated that the electrode operates effectively within a dynamic linear range of 0.1–17.0 μmol/L, highlighting its sensitivity to variations in acesulfame-K concentrations. The research established an LOD at 0.35 μmol/L, indicating that the sensor can reliably identify and quantify low levels of this sweetener. To assess the practical applicability of the sensor in real-world scenarios, the researchers tested its performance in various food products such as cola beverages, candies, and tabletop sweeteners that contain acesulfame-K. The results showed that the sensor is sensitive and highly selective for acesulfame-K, even when analyzed in complex food matrices [117]. Despite the FDA's endorsement of aspartame as a safe tabletop sweetener, emerging research has raised significant concerns regarding its potential health implications. Recent studies have suggested a troubling association between high levels of aspartame consumption and the development of hepatocellular carcinoma, a type of liver cancer [118]. Considering these findings, the International Agency for Research on Cancer (IARC) has categorized aspartame as a possible carcinogen, placing it in Group 2B. In response to these health concerns, the World Health Organization (WHO) has established a recommended acceptable daily intake limit of 40 mg/kg of body weight for aspartame [119]. These guidelines emphasize the importance of being aware of the potential health hazards linked to excessive consumption of aspartame and the need for vigilance against food fraud in the marketplace. As consumers, it is crucial to remain informed about what we eat and its implications for our health. Su et al. developed an MWCNT-MIP sensor to detect aspartame in sports beverages. The sensor showed a linear detection range for aspartame 8 nM to 6 μM, with an LOD of 22 nM [120]. Sha et al. developed a non-enzymatic electrochemical sensor using graphene–polyaniline (PANI) composite to detect urea in tap water and milk. The optimized sensor demonstrated a detection limit of 5.88 μM, excellent reproducibility, selectivity, and stability, and a detection range of 10–200 μM [121].

Electrochemical Detection of Artificial Coloring Agents

Sudan dyes are frequently encountered in various products, including flavoring powders, tangy pepper sauces, zesty soups, eye-catching snacks, and even carbonated beverages. These synthetic colorants, which include compounds such as Sunset Yellow (SY), Sudan Red, and Ponceau 4 R, are often favored by manufacturers due to their low cost and easy accessibility [122]. By utilizing these dyes, producers can enhance the visual appeal of their products and potentially improve the overall flavor and perceived quality. However, it is essential to note that many of these artificial dyes contain an aromatic ring in their chemical structure, which raises concerns regarding health risks when consumed over extended periods. The long-term effects of these substances remain a critical issue for consumer safety. Malachite green is a potent chemical compound commonly used in aquaculture to combat parasite and microbial infections in fish effectively. However, it poses significant health risks due to its tendency to accumulate in the organs of aquatic organisms, potentially leading to long-term health issues. Alarmingly, there have been reports of malachite green being utilized unlawfully within the aquaculture industry, raising concerns about food safety and animal welfare. In response to these challenges, Zhou et al. made a notable advancement: developing an innovative electrochemical sensor crafted from silver/copper metal-organic frameworks (Ag/Cu-MOFs). This sensor was prepared using a one-step solvothermal method, providing a practical solution for detecting malachite green in aquatic environments. The modified electrode exhibited significantly enhanced sensitivity compared to traditional bare electrodes, obtaining an outstanding LOD of just 2.2 nM. Moreover, it operated effectively within a linear concentration range of 10–140 nM, showcasing remarkable anti-interference capabilities to ensure accurate readings even in complex sample matrices [123]. In a separate yet equally significant study, Shabani-Nooshabadi et al. introduced an advanced sensor to detect multiple food dyes, specifically tartrazine, patent blue V, acid violet 7, and ponceau 4RA. This research focused on food products such as jellies, sauces, soft drinks, and candies, where these colorants are commonly found. They employed a modified carbon paste electrode infused with a copper-based metal-organic framework (Cu-BTC MOF) and 1-ethyl-3-methylimidazolium chloride, which acted as an ionic liquid (IL). This combination significantly enhanced the surface area and conductivity of the electrode, thereby improving the electron transfer rate during the detection process. The capabilities of this newly proposed system were thoroughly validated, particularly highlighting its exceptional analytical performance for patent blue V. It achieved a low detection limit of 0.07 μM, with an extensive linear range spanning from 0.08 to 900 μM, alongside impressive recovery rates, making it a reliable tool for monitoring food safety in the presence of these dyes [124]. Ji et al. employed MOFs to detect another azo dye, Sunset Yellow (SY). They highlighted the remarkable electrocatalytic properties of Fe-BTC, a specific MOF featuring an impressive surface area of 637.8 m^2/g. This notable surface area is essential in enhancing the accumulation potential of SY, which, in turn, allows for exceptional electrochemical performance when identifying SY across various complex drink matrices [125]. Another innovative approach was taken by Qi et al., who developed a highly sensitive sensor specifically for detecting SY. This sensor utilized cetyltrimethylammonium bromide (CTAB) combined with graphene (Gr) and Cu/Zr-MOF. The unique hybrid structure of CTAB-Gr-Cu/Zr-MOF was achieved through electrostatic interactions between the components, effectively merging their strengths for optimal results. When tested in soft drinks, this modified sensor demonstrated a remarkable enhancement in electrochemical current, increasing it by an astonishing 6.53-fold compared to bare electrodes. Furthermore, it achieved a detection limit within the nanomolar range, showcasing its potential for precise food and beverage analysis applications [126]. Kumari et al. developed an amperometric sensor based on ternary mixed metal oxide (Ti/Fe/CuO) nanocomposites to detect Sudan-I dye in red chili sauce and chili powder. The sensor exhibited a linear detection range between 0 and 160 μM and 0 and 200 μM for red chili sauce and chili powder, respectively, with an LOD of 0.02 mM [127].

Electrochemical Detection of Meat Adulteration

Meat adulteration is a persistent and worrisome issue in the food industry that raises serious ethical and health concerns. Unscrupulous suppliers may engage in deceitful practices by secretly blending premium cuts of meat, such as succulent beef, with cheaper alternatives like pork or chicken to increase their profit margins [128]. This practice constitutes a form of food fraud and violates important religious dietary laws that require transparency and integrity in food sourcing. Given these implications, developing precise methods is essential for identifying meat adulterants to protect consumers and uphold the standards of food safety and quality. Fu et al. developed an electrochemical sensor with one-dimensional convolutional neural networks (1DCNN) for detecting meat adulteration in beef, lamb, pork, chicken, and lard. The enhanced 1DCNN models demonstrated outstanding performance, achieving accuracies between 95.2% and 98.1% across various meat types, both raw and cooked. Their sensitivity and specificity consistently surpassed 95%, ensuring strong and dependable detection capabilities. The quick processing and measurements from electrochemical techniques, paired with the sophisticated feature extraction and pattern recognition abilities of 1DCNN, allow for identifying adulterants at levels as low as 10% [129]. Pork is frequently included in a wide range of food products because it is more economical than other meat options. This cost advantage has raised awareness of pork potentially being an adulterant in raw meat, which raises issues regarding food integrity and quality. Mandli et al. created two types of electrochemical immunosensors through the electro-entrapment of IgG in a polymer-modified graphite paste electrode. The first was a direct immunosensor, which could detect 0.1% and 1% levels of adulteration in raw and cooked meat within 2 h. The second format could identify 0.01% pork adulteration in just 20 min [130]. Electrochemical genosensors are innovative and efficient tools for rapid diagnostics in food quality control. These sophisticated devices are designed to detect specific DNA or RNA sequences accurately. The process typically hinges on DNA hybridization, wherein an immobilized single-stranded DNA (ssDNA) probe selectively binds to a complementary target DNA sequence. By focusing on specific DNA sequences, these sensors effectively analyze food adulteration. This cutting-edge technology enhances food safety and supports regulatory compliance, ensuring consumers have accurate information about their food [131]. Flauzino et al. developed a graphene-based single-stranded DNA probe, leading to an impedimetric diagnostic device capable of identifying pork adulteration in beef, with a detection limit of 9% w/w [132]. Ruiz-Valdepenas Montiel et al. introduced the first single-use electrochemical sensor to detect horsemeat adulterations based on RNA probes and magnetic microbeads. This biosensor could identify horsemeat by targeting a specific mitochondrial DNA D-loop region fragment. The method demonstrated impressive analytical performance, achieving an LOD of 0.12 ppm (equivalent to 3.0 automobiles) for synthetic target DNA [131]. In a noteworthy study, Zhang et al. developed an innovative electrochemical biosensor to detect unique DNA sequences within various meat products. This cutting-edge technology employs a technique known as toehold-mediated strand displacement, which operates as an enzyme-free isothermal strand displacement amplification method. The researchers rigorously tested their biosensor on 13 meat types: cattle, sheep, pig, horse, donkey, dog, fox, rabbit, mouse, rat, chicken, duck, and goose. This comprehensive assessment allowed them to evaluate the analytical capabilities of the method for accurately identifying different animal species within food products [133]. Faria et al. introduced an electrochemical method for screening horsemeat adulteration. This technique involves immobilizing anti-horse immunoglobulin G on screen-printed carbon electrodes (SPCE) infused with polyaniline to detect horse immunoglobulin G. The sensors were evaluated using PBS homogenates from commercial raw ground samples of horsemeat, pork, and beef. The testing process took around 72 min, achieving a selective response with an LOD of 0.004% [134]. Hartati et al. developed a voltammetric electrochemical sensor without using a DNA amplifier based on reduced graphene oxide-modified SPCE, focusing on the oxidation signal of guanine from pork, chicken, and beef meat. The LOD was determined to be 1.76 μg/mL, with a linear range extending from 0 to 10.0 μg/mL [135]. Lim et al. introduced an innovative label-free electrochemical immunosensor that utilizes a carbon nanofiber-4-carboxyphenyl-modified SPCE. This device detected pork adulteration in raw meat samples, including beef, chicken, and pork. The immunosensor operates by specifically identifying porcine serum albumin, an essential protein indicating pork presence. The researchers reported that this immunosensor exhibited a remarkable linear response across a concentration range of 0.5–500 pg/mL. Furthermore, it has an exceptional LOD of just 0.5 pg/mL when tested in a buffer solution, showcasing its sensitivity and potential for practical food safety and quality control applications [136]. AuNPs feature a high specific surface area, significantly enhancing the flow of electrons. This unique surface property is an effective platform for immobilizing thiolated oligonucleotides onto electrodes by forming robust gold-sulfur (Au-S) bonds. This interaction facilitates improved conductivity and increases sensitivity and selectivity in both the electrode and the overall biosensor performance. As a result, integrating AuNPs into biosensors can lead to more accurate and efficient detection of target molecules [137]. Hartati et al. developed an innovative label-free electrochemical sensor to detect *S. scrofa* Cyt B DNA, utilizing gold-electrodeposited screen-printed carbon electrodes as the platform. In their methodology, they incorporated a sophisticated streptavidin-biotinylated DNA probe system to immobilize target DNA, enhancing the specificity of the sensor (Fig. 4). The researchers employed targeted guanine signals to assess signal transduction, showcasing the sensor's ability to measure DNA concentrations accurately. Notably, this electrochemical sensor demonstrated a detection limit of 0.135 μg/mL and maintained a linear range between 0.5 and 1.5 μg/mL, indicating its effectiveness in identifying low DNA concentrations. Current response analysis conducted as part of the study revealed that the sensor could selectively detect 5% pork DNA amidst various meat samples, highlighting its robust performance in food authenticity testing. The authors proposed that their electrochemical biosensor

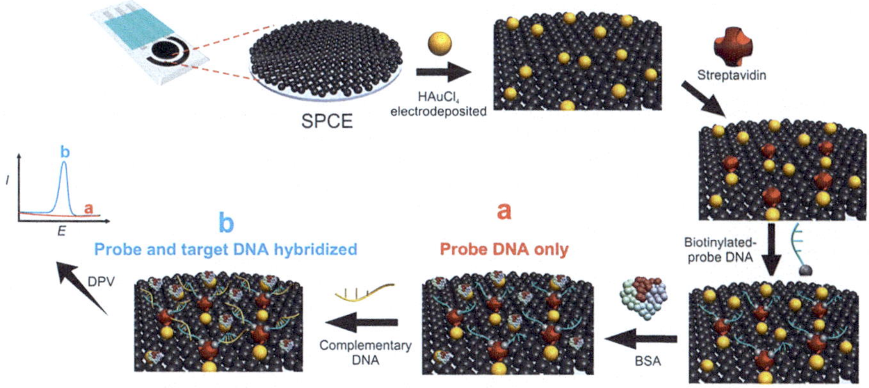

Fig. 4 Preparation of label-free electrochemical sensor to detect *S. scrofa* Cyt B DNA. (Adapted with permission [138])

holds significant potential for further development into a portable detection method, facilitating point-of-care testing for pork or other food adulterants and enhancing food safety and quality assurance [138].

Electrochemical Detection of Oil Substitution

Virgin olive oil's exceptional quality and significant commercial value are mainly due to its impressive nutritional profile and distinct biological benefits. However, virgin olive oil becomes a target for deliberate adulteration due to its limited production, high market prices, and increasing consumer demand for this flavorful and healthful oil. Unscrupulous practices often involve mixing it with lower-quality vegetable oils, driven by the potential for substantial financial gain. Common oils used in these fraudulent mixtures include sunflower oil, rapeseed oil, peanut oil, corn oil, walnut oil, and soybean oil, all of which compromise the integrity of the pure product. Another prevalent method of adulterating virgin olive oil entails combining cold-pressed, filtered oil with refined oil. This practice circumvents crucial stages typical to the oil refining process, such as neutralization, clarification, and smell absorption—resulting in an inferior product that falls short in quality and contains reduced levels of transfatty acids. Reputable organizations, including the EU Commission, the International Olive Council, and the Codex Committee on Fats and Oils, are focused on implementing strict quality standards for virgin olive oil to address this issue. They have established specific limits for various parameters that define olive oil quality. For instance, any rotation of virgin olive oil with soybean or walnut oil must result in fatty acid levels that do not exceed 5% [139]. Detecting adulteration is a sophisticated process that involves assessing specific chemical markers like tocopherols or oleuropein. Additionally, experts analyze the overall chemical composition of the sample to ascertain its authenticity [140]. Through these measures, the integrity of virgin olive oil can be preserved, ensuring that consumers receive a product that meets the high standards they expect.

Harzalli et al. developed an innovative electronic tongue that effectively integrates linear discriminant analysis (LDA) with a meta-heuristic algorithm to identify olive oil adulteration. This cutting-edge approach is particularly valuable for analyzing low-grade olive oil samples, enabling the detection of subtle changes in quality through their associated potentiometric fingerprints. Notably, this method demonstrated a high level of sensitivity, successfully identifying the presence of more than 2.5% rancid olive oil or 5% vinegary olive oil in extra virgin olive oil [141]. Santonico et al. developed an innovative sensor system to analyze gases and liquids, leveraging the unique properties of anthocyanins. This advanced system can differentiate samples originating from various cultivars and geographical backgrounds. By employing a combination of gas sensors, including six quartz crystal microbalances (QMBs) and a sophisticated liquid sensor, which features screen-printed gold (Au) voltammetric electrodes, the platform is adept not only at distinguishing high-quality extra virgin olive oils but also proficient at predicting the potential for adulteration. Notably, this multi-parametric approach effectively evaluates the organoleptic characteristics of the oils. It can detect adulterants at blending levels that traditional methods often fail to identify [142]. Karagözlü et al. conducted an electrochemical study examining alpha-tocopherol oxidation on a pencil graphite electrode (PGE) with rapeseed oil (RO), sunflower oil (SO), and corn oil (CO) through the DPV technique. This research aimed to identify counterfeiting parameters in cold-pressed olive oil (CPOO). Their voltammetric analysis of CPOO and its artificially mixed blends with commercial oils demonstrated a statistically significant ability to distinguish between them, enabling counterfeit detection at levels below 10% by weight [143]. Bougrini et al. created an e-tongue and e-nose to identify argan oil adulterated with sunflower oil utilizing a metal oxide semiconductor sensing platform and pattern recognition methods. They successfully distinguished between unadulterated and adulterated argan oil, attaining a 100% accuracy rate [144]. Rodrigues et al. investigated the impact of storage conditions on the quality of extra virgin olive oils. Their research highlighted a concerning trend: prolonged storage can significantly deteriorate the overall quality of these oils. To explore this phenomenon, the researchers employed cutting-edge electronic tongue technology complemented by the linear discriminant analysis (LDA) chemometric method. This sophisticated approach allowed them to uncover how typical commercial storage conditions, especially exposure to light, and the duration of storage can affect the physicochemical characteristics and the retronasal sensory attributes, including both olfactory and gustatory qualities. The innovative use of e-tongue technology in their electrochemical analysis proved to be highly effective in distinguishing between various samples of extra virgin olive oils [145].

Electrochemical Detection of Adulteration in Alcoholic Beverages

Accurately measuring aliphatic short-chain alcohols, particularly methanol and ethanol, is essential in the food industry. Methanol, specifically, is often recognized as an illicit additive applied in the fabrication of counterfeit spirits and wines, posing a serious risk to public health. When ingested, methanol can lead to severe intoxication because of the buildup of harmful metabolites, including formic acid and formaldehyde, which can have toxic effects on the body [146]. Other hazardous adulterants have been illicitly introduced into alcoholic beverages to enhance flavor, increase alcohol potency, and refine the overall packaging and production processes. Among these dangerous substances are urea, a chemical often used in fertilizers; fuel oil, typically associated with industrial applications; and heavy metals such as lead and arsenic, which pose serious health risks. Additionally, sedatives like benzodiazepines and barbiturates, along with furfural, an industrial solvent, and ethylene alcohol, a toxic compound, are also utilized. These additives not only compromise the safety of the products but can also result in severe health consequences for consumers [147]. Therefore, the development and implementation of straightforward analytical techniques are critical for effectively monitoring methanol levels in various contexts, such as evaluating environmental concentrations, ensuring the safety of alcoholic beverages, and conducting clinical diagnostic assessments. These methods are essential in safeguarding consumers and preserving the integrity of the food and beverage industry. Karimi et al. developed an electrochemical sensor based on palladium nanoparticle-Santa barbara amorphous-15-PrNHEtNH$_2$ (PdNPs@SBA-15-PrEn) to detect methanol via electro-oxidation of methanol in an alkaline solution. The sensor demonstrates a linear detection range between 20 and 1000 μM and an LOD of 12 mM [146]. Liveri et al. developed an innovative electrochemical sensor to detect and quantify apple juice and apple cider levels in white wine samples. Utilizing sophisticated statistical techniques like principal component analysis, the researchers effectively distinguished between the various sample types, successfully isolating distinct clusters representing white wine and apple products. Their advanced class modeling demonstrated a remarkable capability to identify adulteration within white wine, achieving an impressive detection limit of just 5% v/v or lower. This level of sensitivity varied based on the nature of the adulterant, whether it was apple juice or apple cider, showcasing the sensor's effectiveness in discerning even minor contamination levels [148]. In a significant study, Novakowski et al. employed CV techniques utilizing a copper working electrode to investigate the potential for whiskey adulteration. Their research focused on the ability to detect the presence of water and glucose in whiskey samples. By integrating principal component analysis (PCA) into their methodology, they successfully distinguished between high-end whiskey brands and lower-cost alternatives, showcasing the effectiveness of their approach in ensuring product authenticity [149]. Similarly, Wojcik et al. developed a quadruple-disk electrode using an innovative iridium. Their research specifically concentrated on analyzing the faradaic component of the voltammetric signal, which provided more profound insights into the electrochemical characteristics of the beverages being studied. By employing the advanced alternating trilinear decomposition (ATLD) algorithm, they could effectively differentiate between various types of wines and Scotch whiskies, highlighting the potential of their method for classifying spirits based on their unique electrochemical signatures [150]. Rosello et al. used CV with an SPCE alongside support vector machine discriminant analysis (SVMDA) to differentiate between two main types of beer, ales and lagers, based on their fermentation methods. They also employed partial least squares (PLS) regression and artificial neural networks to forecast the alcoholic content of the beers [151]. Lvova et al. used the potentiometric responses obtained from porphyrin-based solvent-polymeric membranes to assess a range of alcoholic beverages, including beer, wine, whisky, and grappa. Employing a partial least squares (PLS) model, they successfully quantified the alcohol content in these diverse drinks, demonstrating the effectiveness of their innovative approach to analyzing alcoholic beverages. Their research highlights the potential of advanced membrane technology in accurately measuring alcohol concentrations [152].

Conclusions and Future Outlook

Electrochemical sensors represent a groundbreaking and highly effective strategy in the ongoing battle against food adulteration, a pervasive issue impacting global food safety. These advanced sensors are celebrated for their remarkable sensitivity, affordability, and capability to deliver real-time analytical results, making them invaluable tools for monitoring food integrity. This chapter details the diverse applications of electrochemical sensors across various food products, underscoring significant advancements in electrode technology, which collectively facilitate the precise detection of harmful adulterants. Moreover, nanotechnology-based sensors and detection systems lead the charge in innovation, providing unprecedented sensitivity, specificity, and real-time analytical capabilities. These innovations aim to create environmentally sustainable and scalable solutions that address urgent food safety challenges worldwide. However, despite these advances, researchers face hurdles, particularly concerning the intricate composition of food matrices and the necessity for rigorous validation compared with traditional testing methods. Future research initiatives should prioritize the development of portable and user-friendly electrochemical devices. Such innovations would enhance accessibility for consumers and food safety inspectors, enabling efficient monitoring in various settings, from local markets to remote regions. Ultimately, electroanalytical techniques hold tremendous promise to bolster food authentication efforts, protect public health, and promote equitable economic practices within international marketplaces, ensuring that consumers receive safe and genuine food products.

References

1. Banti M (2020) Food adulteration and some methods of detection, review. Int J Nutr Food Sci 9(3):86–94
2. Bahrami A, Moaddabdoost Baboli Z, Schimmel K, Jafari SM, Williams L (2020) Efficiency of novel processing technologies for the control of listeria monocytogenes in food products. Trends Food Sci Technol 96:61–78
3. Maduwanthi S, Marapana R (2019) Induced ripening agents and their effect on fruit quality of banana. Int J Food Sci 2019(1):2520179
4. Panghal A, Yadav D, Khatkar BS, Sharma H, Kumar V, Chhikara N (2018) Post-harvest malpractices in fresh fruits and vegetables: food safety and health issues in India. Nutr Food Sci 48(4):561–578
5. Galvin-King P, Haughey SA, Elliott CT (2018) Herb and spice fraud; the drivers, challenges and detection. Food Control 88:85–97
6. Lubis HN, Mohd-Naim NF, Alizul NN, Ahmed MU (2016) From market to food plate: current trusted technology and innovations in halal food analysis. Trends Food Sci Technol 58:55–68
7. Okeke ES, Okagu IU, Okoye CO, Ezeorba TPC (2022) The use of calcium carbide in food and fruit ripening: potential mechanisms of toxicity to humans and future prospects. Toxicology 468:153112
8. Ali A, More TA, Shaikh Z (2021) Artificial sweeteners and their health implications: a review. Biosci Biotechnol Res Asia 18(2):227–237
9. Ghatak B, Naskar H, Pradhan S, Tudu B, Mukherji S, Bandyopadhyay R et al (2021) Electrochemical detection of important biomarker for artificial ripening of mango by polymethacrylic acid imprinted polymer sensor. IEEE Sensors J 21(5):5695–5702
10. Li J, Li S, Wei X, Tao H, Pan H (2012) Molecularly imprinted electrochemical luminescence sensor based on signal amplification for selective determination of trace gibberellin A3. Anal Chem 84(22):9951–9955
11. Islam MN, Imtiaz MY, Alam SS, Nowshad F, Shadman SA, Khan MS (2018) Artificial ripening on banana (Musa Spp.) samples: analyzing ripening agents and change in nutritional parameters. Cogent Food Agric 4(1):1477232
12. Hewajulige I, Premaseela H (2020) Fruit ripening: importance of artificial fruit ripening in commercial agriculture and safe use of the technology for consumer health. Sri Lanka J Food Agric 6(1):57
13. Stephen Inbaraj B, Chen BH (2016) Nanomaterial-based sensors for detection of foodborne bacterial pathogens and toxins as well as pork adulteration in meat products. J Food Drug Anal 24(1):15–28
14. Karabasanavar NS, Singh SP, Kumar D, Shebannavar SN (2014) Detection of pork adulteration by highly-specific PCR assay of mitochondrial D-loop. Food Chem 145:530–534
15. Ali ME, Hashim U, Mustafa S, Che Man YB, Adam T, Humayun Q (2014) Nanobiosensor for the detection and quantification of pork adulteration in meatball formulation. J Exp Nanosci 9(2):152–160
16. Dotel S, Pozo P (2023) Impact of the consumption of adulterated oils and alcoholic beverages. Int J Sci Res Publ 13:14024
17. Yadav S (2018) Edible oil adulterations: current issues, detection techniques, and health hazards. Int J Chem Stud 6:1393–1397
18. Mishra V, Mishra M, Ansari KM, Chaudhari BP, Khanna R, Das M (2012) Edible oil adulterants, argemone oil and butter yellow, as aetiological factors for gall bladder cancer. Eur J Cancer 48(13):2075–2085
19. Raghavan VS, O'Driscoll B, Bloor JM, Li B, Katare P, Sethi J et al (2021) Emerging graphene-based sensors for the detection of food adulterants and toxicants—a review. Food Chem 355:129547

20. Huang S-J, Gokulkumar K, Mani G, Lee Y-Y, Kogularasu S, Chang-Chien G-P (2024) Synthesis and characterization of Bi2S3-embedded carbon nanofibers as a novel electrochemical biosensor for the detection of mycotoxin zearalenone in food crops. FlatChem 45:100652
21. Mahato K, Wang J (2021) Electrochemical sensors: from the bench to the skin. Sensors Actuators B Chem 344:130178
22. Rizzotto F, Khalife M, Hou Y, Chaix C, Lagarde F, Scaramozzino N et al (2023) Recent advances in electrochemical biosensors for food control. Micromachines 14(7):1412
23. Kogularasu S, Lee Y-Y, Chang-Chien G-P, Chen P-Y, Govindasamy M (2024) A novel synthesis of nickel carbide modified glassy carbon electrode for electrochemical investigation of archetypal diabetes biomarker in human serum and urine samples. J Electrochem Soc 171(4):047512
24. Sakthivel K, Balasubramanian S, Chang-Chien G-P, Wang S-F, Ahammad, Billey W et al (2024) Advances in electrochemical sensors: improving food safety, quality, and traceability. ECS Sens Plus 3(2):020605
25. Saleh HM, Hassan AI (2023) Synthesis and characterization of nanomaterials for application in cost-effective electrochemical devices. Sustainability 15(14):10891
26. Ghaani M, Azimzadeh M, Büyüktaş D, Carullo D, Farris S (2024) Electrochemical sensors in the food sector: a review. J Agric Food Chem 72(44):24170–24190
27. Iqbal N, Khan NA, Ferrante A, Trivellini A, Francini A, Khan MIR (2017) Ethylene role in plant growth, development and senescence: interaction with other phytohormones. Front Plant Sci 8:475
28. BelBruno JJ (2018) Molecularly imprinted polymers. Chem Rev 119(1):94–119
29. Zarejousheghani M, Lorenz W, Vanninen P, Alizadeh T, Cämmerer M, Borsdorf H (2019) Molecularly imprinted polymer materials as selective recognition sorbents for explosives: a review. Polymers 11(5):888
30. Ghatak B, Naskar H, Ali SKB, Banerjee S, Chakraborty AK, Das N et al (2023) Dysprosium particles decorated ambroxol imprinted polymer sensor to detect carbide-treated mango. Sensors Actuators A Phys 358:114420
31. Kim W, Noh B-I, Cha YL, Chang Y-H, Dai S, Kim SH et al (2024) Development of molecularly imprinted polymer electrochemical sensors for strawberry sweetness biomarker detection. ACS Appl Polym Mater 6(14):8084–8092
32. Kilic NM, Singh S, Keles G, Cinti S, Kurbanoglu S, Odaci D (2023) Novel approaches to enzyme-based electrochemical Nanobiosensors. Biosensors 13(6):622
33. Yamaguchi H, Miyazaki M (2023) Enzyme-immobilized microfluidic devices for biomolecule detection. TrAC Trends Anal Chem 159:116908
34. Elias Abi-Ramia Silva T, Burisch F, Güntner AT (2024) Gas sensing for space: health and environmental monitoring. TrAC Trends Anal Chem 177:117790
35. Wang L-P, Jin Z, Luo T, Ding Y, Liu J-H, Wang X-F et al (2019) The detection of ethylene using porous ZnO nanosheets: utility in the determination of fruit ripeness. New J Chem 43(8):3619–3624
36. Sekhar PK, Ludwig T, Wilhelm M, Graf D, Riheen MA, Mathur S (2019) Potentiometric ethene sensor for postharvest detection applications. J Electrochem Soc 166(15):B1477
37. Takenaka S (2002) New method for ethephon ((2-chloroethyl)phosphonic acid) residue analysis, and detection of residual levels in the fruit and vegetables of Western Japan. J Agric Food Chem 50(26):7515–7519
38. Zhang Z, Zhai S, Wang M, He L, Peng D, Liu S et al (2015) Electrochemical sensor based on a polyaniline-modified SnO_2 nanocomposite for detecting ethephon. Anal Methods 7(11):4725–4733
39. Zhang N, Casida JE (2002) Novel irreversible butyrylcholinesterase inhibitors: 2-chloro-1-(substituted-phenyl)ethylphosphonic acids. Bioorg Med Chem 10(5):1281–1290
40. Li S, Pang C, Ma X, Li H, Zhao M, Liu C et al (2020) A gold nanoparticle-loaded molecularly imprinted switch sensor with high sensitivity to ethephon. Microchem J 157:105025

41. Wang X, Dong S, Wei H (2023) Recent advances on nanozyme-based electrochemical biosensors. Electroanalysis 35(1):e202100684
42. Tripathi A, Harris KD, Elias AL (2020) Peroxidase-like behavior of Ni thin films deposited by glancing angle deposition for enzyme-free uric acid sensing. ACS Omega 5(16):9123–9130
43. Campuzano S, Pedrero M, Yáñez-Sedeño P, Pingarrón JM (2020) Nanozymes in electrochemical affinity biosensing. Microchim Acta 187(8):423
44. Weerathunge P, Ramanathan R, Torok VA, Hodgson K, Xu Y, Goodacre R et al (2019) Ultrasensitive colorimetric detection of murine norovirus using nanozyme aptasensor. Anal Chem 91(5):3270–3276
45. Ji C, Tang X, Wen R, Xu C, Wei J, Han B et al (2024) A multienzyme reaction-mediated electrochemical biosensor for sensitive detection of organophosphorus pesticides. Biosensors 14(2):62
46. Shu J, Tang D (2020) Recent advances in photoelectrochemical sensing: from engineered photoactive materials to sensing devices and detection modes. Anal Chem 92(1):363–377
47. Wang J, Liu Z (2020) Recent advances in two-dimensional layered materials for photoelectrochemical sensing. TrAC Trends Anal Chem 133:116089
48. Yan X-R, Li J, Kong L-F, Li M-Y, Li H-L, Qian C et al (2021) AuNPs/g-C3N4/BiOCl0.5Br0.5 heterojunction sensitizing sensor for photoelectrochemical sensing of 2-chloroethyl phosphoric acid. Chin J Anal Chem 49(5):798–808
49. Vildanova M, Vishnyakova P, Saidova A, Konduktorova V, Onishchenko G, Smirnova E (2021) Gibberellic acid initiates ER stress and activation of differentiation in cultured human immortalized keratinocytes HaCaT and epidermoid carcinoma cells A431. Pharmaceutics 13(11):1811
50. Ghanbary E, Asiabani Z, Hosseini N, Kiaie SH, Kaki S, Ghasempour H et al (2020) The development of a new modified graphite pencil electrode for quantitative detection of gibberellic acid (GA3) herbal hormone. Microchem J 157:105005
51. Kuthati Y, Kankala RK, Lee C-H (2015) Layered double hydroxide nanoparticles for biomedical applications: current status and recent prospects. Appl Clay Sci 112-113:100–116
52. Zhang J, Zhu J, Chao J (2023) Recent advances in DNA-based electrogenerated chemiluminescence biosensors. Sens Diagn 2(3):582–599
53. Liu H, Yin H, Dong Y, Ding H, Chu X (2020) Electrochemiluminescence resonance energy transfer between luminol and black phosphorus nanosheets for the detection of trypsin via the "off–on–off" switch mode. Analyst 145(6):2204–2211
54. Xie H-Z, Yang B, Li J-P (2020) A molecularly imprinted electrochemical luminescence sensor for detection of gibberellin based on energy transfer. Chin J Anal Chem 48(12):1633–1641
55. Yusaf T, Mahamude ASF, Farhana K, Harun WSW, Kadirgama K, Ramasamy D et al (2022) A comprehensive review on graphene nanoparticles: preparation, properties, and applications. Sustainability 14(19):12336
56. Sheikh-Mohseni MA (2016) Sensitive electrochemical determination of gallic acid: application in estimation of total polyphenols in plant samples. Anal Bioanal Chem Res 3(2):217–224
57. Sparavigna AC (2023) Iron oxide Fe_3O_4 nanoparticles for electromagnetic shielding
58. Neto DMA, da Costa LS, Sousa CP, Becker H, Casciano PNS, Nascimento HO et al (2022) Functionalized Fe_3O_4 nanoparticles for electrochemical sensing of carbendazim. Electrochim Acta 432:141193
59. Mollarasouli F, Zor E, Ozcelikay G, Ozkan SA (2021) Magnetic nanoparticles in developing electrochemical sensors for pharmaceutical and biomedical applications. Talanta 226:122108
60. Pérez-Ferreiro M, Abelairas A, Criado A, Gómez IJ, Mosquera J (2023) Dendrimers: exploring their wide structural variety and applications. Polymers 15(22):4369
61. Fu C, Li J-P (2014) A molecular-imprinted sensor for trace detection of gibberellin based on Ferrocenecarboxylic acid multiply marked dendrimer. Chin J Anal Chem 42(3):315–319
62. Santos LKB, Mendonça PD, Assis LKS, Prudêncio CR, Guedes MIF, Marques ETA et al (2022) A redox-probe-free Immunosensor based on Electrocatalytic Prussian blue nanostructured film one-step-prepared for Zika virus diagnosis. Biosensors 12(8):623

63. Er ZS (2023) Polydopamine-coated hexagonal boron nitride-based electrochemical immunosensing of T-tau as a marker of Alzheimer's disease. Bioelectrochemistry 154:108552
64. Kim E, Winkler TE, Kitchen C, Kang M, Banis G, Bentley WE et al (2017) Redox probing for chemical information of oxidative stress. Anal Chem 89(3):1583–1592
65. Lee H, Rho J, Messersmith PB (2009) Facile conjugation of biomolecules onto surfaces via mussel adhesive protein inspired coatings. Adv Mater 21(4):431–434
66. You Y, Luo B, Wang C, Dong H, Wang X, Hou P et al (2023) An ultrasensitive probe-free electrochemical immunosensor for gibberellins employing polydopamine-antibody nanoparticles modified electrode. Bioelectrochemistry 150:108331
67. Li J, Xiao L-T, Zeng G-M, Huang G-H, Shen G-L, Yu R-Q (2005) Immunosensor for rapid detection of gibberellin acid in the Rice grain. J Agric Food Chem 53(5):1348–1353
68. Prabhu K, Malode SJ, Shetti NP, Pandiaraj S, Alodhayb AN (2024) MCNTs/ZnO nanocomposite-fabricated nanosensor for 1-naphthyleacetic acid electroanalysis and its applications. Microchem J 199:110239
69. Fernández-Argüelles MT, Cañabate B, Segura A, Costa JM, Pereiro R, Sanz-Medel A et al (2005) Flow-through optosensing of 1-naphthaleneacetic acid in water and apples by heavy atom induced–room temperature phosphorescence measurements. Talanta 66(3):696–702
70. Karimi-Maleh H, Cellat K, Arıkan K, Savk A, Karimi F, Şen F (2020) Palladium–Nickel nanoparticles decorated on functionalized-MWCNT for high precision non-enzymatic glucose sensing. Mater Chem Phys 250:123042
71. Wang J (2012) Electrochemical biosensing based on noble metal nanoparticles. Microchim Acta 177(3):245–270
72. Lü S (2003) Voltammetric determination of 1-naphthylacetic acid in soil samples using carbon nanotubes film modified electrode. Anal Lett 36(8):1523–1534
73. Xu X, Li S, Luan X, Xuan C, Zhao P, Zhou T et al (2023) Sensitivity enhancement of a cu (II) metal organic framework-acetylene black-based electrochemical sensor for ultrasensitive detection of imatinib in clinical samples. Front Chem 11:1191075
74. Huang W, Qu W, Zhu D (2008) Electrochemistry and determination of 1-naphthylacetic acid using an acetylene black film modified electrode. Bull Korean Chem Soc 29(7):1323–1325
75. Bruch R, Baaske J, Chatelle C, Meirich M, Madlener S, Weber W et al (2019) CRISPR/Cas13a-powered electrochemical microfluidic biosensor for nucleic acid amplification-free miRNA diagnostics. Adv Mater 31(51):1905311
76. Liang S, Pan J, Yin X, Sui K (2015) Research article the development of an electrochemical Immunosensory for 1-naphthylacetic acid. Adv J Food Sci Technol 9(6):439–443
77. Lourencao BC, Brocenschi RF, Medeiros RA, Fatibello-Filho O, Rocha-Filho RC (2020) Analytical applications of electrochemically pretreated boron-doped diamond electrodes. ChemElectroChem 7(6):1291–1311
78. Janíková L, Chýlková J, Šelešovská R, Sedlák M, Váňa J, Dušek L et al (2020) Electrochemical behavior of plant growth stimulator 1-naphthaleneacetic acid and its voltammetric determination using boron doped diamond electrode. J Electroanal Chem 859:113855
79. Mishra M, Ali S, Das M (2014) Analysis of oxytocin in milk samples and intake pattern in different age groups of Indian population. Toxicol Mech Methods 24(5):342–346
80. Li Y, Pu Z, Sun Q, Pan N (2021) A review on novel activation strategy on carbonaceous materials with special morphology/texture for electrochemical storage. J Energy Chem 60:572–590
81. Salah LS, Ouslimani N, Bousba D, Huynen I, Danlée Y, Aksas H (2021) Carbon nanotubes (CNTs) from synthesis to functionalized (CNTs) using conventional and new chemical approaches. J Nanomater 2021(1):4972770
82. Mehrotra S, Rai P, Gautam K, Saxena A, Verma R, Lahane V et al (2023) Chitosan-carbon nanofiber based disposable bioelectrode for electrochemical detection of oxytocin. Food Chem 418:135965
83. Zhang Z, Bai W-L, Wang K-X, Chen J-S (2020) Electrocatalyst design for aprotic Li–CO_2 batteries. Energy Environ Sci 13(12):4717–4737

84. Thomas R, Balachandran M (2023) Heteroatom engineered graphene-based electrochemical assay for the quantification of high-risk abused drug oxytocin in edibles and biological samples. Food Chem 400:134106
85. Abedeen MZ, Laddha H, Sharma M, Gupta R, Kushwaha HS (2023) Non-enzyme picomolar sensing of acephate by modified glassy carbon electrode using bimetallic Zn-cu metal-organic framework. J Electroanal Chem 948:117810
86. Liu K, Chen Y, Dong X, Hu Y, Huang H (2023) Bimetallic FeCo metal-organic-frameworks anchored multi-walled carbon nanotubes for electrochemical nitrite sensing. Electrochim Acta 456:142441
87. Narayanan M, Singh Chauhan NP, Perumal P (2023) A highly efficient metal oxide incorporated metal organic framework [Nd2O3-MIL(Fe)-88A] for the electrochemical detection of dichlorvos. RSC Adv 13(8):5565–5575
88. Balasubramaniyan NG, Perumal P (2024) Highly efficient electrochemical detection of H2O2 utilizing an innovative copper porphyrinic nanosheet decorated bismuth metal–organic framework modified electrode. Anal Methods 16(4):624–638
89. Abedeen MZ, Sharma M, Kushwaha HS, Gupta R (2025) Tuning the electro-catalytic activity of the Zn–cu MOF/rGO nanocomposite as a novel enzyme-free electrochemical sensor for the detection of the oxytocin hormone. Analyst 150(1):154–167
90. Al-Taee AT, Al-Hyali RH (2021) Electrochemical behavior of oxytocin hormone through its cysteine reduction peak using glassy carbon electrode modified with poly Furfurylamine and multi-walled carbon nanotubes. Egypt J Chem 64(10):5831–5837
91. Rai M, Nema A, Bano M, Dangi SS, Shrivastava K, Khan F (2021) Hierarchical copper nanocomposites and their applications for catalytic reduction of o-nitroaniline to 1,2-benzenediamine and sensor activities against oxytocin. J Mater Sci Mater Electron 32(20):25149–25163
92. Iwuozor KO, Anyanwu VU, Olaniyi BO, Mbamalu PS, Adeniyi AG (2022) Adulteration of sugar: a growing global menace. Sugar Tech 24(3):914–919
93. Grembecka M (2015) Sugar alcohols—their role in the modern world of sweeteners: a review. Eur Food Res Technol 241(1):1–14
94. Puttegowda SKB, Shivaramu M, Manjunath BS, Venkataraman R, Nagarajappa RB (2025) Understanding artificial sweeteners and food colorants-their impact on human health: a review. Indian J Pharm Pract 18(1):8–15
95. Spencer M, Gupta A, Dam LV, Shannon C, Menees S, Chey WD (2016) Artificial sweeteners: a systematic review and primer for gastroenterologists. J Neurogastroenterol Motil 22(2):168–180
96. Ghusn W, Naik R, Yibirin M (2023) The impact of artificial sweeteners on human health and cancer association: a comprehensive clinical review. Cureus 15(12):e51299
97. Dickerson AS, Lee JS, Keshava C, Hotchkiss A, Persad AS (2018) Assessment of health effects of exogenous urea: summary and key findings. Curr Environ Health Rep 5(2):205–212
98. Alabi OA, Ologbonjaye KI, Awosolu O, Alalade OE (2019) Public and environmental health effects of plastic wastes disposal: a review. J Toxicol Risk Assess 5(021):1–13
99. Fakhlaei R, Selamat J, Khatib A, Razis AFA, Sukor R, Ahmad S et al (2020) The toxic impact of honey adulteration: a review. Foods 9(11):1538
100. Naila A, Flint SH, Sulaiman AZ, Ajit A, Weeds Z (2018) Classical and novel approaches to the analysis of honey and detection of adulterants. Food Control 90:152–165
101. Wang S, Guo Q, Wang L, Lin L, Shi H, Cao H et al (2015) Detection of honey adulteration with starch syrup by high performance liquid chromatography. Food Chem 172:669–674
102. Zazoum B, Batoo KM, Khan MAA (2022) Recent advances in flexible sensors and their applications. Sensors 22(12):4653
103. Escobedo P, Bhattacharjee M, Nikbakhtnasrabadi F, Dahiya R (2021) Flexible strain and temperature sensing NFC tag for smart food packaging applications. IEEE Sensors J 21(23):26406–26414

104. Chugh V, Bhattacharjya T, Das C, Tripathy K, Bhattacharjee M (2024) Flexible electrochemical sensing label for the detection of glucose adulteration in honey. IEEE Sensors J 24(9):13823–13830
105. Bougrini M, Tahri K, Saidi T, El Hassani EA, Bouchikhi B, El Bari N (2016) Classification of honey according to geographical and botanical origins and detection of its adulteration using voltammetric electronic tongue. Food Anal Methods 9(8):2161–2173
106. Veloso ACA, Sousa ME, Estevinho L, Dias LG, Peres AM (2018) Honey evaluation using electronic tongues: an overview. Chem 6(3):28
107. Di Rosa AR, Marino AMF, Leone F, Corpina GG, Giunta RP, Chiofalo V (2018) Characterization of Sicilian honeys pollen profiles using a commercial E-tongue and Melissopalynological analysis for rapid screening: a pilot study. Sensors 18(11):4065
108. Podrażka M, Bączyńska E, Kundys M, Jeleń PS, Witkowska NE (2018) Electronic tongue—a tool for all tastes? Biosensors 8(1):3
109. Pérez-Ràfols C, Serrano N, Ariño C, Esteban M, Díaz-Cruz JM (2019) Voltammetric electronic tongues in food analysis. Sensors 19(19):4261
110. Escriche I, Kadar M, Domenech E, Gil-Sánchez L (2012) A potentiometric electronic tongue for the discrimination of honey according to the botanical origin. Comparison with traditional methodologies: physicochemical parameters and volatile profile. J Food Eng 109(3):449–456
111. Ciursa P, Oroian M (2021) Voltammetric E-tongue for honey adulteration detection. Sensors 21(15):5059
112. Revenga-Parra M, Robledo SN, Martínez-Periñán E, González-Quirós MM, Colina A, Heras A et al (2020) Direct determination of monosaccharides in honey by coupling a sensitive new Schiff base Ni complex electrochemical sensor and chemometric tools. Sensors Actuators B Chem 312:127848
113. Song X, She S, Xin M, Chen L, Li Y, Heyden YV et al (2020) Detection of adulteration in Chinese monofloral honey using 1H nuclear magnetic resonance and chemometrics. J Food Compos Anal 86:103390
114. Li Y, Huang Y, Xia J, Xiong Y, Min S (2020) Quantitative analysis of honey adulteration by spectrum analysis combined with several high-level data fusion strategies. Vib Spectrosc 108:103060
115. Berriel V, Perdomo C (2016) Determination of high fructose corn syrup concentration in Uruguayan honey by 13C analyses. LWT 73:649–653
116. Huang T-K, Chuang M-C, Kung Y, Hsieh B-C (2021) Impedimetric sensing of honey adulterated with high fructose corn syrup. Food Control 130:108326
117. Singh R, Singh M (2023) Molecularly imprinted electrochemical sensor for highly selective and sensitive determination of artificial sweetener Acesulfame-K. Talanta Open 7:100194
118. Landrigan PJ, Straif K (2021) Aspartame and cancer—new evidence for causation. Environ Health 20(1):42
119. Manuel M, Kanchi S (2025) Construction of a waste-derived graphite electrode integrated IL/Ni-MOF flowers/Co3O4 NDs for specific enrichment and signal amplification to detect aspartame. Colloids Surf A Physicochem Eng Asp 705:135648
120. Su J, Su X (2024) A multi-walled carbon nanotubes-based molecular imprinting electrochemical sensor for the detection of aspartame in sports beverage. J Food Meas Charact 18(1):618–624
121. Sha R, Komori K, Badhulika S (2017) Graphene–Polyaniline composite based ultra-sensitive electrochemical sensor for non-enzymatic detection of urea. Electrochim Acta 233:44–51
122. Trasande L, Shaffer RM, Sathyanarayana S, Lowry JA, Ahdoot S, Baum CR et al (2018) Food additives and child health. Pediatrics 142:2
123. Zhou Y, Li X, Pan Z, Ye B, Xu M (2019) Determination of malachite green in fish by a modified MOF-based electrochemical sensor. Food Anal Methods 12(5):1246–1254
124. Darabi R, Shabani-Nooshabadi M, Karimi-Maleh H, Gholami A (2022) The potential of electrochemistry for one-pot and sensitive analysis of patent blue V, tartrazine, acid violet 7 and ponceau 4R in foodstuffs using IL/cu-BTC MOF modified sensor. Food Chem 368:130811

125. Ji L, Peng L, Chen T, Li X, Zhu X, Hu P (2022) Facile synthesis of Fe-BTC and electrochemical enhancement effect for sunset yellow determination. Talanta Open 5:100084
126. Lv R, Sun R, Du T, Li Y, Chen L, Zhang Y et al (2022) Cu2+ modified Zr-based metal organic framework-CTAB-graphene for sensitive electrochemical detection of sunset yellow. Food Chem Toxicol 166:113250
127. Kumari R, Kumar H, Sharma R, Yadav A, Kumar G, Tundwal A et al (2023) Highly sensitive amperometric food sensor for Sudan-I dye using nanocomposites modified working electrode. Microchem J 193:109078
128. Grundy HH, Brown LC, Romero MR, Donarski JA (2023) Review: methods to determine offal adulteration in meat products to support enforcement and food security. Food Chem 399:133818
129. Fu J (2024) Research on meat adulteration based on one-dimensional convolutional neural network (1DCNN) combined with electrochemical technology. J Food Meas Charact 18(7):5720–5728
130. Mandli J, El Fatimi I, Seddaoui N, Amine A (2018) Enzyme immunoassay (ELISA/immunosensor) for a sensitive detection of pork adulteration in meat. Food Chem 255:380–389
131. Ruiz-Valdepeñas Montiel V, Gutiérrez ML, Torrente-Rodríguez RM, Povedano E, Vargas E, Reviejo ÁJ et al (2017) Disposable amperometric polymerase chain reaction-free biosensor for direct detection of adulteration with horsemeat in raw lysates targeting mitochondrial DNA. Anal Chem 89(17):9474–9482
132. Flauzino JMR, Nguyen EP, Yang Q, Rosati G, Panáček D, Brito-Madurro AG et al (2022) Label-free and reagentless electrochemical genosensor based on graphene acid for meat adulteration detection. Biosens Bioelectron 195:113628
133. Zhang Y, Wang W, Lin Z, Liu B, Zhou X (2020) Dual-output toehold-mediated strand displacement amplification for sensitive homogeneous electrochemical detection of speciespecific DNA sequences for species identification. Biosens Bioelectron 161:112256
134. de Faria R, Iden H, Bharucha E, Lins V, Messaddeq Y, Matencio T et al (2018) A new tool for the detection of horsemeat adulteration in raw meat. J Biosens Bioelectron 9:264
135. Hartati YW, Setiawati TA, Sofyatin T, Fitrilawati F, Anggraeni A, Gaffar S (2020) Electrochemical DNA biosensor for detection of pork (Susscrofa) using screen printed carbon-reduced graphene oxide electrode. Science 46(1):72–79
136. Lim SA, Ahmed MU (2016) A label free electrochemical immunosensor for sensitive detection of porcine serum albumin as a marker for pork adulteration in raw meat. Food Chem 206:197–203
137. Yudin Kharismasari C, Irkham, Zein MHL, Hardianto A, Nur Zakiyyah S, Umar Ibrahim A et al (2024) CRISPR/Cas12-based electrochemical biosensors for clinical diagnostic and food monitoring. Bioelectrochemistry 155:108600
138. Hartati YW, Irkham I, Sumiati I, Wyantuti S, Gaffar S, Zakiyyah SN et al (2023) The optimization of a label-free electrochemical DNA biosensor for detection of Sus scrofa mtDNA as food adulterations. Biosensors 13(6):657
139. Salah WA, Nofal M (2021) Review of some adulteration detection techniques of edible oils. J Sci Food Agric 101(3):811–819
140. Meenu M, Cai Q, Xu B (2019) A critical review on analytical techniques to detect adulteration of extra virgin olive oil. Trends Food Sci Technol 91:391–408
141. Harzalli U, Rodrigues N, Veloso ACA, Dias LG, Pereira JA, Oueslati S et al (2018) A taste sensor device for unmasking admixing of rancid or winey-vinegary olive oil to extra virgin olive oil. Comput Electron Agric 144:222–231
142. Santonico M, Grasso S, Genova F, Zompanti A, Parente FR, Pennazza G (2015) Unmasking of olive oil adulteration via a multi-sensor platform. Sensors 15(9):21660–21672
143. Karagözlü M, Adun P, Ozsoz M, Aşır S (2024) Detection of adulteration in cold-pressed olive oil by voltammetric analysis of alpha-tocopherol on single-use pencil graphite electrode. Int J Food Sci Technol 59(3):1781–1788

144. Bougrini M, Tahri K, Haddi Z, Saidi T, El Bari N, Bouchikhi B (2014) Detection of adulteration in argan oil by using an electronic nose and a voltammetric electronic tongue. J Sens 2014(1):245831
145. Rodrigues N, Dias LG, Veloso ACA, Pereira JA, Peres AM (2017) Evaluation of extra-virgin olive oils shelf life using an electronic tongue—chemometric approach. Eur Food Res Technol 243(4):597–607
146. Karimi Z, Shamsipur M, Tabrizi MA, Rostamnia S (2018) A highly sensitive electrochemical sensor for the determination of methanol based on PdNPs@SBA-15-PrEn modified electrode. Anal Biochem 548:32–37
147. Thiruvengadam M, Kim J-T, Kim W-R, Kim J-Y, Jung B-S, Choi H-J et al (2014) Safeguarding public health: advanced detection of food adulteration using nanoparticle-based sensors. Crit Rev Anal Chem 13:1–21
148. Liveri M, Tsantili-Kakoulidou A, Tsopelas F (2024) Identification of white wine adulteration with apple juice and apple cider using cyclic voltammetry on a screen-printed electrode aided by chemometric analysis. J Food Compos Anal 136:106751
149. Novakowski W, Bertotti M, Paixão TRLC (2011) Use of copper and gold electrodes as sensitive elements for fabrication of an electronic tongue: discrimination of wines and whiskies. Microchem J 99(1):145–151
150. Wójcik S, Kopeć J, Baś B, Jakubowska M (2019) A new approach in voltammetric profiling of wines and whiskies based on a useful faradaic signal component. Anal Methods 11(47):5984–5996
151. Roselló A, Serrano N, Díaz-Cruz JM, Ariño C (2021) Discrimination of beers by cyclic voltammetry using a single carbon screen-printed electrode. Electroanalysis 33(4):864–872
152. Lvova L, Paolesse R, Di Natale C, D'Amico A (2006) Detection of alcohols in beverages: an application of porphyrin-based electronic tongue. Sensors Actuators B Chem 118(1):439–447

Advanced Liquid Chromatography Techniques for Detecting and Analyzing Food Adulteration: Ensuring Safety and Quality in Food Products

Monika Sindhu and Arpan Tewary

Contents

M. Sindhu (✉)
University Centre for Research and Development, Chandigarh University, Mohali, Punjab, India

A. Tewary
Indian Institute of Solar Energy, Thane, Maharashtra, India

W. Alahmad, M. Ashfaq (eds.), *Advanced Techniques against Food Adulteration*,
https://doi.org/10.1007/978-3-031-99454-8_8

Abstract Food adulteration poses significant risks to public health and safety, necessitating the development of sophisticated analytical techniques for detecting and preventing the introduction of unauthorized substances in food products. Liquid chromatography (LC), particularly its advanced forms such as high-performance liquid chromatography (HPLC) and ultra-high-performance liquid chromatography (UHPLC), has emerged as a powerful tool in the fight against food adulteration. This abstract explores the application of these advanced chromatographic techniques in identifying and quantifying adulterants in various food matrices. HPLC and UHPLC provide high-resolution separation of complex mixtures, enabling the effective analysis of both known and unknown adulterants. These methods offer enhanced sensitivity and specificity, which are crucial for detecting trace levels of contaminants, including synthetic dyes, pesticides, heavy metals, and unapproved additives. The integration of mass spectrometry (MS) with LC techniques further amplifies detection capabilities, allowing for precise identification of adulterants based on their molecular structures. Recent advancements in sample preparation techniques, such as solid-phase extraction (SPE) and microextraction methods, have also improved the reliability and efficiency of LC in food safety applications. These techniques minimize sample matrix interference, ensuring an accurate assessment of food quality. Furthermore, emerging methodologies, including ultra-fast LC and two-dimensional LC, promise increased throughput and comprehensive profiling of food constituents. Overall, the advancement of liquid chromatography techniques plays a vital role in safeguarding food integrity. By enabling rapid and reliable detection of food adulterants, these methods enhance consumer safety, support regulatory compliance, and bolster confidence in food quality. Continuous innovation and integration of LC technologies will be essential for adapting to the evolving landscape of food adulteration practices.

Keywords Food adulteration · Liquid chromatography · Mass spectrometry · Microextraction

Introduction

Food adulteration has emerged as a pressing issue in contemporary food systems, posing significant risks to consumer health and challenging regulatory frameworks worldwide. As global food supply chains become increasingly complex, intentional, or unintentional adulteration opportunities multiply [1]. Adulteration encompasses adding harmful substances, substituting inferior-quality ingredients, or misrepresenting products to deceive consumers and increase profitability [2]. These unethical practices undermine public trust and often lead to severe health consequences, ranging from mild gastrointestinal discomfort to chronic illnesses such as cancer or cardiovascular disorders. In many instances, the economic motives driving food adulteration are exacerbated by inadequate regulatory oversight, insufficient testing mechanisms, and a lack of consumer awareness [3]. Examples of food adulteration are numerous and diverse, spanning categories such as dairy, spices, oils, grains, beverages, and meat products (Fig. 1) [4]. For instance, the addition of melamine in milk to artificially inflate protein content or the substituting of saffron with synthetic dyes are well-documented cases [5]. These instances not only compromise the nutritional value of food products but also pose significant health hazards, particularly to vulnerable populations such as children and the elderly. Addressing this growing concern necessitates developing and implementing robust analytical techniques capable of detecting and quantifying adulterants with high precision and reliability.

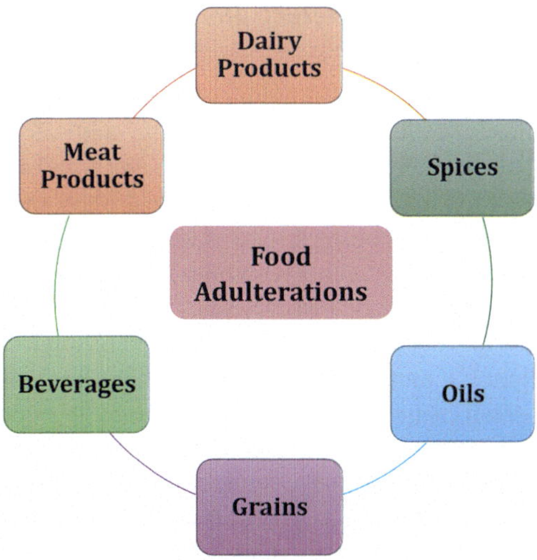

Fig. 1 Illustrating "Food Adulterations" at the center, connected to six categories: "Dairy Products," "Meat Products," "Beverages," "Spices," "Oils," and "Grains." Each category is in a colored rectangle, linked in a circular arrangement around the central topic

Importance of Food Safety and Quality Assurance

Ensuring food safety and quality is paramount in safeguarding public health, sustaining consumer confidence, and supporting economic stability. Quality assurance systems aim to maintain the integrity of food products from production to consumption, ensuring compliance with established standards and minimizing the risk of contamination or adulteration [6]. With the increasing globalization of the food trade, regulatory agencies such as the World Health Organization (WHO) and Food and Agriculture Organization (FAO) emphasize the need for rigorous testing and certification protocols to address food safety concerns [7].

Foodborne illnesses remain a significant public health challenge, with millions of cases reported annually. Contaminated or adulterated food is a major contributor to this burden, underscoring the critical need for effective detection mechanisms. Quality assurance involves not only identifying adulterants but also ensuring compliance with nutritional standards, assessing shelf-life stability, and verifying authenticity claims [8]. For example, labeling organic or non-GMO food products must be substantiated by analytical evidence to maintain consumer trust. The economic implications of food adulteration further highlight the importance of quality assurance. Recalling adulterated products, addressing legal repercussions, and managing reputational damage can impose significant financial strain on food manufacturers. Moreover, consumer preferences are increasingly shifting toward transparency and traceability in food production, further emphasizing the need for sophisticated analytical techniques to meet these expectations. In this context, advancements in LC offer a promising avenue for addressing the multifaceted challenges associated with food adulteration detection and quality assurance.

Role of Chromatographic Techniques in Food Adulteration Detection

Chromatographic techniques have revolutionized analytical chemistry by providing unparalleled precision, sensitivity, and versatility in detecting and quantifying diverse chemical entities. Among these, LC stands out as a cornerstone technology for food analysis, offering a robust platform for identifying adulterants and contaminants in complex food matrices [9]. The ability of LC to separate, identify, and quantify components with high resolution has made it an indispensable tool in food safety laboratories worldwide.

Advanced LC techniques, such as high-performance liquid chromatography (HPLC) and ultra-performance liquid chromatography (UPLC), have further enhanced the efficiency and accuracy of food adulteration detection [10]. These techniques enable the identification of minute quantities of adulterants, such as pesticides, heavy metals, synthetic dyes, and pharmaceutical residues, which might elude conventional detection methods [11]. By coupling LC with advanced detection systems such as mass spectrometry (MS) or fluorescence detectors, analysts can achieve greater specificity and sensitivity, enabling the detection of even trace levels of adulterants, as shown in Fig. 2.

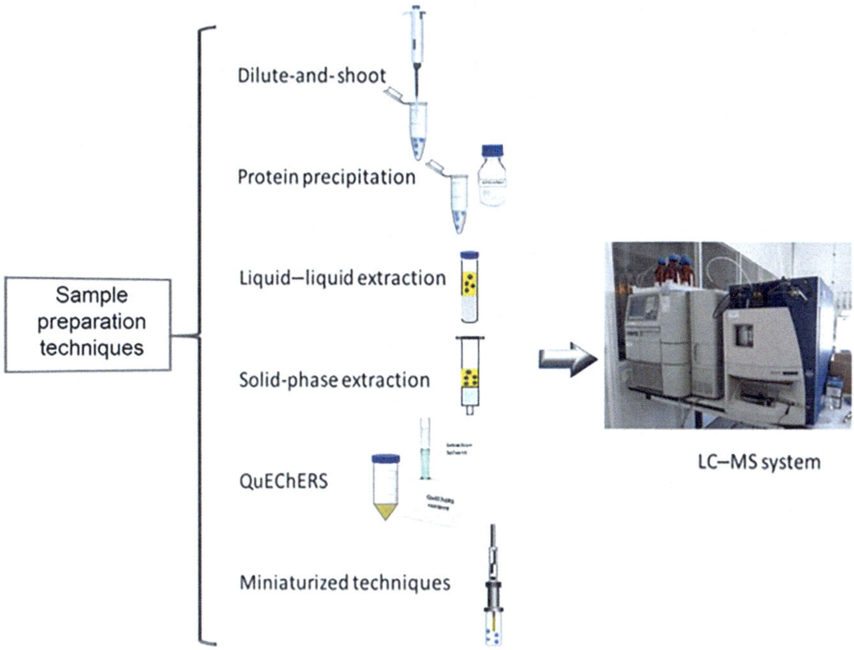

Fig. 2 Illustrating various sample preparation techniques leading to an LC-MS system. Techniques include dilute-and-shoot, protein precipitation, liquid-liquid extraction, solid-phase extraction, QuEChERS, and miniaturized techniques. Each method is depicted with corresponding laboratory equipment icons. An arrow points to an image of an LC-MS system, indicating the final analysis step. Reprinted from [12] with the permission from Elsevier

One of the key advantages of chromatographic techniques is their applicability across a wide range of food products. For instance, HPLC has been effectively employed to detect adulterants in edible oils, identify unauthorized additives in beverages, and confirm the authenticity of high-value products like honey and saffron. Similarly, gas chromatography (GC) combined with mass spectrometry (GC-MS) is widely used to detect volatile and semi-volatile adulterants in spices and essential oils [12]. These methods enhance the reliability of food safety assessments and support regulatory compliance by providing legally defensible evidence of adulteration.

The development of portable chromatography systems has further expanded the applicability of this technology, enabling on-site testing and real-time decision-making in food production and distribution chains. This is particularly valuable in mitigating the risks associated with food fraud and ensuring the timely recall of adulterated products. Additionally, software and data analytics advancements have facilitated the interpretation of complex chromatographic data, enabling more efficient identification of adulterants and trends in food adulteration [13]. In conclusion, the role of advanced LC techniques in detecting and analyzing food adulteration is indispensable for ensuring food safety and quality. By addressing the limitations of conventional methods and providing a high degree of reliability, these techniques contribute to a safer and more transparent food supply chain. The subsequent sections of this chapter will delve deeper into the principles, methodologies, and applications of LC in food adulteration detection, highlighting recent advancements and future directions in this critical area of food science.

Fundamentals of LC

Liquid chromatography (LC) is an indispensable analytical tool in modern food safety and quality assurance, offering unparalleled precision in separating, identifying, and quantifying complex mixtures [14]. Its utility extends to detecting food adulteration, identifying contaminants, verifying authenticity, and ensuring the safety and quality of food products. This section outlines the principles, components, and types of LC, focusing on its advanced applications in food adulteration analysis.

Principles of LC

At its core, LC operates on the principle of differential partitioning, wherein analytes exhibit varying affinities toward a stationary phase and a mobile phase. This differential interaction results in the separation of components as they traverse the chromatographic column at different rates. The separation process depends on polarity, molecular size, ionic charge, and hydrophobic interactions [15]. The stationary phase in LC typically consists of solid or liquid-coated particles with specific chemical properties housed within a column. The mobile phase, a liquid solvent or solvent mixture, transports the analytes through the column. As analytes interact with the stationary phase, their migration slows relative to the flow of the mobile phase, creating distinct separation patterns. The retention time, the time taken for an analyte to elute from the column, is a key parameter for identification and quantification. Gradient elution, a technique involving systematic variation of the mobile phase composition during separation, enhances LC's versatility. It enables the efficient resolution of analytes with widely varying affinities for the stationary phase. Advances in LC technology, including high-pressure systems and refined stationary phases, have further optimized separation efficiency, making LC a cornerstone of food adulteration detection [16].

Components of LC Systems

An LC system comprises several integral components that collectively facilitate high-precision analysis:

1. Mobile phase reservoirs: These containers hold solvents used as the mobile phase. The choice of solvents depends on analyte properties and the nature of the stationary phase. Gradient elution involves multiple reservoirs to achieve compositional changes during analysis.
2. Pump system: The pump delivers the mobile phase to the column at a controlled flow rate and pressure. High-performance systems can withstand pressures exceeding 6000 psi, ensuring consistent flow and optimal separation in HPLC and UPLC.
3. Sample injector: The injector introduces the sample into the mobile phase. Automated systems ensure reproducibility, enabling precise delivery of microliter volumes.
4. Chromatographic column: The column is the core of the LC system, housing the stationary phase. Packed columns contain materials such as silica beads or polymer-based particles, modified with functional groups to optimize interactions with specific analytes.
5. Detector: Detectors identify and quantify analytes as they elute from the column. Common detectors include ultraviolet-visible (UV-Vis) spectrophotometers, fluorescence detectors, and mass spectrometers (MS), each chosen based on analyte characteristics and sensitivity requirements.
6. Data system: Modern LC systems incorporate advanced software for recording, analyzing, and visualizing data. Chromatograms generated by the system provide insights into analyte retention times and peak intensities, facilitating quantitative and qualitative analyses.
7. Waste management: Effluent from the system is either collected for further processing or safely discarded, ensuring compliance with environmental and safety standards.

Types of LC

LC encompasses a variety of techniques, each tailored to specific analytical requirements. The most widely used variants are HPLC and UPLC, complemented by specialized methods for niche applications [17–19].

HPLC

HPLC is a versatile technique that achieves high-resolution separations using high-pressure pumps and columns packed with fine particles. The method is widely employed in food adulteration analysis, offering sensitivity and reproducibility for detecting contaminants such as synthetic dyes, pesticides, and pharmaceutical residues [20]. HPLC remains a cornerstone of modern food analysis due to its robustness, versatility, and accuracy. HPLC operates on the principle of differential partitioning, wherein analytes interact with the stationary and mobile phases, resulting in separation based on their chemical properties.

Applications of HPLC in Food Analysis

1. Detection of pesticides: HPLC is extensively used to quantify pesticide residues in fruits, vegetables, and grains, ensuring compliance with maximum residue limits (MRLs).
2. Adulteration analysis: The technique effectively identifies synthetic dyes in spices and beverages, unauthorized preservatives in processed foods, and contaminants in edible oils.
3. Nutritional profiling: HPLC facilitates the quantification of vitamins, amino acids, and polyphenols, supporting food labeling and nutritional claims.

Advantages of HPLC

1. Versatility: HPLC accommodates diverse stationary phases and detection systems, enabling the analysis of polar, non-polar, and thermally labile compounds.
2. High resolution: The technique's capability to resolve complex mixtures makes it indispensable in detecting closely related compounds.
3. Automation and scalability: Automated sample injection and data analysis enhance reproducibility and throughput.

While HPLC remains widely utilized, its limitations, including longer run times and higher solvent consumption, have spurred the development of advanced techniques such as UHPLC. HPLC has some key features, such as its applicability to both polar and non-polar analytes, compatibility with a wide range of stationary phases and detectors, and high sensitivity for detecting trace contaminants. Despite its advantages, HPLC involves longer analysis times and higher solvent consumption than newer techniques like UPLC.

Ultra-Performance LC (UPLC)

UPLC represents a significant advancement over HPLC, employing sub-2 μm particles and operating at pressures up to 15,000 psi [21]. These enhancements result in faster and more efficient separations, with improved resolution and sensitivity.

Advantages of UHPLC

There are several benefits of UPLC, such as reduced run times, which leads to enhanced efficiency. Also, it helps reduce solvent consumption, operational costs, and environmental impact. It has superior peak resolution, enabling the detection of closely eluting analytes. UPLC is particularly beneficial in high-throughput environments, such as regulatory and industrial food safety laboratories, where rapid detection of adulterants is critical [22].

1. Reduced analysis times: Smaller particle sizes and optimized column packing result in faster separations, increasing analytical throughput.
2. Lower solvent consumption: UHPLC's efficiency reduces solvent usage, offering economic and environmental benefits.
3. Improved resolution and sensitivity: Enhanced peak resolution enables the detection of low-abundance analytes in complex food matrices.

Applications in Food Safety

1. Mycotoxin detection: UHPLC coupled with tandem mass spectrometry (UHPLC-MS/MS) provides sensitive detection of mycotoxins in grains, nuts, and dairy products.
2. Authentication studies: The technique distinguishes authentic products from counterfeit items, such as verifying the purity of honey, olive oil, and saffron.
3. Trace contaminant analysis: UHPLC is ideal for detecting trace levels of contaminants, including heavy metals and pharmaceutical residues.

While UHPLC offers superior performance, challenges such as high initial costs and the need for specialized maintenance persist. Innovations in column technology, mobile phase optimization, and detection systems continue to expand the technique's applicability.

Multi-Dimensional LC (MD-LC)

MD-LC integrates multiple separation mechanisms within a single workflow, enhancing the resolution and selectivity of complex sample analyses [23]. This technique is particularly valuable in untargeted analyses where a single chromatographic dimension may not suffice. MD-LC utilizes orthogonal separation techniques, wherein each chromatographic dimension targets distinct physicochemical properties of analytes. For example, the first dimension may separate compounds based on polarity, while the second dimension resolves components by size or charge.

Applications in Food Analysis

1. Comprehensive profiling: MD-LC excels in analyzing complex food matrices such as coffee, wine, and fermented products, offering insights into chemical composition and authenticity.

2. Adulteration detection: The technique identifies low-abundance adulterants in high-value products, such as spices and essential oils.
3. Allergen analysis: MD-LC provides detailed separation of protein fractions, aiding in detecting allergens in processed foods.

Advantages and Limitations of MD-LC

- Enhanced resolution: The combination of orthogonal separation mechanisms resolves intricate mixtures with unparalleled accuracy.
- Scalability: MD-LC can be tailored for targeted and untargeted analyses, expanding its scope in food safety.

However, MD-LC's complexity and extended analysis times may limit its routine application. Ongoing research aims to simplify workflows and enhance automation.

Specialized LC Techniques

In addition to HPLC and UPLC, several specialized LC methods cater to specific analytical needs in food adulteration analysis:

1. Ion chromatography (IC): Ideal for detecting ionic species, such as anions and cations, in food matrices. Applications include the analysis of additives like phosphates and contaminants such as nitrites.
2. Size-exclusion chromatography (SEC): Separates molecules based on size, making it suitable for analyzing macromolecules like proteins and polysaccharides in food products.
3. Affinity chromatography: Exploits specific interactions between analytes and immobilized ligands on the stationary phase. This technique is widely used for detecting target proteins or enzymes.
4. Chiral chromatography: Employs chiral stationary phases to separate enantiomers, enabling the analysis of chiral contaminants or additives.
5. Preparative chromatography: Focuses on isolating and purifying compounds in larger quantities, often used for authenticity verification or quality control.

Each of these techniques offers unique advantages, expanding the applicability of LC in food adulteration analysis. By integrating advanced detection systems and automation, modern LC methods continue to evolve, addressing the growing demands of food safety and quality assurance [24].

LC in Food Adulteration Analysis

Liquid chromatography (LC) has become an indispensable analytical tool for detecting food adulteration, offering unparalleled precision and sensitivity. Its versatility enables identifying and quantifying a wide array of contaminants and adulterants across diverse food matrices. The applicability of LC in food adulteration analysis is vast, encompassing diverse product categories such as oils, dairy, spices, beverages, and high-value commodities like saffron and honey [25]. Advanced LC techniques enable:

1. Detection of adulterants: Identifying unauthorized additives, synthetic dyes, and harmful contaminants with high precision.
2. Authentication: Verifying the purity and origin of food products, ensuring compliance with labeling claims.
3. Quantification: Measuring trace levels of contaminants, pesticides, and residual solvents to ensure regulatory compliance.

Coupling LC with advanced detection methods, such as mass spectrometry, enhances its sensitivity and specificity, enabling the detection of even trace levels of adulterants. Portable LC systems and automation have further expanded the scope of this technology, allowing on-site testing and real-time decision-making in food production and distribution. By combining advanced separation capabilities with robust detection systems, LC plays a pivotal role in ensuring the safety and quality of food products [26]. Its adaptability and precision make it an indispensable tool in addressing the multifaceted challenges of food adulteration detection.

Advanced LC Techniques for Food Analysis

LC has evolved into a pivotal analytical tool in food safety and quality assurance, enabling precise detection of contaminants, adulterants, and authentic markers in complex food matrices. This section explores the advancements in LC technologies, emphasizing HPLC, UHPLC, MD-LC, and preparative LC.

Preparative LC

Preparative LC isolates and purifies compounds in larger quantities for downstream applications [27]. Unlike analytical LC, which emphasizes detection and quantification, preparative LC prioritizes yield and purity. There are some key features:

1. Scalability: Preparative LC accommodates larger sample volumes, facilitating industrial-scale operations.

2. Customizable columns: Specialized columns optimize separation efficiency for target compounds.
3. High purity levels: The technique's ability to achieve high-purity fractions supports quality assurance and research applications.

Applications in Food Industry

1. Isolation of natural compounds: Preparative LC isolates bioactive compounds, such as antioxidants and flavoring agents, from natural sources like fruits and herbs.
2. Quality control: The technique ensures the purity of additives, supplements, and nutraceuticals, aligning with regulatory standards.
3. Authentication: Preparative LC aids in verifying the authenticity of high-value products, such as extracting pure saffron or vanilla components for comparison studies.

Preparative LC requires significant solvent consumption and may face difficulty separating closely related compounds. Advances in green chromatography and solvent recycling systems aim to address these limitations, making the technique more sustainable and cost-effective [28].

Adulterants in Food Products

Food adulteration remains a significant concern worldwide, posing risks to public health, undermining consumer trust, and challenging food safety authorities. This section explores the common types of adulterants, their sources, associated health risks, and the regulatory frameworks designed to mitigate their presence in the food supply chain.

Common Types of Adulterants and Their Sources

Adulterants are substances intentionally or unintentionally added to food products to alter their properties, increase quantity, or reduce production costs. These substances can be classified into several categories based on their sources and effects on food products [29].

Chemical Adulterants

1. Pesticides and residual chemicals: These chemicals are found in fruits, vegetables, and grains, including organophosphates and carbamates used in agriculture. Their persistence in food products often exceeds permissible limits, leading to contamination.

2. Synthetic colors and dyes: Non-permitted synthetic dyes such as Sudan Red, Metanil Yellow, and Rhodamine B are commonly used to enhance the visual appeal of products like spices, beverages, and sweets.
3. Preservatives and additives: Excessive or unauthorized use of preservatives such as benzoates, sulfites, and formalin can render food unsafe for consumption.
4. Heavy metals: Contaminants like lead, arsenic, cadmium, and mercury often enter the food supply through contaminated soil, water, or packaging materials.

Biological Adulterants

1. Pathogenic microorganisms: Unhygienic practices during production and storage can introduce bacteria such as *Salmonella*, *E. coli*, and *Listeria* into food products.
2. Mycotoxins: Produced by fungi, mycotoxins like aflatoxins and ochratoxins contaminate grains, nuts, and dairy products.

Physical Adulterants

Physical adulterants in food are unwanted foreign materials that may be accidentally added during processing, storage, and transportation or purposely added to increase volume and mislead customers. Dirt, stones, dust, glass, and metal shavings are among the adulterants. Organic contaminants include insect remains, rodent hair, and fungi [30]. Chalk powder in flour, washing powder in pulses, and artificially colored food grains have also been found in food products. Iron filings from milling machines and hazardous substances like lead, arsenic, and mercury are physical adulterants that pose health risks. These pollutants come from poor agriculture and storage methods, food processing cleanliness, and packing and transportation. Physically compromised food can cause bodily harm from sharp particulates, gastrointestinal disturbances from hard impurities, toxicological effects from heavy metal exposure, and infections from contaminated organic materials [31]. Implementing strict food safety measures, including Good Manufacturing Practices (GMP), sieving, metal detection, and tamper-proof packaging, reduces these risks. The FSSAI, FDA, and Codex Alimentarius have developed protocols to ensure food safety and reduce contamination. Compliance with these rules, customer understanding, and industry responsibility are essential for food product integrity and public health. Foreign materials such as stones, glass, metal shards, and plastic fragments often result from inadequate quality control during processing and packaging.

Economic Adulterants

Adulteration motivated by economic gain includes substitution of inferior ingredients (e.g., starch in milk, brick powder in spices), dilution of products (e.g., water in milk), and mislabeling (e.g., selling synthetic honey as natural honey) [32]. These adulterants not only degrade food quality and nutritional value but also pose severe risks to consumer health.

Health Risks Associated with Adulterated Foods

Consumption of adulterated foods can lead to acute and chronic health issues, depending on the type and concentration of adulterants. Some of the notable health risks are discussed below [33].

Toxic Effects

1. Pesticides: Chronic exposure to pesticides is linked to neurological disorders, developmental delays, and cancer.
2. Heavy metals: Lead and mercury are neurotoxic, causing cognitive impairments and developmental delays in children. Arsenic exposure can result in skin lesions, cardiovascular disease, and cancer.
3. Synthetic dyes: Non-permitted dyes are carcinogenic and can cause liver damage, skin rashes, and hyperactivity in children.

Gastrointestinal Issues

Biological adulterants, including bacteria and fungi, often lead to foodborne illnesses such as diarrhea, nausea, and vomiting. Mycotoxins, in particular, are hepatotoxic and can impair liver function.

Allergic Reactions

Adulterants such as undeclared allergens, cross-contaminants, and certain preservatives can trigger allergic reactions ranging from mild skin irritation to life-threatening anaphylaxis.

Nutritional Deficiency

Economic adulteration, such as diluting milk with water or substituting natural ingredients with synthetic counterparts, compromises the nutritional content of food, potentially leading to deficiencies in essential vitamins and minerals.

Table 1 Representation of various types of adulterants, their sources, and the harmful causes of the specific adulterants

S. No	Source of adulterant	Type of adulterant	Harmful effects
1.	Pesticides	Chemical	Neurological disorders, developmental delays, and cancer
2.	Synthetic dye	Chemical	Carcinogenic and can cause liver damage, skin rashes, and hyperactivity in children
3.	Heavy metals	Chemical	Neurotoxic, cognitive impairments and developmental delays in children, cause skin lesions, cardiovascular disease, and cancer
4.	Bacteria and fungi	Biological	Diarrhea, nausea, and vomiting
5.	Mycotoxins	Biological	Hepatotoxic and can impair liver function
6.	Starch in milk, dilution of milk	Economical	Deficiencies of essential vitamins and minerals
7.	Brick powder in spices	Economical	Deficiencies of essential vitamins and minerals

Carcinogenicity and Mutagenicity

Prolonged consumption of certain adulterants, including synthetic dyes and mycotoxins, is associated with DNA damage, increasing the risk of cancer and genetic mutations.

Endocrine Disruption

Hormone-mimicking chemicals, such as certain pesticides and plasticizers (e.g., bisphenol A), interfere with endocrine function, affecting growth, reproduction, and metabolism [34]. The list of adulterants, including their source, variant category, and harmful effects, has been formulated in Table 1. Given these risks, stringent monitoring and detection of adulterants are essential to protect public health and ensure food safety.

Regulatory Standards and Guidelines

To address the challenges posed by food adulteration, international and national regulatory bodies have established stringent standards and guidelines to ensure the safety and quality of food products [35]. Key frameworks include,

International Standards

1. Codex Alimentarius: Developed by the Food and Agriculture Organization (FAO) and World Health Organization (WHO), the Codex sets global food standards, guidelines, and codes of practice to protect consumer health and promote fair trade practices [36].
2. European Union Regulations: The EU enforces stringent food safety regulations, including the General Food Law Regulation (EC 178/2002) and directives on contaminants and food additives [37].
3. United States Food and Drug Administration (FDA): The FDA regulates food safety under the Food Safety Modernization Act (FSMA), emphasizing preventive controls and stringent inspection protocols.

National Standards

1. India: The Food Safety and Standards Authority of India (FSSAI) oversees food safety through regulations such as the Food Safety and Standards Act 2006 [38]. It mandates regular inspections, licensing, and testing of food products.
2. China: The National Food Safety Standard and related laws address food contamination and adulteration issues, emphasizing stringent quality controls.
3. Australia and New Zealand: Food Standards Australia New Zealand (FSANZ) establishes safety standards and labeling requirements to ensure transparency and consumer protection.

Analytical Methods and Monitoring

Regulatory agencies mandate the use of advanced analytical techniques, such as liquid chromatography (LC), gas chromatography (GC), and mass spectrometry (MS), for detecting adulterants with high accuracy and sensitivity. Sampling plans, risk-based monitoring, and traceability systems are integral to ensuring compliance with safety standards.

Enforcement and Penalties

Food adulteration is often penalized through fines, product recalls, and license revocations. Repeat offenders may face stricter consequences, including criminal prosecution.

Consumer Awareness and Education

Public awareness campaigns and educational initiatives are crucial in empowering consumers to identify and report adulterated foods. Labeling standards and certifications, such as organic and non-GMO labels, play a significant role in fostering trust and transparency [39].

Detection of Pesticide Residues in Fruits and Vegetables

Pesticide residues in fruits and vegetables are a significant public health concern due to their potential toxicity and persistence in the food supply chain [40]. LC techniques, particularly HPLC and UPLC, coupled with tandem mass spectrometry (MS/MS), have revolutionized pesticide residue analysis.

Methodology

1. Sample preparation: Fruits and vegetables undergo extraction using solvents like acetonitrile, followed by clean-up processes such as QuEChERS (Quick, Easy, Cheap, Effective, Rugged, and Safe).
2. Chromatographic separation: The choice of stationary and mobile phases ensures the effective separation of multiple pesticide residues within a single run.
3. Detection: LC-MS/MS enables quantification at parts per billion (ppb) levels, meeting regulatory requirements for maximum residue limits (MRLs).

Identification of Synthetic Additives in Beverages

Synthetic additives in beverages, such as artificial sweeteners, preservatives, and coloring agents, are often subject to regulatory scrutiny. LC techniques are essential for verifying compliance and detecting unauthorized additives.

Analytical Approach

1. Targeted analysis: In soft drinks and juices, HPLC with UV or fluorescence detection is commonly used to identify synthetic dyes, including Tartrazine and Sunset Yellow.

2. Non-targeted analysis: UHPLC-MS/MS facilitates comprehensive screening for unapproved additives in energy drinks and alcoholic beverages.

Analysis of Heavy Metals and Contaminants in Processed Foods

Heavy metal contamination in processed foods poses severe health risks, including neurotoxicity and carcinogenicity. It applies to monitoring heavy metals in baby food, canned goods, and ready-to-eat meals [41]. Evaluating contamination from packaging materials and processing equipment and ensuring compliance with international standards, such as those established by the Codex Alimentarius. In combination with inductively coupled plasma mass spectrometry (ICP-MS), LC methods provide a robust platform for heavy metal analysis.

Techniques

1. Speciation analysis: LC separates different oxidation states or species of metals, such as arsenic (As(III) vs. As(V)), before detection.
2. Coupled techniques: HPLC-ICP-MS offers high sensitivity for detecting trace levels of lead, cadmium, mercury, and arsenic.

Detection of Antibiotics and Hormones in Meat and Dairy Products

The use of antibiotics and hormones in livestock farming raises concerns about residues in meat and dairy products, potentially leading to antimicrobial resistance and hormonal imbalances in consumers. Practical application for monitoring veterinary drug residues in meat products to comply with regulatory limits. Also, hormone residues in milk powder and infant formula should be detected to ensure safety and identify non-compliance in organic meat and dairy production. LC-based methods are instrumental in residue detection and quantification [42].

Analytical Strategies

1. Antibiotic residues: HPLC-MS/MS detects sulfonamides, tetracyclines, and fluoroquinolones in milk, cheese, and meat samples.
2. Hormonal residues: UHPLC-MS/MS provides high-resolution analysis of growth promoters like clenbuterol and synthetic estrogens.

Identification of Melamine and Other Protein Adulterants

Melamine adulteration, highlighted by food scandals involving infant formula and dairy products, underscores the need for advanced detection methods [43]. Several applications include screening milk and dairy products for melamine contamination. Identifying urea and cyanuric acid as adulterants in protein supplements. Ensuring compliance with zero-tolerance policies for melamine in food products. LC techniques play a critical role in identifying melamine and other protein adulterants.

Detection Techniques

1. Sample extraction: Protein-rich matrices undergo acid hydrolysis or solvent extraction to release melamine and its analogs.
2. Chromatographic analysis: HPLC coupled with diode-array detection (DAD) or MS detects melamine at trace levels.
3. Confirmatory methods: UHPLC-MS/MS enhances specificity for distinguishing melamine from structurally similar compounds.

Detection and Analysis of Food Adulterants Using LC

Liquid chromatography (LC) has become a cornerstone analytical technique for detecting and analyzing food adulterants. Its unparalleled versatility, sensitivity, and precision make it indispensable for ensuring food safety and quality [44]. This section delves into the identification of chemical adulterants, the analysis of food colors, preservatives, and additives, the detection of heavy metals and pesticides, and the methods for authenticating food products.

Identification of Chemical Adulterants

Chemical adulterants are intentionally or unintentionally introduced into food products to manipulate their quality, extend shelf life, or reduce production costs [45]. LC techniques, particularly HPLC and UHPLC, are crucial in identifying these substances. There are some common chemical adulterants and their analysis.

1. Melamine can be identified using HPLC coupled with diode-array detection (DAD) or mass spectrometry (MS) to falsely elevate protein content in dairy products. Sample preparation involves acid hydrolysis or solvent extraction, followed by chromatographic separation.

2. Urea: Adulteration of milk with urea is detected using HPLC methods that exploit its distinct retention behavior. Detection is often paired with UV or fluorescence detectors.
3. Synthetic sweeteners: Artificial sweeteners, such as aspartame and saccharin, are analyzed using reversed-phase HPLC with UV detection. These methods ensure compliance with regulatory thresholds.

Advantages of LC in Identifying Chemical Adulterants

1. High sensitivity and specificity for trace-level detection.
2. Applicability across diverse food matrices, including dairy, oils, and beverages.
3. Robustness in distinguishing target adulterants from complex sample backgrounds.

Methods for Authentication of Food Products

Authentication of food products is vital to prevent fraud and maintain consumer trust. LC techniques provide robust tools for verifying product authenticity based on chemical composition.

Authenticity of High-Value Products

1. Honey: Authentic honey is distinguished from adulterated versions using HPLC analysis of carbohydrate profiles. Adulterants like high-fructose corn syrup are identified based on their distinct retention times.
2. Olive oil: LC-MS/MS profiles phenolic compounds and detects blending with cheaper oils.
3. Wine: UHPLC separates and identifies phenolic and flavor compounds, distinguishing genuine wines from counterfeit products.

Origin and Geographical Indication

1. Saffron: The authenticity of saffron is verified by analyzing its characteristic components, such as crocin, picrocrocin, and safranal, using HPLC-DAD.
2. Tea: LC-MS/MS profiles polyphenols and amino acids to confirm the geographical origin of premium tea varieties.

Detection of Adulteration in Protein-Rich Foods

1. Milk and dairy products: LC-MS/MS identifies melamine and its analogs in milk powder and infant formula.
2. Meat products: HPLC quantifies protein hydrolysates and non-meat fillers, ensuring compliance with labeling claims.

A brief outline of some additives and their detection techniques is mentioned in Table 2.

Table 2 Some common adulterations with their detection methods and applications

S. No.	Commodity	Detection technique	Application	References
1.	Pesticide residues in fruits and vegetables	LC-MS/MS	Identifying illegal pesticide use in organic and conventional farming	[46]
2.	Synthetic additives in beverages	HPLC with UV	Identification of non-permitted dyes, such as rhodamine B, in fruit-based beverages	[47]
3.	Heavy metal contamination in processed foods	HPLC-ICP-MS	Monitoring heavy metals in baby food, canned goods, and ready-to-eat meals	[48]
4.	Antibiotic residues in dairy and meat products	HPLC-MS/MS	Monitoring veterinary drug residues in meat products to comply with regulatory limits	[49]
5.	Hormonal residues in dairy and meat products	UHPLC-MS/MS	Detecting hormone residues in milk powder and infant formula to ensure safety	[50]
6.	Melamine adulteration	HPLC coupled with diode-array detection	Screening milk and dairy products for melamine contamination	[51]
7.	Analysis of food colors	UHPLC-MS/MS	Employed to authenticate natural food colors,	[52]
8.	Heavy metals	HPLC coupled with inductively coupled plasma mass spectrometry (ICP-MS)	Monitor contamination in baby foods, seafood, and beverages	[53]
9.	Adulteration in protein-rich foods	LC-MS/MS	Identifies melamine and its analogs in milk powder and infant formula	[54]
10.	Authentication of food products such as honey	HPLC	Adulterants such as high-fructose corn syrup are identified based on their distinct retention times	[55]

Challenges and Limitations of Advanced LC Techniques

High-performance and ultra-high-performance LC have revolutionized food adulteration detection and analysis by identifying contaminants, additives, and adulterants with exceptional sensitivity, precision, and efficiency. However, these new analytical approaches face many obstacles that limit their applicability in food safety monitoring. In resource-constrained situations, small-scale food producers and regulatory agencies cannot afford LC systems due to their high procurement, maintenance, and operation costs [56]. Sample preparation, which sometimes requires comprehensive extraction, purification, and derivatization, can be laborious and inaccurate, affecting consistency and dependability. Complex food matrices, such as those high in fats, proteins, or fibers, complicate chromatographic separation and detection of adulterants, causing interferences and co-elution issues [57]. LC instruments and data analysis require competent personnel, as improper handling and interpretation can lead to inaccurate results and adulterant detection. LC-MS has improved the specificity and sensitivity of LC, but it requires expensive mass spectrometers and extensive data processing software, adding complexity. The continuing evolution of deceptive procedures and the introduction of novel adulterants in food adulteration demands advancements in LC, reference databases, and detection systems. Given these limits, LC techniques remain vital for food quality and safety evaluation. To effectively monitor and regulate food adulteration worldwide, affordable, accessible, and fast chromatographic methods are needed. One of the most significant barriers to adopting advanced LC techniques, such as UHPLC and MD-LC, is the high cost of instrumentation [23]. State-of-the-art LC systems, especially with advanced detectors like mass spectrometers (MS), represent a substantial laboratory investment.

Future Trends and Innovations in LC

As food adulteration practices become increasingly sophisticated, the LC field is poised to advance in response. Future trends and innovations in LC are focused on improving instrumentation, enhancing portability, and integrating with emerging technologies like mass spectrometry (MS) and artificial intelligence (AI). This section explores these developments in detail, highlighting their implications for food safety and quality assurance as depicted in Fig. 3.

Advances in LC Instrumentation

Advancements in LC instrumentation aim to enhance resolution, speed, and sensitivity while reducing environmental impact and operational costs [58]. These improvements are driving the evolution of LC systems to meet the growing demands of modern analytical laboratories.

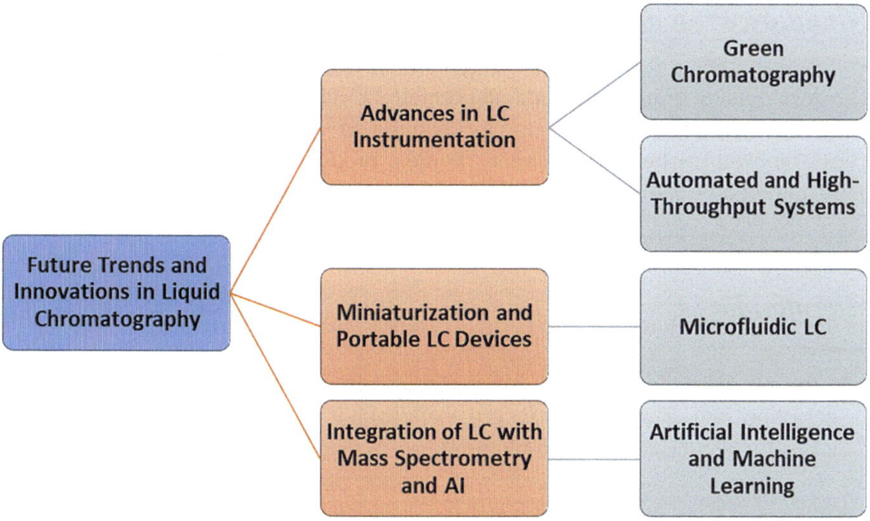

Fig. 3 Flow chart illustrating future trends and innovations in liquid chromatography. The central node is titled "Future Trends and Innovations in Liquid Chromatography." It branches into three main categories: "Advances in LC Instrumentation," "Miniaturization and Portable LC Devices," and "Integration of LC with Mass Spectrometry and AI." Each category further branches into specific areas: "Advances in LC Instrumentation" leads to "Green Chromatography" and "Automated and High-Throughput Systems"; "Miniaturization and Portable LC Devices" leads to "Microfluidic LC"; "Integration of LC with Mass Spectrometry and AI" leads to "Artificial Intelligence and Machine Learning"

Green Chromatography

1. Low-solvent consumption: Instruments designed for minimal solvent use, combined with water-based mobile phases, reduce LC techniques' environmental footprint.
2. Recyclable materials: Advances in column and tubing materials aim to make LC systems sustainable without compromising performance.

Automated and High-Throughput Systems

1. Multi-injection platforms: Automation of sample injection increases throughput and reproducibility, enabling laboratories to process more samples in less time.
2. Integration with robotics: Robotic systems are employed to handle sample preparation and injection, reducing manual errors and enhancing efficiency.

Miniaturization and Portable LC Devices

The development of miniaturized and portable LC devices is transforming the landscape of food adulteration detection by enabling on-site and real-time analysis [59]. These innovations benefit regulatory bodies, producers, and consumers seeking rapid and reliable testing solutions.

Microfluidic LC

1. Lab-on-a-Chip Technology**:** Microfluidic LC platforms integrate separation, detection, and data processing onto a single chip, significantly reducing the volume of reagents and samples required.

Integration of LC with Mass Spectrometry and AI

The integration of LC with mass spectrometry (LC-MS) and artificial intelligence (AI) is paving the way for more sophisticated and automated analytical workflows [60]. These technologies enhance food adulteration detection's specificity, sensitivity, and efficiency.

Artificial Intelligence and Machine Learning

1. Data analysis: AI algorithms process large LC-MS datasets, identifying patterns and correlations that might be missed by manual interpretation. This capability is critical for non-targeted screening approaches.
2. Predictive models: Machine learning models predict adulteration trends, helping regulatory agencies and manufacturers preemptively address emerging threats.
3. Automated method development: AI assists in optimizing LC parameters, such as gradient profiles and column selection, to achieve the best separation for specific analytes.

Conclusion

Advanced LC methods like HPLC and UHPLC can detect and analyze food adulteration, ensuring food safety and quality. These methods detect several pollutants, adulterants, and food fraud signs with high accuracy, sensitivity, and specificity. These methods have many advantages, but high operational costs, complicated sample preparation, the need for trained staff, and matrix interference complications make them difficult to implement, especially in low-resource environments. As food adulteration tactics evolve, more resilient, inexpensive, and fast analytical approaches are needed. Miniaturized chromatography devices, automation, and machine learning-assisted data processing will make LC for routine food safety monitoring more efficient and accessible. Regulatory agencies, research groups, and the food sector must collaborate to harmonize methods, improve reference databases, and deploy more accessible technology. LC is essential to food safety analysis, but spectroscopy and biosensor-based platforms could improve food fraud and adulteration prevention. These solutions must be consistent, cost-effective, and expandable to maintain public health and food consumer confidence. While advanced LC techniques offer unparalleled capabilities in food adulteration detection, their widespread adoption is hindered by several challenges. High instrumentation costs, complex sample preparation, technical expertise requirements, and limitations in detecting novel adulterants represent significant obstacles. Addressing these issues through collaborative efforts, technological innovations, and targeted training programs will be crucial for maximizing the impact of LC technologies on global food safety and quality assurance.

References

1. Thiruvengadam M, Kim J-T, Kim W-R, Kim J-Y, Jung B-S, Choi H-J, Chi H-Y, Govindasamy R, Kim S-H (2024) Safeguarding public health: advanced detection of food adulteration using nanoparticle-based sensors. Crit Rev Anal Chem. https://doi.org/10.1080/10408347.2024.2399202
2. Thangaraju S, Modupalli N, Natarajan V (2021) Food adulteration and its impacts on our health/balanced nutrition. In: Food chemistry: the role of additives, preservatives and adulteration. Wiley, Hoboken, pp 189–216
3. Chukwugozie DC, Njoagwuani EI, David K, Okonji BA, Milovanova N, Akinsemolu AA, Mazi IM, Onyeaka H, Winnall L, Ghosh S (2024) Combatting food fraud in sub-Saharan Africa: strategies for strengthened safety and security. Trends Food Sci Technol 150:104575
4. Haider A, Iqbal SZ, Bhatti IA, Alim MB, Waseem M, Iqbal M, Khaneghah AM (2024) Food authentication, current issues, analytical techniques, and future challenges: a comprehensive review. Compr Rev Food Sci Food Saf 23(3):e13360
5. Seddaoui N, Amine A (2023) Recent advances in sensor and biosensor technologies for adulteration detection. Adv Sens Technol 2023:699–739
6. Okpala, Charles Odilichukwu R, Korzeniowska M (2023) Understanding the relevance of quality management in agro-food product industry: from ethical considerations to assuring food hygiene quality safety standards and its associated processes. Food Rev Int 39(4):1879–1952
7. Savelli CJ, Bradshaw A, Embarek PB, Mateus C (2019) The FAO/WHO international food safety authorities network in review, 2004–2018: learning from the past and looking to the future. Foodborne Pathog Dis 16(7):480–488
8. Bhat MA, Rather MY, Singh P, Hassan S, Hussain N (2024) Advances in smart food authentication for enhanced safety and quality. Trends Food Sci Technol 155:104800
9. Sarkar J, Singh R, Chandel S (2025) Understanding LC/MS-based metabolomics: a detailed reference for natural product analysis. PROTEOMICS Clin Appl 19(1):e202400048

10. Pandreka MK, Yamini M, Abhishek G, Gope ER, Raghava D (2024) Advances in high-performance liquid chromatography (HPLC) and ultra-performance liquid chromatography (UPLC). J Pharma Insights Res 2(6):39–46
11. Aslam N, Fatima R, Altemimi AB, Ahmad T, Khalid S, Hassan SA, Aadil RM (2024) Overview of industrial food fraud and authentication through chromatography technique and its impact on public health. Food Chem 460:140542
12. Antunes M, Rosado T, Simão AY, Gonçalves J, Soares S, Barroso M, Gallardo E (2021) Liquid chromatography–mass spectrometry as a tool to identify adulteration in different food industries. In: Food toxicology and forensics. Academic Press, New York, pp 123–180
13. Vinothkanna A, Dar OI, Liu Z, Jia A-Q (2024) Advanced detection tools in food fraud: a systematic review for holistic and rational detection method based on research and patents. Food Chem 446:138893
14. Siddique I (2023) High-performance liquid chromatography: comprehensive techniques and cutting-edge innovations. Eur J Adv Eng Technol 10(9):66–70
15. McCalley DV (2017) Understanding and manipulating the separation in hydrophilic interaction liquid chromatography. J Chromatogr A 1523:49–71
16. Neole NG (2024) Analytical techniques in forensic science: spectroscopy and chromatography. In: Cases on forensic and criminological science for criminal detection and avoidance. IGI Global, Hershey, pp 188–240
17. Cortés-Herrera C, Artavia G, Leiva A, Granados-Chinchilla F (2018) Liquid chromatography analysis of common nutritional components, in feed and food. Foods 8(1):1
18. Gupta MK, Ghuge A, Parab M, Al-Refaei Y, Khandare A, Dand N, Waghmare N (2022) A comparative review on high-performance liquid chromatography (HPLC), ultra performance liquid chromatography (UPLC) & high-performance thin layer chromatography (HPTLC) with current updates. Curr Issues Pharm Med Sci 35(4):224–228
19. Swapna M, Rongali Indu PV, Latha M, Rama Madhuri B (2024) A comprehensive review on UHPLC and UPLC: advancements, comparison, and applications. Int J Sci R Tech 1(11):240211045
20. Moldoveanu SC, David V (2022) Essentials in modern HPLC separations. Elsevier, Amsterdam
21. Dhull P, Dunuweera S, Bietsch J, Bandu R, Wannere C, Achanta S, Krishnamurthy D, Bo Q, Senanayake C (2025) Recent advances and application of liquid chromatography in pharmaceutical industry. J Liq Chromatogr Relat Technol 48:1–20
22. Fekete S, Schappler J, Veuthey J-L, Guillarme D (2014) Current and future trends in UHPLC. TrAC Trends Anal Chem 63:2–13
23. Bouvarel T, Camperi J, Guillarme D (2024) Multi-dimensional technology–recent advances and applications for biotherapeutic characterization. J Sep Sci 47(5):2300928
24. Medina S, Perestrelo R, Silva P, Pereira JAM, Câmara JS (2019) Current trends and recent advances on food authenticity technologies and chemometric approaches. Trends Food Sci Technol 85:163–176
25. Kotsanopoulos K (2022) Authenticity of foods of plant origin. CRC Press, Boca Raton
26. Siddique I (2021) Unveiling the power of high-performance liquid chromatography: techniques, applications, and innovations. Eur J Adv Eng Technol 8(9):79–84
27. Speybrouck D, Lipka E (2016) Preparative supercritical fluid chromatography: a powerful tool for chiral separations. J Chromatogr A 1467:33–55
28. Peyrin E, Lipka E (2024) Preparative supercritical fluid chromatography as green purification methodology. TrAC Trends Anal Chem 171:117505
29. Choudhary A, Gupta N, Hameed F, Choton S (2020) An overview of food adulteration: concept, sources, impact, challenges and detection. Int J Chem Stud 8(1):2564–2573
30. Peariso D (2008) Preventing foreign material contamination of foods. Wiley, Hoboken
31. Dehdashti B, Amin MM, Chavoshani A (2020) Other trace elements (heavy metals) and chemicals. In: Micropollutants challenges: emerging in the aquatic environments and treatment processes. Elsevier, Amsterdam, p 215

32. Momtaz M, Bubli SY, Khan MS (2023) Mechanisms and health aspects of food adulteration: a comprehensive review. Foods 12(1):199
33. Gizaw Z (2019) Public health risks related to food safety issues in the food market: a systematic literature review. Environ Health Prev Med 24:1–21
34. Ason B. Contaminant fate and human exposure to endocrine disrupting chemicals in selected contaminated soils in Ghana. PhD diss., University of Cape Coast, 2022
35. Eruaga MA (2024) Enhancing global food safety standards through international collaboration and policy harmonization. Int J Sch Res Multidiscip Stud 4(1):20–32
36. Fortin ND (2023) Global governance of food safety: the role of the FAO, WHO, and codex Alimentarius in regulatory harmonization. In: Research handbook on international food law. Edward Elgar Publishing, Cheltenham, pp 227–242
37. Pisanello D (2014) Chemistry of foods: EU legal and regulatory approaches. Springer, Berlin
38. Baliyan N, Kumari R, Kumari A, Kumar A (2024) Food safety and detection of bacterial foodborne pathogens in India. In: Handbook of public health nutrition: international, national, and regional perspectives. Springer, Cham, pp 1–24
39. Jesada R (2011) Buyer beware: an exploration of health risks and legal policies in favor of a labeling requirement for genetically modified organisms. J Health Care 14:30
40. Keikotlhaile BM, Spanoghe P (2011) Pesticide residues in fruits and vegetables. Pesticides 2011:243–252
41. Żmudzińska A, Puścion-Jakubik A, Bielecka J, Grabia M, Soroczyńska J, Mielcarek K, Socha K (2022) Health safety assessment of ready-to-eat products consumed by children aged 0.5–3 years on the polish market. Nutrients 14(11):2325
42. Al-Amri I, Kadim IT, AlKindi A, Hamaed A, Al-Magbali R, Khalaf S, Al-Hosni K, Mabood F (2021) Determination of residues of pesticides, anabolic steroids, antibiotics, and antibacterial compounds in meat products in Oman by liquid chromatography/mass spectrometry and enzyme-linked immunosorbent assay. Vet World 14(3):709
43. Liang W, Wei Y, Gao M, Yan X, Zhu X, Guo W (2020) Detection of melamine adulteration in milk powder by using optical spectroscopy technologies in the last decade—a review. Food Anal Methods 13:2059–2069
44. Sorapalli SK, Kagitha KB, Doddi NS, Alla M, Nagidi M, Gope ER, Raghava D (2024) A review of liquid chromatography-mass spectrometry and its applications in chemical analysis. J Pharma Insights Res 2(6):025–032
45. Sen M (2021) Food chemistry: role of additives, preservatives, and adulteration. In: Food chemistry: the role of additives, preservatives and adulteration. Scrivener Publishing LLC, Austin, pp 1–42
46. Stachniuk A, Szmagara A, Czeczko R, Fornal E (2017) LC-MS/MS determination of pesticide residues in fruits and vegetables. J Environ Sci Health B 52(7):446–457
47. Sun X-D, Hai-Long W, Liu Z, Chen Y, Liu Q, Ding Y-J, Ru-Qin Y (2019) Rapid and sensitive detection of multi-class food additives in beverages for quality control by using HPLC-DAD and chemometrics methods. Food Anal Methods 12:381–393
48. Saraiva M, Chekri R, Guérin T, Sloth JJ, Jitaru P (2021) Chromium speciation analysis in raw and cooked milk and meat samples by species-specific isotope dilution and HPLC-ICP-MS. Food Add Contamin Part A 38(2):304–314
49. Tuzimski T (2018) HPLC–MS (MS/MS) as a method of identification and quantification of drugs and veterinary drug residues in food, feed, environmental and biological samples. In: Determination of target xenobiotics and unknown compound residues in food, environmental, and biological samples. CRC Press, Boca Raton, pp 73–102
50. Zheng J, Xi C, Wang G, Cao S, Tang B, Zhaode M (2018) Rapid screening and determination of the residues of hormones and sedatives in milk powder using the UHPLC-MS/MS and SPE. Food Anal Methods 11(12):3435–3451
51. Chen Z, Yan X (2009) Simultaneous determination of melamine and 5-hydroxymethylfurfural in milk by capillary electrophoresis with diode array detection. J Agric Food Chem 57(19):8742–8747

52. Sadok I, Szmagara A, Krzyszczak A (2023) Validated QuEChERS-based UHPLC-ESI-MS/MS method for the postharvest control of patulin (mycotoxin) contamination in red-pigmented fruits. Food Chem 400:134066
53. Letsoalo MR, Mamo MA, Ambushe AA (2021) Simultaneous quantitative speciation of selected toxic elements in water using high performance liquid chromatography coupled to inductively coupled plasma-mass spectrometry (HPLC-ICP-MS). Phys Chem Earth 124:103011
54. Valletta M, Ragucci S, Landi N, Di Maro A, Pedone PV, Russo R, Chambery A (2021) Mass spectrometry-based protein and peptide profiling for food frauds, traceability and authenticity assessment. Food Chem 365:130456
55. Zhang G, Abdulla W (2022) On honey authentication and adulterant detection techniques. Food Control 138:108992
56. Clarke W, Rhea JM, Molinaro R (2013) Challenges in implementing clinical liquid chromatography–tandem mass spectrometry methods–seeing the light at the end of the tunnel. J Mass Spectrom 48(7):755–767
57. Cafarella C (2024) Use of liquid chromatography and mass spectrometry methods for the detailed characterization of food and biological samples
58. Wilson S, Steele S, Adeli K (2022) Innovative technological advancements in laboratory medicine: predicting the lab of the future. Biotechnol Biotechnol Equip 36(1):S9–S21
59. Hemida M, Ghiasvand A, Macka M, Gupta V, Haddad PR, Paull B (2023) Recent advances in miniaturization of portable liquid chromatography with emphasis on detection. J Sep Sci 46(15):2300283
60. Yee WLS, Drum CL (2020) Increasing complexity to simplify clinical care: high resolution mass spectrometry as an enabler of AI guided clinical and therapeutic monitoring. Adv Therap 3(4):1900163

Emerging Nanotechnology Applications in Food Safety Analysis

Rahul Gogoi, Madhumita Barooah, and Samim Dullah

Contents

R. Gogoi · M. Barooah
Department of Agricultural Biotechnology, Assam Agricultural University, Jorhat, Assam, India

S. Dullah (✉)
Department of Zoology and DBT-IBT-HUB, Nanda Nath Saikia College, Titabar, Jorhat, Assam, India

W. Alahmad, M. Ashfaq (eds.), *Advanced Techniques against Food Adulteration*,
https://doi.org/10.1007/978-3-031-99454-8_9

Abstract According to the World Health Organization (WHO), contaminated food is the leading cause of morbidity and mortality worldwide, with millions of people suffering from foodborne illnesses. Ensuring the safety and quality of our food has grown more difficult due to the food industry's rapid growth and the diversification of its resources. While traditional food safety analysis techniques have some advantages, they also have serious drawbacks, such as long processing times, high costs, and limited sensitivity and specificity. These difficulties are made worse by the intricacy of foodborne illnesses, newly discovered contaminants, and the global reach of food supply systems. Given these challenges, there is a growing need for effective, accurate, and scalable solutions that may resolve these problems and give consumers trust in the foods they choose. The advancement of nanotechnology is viewed favorably in this context since it provides several advantages for food safety analysis, including intelligent packaging options, enhanced food quality monitoring, and quick identification of foodborne viruses. Nanomaterials are ideal for food safety analysis because of their unique characteristics, such as their large surface area and high surface-to-volume ratio, which facilitate efficient interaction with analytes and raise detection sensitivity. Significant applications of nanotechnology have emerged in the field of food safety, notably in the areas of monitoring, detection, and intervention. These developments increase food safety analyses' scalability, precision, and efficiency by utilizing state-of-the-art discoveries. This chapter aims to outline the many kinds of nanomaterials utilized in food safety and their uses, such as identifying food pollutants and foodborne pathogens. It draws attention to the developments in nanotechnology and the difficulties posed by safety, legal, and moral issues.

Keywords Contamination · Food · Foodborne · Illnesses · Challenges · Nanomaterials

Introduction

Food safety and quality assurance are increasingly urgent concerns in our global landscape. With the fast expansion of the food industry and the diversification of sources, guaranteeing the safety and quality of our food has become increasingly complex [1]. The World Health Organization (WHO) reports that millions globally suffer from foodborne illnesses, with contaminated food being a primary contributor to morbidity and mortality. This underscores the necessity for proficient food safety analysis to detect dangers, including infections, pollutants, and spoiling agents in food products [2].

Although conventional food safety analysis methods provide certain benefits, they also exhibit significant disadvantages, including prolonged processing durations, elevated prices, and restricted sensitivity and specificity. The complexity of foodborne infections, emerging pollutants, and the worldwide scope of food supply chains exacerbate these challenges [3]. Considering these obstacles, there is an increasing desire for efficient, precise, and sufficiently scalable solutions to address these issues while instilling consumer confidence in their food selections.

Nanotechnology, which includes manipulating matter on an atomic or molecular level (usually at scales less than 100 nanometers), has emerged as a revolutionary force in several sectors, including health care, environmental science, and food safety. Nanomaterials (NMs) stand out because of their unique features, like a high surface area and reactivity, and their applicability for specialized applications, making them particularly beneficial in food safety analysis [4]. For example, nanoparticles (NPs) and nanostructures can be developed to detect exceedingly minute levels of infections, pollutants, and spoiling signs in food. This leads to quick and extremely sensitive testing methods that often surpass older techniques [5].

Nanotechnology offers various benefits in food safety analysis, such as developing innovative packaging solutions, increased food quality monitoring, and rapid detection of foodborne pathogens. Nanosensors can identify infections such as *Salmonella*, *Listeria*, and *E. coli* with superior sensitivity to traditional approaches [6]. Additionally, assays utilizing NMs can correctly identify pollutants such as heavy metals, pesticide residues, and allergies. This novel method can revolutionize food safety processes by facilitating real-time detection and improving the overall efficacy of food safety systems [7]. This chapter will look into the expanding significance of nanotechnology in food safety analysis, demonstrating its ability to handle significant difficulties in the field, such as pathogen identification, contamination assessment, and food quality management. We will also look at how nanotechnology can be coupled with artificial intelligence (AI) to improve food safety systems, wearable nanodevices for field testing, and high-throughput nano-enabled systems for large-scale applications. While the benefits of NMs are enormous, it is vital to address concerns surrounding their toxicity and environmental impact, particularly if they build up in the food chain. Therefore, regulatory standards governing NPs in food products will be the main topic of our discussion. We will also examine consumer views and ethical implications, as the success of nanotechnology-based food safety measures rests not only on their effectiveness but also on obtaining public trust and knowledge. Finally, we will discuss safety and regulatory issues surrounding the use of NPs in food safety.

Overview of Nanotechnology

Nanotechnology is fundamentally concerned with manipulating matter at the atomic and molecular levels. This topic concerns the development of nanostructures and NMs tailored to specific functions. Because of their increased surface area, NMs are significantly more reactive and may quickly interact with target molecules, making them extremely useful for detection and analysis activities [8]. The foundations of nanotechnology are built on the collaboration of various scientific fields, such as chemistry, biology, physics, and materials science. Techniques such as bottom-up synthesis, which includes self-assembly, and top-down approaches, such as lithography, are often used to create nanostructures with precise sizes and capabilities.

Types of NMs Used in Food Safety

NMs can be classified according to their composition, structure, and functionality (Fig. 1). Here is an overview of some of the NMs often utilized in food safety research:

NPs

The three main NMs used in food safety—gold, silver, and magnetic—have unique properties and uses. Au-NPs are highly valued for their superior optical qualities and compatibility with biological systems, which make them perfect for biosensors and colorimetric tests that identify toxins and pathogens. They can also use localized surface plasmon resonance (LSPR), which allows for a highly sensitive detection method [9]. However, Ag-NPs are notable for their potent anti-bacterial qualities and are frequently added to surface coatings and packaging materials to prevent microbial contamination. Additionally, they are essential to water filtration systems since they efficiently eliminate infections [10]. The main application of magnetic NPs is the rapid isolation and concentration of particular target analytes from intricate food matrices, such as pathogens and heavy metals. They are useful in food safety monitoring and diagnostics because of their magnetic qualities, which make separation and analysis simple [11].

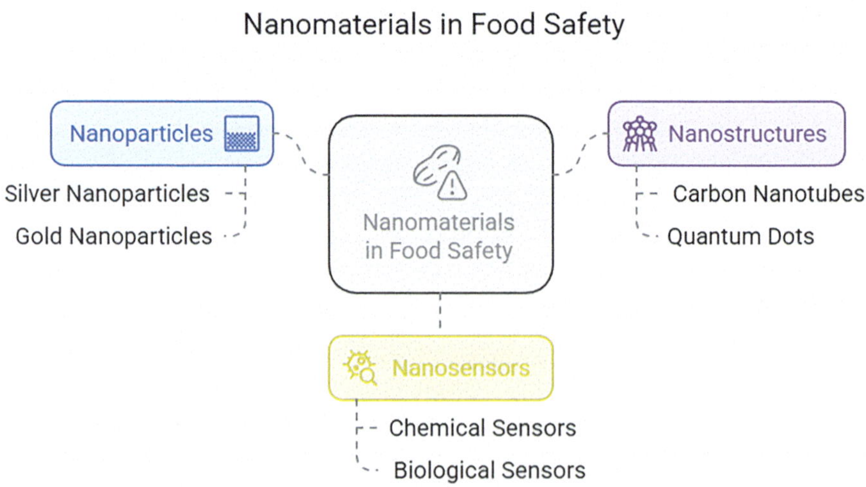

Fig. 1 Various NMs used in food safety

Nanostructures

Nanotubes, nanowires, and nanosheets are multifunctional nanomaterials (NMs), each demonstrating unique features advantageous in food safety and detection. Carbon nanotubes, for example, are known for their outstanding electrical conductivity and good strength-to-weight ratio, making them excellent for sensor platforms focused on detecting chemical pollutants. Multi-walled and single-walled nanotubes can be functionalized, leading to bespoke solutions for specific detection needs [12]. Nanowires are great for real-time monitoring applications because of their one-dimensional structure, giving them extraordinary sensitivity to environmental changes. They are frequently found in chemical sensors and electronic noses that detect volatile substances. In the meantime, nanosheets such as graphene and other two-dimensional materials are prized for their substantial surface area and improved electrical characteristics, which increase their usefulness in sophisticated sensor design [13]. Nanosheets can be functionalized to bind to target analytes selectively, increasing the sensitivity and specificity of detection. These materials promise to improve food safety by developing more effective and sensitive detection systems [14].

Nanosensors

Because of their remarkable sensitivity, selectivity, and fast reaction times, nanosensors, advanced analytical instruments that combine sensing elements with NMs to detect specific chemical or biological substances, are essential for food safety because they can identify contaminants, pathogens, and toxins, guaranteeing food safety and quality [15]. There are several types of nanosensors, including piezoelectric, optical, and electrochemical sensors. Electrochemical sensors detect chemical or biological contaminants by detecting changes in electrical properties. Conversely, optical sensors assess environmental changes using light interactions [16]. Meanwhile, piezoelectric sensors use frequency shifts to identify changes in mass or force. Each type of nanosensor offers unique advantages, enabling customized solutions tailored to the different detection demands in food safety monitoring [17].

Properties of NMs

NMs are perfect for analyzing food safety because of their special properties. Because of their vast surface area and high surface-to-volume ratio, they interact with analytes more effectively and increase detection sensitivity. Porous NPs, for instance, can efficiently separate and concentrate trace amounts of pollutants, increasing detection accuracy [18]. Maintaining food safety requires the ability to detect even the smallest amounts of contaminants, pathogens, or spoiling indicators, and the great sensitivity of NPs facilitates this process. They can be functionalized with biomolecules like enzymes, antibodies, or nucleic acids to increase their selectivity towards target analytes [19].

Two other crucial characteristics of NMs are stability and reactivity. While reactivity can be adjusted for diverse applications, stability guarantees constant performance under various environmental conditions, supporting real-time monitoring operations. Novel sensing platforms are also made possible by NMs' optical and electrical characteristics, such as conductivity, plasmon resonance, and fluorescence [20]. Quantum dots, for example, have size-dependent fluorescence, making them especially helpful in multiplex assays, which detect several targets simultaneously. These special qualities allow NPs to greatly improve food safety and aid in the creation of more dependable and efficient monitoring systems [21].

Nanotechnology in Food Safety Analysis

The methods used to identify and track pollutants, infections, and general food quality have changed due to the application of nanotechnology in food safety analysis. Compared to traditional procedures, nanotechnology-derived solutions show speed, sensitivity, and cost-effectiveness advantages. The main areas where nanotechnology significantly impacts are examined in the following sections (Fig. 2).

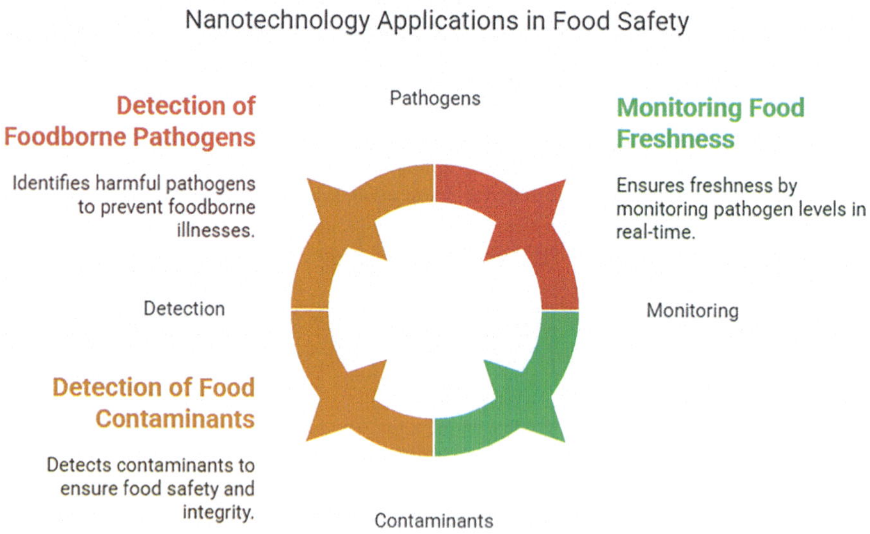

Fig. 2 Applications of nanotechnology in food safety analysis

Detection of Foodborne Pathogens

Salmonella, Escherichia coli, and *Listeria monocytogenes* are the main bacteria that cause foodborne illnesses, a serious worldwide problem. To stop outbreaks and guarantee food safety, detecting these illnesses as soon as possible is crucial. By fusing biological recognition components with NMs, nanosensors enable highly sensitive and precise pathogen identification, an essential step in this detection process [22] (Table 1). For example, Au-NPs coated with DNA probes or antibodies are frequently used in colorimetric testing. In these assays, the presence of a pathogen is indicated by a discernible color shift [40]. Furthermore, nanostructured biosensors use materials like graphene, carbon nanotubes, and quantum dots to improve sensitivity and accuracy in pathogen detection. To precisely detect infections, these

Table 1 Different types of NPs, their effects on foodborne pathogens, and the diseases caused by these pathogens

Nanoparticle type	Foodborne pathogen	Diseases caused	References
Ag-NPs	*Escherichia coli* (EHEC, ETEC)	Diarrhea, hemolytic uremic syndrome	[23]
	Salmonella spp.	Salmonellosis (gastroenteritis, typhoid fever)	[24]
	Vibrio parahaemolyticus	Acute gastroenteritis	[25]
ZnO-NPs	*Listeria monocytogenes*	Listeriosis (meningitis, sepsis)	[26]
	Bacillus cereus	Emetic and diarrheal syndromes	[27]
TiO_2-NPs	*Staphylococcus aureus*	Food poisoning (enterotoxin-mediated illness)	[28]
	Clostridium botulinum	Botulism (paralysis, respiratory failure)	[29]
Cu-NPs	*Campylobacter jejuni*	Campylobacteriosis (gastroenteritis, Guillain-Barré syndrome)	[30]
	Shigella spp.	Shigellosis (bacillary dysentery)	[31]
Chitosan-NPs	*Vibrio cholerae*	Cholera	[32]
	Helicobacter pylori	Gastritis, peptic ulcers, gastric cancer	[33]
GO-NPs	*Clostridium perfringens*	Foodborne toxicoinfection (gas gangrene, diarrhea)	[34]
Fe_3O_4-NPs	*Yersinia enterocolitica*	Yersiniosis (gastroenteritis, septicemia)	[35]
Au-NPs	*Norovirus*	Viral gastroenteritis	[36]
	Rotavirus	Severe diarrhea, dehydration	[37]
CNTs	*Pseudomonas aeruginosa*	Opportunistic infections (UTIs, pneumonia)	[38]
SiO_2-NPs	*Enterococcus faecalis*	Endocarditis, urinary tract infections	[39]

sensors usually use detection techniques like surface plasmon resonance, electrochemical signals, or fluorescence [41]. Because they are particularly made to target specific diseases, Au-NPs are frequently employed in lateral flow assays for rapid, on-the-spot pathogen identification. Quantum dots also make it easier to multiplex in fluorescence-based detection, which enables the identification of several diseases in a single test [42].

Detection of Food Contaminants

Pollutants, including pesticides, heavy metals, and allergies, pose health hazards. Nanotechnology provides new tools to address this problem, enabling us to precisely identify these dangerous compounds [43] (Table 2). For example, chelating agent-tailored magnetic NPs can detect heavy metals and poisons like lead, cadmium, and mercury. These unique NPs are excellent at separating and absorbing heavy metals, making it possible to create sensors, especially for quick field testing [63]. Furthermore, pesticide residues can be effectively detected using assays based on NMs that use elements like carbon nanotubes (CNTs) or Ag-NPs. It is simpler to find minute levels of pesticide residues in food with these techniques, which usually rely on optical or electrochemical signals. This method provides a sensitive and trustworthy technique to monitor food safety [64]. Additionally, allergens, including gluten, peanuts, and shellfish proteins, can be swiftly and precisely identified using DNA-based sensors and nano-enhanced immunoassays. Developing these cutting-edge instruments is essential to protecting individuals vulnerable to food allergies and guaranteeing that food items fulfill safety requirements for all [65].

Monitoring Food Freshness and Quality

With real-time sensor systems and innovative packaging, nanotechnology is opening the door to some truly remarkable ways to monitor the freshness and quality of food. These innovative packaging solutions use nanosensors and nano-based materials to check crucial spoiling indicators like pH levels, gas emissions, or bacterial growth. Colorimetric nanosensors, for example, can change color when food begins to degrade, providing producers and consumers with a straightforward visual cue [66]. Real-time monitoring is made possible by incorporating these nanosensors into supply chain tracking systems. This reduces waste and spoiling by constantly monitoring food quality, ensuring that goods remain fresh from preparation until they are placed on your plate. Additionally, nanosensors that identify gases like ammonia or volatile organic compounds (VOCs) offer crucial information on the deterioration of food [67]. These sensors are essential for identifying early spoiling indicators, which reduces waste and improves food safety by guaranteeing that goods are consumed at their freshest. These nanotechnology-based solutions significantly improve food sustainability, safety, and preservation [68].

Table 2 Different types of NPs, their impact on food contaminants, and the diseases they can cause

Nanoparticle type	Food contaminant	Diseases caused	References
Ag-NPs	Heavy metals (lead, mercury, cadmium)	Neurotoxicity, kidney damage, developmental disorders	[44]
	Mycotoxins (*Aspergillus* spp., *Fusarium* spp.)	Liver cancer, immunosuppression	[45]
	Pesticide residues (organophosphates, carbamates)	Endocrine disruption, neurological disorders	[46]
ZnO-NPs	*Bacillus cereus* toxins	Food poisoning, diarrhea	[47]
	Mycotoxins (*aflatoxins, ochratoxins*)	Hepatotoxicity, nephrotoxicity	[48]
	Food preservatives (sodium nitrite)	Carcinogenesis, DNA damage	[49]
TiO_2-NPs	Synthetic dyes (Tartrazine, sunset yellow, Allura red)	Allergies, hyperactivity, DNA damage	[50]
	Nanoplastic contaminants	Gut microbiota imbalance, inflammation	[51]
Cu-NPs	Pathogenic bacteria (*Salmonella, E. coli, Listeria*)	Gastroenteritis, typhoid fever, listeriosis	[52]
	Antibiotic residues	Antibiotic resistance, dysbiosis	[53]
Chitosan NPs	Pesticide residues (organochlorines, neonicotinoids)	Endocrine disruption, cancer risk	[54]
	Heavy metals (arsenic, chromium)	Carcinogenesis, liver toxicity	[55]
GO-NPs	Heavy metals (cadmium, arsenic, lead)	Carcinogenic effects, organ damage	[56]
	Toxins from *Clostridium botulinum*	Botulism (paralysis, respiratory failure)	[57]
Fe_3O_4-NPs	Foodborne parasites (*Toxoplasma gondii, Cryptosporidium*)	Neurological disorders, gastrointestinal illness	[58]
Au-NPs	Foodborne viruses (*Norovirus, rotavirus, Hepatitis A virus*)	Viral gastroenteritis, hepatitis	[36]
	Biogenic amines (histamine, putrescine)	Allergic reactions, hypertension	[59]
CNTs	Polycyclic aromatic hydrocarbons (PAHs)	DNA damage, carcinogenesis	[60]
SiO_2-NPs	Microplastics	Hormonal imbalance, gut inflammation	[61]
	Foodborne fungi (*Penicillium* spp., *Fusarium* spp.)	Mycotoxicosis, respiratory issues	[62]

Advances in Nanotechnology for Food Safety

The development of nanotechnology has led to essential uses in the food safety sector, particularly addressing problems related to monitoring, detection, and intervention. These advancements use cutting-edge breakthroughs to improve the scalability, accuracy, and efficiency of food safety analyses.

Integration of Artificial Intelligence with Nanosensors

Nanosensor technology's integration of artificial intelligence (AI) has greatly enhanced data processing and decision-making capabilities. To identify trends, evaluate dangers, and provide instant responses, artificial intelligence (AI) algorithms are used to analyze the massive amounts of data generated by nanosensors. AI-driven nanosensor systems, which can predict the shelf life of perishable commodities by considering environmental parameters like temperature and humidity, are an example of this use [69].

Wearable and Portable Nanodevices

The development of portable nanodevices has made testing and monitoring food safety on-site easier. These gadgets are suitable for usage in remote areas or by people without specific training because of their lightweight design, low cost, and simplicity of use. Food workers can evaluate contamination concerns in real time with wearable nanosensors, such as those built into gloves or wristbands. These advancements have enhanced the ability to address food safety concerns in quickly evolving contexts [70].

High-Throughput Nano-Enabled Systems

Nano-enabled technologies have been devised to facilitate comprehensive testing of food products. These systems combine robotics, automation, and nanotechnology, providing high-throughput analysis. Testing methods require less time and labor due to this integration [71]. For instance, the simultaneous examination of many food samples is made possible by microfluidic systems that integrate nanosensors. Thus, this development strengthens safety assurance in large-scale food production and increases the effectiveness of quality control procedures [72].

Multifunctional NMs

Multifunctional NMs have been developed to perform various tasks, such as contamination detection and cleanup. An example is provided by photocatalytic NMs, which may both provide analytical feedback and break down pesticide residues.

Because of this dual functionality, there is less need for separate processes, which saves time and money. Furthermore, this strategy helps to reduce environmental hazards [73].

Green Nanotechnology

Green nanotechnology is focused on the development of sustainable and environmentally friendly NMs. The synthesis of these materials involves procedures that are benign to the environment and result in low toxicity levels. This topic addresses concerns linked to safety and corresponds with global aims for sustainability, presenting a promising option for future research endeavors. For example, manufacturing NPs using plant-based technologies has emerged as a potential way to limit environmental impact while maintaining effectiveness [74].

Safety and Regulatory Considerations

The advancement of nanotechnology is considerably revolutionizing food safety; nonetheless, examining the accompanying safety and regulatory difficulties that come from its implementation is vital. Although NPs provide significant benefits for detection and monitoring purposes, their particular properties may also threaten human health and the environment.

Potential Risks of Nanotechnology in Food Applications

The expanding utilization of NMs has produced substantial anxiety over their toxicity, environmental implications, and capacity for bioaccumulation. The small size and heightened reactivity of NPs, such as silver and titanium dioxide, may result in unanticipated biological interactions. These interactions can generate oxidative stress, DNA damage, and cellular toxicity, with the effects determined by various parameters, including particle size, shape, concentration, and surface chemistry. Furthermore, the extensive application of these materials may lead to their buildup in the environment, particularly through waste disposal methods and food packaging. This accumulation might result in their persistence in soil and water, which may damage natural processes [75]. The prospect of bioaccumulation in organisms, especially within aquatic species, exacerbates these concerns since NPs may integrate into the food chain and harm higher trophic levels, including humans. These challenges underline the importance of rigorous research to understand better the fate, behavior, and enduring implications of NMs on public health and the ecosystem.

Regulatory Frameworks for NMs in Food Safety

Regulatory bodies such as the Food and Drug Administration (FDA), the European Food Safety Authority (EFSA), and the Codex Alimentarius Commission have established global standards and guidelines for using NMs, especially in the food industry. The safe use of NMs depends on risk assessment, safety evaluation, and label transparency, all prioritized in these frameworks. Nanomaterial risk assessment procedures include a detailed analysis of the materials' physicochemical characteristics, toxicity, exposure levels, and environmental impacts [76]. Following standardized testing procedures, like those offered by the Organization for Economic Co-operation and Development (OECD), guarantees uniformity and dependability in these evaluations. However, current regulatory frameworks face significant issues due to the continuously changing nature of nanotechnology. Traditional risk assessment techniques may not adequately address the unique properties of NPs, necessitating the creation of novel and more potent instruments for their impact analysis and safety assessment [77].

Consumer Perceptions and Ethical Considerations

The widespread implementation of nanotechnology greatly depends on public awareness and acceptance. Consumer attitudes, which are commonly impacted by disinformation or a lack of understanding regarding the safety and benefits of NMs, may limit their acceptance. Therefore, good communication and involvement with the public are crucial for developing trust and confidence in nanotechnology [78]. It is also critical to address ethical issues, especially those about unequal access to nanotechnology and the differences in its advantages. The responsible development and use of nanotechnology must be guided by ethical frameworks, highlighting the significance of social justice, environmental sustainability, and public health in applying these cutting-edge technologies [79].

Mitigation Strategies

Because safety-by-design methods make integrating safety concerns into the design and synthesis of NMs easier, they are essential for lowering the associated risks. Using surface changes, for instance, may reduce toxicity and increase biocompatibility, making NPs safer to use. Addressing environmental challenges can also be significantly aided by developing sustainable practices, such as green nanotechnology and eco-friendly synthesis techniques [80]. Adopting sustainable practices is further encouraged by implementing lifecycle assessments for products enabled by nanotechnology. Furthermore, addressing safety and regulatory issues requires cooperative efforts involving researchers, legislators, and business representatives. The responsible development of nanotechnology can be facilitated by interdisciplinary initiatives that promote innovation while guaranteeing adherence to safety regulations and standards [81].

Challenges and Future Directions in Nanotechnology for Food Safety

Nanotechnology has shown itself to be a revolutionary approach to tackling important food safety challenges with more sensitive, effective, and efficient solutions than conventional techniques. Nevertheless, despite its enormous promise, several obstacles exist in commercializing and broad use of nanotechnology in food safety. Unlocking nanotechnology's full potential in guaranteeing global food safety requires addressing these issues and determining future approaches.

Challenges in Nanotechnology for Food Safety

Safety Concerns and Toxicity

One of the biggest hurdles encountered in implementing nanotechnology within food safety is the possible toxicity associated with NMs. The particular properties of these materials, especially their higher surface-to-volume ratio and strong reactivity, may result in unexpected interactions with biological systems. Research reveals that particular NMs, such as silver NPs, could represent cytotoxicity, genotoxicity, and bioaccumulation hazards in the human organism [82]. Additionally, the environmental ramifications of NMs, especially their possible accumulation within ecosystems, represent an ongoing worry. Consequently, a compelling need exists for substantial toxicological research and risk assessments to guarantee the safe deployment of nanotechnology in food-related applications [83].

Regulatory and Standardization Challenges

One major barrier to the commercialization of nanotechnology is the lack of standardized regulatory frameworks controlling its use in food safety. Regulatory organizations like the European Food Safety Authority and the US Food and Drug Administration have not yet developed clear standards for evaluating and approving food safety projects based on nanotechnology [84]. This ambiguity makes producers uneasy and more challenging to incorporate nanotechnology into the food industry. Moreover, the regulatory environment is made more complex by the lack of standardized testing procedures for assessing the efficacy and safety of NMs [85].

Cost and Scalability

Incorporating nanotechnology into current food safety systems may necessitate careful consideration. Nanomaterial synthesis and the development of nano-enabled devices frequently incur significant costs, limiting their availability and scalability for widespread applications.

Public Perception and Consumer Acceptance

Consumer skepticism and negative attitudes surrounding the implementation of nanotechnology in the food business represent a considerable hurdle. Numerous customers express apprehensions concerning NMs' safety and ethical consequences in food production and packaging operations. This resistance typically stems from a shortage of understanding and awareness concerning both the advantages and concerns of nanotechnology. Establishing public trust involves clear communication, education, and active involvement to clarify misconceptions and underline the benefits of nanotechnology for food safety [86].

Technical Challenges

Introducing nanotechnology into applications linked to food safety confronts various technological hurdles. One of the significant obstacles is the complexity of food matrices, wherein the detection of contaminants may prove challenging due to the presence of compounds that interfere with the performance of nanosensors and diagnostic devices. Furthermore, it is necessary to ensure the longevity and stability of nano-enabled sensors and packaging materials; these components must keep their dependability and functioning despite variations in environmental conditions to achieve efficacy in actual applications [87]. Moreover, the application of nanosensors and systems enabled by the Internet of Things (IoT) in food safety results in the collection of considerable data. Consequently, a compelling demand arises for robust data management and analytical systems capable of processing and understanding this information proficiently. Such problems underscore the demand for constant innovation and refinement in nanotechnology's technological and data management elements concerning food safety [88].

Future Directions in Nanotechnology for Food Safety

Given the problems mentioned above, the prospects of nanotechnology in the realm of food safety appear to be optimistic. Continuous advancements are helping the development of fresh remedies.

Development of Safer and Biodegradable NMs

Current research efforts are increasingly aimed at producing NSMs that offer minimum dangers to human health and the environment. Biodegradable NPs, particularly those generated from natural polymers such as chitosan and cellulose, present a sustainable alternative to standard NMs. These compounds minimize toxicity concerns and match the rising focus on sustainability and environmental responsibility within food systems [89].

Advancements in Smart Packaging

Using nanotechnology in innovative packaging will contribute considerably to food safety and quality assurance. Innovative packaging solutions are anticipated to incorporate nanosensors that can monitor environmental conditions and indicators of deterioration in real time. Furthermore, when these technologies are paired with Internet of Things (IoT) connectivity, they will enable stakeholders to access actionable information to boost supply chain transparency and lower food waste [90].

Integration with Artificial Intelligence (AI) and Machine Learning

Combining nanotechnology, artificial intelligence (AI), and machine learning could revolutionize food safety analysis. Artificial intelligence (AI)-powered algorithms can analyze large datasets generated by nanosensors, which can help identify pollution trends, forecast related hazards, and improve detection techniques. Additionally, machine learning models can increase the sensitivity and precision of nanotechnology-enhanced diagnostic instruments, enabling more accurate contamination diagnosis [91].

Portable and Cost-Effective Solutions

To reduce reliance on centralized laboratories and improve response times, future research efforts will focus on developing portable, user-friendly, and cost-effective nanotechnology-based devices for field testing applications, such as handheld nanosensors and lateral flow assays, which will allow food producers, distributors, and consumers to conduct on-site testing [92].

Harmonized Regulatory Frameworks

To encourage the commercialization of nanotechnology in food safety, regulatory bodies must provide coherent frameworks and clear instructions on assessing, approving, and labeling goods that use nanotechnology. Additionally, it will be crucial for governments, industry stakeholders, and international organizations to work together to create standardized procedures.

Public Engagement and Education

Effective communication tactics that address safety concerns and emphasize the benefits of nanotechnology are crucial for resolving consumer concerns and fostering public trust in the technology. To close the gap between public perception and scientific advancement, tactics including educational campaigns, open labeling, and active stakeholder involvement are essential. Such initiatives will make the increased adoption of nanotechnology concerning food safety possible.

Collaborative and Interdisciplinary Research

Collaboration and multidisciplinary research projects involving scientists, engineers, regulatory agencies, and industry players are essential to developing nanotechnology in food safety. Freire and Gonçalves state that these cooperative efforts are crucial for accelerating innovation, overcoming technological and legal challenges, and guaranteeing the moral development of nano-enabled food safety solutions [93].

Conclusion

Nanotechnology offers the capacity to dramatically improve food safety analysis by delivering imaginative solutions to satisfy the rising issues experienced by the food sector. The special qualities of NMs, which include raised surface area, heightened sensitivity, and reactivity, permit the creation of highly effective detection systems to recognize foodborne pathogens, pollutants, and signs of degradation. Advancements coming from nanotechnology, such as nanosensors and nano-enabled packaging, allow rapid, cost-efficient, and more dependable techniques for preserving food safety across the supply chain. Evidence indicates that NPs can detect pathogens such as *E. coli*, *Salmonella*, and *Listeria* with improved sensitivity and speed compared to current techniques, allowing real-time monitoring inside food production systems. Furthermore, devices based on nanotechnology are growing as essential resources for identifying contaminants such as heavy metals, herbicides, and allergies, which typically pose challenges for conventional monitoring systems. Integrating nanotechnology with artificial intelligence (AI) and wearable devices dramatically increases food safety systems by permitting intelligent, quick testing. Nevertheless, the widespread implementation of nanotechnology in food safety has several difficulties. It is vital to adequately assess NMs' safety, toxicity, and environmental consequences, alongside the necessity to construct regulatory frameworks that oversee their right utilization in food goods. Additionally, public perception and ethical issues regarding the employment of NMs in food remain crucial variables impacting the acceptability of these technologies. In summary, nanotechnology offers enormous potential for the progress of food safety analysis and the enhancement of food quality assurance. As research improves and regulatory needs increase, nanotechnology is anticipated to become a key aspect of contemporary food safety systems, contributing to the protection of public health while promoting innovation within the food sector. The future landscape of food safety is predicted to be influenced by the continual investigation and application of nanotechnology into global food systems.

References

1. Fung F, Wang HS, Menon S (2018) Food safety in the 21st century. Biom J 41(2):88–95
2. Varghese SM, Parisi S, Singla RK, Begum AA (2022) Food safety and quality control in food industry. In: Trends in food chemistry, nutrition and technology in Indian sub-continent. Springer, Cham, pp 31–44
3. Lehotay SJ (2018) Food safety analysis. Anal Bioanal Chem 410:5329–5330
4. Madkour LH, Madkour LH (2019) Introduction to nanotechnology (NT) and nanomaterials (NMs). In: Nanoelectronic materials: fundamentals and applications. Springer, Cham, pp 1–47
5. Banerjee A, Maity S, Mastrangelo CH (2021) Nanostructures for biosensing, with a brief overview on cancer detection, IoT, and the role of machine learning in smart biosensors. Sensors 21(4):1253
6. Bajpai VK, Kamle M, Shukla S, Mahato DK, Chandra P, Hwang SK, Han YK (2018) Prospects of using nanotechnology for food preservation, safety, and security. J Food Drug Anal 26(4):1201–1214
7. Yang T, Huang H, Zhu F, Lin Q, Zhang L, Liu J (2016) Recent progresses in nanobiosensing for food safety analysis. Sensors 16(7):1118
8. Jain K, Patel AS, Pardhi VP, Flora SJS (2021) Nanotechnology in wastewater management: a new paradigm towards wastewater treatment. Molecules 26(6):1797
9. Karnwal A, Kumar Sachan RS, Devgon I, Devgon J, Pant G, Panchpuri M, Kumar G (2024) Gold nanoparticles in nanobiotechnology: from synthesis to biosensing applications. ACS Omega 9(28):29966–29982
10. Bruna T, Maldonado-Bravo F, Jara P, Caro N (2021) Silver nanoparticles and their antibacterial applications. Int J Mol Sci 22(13):7202
11. Franci G, Falanga A, Galdiero S, Palomba L, Rai M, Morelli G, Galdiero M (2015) Silver nanoparticles as potential anti-bacterial agents. Molecules 20(5):8856–8874

12. Gupta N, Gupta SM, Sharma SK (2019) Carbon nanotubes: synthesis, properties and engineering applications. Carbon Lett 29:419–447
13. Salhi B, Hossain MK, Mukhaimer AW, Al-Sulaiman F (2016) Nanowires: a new pathway to nanotechnology-based applications. J Electroceram 37:34–49
14. Gu J, Shen Y, Tian S, Xue Z, Meng X (2023) Recent advances in nanowire-based wearable physical sensors. Biosensors 13(12):1025
15. Sargazi S, Fatima I, Kiani MH, Mohammadzadeh V, Arshad R, Bilal M, Behzadmehr R (2022) Fluorescent-based nanosensors for selective detection of a wide range of biological macromolecules: a comprehensive review. Int J Biol Macromol 206:115–147
16. Ramesh M, Janani R, Deepa C, Rajeshkumar L (2022) Nanotechnology-enabled biosensors: a review of fundamentals, design principles, materials, and applications. Biosensors 13(1):40
17. Toydemir G, Cekic SD, Ozkan G, Uzunboy S, Avan AN, Capanoglu E, Apak R (2020) Nanosensors for foods. In: Nano-food engineering, vol 1. Springer, Cham, pp 327–375
18. Pattnaik BK, Sahu C, Moulick D (2024) Nanoparticles in soil–water–plant inference. In: The nanotechnology driven agriculture. CRC Press, pp 85–105
19. Torres-Giner S, Prieto C, Lagaron JM (2020) Nanomaterials to enhance food quality, safety, and health impact. Nano 10(5):941
20. Hafeez A (2024) Two-dimensional nanomaterials. In: Engineering materials: fundamentals, processing and properties. Springer, Cham, pp 181–204
21. Phan HT, Haes AJ (2019) What does nanoparticle stability mean? J Phys Chem C 123(27):16495–16507
22. Bintsis T (2017) Foodborne pathogens. AIMS Microbiol 3(3):529
23. Gabrielyan L, Badalyan H, Gevorgyan V, Trchounian A (2020) Comparable anti-bacterial effects and action mechanisms of silver and iron oxide nanoparticles on Escherichia coli and salmonella typhimurium. Sci Rep 10(1):13145
24. Losasso C, Belluco S, Cibin V, Zavagnin P, Mičetić I, Gallocchio F, Ricci A (2014) Antibacterial activity of silver nanoparticles: sensitivity of different salmonella serovars. Front Microbiol 5:227
25. Marukurti A, Reddy AM, Medapalli SR, Kasi PB, Rao PT, Anupoju R, Nirmala PV (2025) Evaluation of anti-vibriocidal, antioxidant properties and cytotoxicity of bio fabricated/green synthesised silver nanoparticles using euphorbia hitra L. leaf extract. Next Mater 7:100355
26. Jin T, Sun D, Su JY, Zhang HW, Sue HJ (2009) Antimicrobial efficacy of zinc oxide quantum dots against listeria monocytogenes, salmonella enteritidis, and Escherichia coli O157: H7. J Food Sci 74(1):M46–M52
27. Jin SE, Jin HE (2021) Antimicrobial activity of zinc oxide nano/microparticles and their combinations against pathogenic microorganisms for biomedical applications: from physicochemical characteristics to pharmacological aspects. Nano 11(2):263
28. Durango-Giraldo G, Cardona A, Zapata JF, Santa JF, Buitrago-Sierra R (2019) Titanium dioxide modified with silver by two methods for bactericidal applications. Heliyon 5(5):e01608
29. de Dicastillo CL, Patiño C, Galotto MJ, Vásquez-Martínez Y, Torrent C, Alburquenque D, Escrig J (2019) Novel hollow titanium dioxide nanospheres with antimicrobial activity against resistant bacteria. Beilstein J Nanotechnol 10(1):1716–1725
30. Benoit SL, Maier RJ (2022) Copper toxicity towards *campylobacter jejuni* is enhanced by the nickel chelator dimethylglyoxime. Metallomics 14(1):mfab076
31. Shabatina TI, Vernaya OI, Melnikov MY (2023) Hybrid nanosystems of antibiotics with metal nanoparticles—novel anti-bacterial agents. Molecules 28(4):1603
32. Saberpour M, Bakhshi B, Najar-Peerayeh S (2020) Evaluation of the antimicrobial and antibiofilm effect of chitosan nanoparticles as carrier for supernatant of mesenchymal stem cells on multidrug-resistant *vibrio cholerae*. Infect Drug Resistance 13:2251–2260
33. Moosazadeh Moghaddam M, Bolouri S, Golmohammadi R, Fasihi-Ramandi M, Heiat M, Mirnejad R (2023) Targeted delivery of a short antimicrobial peptide (CM11) against helicobacter pylori gastric infection using concanavalin A-coated chitosan nanoparticles. J Mater Sci Mater Med 34(9):44

34. Kumar P, Huo P, Zhang R, Liu B (2019) Anti-bacterial properties of graphene-based nanomaterials. Nanomaterials 9(5):737
35. Gabrielyan L, Hovhannisyan A, Gevorgyan V, Ananyan M, Trchounian A (2019) Anti-bacterial effects of iron oxide (Fe3O4) nanoparticles: distinguishing concentration-dependent effects with different bacterial cells growth and membrane-associated mechanisms. Appl Microbiol Biotechnol 103(6):2773–2782
36. Broglie JJ, Alston B, Yang C, Ma L, Adcock AF, Chen W, Yang L (2015) Antiviral activity of gold/copper sulfide core/shell nanoparticles against human norovirus virus-like particles. PLoS One 10(10):e0141050
37. Ayipo YO, Bakare AA, Badeggi UM, Jimoh AA, Lawal A, Mordi MN (2022) Recent advances on therapeutic potentials of gold and silver nanobiomaterials for human viral diseases. Curr Res Chem Biol 2:100021
38. Banihashemi K, Sobouti B, Mehregan I, Bakhtiari R, Amirmozafari N (2020) The construction of carbon nanotubes containing an anti-bacterial chemical component and its effect on MDR and XDR isolates of *Pseudomonas aeruginosa*. Rep Biochem Mol Biol 9(1):89
39. Hameed S, Antony DP, Shanmugam R, Raghu S, Adimulapu HS (2024) Enhancing antimicrobial efficacy and synergistic effects of nano-silica-based combinations with doxycycline, metronidazole, and ciprofloxacin against *enterococcus faecalis* biofilms. Cureus 16(2):e54668
40. Aladhadh M (2023) A review of modern methods for the detection of foodborne pathogens. Microorganisms 11(5):1111
41. Kaya HO, Cetin AE, Azimzadeh M, Topkaya SN (2021) Pathogen detection with electrochemical biosensors: advantages, challenges and future perspectives. J Electroanal Chem 882:114989
42. Banakar M, Hamidi M, Khurshid Z, Zafar MS, Sapkota J, Azizian R, Rokaya D (2022) Electrochemical biosensors for pathogen detection: an updated review. Biosensors 12(11):927
43. Alengebawy A, Abdelkhalek ST, Qureshi SR, Wang MQ (2021) Heavy metals and pesticides toxicity in agricultural soil and plants: ecological risks and human health implications. Toxics 9(3):42
44. Sudarman F, Shiddiq M, Armynah B, Tahir D (2023) Silver nanoparticles (AgNPs) synthesis methods as heavy-metal sensors: a review. Int J Environ Sci Technol 20(8):9351–9368
45. Hashem AH, Saied E, Amin BH, Alotibi FO, Al-Askar AA, Arishi AA, Elbahnasawy MA (2022) Antifungal activity of biosynthesised silver nanoparticles (AgNPs) against aspergilli causing aspergillosis: ultrastructure study. J Funct Biomater 13(4):242
46. Albadawi EA, Musa ENA, Ghaban HM, Ebrahim NA, Albadrani MS, El-Tokhy AI (2024) Eco-friendly green synthesis of silver NPs from guajava leaves extract for controlling organophosphorus pesticides hazards, characterisation, and in-vivo toxicity assessment. BMC Pharmacol Toxicol 25(1):1–27
47. Sirelkhatim A, Mahmud S, Seeni A, Kaus NHM, Ann LC, Bakhori SKM, Mohamad D (2015) Review on zinc oxide nanoparticles: anti-bacterial activity and toxicity mechanism. Nanomicro Lett 7:219–242
48. Nxumalo KA, Adeyemi JO, Leta TB, Pfukwa TM, Okafor SN, Fawole OA (2024) Antifungal properties and molecular docking of ZnO NPs mediated using medicinal plant extracts. Sci Rep 14(1):18071
49. Gökmen GG, Mirsafi FS, Leißner T, Akan T, Mishra YK, Kışla D (2024) Zinc oxide nanomaterials: safeguarding food quality and sustainability. Compr Rev Food Sci Food Saf 23(6):e70051
50. Mandal AK, Katuwal S, Tettey F, Gupta A, Bhattarai S, Jaisi S, Parajuli N (2022) Current research on zinc oxide nanoparticles: synthesis, characterisation, and biomedical applications. Nano 12(17):3066
51. Kumar AA, Jain RK (2024) Synthesis and characterisation of the zinc-oxide: tin-oxide nanoparticle composite and assessment of its anti-bacterial activity: an in vitro study. Cureus 16(1):e53016

52. Hesabizadeh T, Sung K, Park M, Foley S, Paredes A, Blissett S, Guisbiers G (2023) Synthesis of anti-bacterial copper oxide nanoparticles by pulsed laser ablation in liquids: potential application against foodborne pathogens. Nano 13(15):2206
53. Ma X, Zhou S, Xu X, Du Q (2022) Copper-containing nanoparticles: mechanism of antimicrobial effect and application in dentistry-a narrative review. Front Surg 9:905892
54. Zhang Z, Yang N, Yu J, Jin S, Shen G, Chen H, Qian K (2023) Research progress of a pesticide polymer-controlled release system based on polysaccharides. Polymers 15(13):2810
55. Almutairi FM, El Rabey HA, Alalawy AI, Salama AA, Tayel AA, Mohammed GM, Zayed MM (2021) Application of chitosan/alginate nanocomposite incorporated with phycosynthesized iron nanoparticles for efficient remediation of chromium. Polymers 13(15):2481
56. Husein DZ, Hassanien R, Khamis M (2021) Cadmium oxide nanoparticles/graphene composite: synthesis, theoretical insights into reactivity and adsorption study. RSC Adv 11(43):27027–27041
57. Shi J, Guo J, Bai G, Chan C, Liu X, Ye W, Yang M (2015) A graphene oxide based fluorescence resonance energy transfer (FRET) biosensor for ultrasensitive detection of botulinum neurotoxin a (BoNT/a) enzymatic activity. Biosens Bioelectron 65:238–244
58. Sharaf M, Sewid AH, Hamouda HI, Elharrif MG, El-Demerdash AS, Alharthi A, Saber T (2022) Rhamnolipid-coated iron oxide nanoparticles as a novel multitarget candidate against major foodborne *E. coli* serotypes and methicillin-resistant *S. aureus*. Microbiol Spectr 10(4):e00250–e00222
59. Bi J, Tian C, Zhang GL, Hao H, Hou HM (2020) Detection of histamine based on gold nanoparticles with dual sensor system of colorimetric and fluorescence. Foods 9(3):316
60. Shrestha B, Anderson TA, Acosta-Martinez V, Payton P, Cañas-Carrell JE (2015) The influence of multi-walled carbon nanotubes on polycyclic aromatic hydrocarbon (PAH) bioavailability and toxicity to soil microbial communities in alfalfa rhizosphere. Ecotoxicol Environ Saf 116:143–149
61. Pu J, Ma J, Li J, Wang S, Zhang W (2023) Organosilicon and inorganic silica inhibit polystyrene nanoparticles uptake in rice. J Hazard Mater 442:130012
62. Ding C, Zhang Y, Chen C, Wang J, Qin M, Gu Y, Luo Y (2024) Hollow mesoporous silica nanoparticles as a new nanoscale resistance inducer for fusarium wilt control: size effects and mechanism of action. Int J Mol Sci 25(8):4514
63. Shetty SS, Deepthi D, Harshitha S, Sonkusare S, Naik PB, Madhyastha H (2023) Environmental pollutants and their effects on human health. Heliyon 9(9):e19496
64. Zhang C, Qiu M, Wang J, Liu Y (2023) Recent advances in nanoparticle-based optical sensors for detection of pesticide residues in soil. Biosensors 13(4):415
65. Jia ME, Zhai F, Bing X (2020) Rapid multi-residue detection methods for pesticides and veterinary drugs. Molecules 25(16):3590
66. Bhatlawande AR, Ghatge PU, Shinde GU, Anushree RK, Patil SD (2024) Unlocking the future of smart food packaging: biosensors, IoT, and nano materials. Food Sci Biotechnol 33(5):1075–1091
67. George J, Kumar R, Aaliya B, Sunooj KV (2023) Packaging solutions for monitoring food quality and safety. In: Engineering aspects of food quality and safety. Springer, Cham, pp 411–442
68. Chhetri KB (2024) Applications of artificial intelligence and machine learning in food quality control and safety assessment. Food Eng Rev 16(1):1–21
69. Liu Q, Li S, Li Z, Zou C, Feng S, Song J, Li X (2025) A novel multimodal nanosensor detection system based on artificial intelligence and two-dimensional Mxenes for Ochratoxin a in food. Food Control 170:111055
70. Hassan RY (2022) Advances in electrochemical nano-biosensors for biomedical and environmental applications: from current work to future perspectives. Sensors 22(19):7539
71. He X, Hwang HM (2016) Nanotechnology in food science: functionality, applicability, and safety assessment. J Food Drug Anal 24(4):671–681
72. Nile SH, Baskar V, Selvaraj D, Nile A, Xiao J, Kai G (2020) Nanotechnologies in food science: applications, recent trends, and future perspectives. Nano-micro Lett 12:1–34

73. Gupta S, Kaur R, Bhardwaj A, Parashar D (2024) Multifunctional nanomaterials: recent advancements in cancer therapeutics and vaccines. Indian J Microbiol 65(1):51–68
74. Hano C, Abbasi BH (2021) Plant-based green synthesis of nanoparticles: production, characterisation and applications. Biomolecules 12(1):31
75. Ray PC, Yu H, Fu PP (2009) Toxicity and environmental risks of nanomaterials: challenges and future needs. J Environ Sci Health C 27(1):1–35
76. Soltani AM, Pouypouy H (2019) Standardisation and regulations of nanotechnology and recent government policies across the world on nanomaterials. In: Advances in phytonanotechnology. Academic Press, New York, pp 419–446
77. Katsiadaki I, Ellis T, Andersen L, Antczak P, Blaker E, Burden N, Jacobs MN (2021) Dying for change: a roadmap to refine the fish acute toxicity test after 40 years of applying a lethal endpoint. Ecotoxicol Environ Saf 223:112585
78. Cormick C (2009) Why do we need to know what the public thinks about nanotechnology? NanoEthics 3(2):167–173
79. Giles EL, Kuznesof S, Clark B, Hubbard C, Frewer LJ (2015) Consumer acceptance of and willingness to pay for food nanotechnology: a systematic review. J Nanopart Res 17:1–26
80. Trump BD, Antunes D, Palma-Oliveira J, Nelson A, Hudecova AM, Rundén-Pran E, Linkov I (2024) Safety-by-design and engineered nanomaterials: the need to move from theory to practice. Environ Syst Decis 44(1):177–188
81. van Gelder P, Klaassen P, Taebi B, Walhout B, van Ommen R, van de Poel I, Jung D (2021) Safe-by-design in engineering: an overview and comparative analysis of engineering disciplines. Int J Environ Res Public Health 18(12):6329
82. Gupta R, Xie H (2018) Nanoparticles in daily life: applications, toxicity and regulations. J Environ Pathol Toxicol Oncol 37(3):209–230
83. Ramanathan A (2019) Toxicity of nanoparticles_ challenges and opportunities. Appl Microsc 49(1):2
84. Gottardo S, Mech A, Drbohlavová J, Małyska A, Bøwadt S, Sintes JR, Rauscher H (2021) Towards safe and sustainable innovation in nanotechnology: state-of-play for smart nanomaterials. NanoImpact 21:100297
85. Allan J, Belz S, Hoeveler A, Hugas M, Okuda H, Patri A, Anklam E (2021) Regulatory landscape of nanotechnology and nanoplastics from a global perspective. Regul Toxicol Pharmacol 122:104885
86. Gupta N, Fischer AR, Frewer LJ (2015) Ethics, risk and benefits associated with different applications of nanotechnology: a comparison of expert and consumer perceptions of drivers of societal acceptance. NanoEthics 9:93–108
87. Duncan TV (2011) Applications of nanotechnology in food packaging and food safety: barrier materials, antimicrobials and sensors. J Colloid Interface Sci 363(1):1–24
88. Ali O, Ishak MK, Bhatti MKL (2021) Emerging IoT domains, current standings and open research challenges: a review. PeerJ Comput Sci 7:e659
89. Choudhary A (2024) Internet of things: a comprehensive overview, architectures, applications, simulation tools, challenges and future directions. Discov Internet Things 4(1):31
90. Primožič M, Knez Ž, Leitgeb M (2021) (bio) nanotechnology in food science—food packaging. Nano 11(2):292
91. Liu Z, Wang S, Zhang Y, Feng Y, Liu J, Zhu H (2023) Artificial intelligence in food safety: a decade review and bibliometric analysis. Foods 2023(12):1242
92. Wolford T, Sutton J, Mangal CN (2023) Laboratory response to pandemic threats: challenges, needs, and solutions. Health Secur 21(S1):S56–S59
93. Freire JAF, Gonçalves E (2022) Co-operation in innovative efforts: a systematic literature review. J Knowl Econ 13(4):3364–3400

Enzyme-Linked Immunosorbent Assay (ELISA) for Food Adulteration Screening

Alok Raghav, Saurabh Mishra, Rachna Dixit, and Goo-Bo Jeong

Contents

Alok Raghav, Saurabh Mishra and Rachna Dixit contributed equally with all other contributors.

A. Raghav (✉)
Department of Anatomy and Cell Biology, Lee Gill Ya Cancer and Diabetes Institute, Incheon, South Korea

University Center for Research and Development (UCRD), Chandigarh University, Chandigarh, India

Center for Global Health Research, Saveetha Medical College and Hospital, Chennai, India
e-mail: alokalig@gachon.ac.kr

S. Mishra · R. Dixit
Department of Life Sciences and Biotechnology, Chhatrapati Shahu Ji Maharaj University, Kanpur, India

G.-B. Jeong
Department of Anatomy and Cell Biology, Lee Gill Ya Cancer and Diabetes Institute, Incheon, South Korea

W. Alahmad, M. Ashfaq (eds.), *Advanced Techniques against Food Adulteration*,
https://doi.org/10.1007/978-3-031-99454-8_10

Abstract Controlling harmful contaminants and spotting fraud in the food business is crucial for preserving consumer health and safety. Detecting and screening for toxins and pollutants in food, antibiotics, and pathogens in the food processing business using traditional methods will be very expensive and time-consuming. As a result, the use of immunological techniques like ELISA, which are quick, inexpensive, accurate, and do not require sophisticated equipment, was introduced to improve food quality and safety. Indirect ELISA and sandwich ELISA are two in vitro methods for food confirmation that allow for the simultaneous analysis of numerous samples. Generally used as a diagnostic tool in pathology and medicine and for quality control testing in many industries, particularly the food industry, this immunology technique finds antibodies or antigens in food samples, pesticide residues, pathogenic microorganisms, and eventually natural materials. The use of commercial ELISA kits for identifying food ingredients and their changes throughout function and processing for safety evaluation has been debated as the technology for screening and detection has advanced in recent years. The significance of the ELISA technology and its use in identifying genetically modified foods, allergies, and eventually food fraud were also discussed in this chapter.

Keywords ELISA · Genetically modified food · Chromatography · Spectroscopy

Introduction

Food adulteration is a burden on humanity that has been occurring for ages and is still happening more recently. Numerous technologies have been developed to control such issues. It puts human health at risk, damages consumer confidence, messes up supply chains, and jeopardizes regulatory systems. The history of food adulteration is integrally linked to economic, societal, and technological advances, illustrating the dynamic interaction of changing consumer behavior, technological improvements, and financial benefits [1]. Adulteration incidents from the 1850s to the present are therefore considered contemporary. It was believed that the German scientist Frederick Accum was the first to study adulteration in 1820 after he discovered numerous harmful metal colorings in foods and beverages. Later, in the early 1850s, doctor Arthur Hill Hassall conducted much research that resulted in the 1860 Food Adulteration Act and other laws [2, 3]. Food safety standards in the nation were established by the Prevention of Food Adulteration Act of 1954, which also specified what constitutes an adulterated food item [4]. Techniques and procedures used in food adulteration have become increasingly sophisticated over time. Traditionally, the only way to adulterate food was to add basic ingredients, such as water to milk or chalk powder to wheat. However, due to scientific and technological breakthroughs, adulteration is becoming more complicated and challenging to identify. Today's adulteration tactics utilize synthetic ingredients, dangerous chemicals, and sophisticated ways to imitate real food products' texture, taste, or look. Together with globalization and the growth of food supply networks, these developments have made it more challenging to identify and stop adulteration. Economic incentives, poor regulatory frameworks, and customer demand are some of the causes that lead to the prevalence of food adulteration. Producers and suppliers are frequently influenced by economic incentives to reduce expenses and increase profits by utilizing poor supplies, diluting goods, or replacing costly items with less expensive ones. Identifying the source of contaminated products is more difficult due to the intricacy of global supply chains. The absence of strict enforcement of food safety regulations and unregulated marketplaces fosters an atmosphere favorable to adulteration in many developing nations, including India. Furthermore, consumer attitudes and the desire for inexpensive or aesthetically pleasing goods can unintentionally contribute to the cycle of adulteration. The World Health Organization (WHO) estimates that each year, contaminated food results in 42,000 deaths and 600 million illnesses worldwide [5]. Three categories of contaminants can purposefully or inadvertently contaminate foods in the food supply chain: (1) physical contaminants (such as metals, plastic, glass, rocks and human hairs); (2) chemicals (such as toxins, pesticides, adulterants, and allergens); and (3) microbes (such as bacteria, viruses, and fungi) [6]. We concentrate on identifying particles consisting of plastic because they are frequently encountered in packing materials, and visual inspection, sieves, or screening devices easily discover significant physical pollutants.

India's food business is big and diverse, making it challenging to tackle food adulteration. The Food Safety and Standards Authority of India (FSSAI) guarantees food safety and is the main regulatory agency. Despite several efforts to improve public knowledge and enforce food safety laws, regional differences, resource shortages, and infrastructure restrictions cause enforcement to be inconsistent. In India, it is common practice to adulterate food by adding dangerous substances like detergents to milk, formalin to fish, and artificial pigments to spices. These techniques present significant health hazards, including gastrointestinal disorders, organ damage, and prolonged exposure to carcinogenic compounds, compromising the quality of the food. On the other hand, the USA has a stronger food safety regulation system, which is supervised by organizations like the Food and Drug Administration (FDA) and the US Department of Agriculture (USDA). To ensure the safety of food items, these organizations maintain strict standards and enforcement procedures, paying special attention to processed foods. Although there is less food adulteration in the USA than in India, problems, including fraud, contamination, and mislabeling of inexpensive goods like seafood and organic products, still exist. Preventing foodborne illnesses, which are frequently connected to contaminated or adulterated foods, is another top priority in the USA [1]. The 2013 incident in China, where a group of more than 900 individuals marketed meat from martens, rats, foxes, and mink as lamb, is an example of a major infringement on a global scale. The meat's origin was successfully hidden by adding gelatin and treating it properly. Mechanically separated meat, or MSM, is becoming a more common acronym among the components stated on the labels of different kinds of meat products. Mechanically separated meat is a paste made from fat and animal meat. Because it is prone to oxidation and has far higher lipid content than real meat, antioxidants must be used during manufacture. Exceeding the allowable amount of MSM in meat products is a common violation [7]. Food adulteration has become even more complex due to globalization, bringing up new hazards and shortcomings. The risk of contamination and adulteration has increased due to the expansion of supply chains across borders prompted by the international trade in food items. Furthermore, there are currently more chances for fraud and mislabeling due to the growing demand for health-conscious goods like organic or fortified meals. Concern over adulteration in high-value products, such as olive oil, honey, and premium spices, is developing since these goods frequently fetch higher costs and are more vulnerable to dishonest business practices. A comprehensive and cooperative strategy is needed to combat food adulteration. Technical developments must be combined with regulatory reforms to increase the detection and prevention of adulteration. Food testing innovations, including blockchain technology for traceability and quick diagnostic tests, can significantly improve food safety. Campaigns to raise consumer awareness of the dangers of contaminated food and to empower people to make wise decisions are equally crucial because they can help lower the demand for such goods.

Physical Methods for Screening of Food Adulterations

For adulteration, detection several physical techniques, such as visual structural analysis at the microscopic and macroscopic levels. The initial approaches to food identification focused on morphological traits, including shape, appearance, taste, color, and smell. In the past, weight and volume were the most significant indicators. Because these methods relied solely on visual inspection, they were incredibly unreliable for detecting food adulterations. For example, specialized honey inspectors performed physical inspections to verify the honey's quality and purity. The use of analytical scales and microscopes for identification was added to the list of inspection techniques in the nineteenth century [8]. Detecting elements of animals and plants origin in feed and food is still done using this kind of testing in Germany and other European nations. However, the species origin of feed or food components is sometimes not detectable by microscopic means. However, these techniques do not ensure the qualitative detection of adulterants [9]. For example, when examined under a microscope, it is simple to identify extraneous starch in powdered spices like coriander, cumin, chilies, and cloves. Since pure food grains would not contain any of these impurities, it is possible to detect adulteration in grains by taking a small sample on a glass plate and visually examining the impurities (dust, straw, weed seeds, stones, weevil grain, rodent hair, damaged grain, insects, and excreta, etc.) [2].

Microscopic

Efforts to increase food quality have been made since the end of the nineteenth century. Because it is quick and accurate to describe the composition of food and plant pollutants, the microscopic detection of plant species utilized in nature or the preparation of food items is essential. Physical, biological, and microbiological pollutants can be observed by microscopy, which also enables assessments of the food's hygienic purity. Despite being less expensive than other chemical or molecular methodologies, it depends on the analyst's skill and understanding, particularly in plant anatomy, as it primarily relies on comparing food products [2]. Most pharmacy programs stopped teaching pharmacognosy in the 1960s. Thus, the majority of pharmacognosists who are new to the field of herbal medicines today, as well as many seasoned pharmacists, are frequently unaware of the possible applications of advancements in microscopy techniques for evaluating the quality of botanical materials. Computer-based image processing and improvements in microscope designs have led to a renewed interest in light microscopy. By combining optical elements, digital imaging, and computerized controls, fluorescence microscopy has made it possible to obtain electronic images at low light levels or invisible wavelengths to the human eye. This allows for a great degree of sophistication and image analysis. These imaging methods have significantly increased microscopy's usefulness [10].

Spectroscopic Techniques-Based Methods to Identify the Adulterants in Food

The most excellent practical application of the IR-spectroscopic approach appears in numerical plant taxonomy. This flexible, non-destructive method can analyze samples in all physical states. The IR spectra obtained include details on combination bands and overtones and provide inferences regarding the chemical structure of the primary elements. The distinct bonding behavior and atom-to-atom interactions are unique and can be utilized for fingerprinting analysis, which offers quick identification of the plant material under examination. Plant material analysis uses near-infrared (NIR), mid-infrared (MIR), Fourier-transform infrared (FTIR), and attenuated total reflection (ATR) spectroscopy. Chemometric analysis and infrared data combine to create a potent instrument for quickly and accurately identifying and differentiating the plants under study. Raman spectroscopy is another crucial method for examining molecular rotary-vibrational spectra. This approach has several crucial qualities. The potential for registering the Raman spectrum with visible light is one of them. Moreover, Raman scattering, like infrared spectroscopy, offers details on molecules' polarizability, structure, and interatomic interactions. When analyzing molecules without a dipole moment, it works very well. It has been used as an analytical tool for characterizing plant materials and is complementary to the infrared approach. Due to its high cost, saffron is the most frequently faked spice. Calendula or staff lower petals, powdered turmeric rhizomes, gardenia fruit extract, and buddleia (butterfly bush) buds, which contain pigments resembling the crocetin esters found in saffron, are utilized for this purpose. Diffuse reflectance infrared Fourier transform spectroscopy (DRIFTS) in conjunction with partial least squares discrimination analysis (PLS-DA) were combined to create a straightforward and thorough research methodology for identifying six saffron contaminants (i.e., *Crocus sativus* stamens, calendula, safflower, turmeric, buddleia, and gardenia) [11, 12]. Nuclear magnetic resonance (NMR) spectroscopy is an effective technique providing comprehensive details of a molecule's structure. These days, metabolomics, lipidomics, and other "omics" analysis fields use their identifying potential. Therefore, a crude extract's NMR spectrum can reveal a great deal about its composition and be used to identify different plants. Some issues related to signal overlapping may occur in the event of extremely complicated fusions. The 2D NMR approach can get around this obstacle [12, 13]. There aren't many examples of using UV-Vis spectrophotometry to construct chemometric models. Particular aspects of this method bring this on. Because the acquired UV-Vis spectrum is complicated (rotary + vibrational + electronic), it is challenging to interpret. Furthermore, the analyzed extract's spectral features are the sum of the gradients inside it. Therefore, the absorption bands describe a family of structurally related components rather than the existence of a single, well-defined substance. Chemical groups with conjugated multiple carbon-heteroatoms or carbon-carbon bonds that may absorb UV light are found via UV-VIS spectroscopy. Alkenes, arenes, and several heterocyclic compounds are examples of compounds that contain such groups. This group also includes secondary plant metabolites. The UV-Vis spectrum is complicated, making it challenging to categorize or differentiate samples from the set under study. Despite these challenges, certain attempts were made to use UV-Vis spectrophotometry in conjunction with chemometric techniques to categorize, authenticate, and investigate similarities and differences in herbal material [12].

Advantages and Disadvantages of Physical Approaches

Performing microscopic examinations and verifying the components indicated on the label are the primary methods of physical food identification. The primary drawback of labels is that they merely provide the manufacturer's word about the composition without any scientific validation. Inaccurate labels are not checked and could disappear while being stored or transported. Microscopic analysis may result in erroneous sample identification due to processing-related alterations in the final output. Additionally, samples may be lost, and microscopic analysis is time-consuming, challenging, and expensive because specialist microscopes are required [14].

Chemical-Based Methods for Identification of Food Adulterants

The practice of dishonest producers and dealers purposefully lowering vital nutrients in food products to boost profit margins by adding chemicals and cheaper foods is known as economic adulteration. Adding components like wheat, starch, vegetable oils, cane sugar, water, chalk powder, skim milk, sand, molasses, brick, and stone powder also increases the volume of food. This approach is risky because it ignores the humanity underneath the profit-driven mindset while adding unnecessary substances and reducing nutrition. Previous food fraud disasters, such as the European horsemeat scandal, the indirect increase in protein content due to melamine in wheat gluten, and the addition of illegal Sudan dyes to paprika and pepper, have had a significant influence [15] (Fig. 1). Traditional chemical testing analyses in food adulteration are colorimetric and titrimetric analysis, for the preliminary screening and field testing. Colorimetric tests rely on specific color reactions and visual indicators. For example, for milk and dairy products lactometer test was performed. A lactometer measures the specific gravity (1.026–1.032), and if its value is low, it indicates water addition. The presence of starch is indicated by the methylene blue test in this mix of the milk sample with a few drops of methylene blue solution, and the blue color indicates starch presence. The same test is performed for detergents present in milk products. For spices commonly using turmeric, an HCL test is performed. In this, add concentrated HCL to turmeric powder, pink/magenta indicates metanil yellow, etc. [17].

According to studies, FA concentration varies significantly, from 0.05 to 267.3 mg/kg. Cooking techniques like pan-frying, stir-frying, and boiling can lower FA levels. However, frying with a lot of heat energy can raise FA levels. Fresh fruits, vegetables, meat, poultry, mutton, processed foods, and commercial milk all contain FA. Commercial milk's high FA content may concern young consumers, particularly

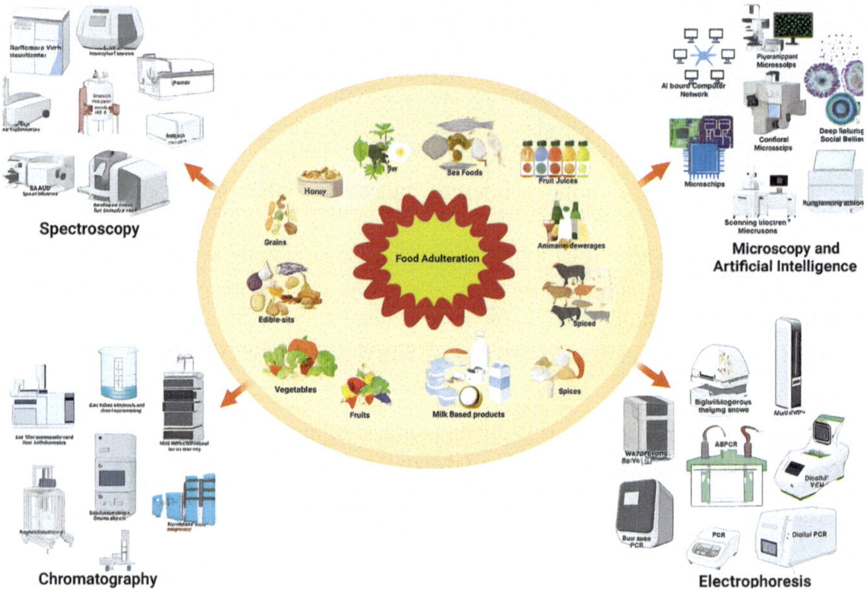

Fig. 1 Various cutting-edge tools and techniques are employed to identify food fraud or product counterfeiting. (Adopted from [16] under Creative Commons CC-BY 4.0)

kids and teenagers. Fruits and vegetables that are grown nearby pose no health hazards. According to the research, the formaldehyde (FA) amount in raw poultry flesh varied between 0.063 and 7.47 mg/kg. The FA content was lowered by 60% by boiling the meat. Stored samples had greater FA concentrations than fresh food samples in Thailand, with values within natural limits. High quantities of FA were discovered in shiitake mushrooms, which impacted consumer health. They advise looking into the source and developing strategies to control FA content [18]. In most cases, the adulterants are so similar to their replaced counterparts in appearance, chemical makeup, and organoleptic characteristics that even the analyst has trouble identifying them. The specificity of the immunological reaction makes immunoassay an appropriate technique for detecting food contaminants. Two common immunoassays for determining whether less acceptable or undesirable species are present in meat products are agar gel immune diffusion and ELISA.

Advantages and Disadvantages of Chemical Methods

Chemical colors and adulterants frequently turn out to be lethal. Since they are carcinogens, they pose a health risk. Certain contaminated foods can also directly impact our internal organs, causing problems and failures with the heart, kidneys, liver, and many other organs. The potential of hyperspectral (HIS) and multispectral imaging (MSI) techniques to obtain one or more images at specific wavelengths in the NIR, MIR, and Vis range has increased their application in food analysis. These methods can identify certain characteristics or attributes of quality in various products and raw materials used in food production. The most alluring benefit of these methods is their capacity to gather multiconstituent, spectral, and geographical information about the sample under study. Food analysis has also used other vibrational spectroscopy technologies, such as fingerprinting methods. These techniques have been widely reintroduced in recent years by researchers in various applications (feasibility studies) or real-world applications. These approaches' timeliness and ease of use in normal operations are their primary advantages over conventional chemical and chromatographic methods. However, their global use in research and the food business is hampered by factors like reliable and commercially available equipment, the huge amount of data produced, the requirement for sophisticated algorithms and data processing, and the lack of academic training [19].

Analytical Methods for Screening of Food Adulterations

An analytical technique for identifying food adulteration is the identification of lipids and volatile chemical molecules. Lipid biomarkers are used to detect lipids. In short, the study involves identifying the species by examining the positions of fatty acids in 4-monoacylglycerol (2-MAG) and triacylglycerol (TAG) molecules. Principal component analysis (PCA), gas chromatography-mass spectrometry (GC-MS), and Fourier-transform infrared spectroscopy (FTIR) are examples of identification techniques. According to reports, food adulteration can be detected using various analytical techniques. Detection methods were ranked based on how frequently they were used in the literature. With 20.6%, mass spectroscopy had the most significant share, followed by polymerase chain reaction (18.5%) and liquid chromatography (11.6%). In South Korea and Asian nations, MS comprised the most significant percentage (20.7% and 38.1%, respectively). It is noteworthy that MS is the most widely used approach for analyzing spices, extracts, cereals, grains, and pulses and is utilized extensively in most food categories. However, organic foods, oils, and spices also use LC and GC extensively. Fruit juices, oils, cereals, grains, and alcoholic beverages can all be authenticated using NMR. PCR is the most often utilized detection technology for the food categories of fish and seafood, milk and milk products, and meat and meat products. For fruits, fruit juices, and sweeteners, HPLC and LC are frequently utilized in addition to MS. Biosensors, immunosorbent tests (like ELISA), Raman, and IR spectroscopy are less common than the other detection techniques [20].

Chromatographic-Based Methods to Identify the Food Adulterants

Chromatography is an approach that relies on the idea that various chemical compounds have varying degrees of solubility. Affinity chromatography (AC), gas chromatography, liquid chromatography, and ion exchange chromatography (IEC) are the broad areas where this technology occurs. However, compared to other chromatographic techniques, GC-MS and LC-MS are employed more often for food adulteration detection [16]. Non-targeted metabolomics based on gas chromatography-mass spectroscopy has been used to differentiate between different types of lettuce, shrimp and tempeh that grow under various conditions and finely clustered or cultivars of melons and chilies [21]. Another study used non-targeted metabolomics and GC-MS to detect meat contamination [22]. Samples of pork and beef blended in various ratios were examined in this study to determine elements that can be utilized as markers for detecting pig adulteration. Additionally, cooking oils have been screened for adulteration using GC-MS. Pure oils can be separated from mixes, and distinct seed oils used in adulteration can be distinguished thanks to GC-MS. For example, GC-MS-based profiling was used to identify chemical indicators of used and adulterated cooking oil and 6-monoglyceride markers were suggested to differentiate between fresh and used edible oil. Accordingly, it has been demonstrated that GC-MS is an effective conventional technique for identifying food scams in market oil products. Using an atmospheric pressure chemical ionization source, a screening technique combining gas chromatography and quadrupole time-of-flight MS was developed to identify and measure pesticide residues in fruit and vegetable samples. The samples only contained 15 of the 104 pesticides used to treat 34 different kinds of fruits and vegetables. The quantitation capabilities were then assessed using a quantitative validation of the pesticides detected in the samples. Five matrices: carrot, lettuce, tomato, orange, and apple [23]. LC-MS is the most effective method for food fraud analysis because of its excellent sensitivity and specificity in detecting adulteration in various foods, including meat products, fruits, vegetables, seafood, and edible oils [24, 25]. Samples of fox meat were mixed in different proportions with raw components for this investigation. Multivariate statistical analysis was used on the mixed meat samples to identify meat items with varying contents of fox and demonstrate the viability of quantitative analysis. A lot of research has been done on milk metabolites utilizing LC-MS-based metabolomics. Human, equine, bovine, and goat milk composition of milk metabolites was examined [26]. Food chemical analysis now requires LC/MS. Carotenoid and polyphenol profiles have been used to characterize specific foods of interest due to their possible biological activity, and the technique has also been used to identify and measure bioactive substances, such as water-soluble and fat-soluble vitamins. Liquid chromatography is a powerful tool in food safety that is widely utilized in routine analysis and research labs. Monitoring for food contamination is expected to involve the use of LC-MS.

Environmental contaminants and naturally occurring poisons are the most health-threatening. Toxins include cyanogenic glycosides, mycotoxins, marine biotoxins, and toxins from toxic mushrooms (such as ustalic acid and γ-amanitin). In biological samples, including serum, plasma, and urine, toxins are usually tested as an exposure indicator. Furthermore, antimicrobials—particularly those with agronomical uses—are frequently examined in animal tissue, other necessary commodities, non-medicated feed and medicated feed (because they are employed for growth promotion, prophylaxis, and metaphylaxis; see, for example, eggs and vegetables). Detecting pesticide residues in food is another common use for LC-MS since the buildup of these compounds can result in harmful and allergic reactions when contaminated goods are consumed [27].

Advantages and Disadvantages of Analytical Methods

The primary drawback of these methods is that if the lipid acids, as well as quantity and kind of lipid, are altered during technological treatments, the outcome is less accurate. A complex statistical analysis is necessary for getting accurate conclusions. Using expensive technology and requiring skilled personnel makes this research costly [28]. Gas chromatography, mass spectrometry, and spectroscopy identify volatile organic molecules. Western blotting, ELISA, isoelectric focusing, 2D electrophoresis, and HPLC (high-performance liquid chromatography) are methods for identifying proteins. Histidine is one species-specific protein biomarker that can be found and measured using HPLC. The origin of animal proteins cannot be ascertained using this method. Animals' environment, age, and health can be determined by isoelectric focusing by analyzing structural proteins like actin, myosin, or tropomyosin. One hundred protein mixtures can be analyzed using 2D electrophoresis. Both methods are exceedingly expensive, labor-intensive, and need specialized knowledge. Furthermore, extremely complicated combinations cannot be used for them [29]. An enzyme immunoassay called the ELISA test finds antigens and antibodies unique to a species. It can identify the type of protein present in animals as well as the origin of their tissues. This test's high sensitivity and low specificity are drawbacks, meaning confirmation is necessary for the result [30]. GC/MS is, therefore, one of the most widely used instruments due to its low cost, small sample size, ability to detect compounds at very low concentrations, stability, reproducibility, and convenience of data processing. However, derivatization might become a disadvantage of this technique in the case of gas chromatography/mass spectroscopy analysis of non-volatile chemicals because of the potential for unstable functional group alteration, limiting precise measurement and identification. The primary limitation of GC/MS is that all analytes need to be thermally stable after volatilizing. Certain carrier gases may also be costly. One of LC-MS's many benefits is that it may be used with a wide range of analytes, including complex compounds like peptides and those with low molecular weight. It is possible to simultaneously determine several analytes of the same type, such as the allergen-causing peptides found in whole eggs, skim milk, ground peanuts, soy flour, and hazelnuts in cookies. Even analytes of different types could be analyzed using LC-MS, as demonstrated by the simultaneous detection of hydroxymethyl furfural and acrylamide in extruded goods. Costly equipment and time-consuming, complex data analysis are two of the drawbacks. High-purity solvents and inert polymers, such as PTFE, must be used when preparing samples since mass detectors are extremely sensitive. Last but not least, solvent waste from some chromatographic separations may be significant or solvents deemed environmentally harmful (such as CH2Cl2, MTBE) may be needed [7].

Immuno-Based Methods to Identify the Food Adulterants

Milk adulteration affects most of the population because toddlers and older people depend on milk, resulting in serious health and economic issues. There are safety concerns when selling breast milk (BM) online in the UK because it may lose its nutritional content and passive immunity properties. Concerns regarding the advantages of breastfeeding were raised by a 10% investigation of 102 breast milk products, which revealed 10% of them were contaminated with cow milk [31]. Several techniques have been developed to date to identify adulteration of dairy products. Infrared spectroscopy was applied to produce a quantitative method for identifying the various types of milk. Milk from various sources was separated using polypropylene gel electrophoresis. The low-field NMR technique and liquid chromatography were used to measure the adulteration of camel milk and dairy products. Even with the advancement of these techniques, there are still drawbacks, such as the high cost of testing and the need for costly, bulky experimental equipment. Ultimately, ELISA was chosen for this investigation due to its sensitivity, a fast and highly specific technique widely used in the food industry. The biomarker's low concentration is quite enough to perform ELISA because this technique uses the enzyme's amplification effect and usually has a better sensitivity than HPLC. Compared to HPLC, ELISA is typically more cost-effective and has a better throughput. A characteristic protein must be discovered before creating an ELISA to identify camel milk that has been tampered with by cow milk. HPLC analysis was used in the current investigation to analyze the chromatogram of camel milk with various adulteration percentages. After that, principal component analysis (PCA) was used to examine the peak region of liquid chromatography. β-Lg was screened as a distinctive protein following analysis. Due to the absence of β-Lg in camel milk whey protein, β-Lg can be utilized as a crucial signal to differentiate camel milk from cow milk. The current work developed an ELISA approach using β-Lg as a marker antigen to identify camel milk adulteration. It produced a particular and sensitive monoclonal antibody against β-Lg. A novel technique for quantifying camel milk tainted with cow milk is presented in this paper [32]. According to Borková and Snášelová (2005), polyclonal antibodies against β-CN, αS1-CN, β-LG, or κ-CN are being utilized in immunological approaches to identify milk adulteration [33]. According to Manzo et al., these tests have certain drawbacks when distinguishing between closely related species, such as cow and buffalo. When detecting cow milk adulteration, using bovine IgG as an analyte is more sensitive and specific. Two immunoassays targeting cow IgG (indirect sandwich ELISA and indirect competitive ELISA) were developed to identify adulteration of dairy products. They demonstrated detection limits of 0.01 and 0.1 μg/mL of bovine IgG, respectively [34–36]. Caseins and other milk proteins are poorly immunogenic and vulnerable to proteolysis [35, 37, 38]. Numerous immunoassays have been developed to detect cow milk. Since IgG is abundant in blood, meat, milk, and other tissues and secretions and has species-specific epitopes and strong immunogenicity, we chose it as the target protein for the developed direct sandwich ELISA technique [39]. In addition, prior studies have indicated that IgG may be employed as a species-specific target antigen in tests for milk product authenticity [35]. Any ELISA's performance should be assessed in terms of sensitivity and specificity. Hence, assay reagent tuning was done to improve both metrics. To avoid inaccurate results, optimization also aims to avoid cross-reaction between the detection and capture antibodies [40]. Following optimization, the proposed assay demonstrated a limit of detection of 0.01 μg/mL for bovine immunoglobulin G (IgG) detection without any cross-reaction with human IgG or caprine stock samples. This indicates that the assay is specifically specific and sensitive to bovine IgG, at least for the two species tested. Additionally, it has identified bovine IgG in fresh whole cow milk without any cross-reaction with goat or human BM, with a limit of detection of 1/10,000 dilution. In terms of detecting cow milk adulteration, this assay demonstrated a detection limit of 0.001% for goat milk adulteration with whole cow milk and 0.001, 0.001, and 0.001% for human BM adulteration with skimmed, semi-skimmed, and whole cow milk, respectively. According to these results, the sensitivity of cow milk detection is higher than that of both the indirect competitive ELISA with a detection limit of detection is 1% using anti-bovine β-CN monoclonal antibody and the established indirect competitive ELISA cow milk detection limit of 0.1% using monoclonal anti-bovine IgG antibody [35, 37].The direct sandwich anti-bovine immunoglobulin (IgG) ELISA could not detect bovine immunoglobulin G (IgG) when tested with dried skimmed cow milk and long-life cow milk. This restriction may be due to the denatured IgG that results from pasteurizing milk at very high temperatures [41]. This issue might be resolved by heat-denatured epitope-specific antibodies or another technique, such as PCR; however, long-life cow and dried skim milk are expensive to produce and are not ideal for adulteration [42]. With a detection limit of 0.001%, the direct sandwich anti-bovine IgG ELISA detectable in adulterated human BM included skimmed, semi-skimmed, and whole cow milk. The test also detected adulterated goat milk with cow milk at a detection limit of 0.001%. According to these results, the assay is highly sensitive and specific to finding human breast milk adulterated in cow's milk. According to a study conducted by scientists, the assay can also be used to detect cow milk adulteration in other commercial milk products. Developing a kit for routine human breast milk monitoring using ELISA is possible, particularly in milk banks. With more advancements, the assay might also be used to identify cow milk adulteration in dairy products produced from goat, buffalo, sheep, camel milk, almond, soy, and oat milk [43]. Food manufacturers in most countries use lard as an appropriate oil ingredient since it is readily available and reasonably priced. Some religious groups (Judaism and Islam) consider lard and pork to be serious health problems. Daily use may be linked to health hazards and biological issues. A family of globular proteins with a range of protective bioactivities, immunoglobulins (Ig) are found in the sera of all vertebrate species and primarily offer immunological protection. They are sufficiently conserved as an essential component of the humoral immune system to allow comparisons across various species. According to their structural composition, immunoglobulins (IGs) are divided into three main classes: IgG, IgM, and IgA. IgG antibodies are the primary neutralizing antibodies for toxins found in tissues because of their ability to diffuse easily into the peripheral muscles through the extracellular fluid. The recommended approach makes it easy to separate and purify IgG, the most common antibody in all body fluids. Furthermore, the enzyme-conjugated antibody, IgG-species (immune-reactants), and anti-species IgG antibody are easily accessible, reasonably priced, and used to authenticate any meat species, not just pig, by applying the same principle with an immune-sensor or the developed ELISA. Accordingly, the current work uses immuno-enzymatic techniques (immune sensor and ELISA) and the facile extraction of immunoglobulin G (IgG) directly from muscle for meat verification. Pork is frequently used in food products because it is less expensive than beef and is a possible adulterant in raw meat. For food control labs to identify animal species in food items, they must be able to differentiate between species used in raw materials. The most widely used analytical methods for halal or adulterated pork are based on DNA or protein analysis. These include chromatography, electrophoresis, single-strand conformational polymorphism, mitochondrial DNA polymerase chain reaction, and real-time polymerase chain reaction analysis. However, these techniques are not practical for regular sample analysis due to their high cost, labor-intensive nature, time consumption, and complexity. Contrarily, immune techniques stand out for their high output of serial testing, convenience of sample preparation, and not requiring specialized staff and expensive equipment. Since electrochemical immune sensors are a different detection method for food authenticity and are highly useful for on-site use, only one immune sensor for meat verification has been reported previously. An anti-swine immunoglobulin G polyclonal antibody-based immune-strip test was recently developed to identify pig meat adulteration. Assessing a high-affinity anti-pig immunoglobulin G (IgG) polyclonal antibody allowed for developing a simple, rapid, and sensitive ELISA (competitive and direct ELISA) for identifying pork adulteration in beef meat samples. Furthermore, the suggested work developed a sensitive electrochemical immune sensor (competitive and direct immune sensor) based on pig-IgG detection for the first time using electro-polymerized polypyrrole-modified graphite paste electrodes. Using peroxidase-conjugated anti-pig immunoglobulin G (IgG) polyclonal antibody (anti-pig IgG-HRP), pig-IgG was collected, and the detection was monitored using chrono-amperometry [44] (Fig. 2).

Food Control Laboratories: Sensitivity, Selectivity, and Accuracy

Keeping food items safe and high-quality is crucial for consumer confidence, public health, and regulatory compliance. Before they are used, the laboratory must thoroughly evaluate or verify the "fitness of purpose" of all food testing techniques (microbiological and chemistry). This validation or verification will follow processes that have been thoroughly documented. When applicable, the following characteristics must be included in the validation of quantitative methods: range, selectivity, estimation of measurement uncertainty, and robustness, limit of quantitation, reporting limits, reproducibility (precision), accuracy, repeatability, etc.

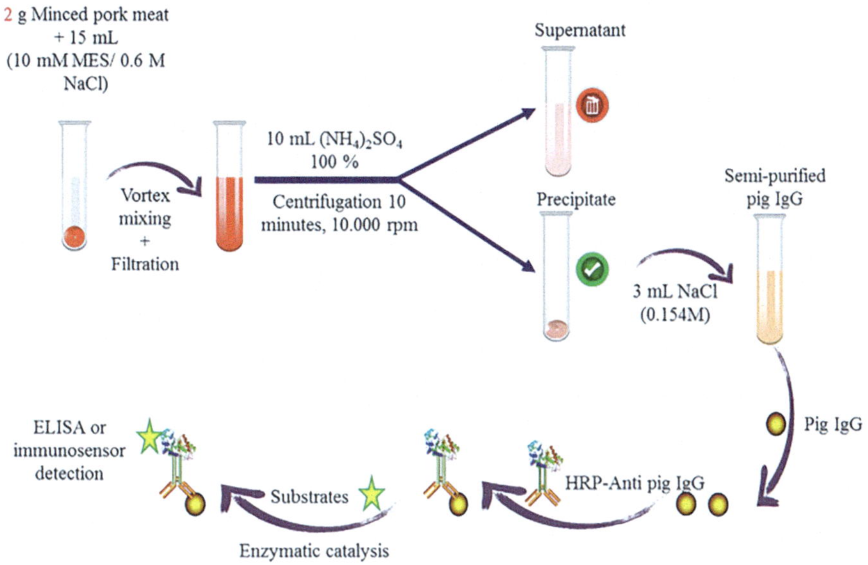

Fig. 2 Diagrammatic illustration of how enzyme immunoassay (ELISA or immune sensor) was used to identify pork adulteration in meat. (Adopted from [44] with permission)

Additionally, when appropriate, the following factors will be considered throughout the validation process: interference, matrix effect, and sample homogeneity.

Application of ELISA in the Identification of Allergens and Contaminants

In the United States, milk, wheat, eggs, fish, tree nuts, shellfish, peanuts, sesame, and soybeans are the top nine food allergy causes. Laws enforced by the USDA's Food Safety and Inspection Service and US Food and Drug Administration mandate that the food label list every ingredient in a product. More than 90% of the recorded cases of food allergies are caused by these allergens (Table 1).

ELISA and PCR are the two most widely used and favored techniques for allergen detection in the food business. Currently, testing techniques can identify these allergies at extremely low concentrations in final products. Low parts per million (ppm) of these pollutants can be detected using enzyme-linked immunosorbent assay and polymerase chain reaction. With the help of these techniques, which identify food allergies at the molecular level, manufacturers may quickly and conclusively throw away or re-label contaminated products before their release. Furthermore, it informs them of areas of their processing facilities that need to be decontaminated or production lines that need to be reused for other projects [45, 46].

Table 1 A list of foods, their adulterants, their uses, and their analysis methods. (Adopted from 20 under Creative Commons Attribution-NonCommercial 4.0 License)

Food products	Use of adulterants	Purpose	Analytical techniques
Milk	Melamine, reconstituted milk, skim milk powder, starch flour, glucose Water (H_2O) Sugarcane, vegetable oil, or urea H_2O_2, carbonates, bicarbonates, antibiotics, caustic soda Detergents or urea and thioglycolic acid calcium salts	Thickening agents, enhanced volume, and act as a sweetener More volume Compensate the carbohydrate, fat, or protein content The shelf life increased Improve the whitening and cosmetic qualities	Hyperspectral imaging, HPLC, gas chromatography Gas chromatography-mass spectroscopy, heat treatments FAA, liquid chromatography-mass spectroscopy, heat treatments Gas chromatography, heat treatments FAA, HPLC Heat treatments, FAA
Flour of wheat	Alloxan monohydrate and benzoyl peroxide	Whitening agent	polymerase chain reaction, Raman hyperspectral imaging
Spice beef	Paprika and chili powders, horse or donkey meat	Increase the amount. Replace expensive food with cheaper items	UV-visible and PLS-DA HPLC/MS, NMR, IR
Powder of meat	Vegetable fat, foreign proteins	Bulking agents	Meat PCR testing
Honey	Beet inverted syrup, sugar, HFCS, rice molasses or rice syrups	Increase volume and sweetener	GC-MS, HPLC, Raman spectroscopy
Coffee	Corn, wheat, chickpeas	To increase volume	LIBS
Butter	Fatty acid, sterol, TAG, vegetable oil, banana, oleomargarine	Volume increases and makes yellowish	Gas chromatography-mass spectroscopy and RP-HPLC

Microbes in Foods

Foods may be thought of as a medium for the growth of microorganisms. Both types of microbes grow, whether good or bad. Harmful microbes cause food to spoil, and beneficial microorganisms enhance food quality. Foods that include protein, especially meats, are putrefied by microorganisms like *Proteus*, *Clostridium*, and *Pseudomonas* bacteria, which break down the lengthy peptide chains of proteins into amino acids and odorous substances including hydrogen sulfide (H_2S), amines, and ammonia. When carbohydrates (starches and sugars) ferment, they produce gases, alcohols, particularly carbon dioxide, and acids (such as acetic acid in vinegar). This is the mechanism by which yeasts cause spoiled chocolate cream candies to break down. Molds, yeasts, and bacteria can also cause improperly canned food to decay. The high level of resilience that the spores of bacteria like *Clostridium* and *Bacillus* possess makes them particularly important in the canning industry. "Sulfide spoilage," which is brought on by *Carnimonas nigrificans* and results in blackened contents and smells like rotten eggs, is an example of microbiological spoiling of canned goods. Another example is "flat sour," where the ruined product tastes sour due to its reduced pH, hazy appearance, and unusual odor. A can may bulge and burst due to putrefaction from *Clostridium sporogenes*, expelling its partially digested contents along with a foul odor [47].

Herbicides in Foods

Three classes comprise the majority of pesticides used in food: herbicides, fungicides, and insecticides. Weeds compete with crop plants for sunshine, water, nutrients, and space. Thus, herbicides are employed to keep them under control. Crop quality is improved, and agricultural labor requirements are reduced through weed control. Herbicides use several different mechanisms of action. They can change photosynthesis or nutrient intake or, harm leaf cells and cause the plant to dry out. Certain pesticides inhibit seed germination and seedling growth. Others destroy leaf and stem tissues when they come into contact with foliage. The herbicides 2,4-D [(2,4-dichlorophenoxy) acetic acid], simazine, atrazine, alachlor, dacthal, glyphosate and metolachlor are among the most commonly used ones [48].

Antibiotics in Foods

Antibiotic residues pose a threat to human health because they cause allergies (penicillin), accelerate the development of antibiotic resistance, facilitate the spread of bacteria resistant to antibiotics (Table 2), and cause other serious illnesses like cancer (sulfamethazine, furazolidone, and oxytetracycline), mutagenic effects, anaphylactic shock, nephropathy (gentamicin), reproductive disorders (chloramphenicol), and bone marrow toxicity [49].

Table 2 Contamination with antibiotics in a variety of foods consumed across different countries

Antibiotics	Countries	Contaminated food stuff
Tetracyclines	America, Cameroon, China, Egypt, South Africa	Chicken meat/beef/pork
Tetracyclines	India, Tanzania, Kenya, Ghana, China	Milk/raw milk
Tetracyclines	Nigeria, Tanzania	Eggs
Sulfonamides	China, Malaysia	Chicken/chicken giblets/eggs
Nitrofuran	Italy, Greece	Pork

Principles of ELISA

In 1971, Peter Perlmann and Eva Engvall developed the ELISA technology at the University of Stockholm in Sweden [50]. Peptides, antibodies, enzymes, and hormones found in various samples are frequently quantified using this technique. On the other hand, ELISA is a plate-based technique that produces a versatile system for scientific research by combining an antibody-enzyme conjugate with a chromogenic substrate [51]. Food analysis and immunology research can both benefit from the powerful and adaptable ELISA technology [52].

Classification of ELISA

ELISA is a well-known technique for identifying many targets. It is frequently used in research and diagnostic settings because of its ease of use and speedy outcomes. As the name implies, ELISA considers particular interactions between antigens, antibodies, and enzymes. Direct, indirect, sandwich, and competitive are the four primary forms of the approach, and each has a specific function in the analysis and detection procedures [53] (Fig. 3).

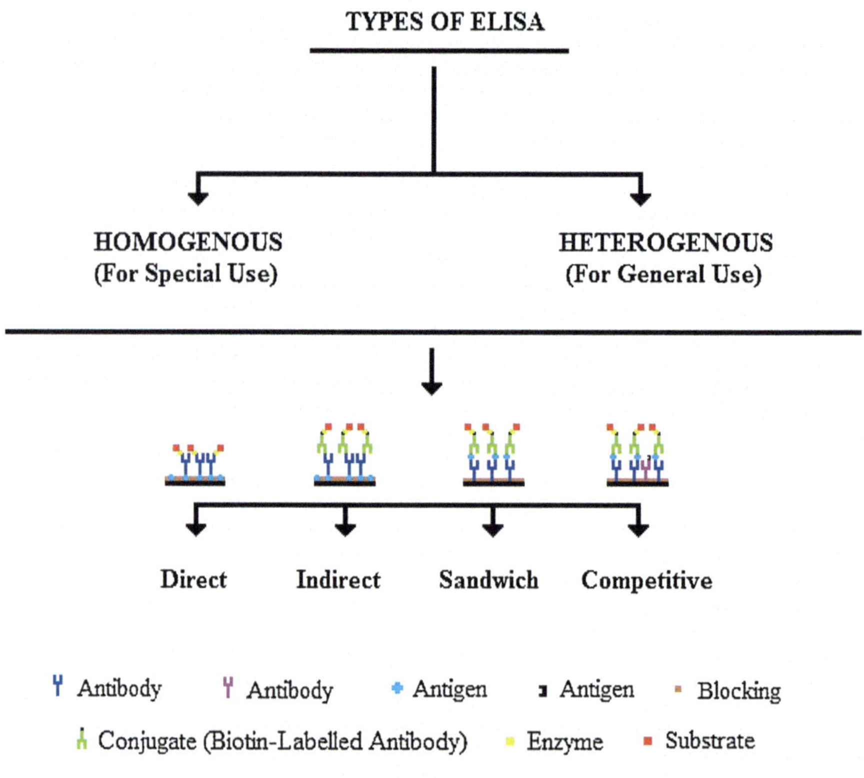

Fig. 3 Type of ELISA. (Adopted from [52] with permission)

Direct ELISA

Direct ELISA eliminates the need for a secondary antibody by using an enzyme-labeled primary antibody. An ELISA plate or solid surface is used to immobilize the target (antigen), which is then immediately recognized by the primary antibody-enzyme combination. Second, the primary antibody-coupled enzyme reacts with its substrate to produce a detectable signal whose strength may be ascertained. This reveals the target antigen [53, 54].

Indirect ELISA

Two types of antibodies are utilized in the indirect ELISA: primary and secondary. The first antibody, in this case, is not enzyme-labeled. Conversely, a secondary antibody is labeled with an enzyme [55]. Enzyme-linked secondary antibodies react with the immobilized antigen on the plate once the primary antibody binds to it. In conclusion, the secondary antibody-enzyme generates a visible reaction signal that can be measured about its substrate [55].

Sandwich ELISA

The sandwich approach, the most popular ELISA technique, involves sandwiching an antigen between two particular antibodies. This method involves the antibody attaching to a solid phase [56]. After the secondary antibody is washed with the enzyme, a specific quantity of the antigen-attached sample is added. The enzyme reacts with the leftover antigens and then binds to the fixed antibody. This method is pretty straightforward. Additionally, higher sensitivity and specificity are linked to it [57].

Competitive ELISA or Inhibitory ELISA

Other popular varieties of this method are competitive or inhibitory ELISA. The competitive approach relies on two antigens or two antibodies competing to bind the ligand in a restricted quantity [52]. The labeled antigen and the standard or unlabeled test antigen compete during incubation. Following incubation, the reaction wells are cleaned with an appropriate buffer to eliminate unbound reactants. The compound's enzyme then breaks down the substrate solution, which results in a colorful product [58]. The primary benefit of competitive ELISA is its increased sensitivity to changes within the mixed antigen complex, particularly when a minimal amount of the specific detection antibody is present. Thus, it is referred to as "competitive" since both analytes are introduced simultaneously. On the other hand, the approach may be referred to as inhibitory when inhibitory competition occurs, wherein an analyte is supplied initially, and a tagged one is added after the incubation period [59].

Assay Development and Examples of Sample Preparation of ELISA

In ELISA, assay development and sample preparation are the crucial steps for getting accurate results. The significant steps to be considered accurately are centrifugation, sample handling, and storage conditions. One of the reaction components in this kind of assay is covalently or non-specifically adsorbed to the surface of a solid phase, like a plastic bead, a magnetic particle, or a microtiter well. This association makes the separation of reactants with and without labels easier. An aliquot of sample or calibrator containing the antigen (Ag) to be measured is introduced to a solid-phase antibody (Ab) and allowed to bind in the most popular ELISA technique method. After washing, an enzyme-labeled antibody is added, forming a solid-phase Ab-Ag-Ab "sandwich complex." After the unbound antibody has been removed by washing, the enzyme substrate is added. The amount of antigen in the sample directly correlates with the product produced. The main pitfalls that majorly affect the reliability and accuracy of the desired results are sample contamination, inconsistent sample dilutions, inadequate mixing, storage conditions not properly avoiding the repeated freeze-thaw cycles, and delay in the processing time of the sample, which can lead to concentration changes due to the adsorption or degradation to container surfaces, incompatible buffer selection which affect antigen-antibody interaction, insufficient washing unable to remove non-specific binding which leads to poor signal to noise ratios and high background, matrix effects ignoring, and validate process failure [60–64].

Lateral-Flow Immuno-chromatographic Strips: A Highly Applicable ELISA-Based Technique for On-Site Food Detection

One-step immuno-chromatographic lateral flow strip testing makes it possible to quickly and affordably identify a variety of analytes at the point of care. We have created lateral flow strip assays for the precise qualitative or semi-quantitative detection of antibodies, antigens, and haptens, including drug residues. Depending on the molecular sizes and immunological characteristics of the target molecules, there are two types of lateral flow strips: competitive solid-phase immunoassay and direct solid-phase (sandwich) immunoassay. Pork adulteration in processed meats can be visually detected with a quick immunostain test. A polyclonal antibody against swine IgG was coupled with gold nanoparticles. A lateral flow immunosensor has been developed to screen for pork tampered with in processed meats, such as beef meatball samples. The simulated samples were pork-in-beef meatballs extracted in phosphate-buffered solution at adulteration levels ranging from 0 to 100 (%w/w). As controls, there were pure beef meatballs devoid of pork. Five minutes after the incubation period, the response time was finished. 0.1% (w/w) was the immunosensor strip's detection limit. Low amounts of pig adulteration in beef meatballs can be found with these immunosensor assays. The assay's reliability was further examined by contrasting the outcomes with those of ELISA kits that are sold commercially [64]. A dipstick test for immunochromatographic lateral flow has been developed to identify bovine rennet whey in milk powder and liquid quickly. Based on the binding of casein glycomacropeptide (cGMP) by two particular anti-bovine κ-casein monoclonal antibodies, the test has a visual detection limit of approximately 15 ng mL^{-1} for cGMP and 1% (v/v) for rennet whey in milk utilizing standards and spiked samples. In a collaborative experiment, raw, pasteurized, UHT milk and skim milk powder were used to assess the dipstick performance. Through the analysis of 60 raw milk samples gathered from Brazilian farms, the applicability of the dipstick was confirmed in comparison to gel permeation-high-performance liquid chromatography and a colorimetric approach. More dependable than the colorimetric approach, the dipstick results strongly correlated with the HPLC results. A more than 4% rennet whey level was accurately detected in all raw milk samples using the dipstick test [65–70].

Future Perspective of ELISA as the point of Detection Devices and Integration with AI

AI provides state-of-the-art approaches for detecting food adulteration more quickly, accurately, and efficiently than traditional techniques. AI is revolutionizing food adulteration detection in the following ways. (a) Data analysis from sophisticated spectroscopic methods, such as Raman and Near-Infrared (NIR) spectroscopy, is improved by AI. AI uses machine learning methods to identify minute variations in the spectral data that can indicate adulterants' presence. This makes it possible to evaluate food products quickly and without causing any damage. (b) Pattern recognition in complex datasets is one area in which AI specializes. Large datasets with details on both real and manipulated food samples can be used to train AI models for food adulteration detection. After training, these models can detect deviations from known patterns to reliably detect adulteration in fresh samples. (c) Food production lines can incorporate AI-powered sensors and Internet of Things devices to check for adulteration indications continuously. These systems are capable of real-time anomaly detection, including sudden changes in physical characteristics or chemical composition, which encourages prompt alerts and preventive measures to keep compromised products off the market, in short, real-time monitoring. (d) The food sector produces enormous volumes of data from various sources, such as market statistics, supply chain records, and quality control inspections. AI can analyze this large amount of data to find patterns, connections, and possible dangers associated with food adulteration. This helps authorities and businesses to anticipate and reduce adulteration hazards before they are marketed and become a health hazard. (e) Blockchain technology and artificial intelligence (AI) can improve food product traceability throughout the supply chain. Blockchain ensures that every food product movement and transaction is documented in a permanent ledger, and artificial intelligence (AI) may examine this data to find irregularities or questionable activity that could potentially point to adulteration. (f) With the help of smartphone cameras or barcode scanning, consumers can access AI-powered databases that analyze the product's ingredients and history, giving them immediate information on its safety and authenticity. Ultimately, it provides the consumer with the power to detect food adulteration without any hurdles. AI-driven mobile applications are increasingly being developed to empower consumers to detect food adulteration.

Artificial intelligence greatly improves food safety and legal frameworks in many aspects of the food supply chain, such as production, processing, distribution, and consumption. Integrating high-resolution photography with AI can detect foreign objects, spoilage, and faults beyond what is now possible, vital in recognizing and preventing dangerous compounds. Artificial intelligence (AI)-powered "radio-frequency identification" (RFID) and Internet of Things (IoT) sensors track and document environmental conditions, guaranteeing compliance with safety standards and preserving product quality. Because of AI's Natural Language Processing (NLP) capabilities, it can evaluate and understand new laws, keeping businesses informed of developments and allowing for quick adjustments. Apps with AI capabilities are helping consumers make safer food selections and remain updated about issues related to food safety. Mobile apps powered by AI give users access to the most recent information on food recalls, allergy alerts, and nutritional facts, enabling them to make educated decisions. The impact of artificial intelligence on food safety is expected to grow in the future. Biological dangers might appear accidentally or on purpose. Public health benefits from biomonitoring by reducing foodborne illness, while public safety benefits from preventing intentional food adulteration and bioterrorism. Pre-harvest and post-harvest phases of the food chain might all be integrated with AI-assisted bio-monitoring systems through logistical networks. Data from each sensor or analyzer that comes into contact with an agricultural product that humans consume is distinctive. AI may then combine and examine this information. To detect tainted product molecules, we might build a program to look for particular patterns and correlations. It is also capable of identifying previously undetected anomalies in processes or outputs. Subject Matter Experts (SMEs) and system operators would then have quick access to and evaluation of the real-time AI discoveries. Human involvement is crucial for analysis based on judgment, experience, and reasoning. The AI-powered analytical system can combine semi-structured, unstructured, and structured data to facilitate analysis. In conclusion, we can confidently say that AI-assisted analytics is a quickly developing field that, with the proper focus, can safeguard public health, lower the risk of foodborne illnesses, and preserve brand excellence and profitability. Adopting this technology requires attention [66].

Conclusion

Global public health and food security are seriously threatened by food adulteration, and financial rewards, inadequate laws regulations, and consumer perceptions could heighten the risks. In most cases, the adulterants are so similar to their replaced counterparts in appearance, chemical makeup, and organoleptic characteristics that even the analyst has trouble identifying them. The specificity of the immunological reaction makes immunoassay an appropriate technique for detecting food contaminants. Two common immunoassays for determining whether less acceptable or undesirable species are present in food products are agar gel immunodiffusion and ELISA. The impact of artificial intelligence on food safety is expected to grow in the future. Public health benefits from biomonitoring by reducing foodborne illness, while public safety benefits from preventing intentional food adulteration and bioterrorism.

Acknowledgments The authors AR and GBJ thank the National Research Foundation, South Korea, for providing financial assistance.

Funding National Research Foundation, South **Korea.**

References

1. Gopalan KR (2025) Contaminated consumption: unveiling the health hazards of food adulteration and its profound impact on public health in India. J Pioneer Med Sci 13:75–88
2. Banti M (2020) Food adulteration and some methods of detection, review. Int J Nutr Food Sci 9(3):86–94
3. Manasha S, Janani M (2016) Food adulteration and its problems (intentional, accidental and natural food adulteration). Int J Res Finance Market 6(4):131–140
4. He Y, Bai X, Xiao Q, Liu F, Zhou L, Zhang C (2021) Detection of adulteration in food based on non-destructive analysis techniques: a review. Crit Rev Food Sci Nutr 61(14):2351–2371
5. World Health Organization. Food safety. https://www.who.int/westernpacific/health-topics/food-safety. Accessed 5 May 2021
6. Yang T, Wang P, Guo H, He L (2017) Surface-enhanced Raman spectroscopy: a tool for all classes of food contaminants. Elsevier, Amsterdam
7. Habza-Kowalska E, Grela M, Gryzińska M, Listos P (2019) Molecular techniques for detecting food adulteration. Med Weter 75(5):6260–2020. https://doi.org/10.21521/mw.6261
8. Van Raamsdonk LWD, von Holst C, Baeten V, Berben G, Boix A, de Jong J (2007) New developments in the detection and identification of processed animal proteins in feeds. Anim Feed Sci Technol 133:63–83

9. Bansal S, Singh A, Mangal M, Mangal AK, Kumar S (2017) Food adulteration: sources, health risks, and detection methods. Crit Rev Food Sci Nutr 57(6):1174–1189
10. Joshi VC, Khan IA (2005) Microscopy techniques for the identification and authentication of botanicals. In: IV international conference on quality and safety issues related to botanicals, vol 720. Elsevier, Amsterdam, pp 73–80
11. Petrakis EA, Polissiou MG (2017) Assessing saffron (Crocus sativus L.) adulteration with plant-derived adulterants by diffuse reflectance infrared Fourier transform spectroscopy coupled with chemometrics. Talanta 162:558–566
12. Kucharska-Ambrożej K, Karpinska J (2020) The application of spectroscopic techniques in combination with chemometrics for the detection of adulteration of some herbs and spices. Microchem J 153:104278
13. Sidhu OP, Annarao S, Chatterjee S, Tuli R, Roy R, Khetrapal CL (2011) Metabolic alterations of Withania somnifera (L.) Dunal fruits at different developmental stages by NMR spectroscopy. Phytochem Anal 22(6):492–502
14. Ballin NZ, Vogensen FK, Karlsson AH (2009) Species determination—can we detect and quantify meat adulteration? Meat Sci 83:165–174
15. Anagaw YK, Ayenew W, Limenh LW, Geremew DT, Worku MC, Tessema TA, Simegn W, Mitku ML (2024) Food adulteration: causes, risks, and detection techniques-review. SAGE Open Med 12:20503121241250184. https://doi.org/10.1177/20503121241250184
16. Vinothkanna A, Dar OI, Liu Z, Jia AQ (2024) Advanced detection tools in food fraud: a systematic review for holistic and rational detection method based on research and patents. Food Chem 4469:138893
17. https://www.fssai.gov.in/book-details.php?bkid=201. Accessed 14 January 2025
18. Rahman MB, Hussain M, Kabiraz MP, Nordin N, Siddiqui SA, Bhowmik S, Begum M (2023) An update on formaldehyde adulteration in food: sources, detection, mechanisms, and risk assessment. Food Chem 427:136761. https://doi.org/10.1016/j.foodchem.2023.136761
19. Roberts J, Power A, Chapman J, Chandra S, Cozzolino D (2018) A short update on the advantages, applications and limitations of hyperspectral and chemical imaging in food authentication. Appl Sci 8(4):505. https://doi.org/10.3390/app8040505
20. Hong E, Lee SY, Jeong JY, Park JM, Kim BH, Kwon K, Chun HS (2017) Modern analytical methods for the detection of food fraud and adulteration by food category. J Sci Food Agric 97(12):3877–3896
21. Moing A, Allwood JW, Aharoni A, Baker J, Beale MH, Ben-Dor S, Schaffer AA (2020) Comparative metabolomics and molecular phylogenetics of melon (Cucumis melo, Cucurbitaceae) biodiversity. Meta 10(3):121
22. Selamat J, Rozani NAA, Murugesu S (2021) Application of the metabolomics approach in food authentication. Molecules 26(24):7565
23. Cheng Z, Zhang X, Geng X, Organtini KL, Dong F, Xu J, Zheng Y (2018) A target screening method for the detection of organic pollutants in fruits and vegetables by atmospheric pressure gas chromatography quadrupole-time-of-flight mass spectrometry combined with an informatics platform. J Chromatogr A 1577:82–91
24. Kang C, Zhang Y, Zhang M, Qi J, Zhao W, Gu J, Li Y (2022) Screening of specific quantitative peptides of beef by LC-MS/MS coupled with OPLS-DA. Food Chem 387:132932
25. Ajani PA, Sarowar C, Turnbull A, Farrell H, Zammit A, Helleren S, Murray SA (2021) A comparative analysis of methods (LC-MS/MS, LC-MS and rapid test kits) for the determination of diarrhetic shellfish toxins in oysters, mussels and pipis. Toxins 13(8):563
26. Wu R, Chen J, Zhang L, Wang X, Yang Y, Ren X (2021) LC/MS-based metabolomics to evaluate the milk composition of humans, horses, goats and cows from China. Eur Food Res Technol 247:663–675
27. Artavia G, Cortés-Herrera C, Granados-Chinchilla F (2021) Selected instrumental techniques applied in food and feed: Quality, safety and adulteration analysis. Foods 10(5):1081
28. Rohman A, Sismindari EY, Cheman YB (2011) Analysis of pork adulteration in beef meatball using Fourier transform infrared (FTIR) spectroscopy. Meat Sci 88:91–95

29. Van der Spiegel M, Van der Fels-Klerx HJ, Sterrenburg P, Ruth SM, Scholtens-Toma IMJ, Kok EJ (2012) Halal assurance in food supply chains: verification of halal certificates using audits and laboratory analysis. Trends Food Sci Technol 27:109–119
30. Mecca JN, Meireles R, De Andrade HF (2011) Quality control of *Toxoplasma gondii* in meat packages: standardization of an ELISA test and its use for detection in rabbit meat cuts. Meat Sci 88:584–589
31. Keim SA, Kulkarni MM, McNamara K, Geraghty SR, Billock RM, Ronau R, Hogan JS, Kwiek JJ (2015) Cow's milk contamination of human milk purchased via the internet. Pediatrics 135(5):e1157–e1162. https://doi.org/10.1542/peds.2014-3554
32. Chi S, Liu B, Zhang B, Wang B, Zhou J, Li L, Zhang Y, Mu Z (2024) Development of an ELISA method to determine adulterated cow milk in camel milk. Int Dairy J 155:105953. https://doi.org/10.1016/j.idairyj.2024.105953
33. Borková M, Snášelová J (2005) Possibilities of different animal milk detection in milk and dairy products - a review. Czech J Food Sci 23(2):41–50. https://doi.org/10.17221/3371-cjfs
34. Manzo N, Pizzolongo F, Montefusco I, Romano A, Masi P, Romano R (2017) Using whey proteins to detect the addition of bovine milk fat in buffalo cream destined for the butter-making process. Food Control 81:164–167. https://doi.org/10.1016/j.foodcont.2017.06.002
35. Hurley I, Coleman R, Ireland H, Williams JH (2004) Measurement of bovine IGG by indirect competitive ELISA as a means of detecting milk adulteration. J Dairy Sci 87(3):543–549. https://doi.org/10.3168/jds.s0022-0302(04)73195-1
36. Hurley IP, Coleman RC, Ireland HE, Williams JH (2006) Use of sandwich IgG ELISA for the detection and quantification of adulteration of milk and soft cheese. Int Dairy J 16(7):805–812. https://doi.org/10.1016/j.idairyj.2005.07.009
37. Ren Q, Zhang H, Guo H, Jiang L, Tian M, Ren F (2014) Detection of cow milk adulteration in yak milk by ELISA. J Dairy Sci 97(10):6000–6006. https://doi.org/10.3168/jds.2014-8127
38. Cosenza G, Iannaccone M, Gallo D, Pauciullo A (2019) A fast and reliable polymerase chain reaction method based on short interspersed nuclear elements detection for the discrimination of buffalo, cattle, goat, and sheep species in dairy products. Asian Australas J Anim Sci 32(6):891–895. https://doi.org/10.5713/ajas.18.0459
39. Vidarsson G, Dekkers G, Rispens T (2014) IGG subclasses and allotypes: from structure to effector functions. Front Immunol 5:520. https://doi.org/10.3389/fimmu.2014.00520
40. Mousavi SF, Fatemi S, Siadat SD, Zahraei SM, Nikanpour E, Malekan MA, Khabiri AR, Janani AR (2016) Development and optimization of a homemade ELISA kit for detection of antibodies against Haemophilus influenzae type B. Jundishapur J Microbiol 9(5):30629. https://doi.org/10.5812/jjm.30629
41. Krishna TC, Najda A, Bains A, Tosif MM, Papliński R, Kapłan M, Chawla P (2021) Influence of ultra-heat treatment on properties of milk proteins. Polymers 13(18):3164. https://doi.org/10.3390/polym13183164
42. Azad T, Ahmed S (2016) Common milk adulteration and their detection techniques. Int J Food Contam 3(1):22. https://doi.org/10.1186/s40550-016-0045-3
43. Eldahshoury MK, Hurley IP (2023) Direct sandwich ELISA to detect the adulteration of human breast milk by cow milk. J Dairy Sci 106(9):5908–5915. https://doi.org/10.3168/jds.2022-22589
44. Mandli J, Fatimi IE, Seddaoui N, Amine A (2018) Enzyme immunoassay (ELISA/immunosensor) for a sensitive detection of pork adulteration in meat. Food Chem 255:380–389. https://doi.org/10.1016/j.foodchem.2018.01.184
45. https://www.fsis.usda.gov/food-safety/safe-food-handling-and-preparation/food-safety-basics/food-allergies-big-9. Accessed 14 January 2025
46. Van Hengel AJ (2007) Food allergen detection methods and the challenge to protect food-allergic consumers. Anal Bioanal Chem 389(1):111–118. https://doi.org/10.1007/s00216-007-1353-5
47. Pelczar MJ, Pelczar RM (2025) Microbiology. Encyclopedia Britannica. https://www.britannica.com/science/microbiology. Accessed 14 January 2025

48. National Research Council (US) Committee on Pesticides in the Diets of Infants and Children (1993) Pesticides in the diets of infants and children. National Academies Press, Washington, p 1. Available from https://www.ncbi.nlm.nih.gov/books/NBK236265/. Accessed 14 January 2025
49. Arsène MMJ, Davares AKL, Viktorovna PI, Andreevna SL, Sarra S, Khelifi I, Sergueïevna DM (2022) The public health issue of antibiotic residues in food and feed: causes, consequences, and potential solutions. Vet World 15(3):662–671. https://doi.org/10.14202/vetworld.2022.662-671
50. Voller A, Bartlett A, Bidwell DE (1978) Enzyme immunoassays with special reference to ELISA techniques. J Clin Pathol 31(6):507–520. https://doi.org/10.1136/jcp.31.6.507
51. Lequin RM (2005) Enzyme immunoassay (EIA)/enzyme-linked immunosorbent assay (ELISA). Clin Chem 51(12):2415–2418. https://doi.org/10.1373/clinchem.2005.051532
52. Aydin S (2015) A short history, principles, and types of ELISA, and our laboratory experience with peptide/protein analyses using ELISA. Peptides 72:4–15. https://doi.org/10.1016/J.PEPTIDES.2015.04.012
53. Hayrapetyan H, Tran T, Tellez-Corrales E, Madiraju C (2023) Enzyme-linked immunosorbent assay: types and applications. Methods Mol Biol 2612:1–17. https://doi.org/10.1007/978-1-0716-2903-1_1
54. Kurbanov M, Mukhamadiev B, Kalanova D, Muzafarova K, Kurbanova S (2021) Immune-enzyme methods of food safety analysis. IOP Conf Ser Earth Environ Sci 848(1):012185. https://doi.org/10.1088/1755-1315/848/1/012185
55. Sakamoto S et al (2018) Enzyme-linked immunosorbent assay for the quantitative/qualitative analysis of plant secondary metabolites. J Nat Med 72(1):32–42. https://doi.org/10.1007/s11418-017-1144-z
56. Chaya D, Parija SC (2013) Evaluation of a newly designed sandwich enzyme-linked immunosorbent assay for the detection of hydatid antigen in serum, urine and cyst fluid for diagnosis of cystic echinococcosis. Trop Parasitol 3(2):125–131. https://doi.org/10.4103/2229-5070.122131
57. Zhang X et al (2021) A double-antibody sandwich ELISA for sensitive and specific detection of swine fibrinogen-like protein 1. Front Immunol 12:1–13. https://doi.org/10.3389/fimmu.2021.670626
58. Khan MS, Rahman MS (2021) Techniques to measure food safety and quality: microbial, chemical, and sensory. https://doi.org/10.1007/978-3-030-68636-9
59. Xiong Y, Leng Y, Li X, Huang X, Xiong Y (2020) Emerging strategies to enhance the sensitivity of competitive ELISA for detection of chemical contaminants in food samples. TrAC Trends Anal Chem 126:115861. https://doi.org/10.1016/j.trac.2020.115861
60. Alhajj M, Zubair M, Farhana A (2025) Enzyme-linked immunosorbent assay. In: StatPearls. StatPearls Publishing, Treasure Island. Available from https://www.ncbi.nlm.nih.gov/books/NBK555922/
61. Engvall E (2010) The ELISA, enzyme-linked immunosorbent assay. Clin Chem 56(2):319–320
62. https://www.phoenixpeptide.com/. Accessed 14 January 2025
63. Puchades R, Maquieira Á (2013) ELISA tools for food PDO authentication. In: Comprehensive analytical chemistry, vol 60. Elsevier, Amsterdam, pp 145–193
64. Kuswandi B, Gani AA, Ahmad M (2017) Immuno strip test for detection of pork adulteration in cooked meatballs. Food Biosci 19:1–6. https://doi.org/10.1016/j.fbio.2017.05.001
65. Martín-Hernández C, Muñoz M, Daury C, Weymuth H, Kemmers-Voncken AE, Corbatón V, Toribio T, Bremer MG (2008) Immunochromatographic lateral-flow test strip for the rapid detection of added bovine rennet whey in milk and milk powder. Int Dairy J 19(4):205–208. https://doi.org/10.1016/j.idairyj.2008.10.016
66. Parul A et al (2023) Food adulteration and its detection using artificial intelligence. Educ Admin 29(4):2576–2583

67. Joia R, Modaqeq T (2024) Unlocking precision: a dive into Elisa methods and significance in ensuring food safety and quality. Acad Res Educ Sci 5:163–176
68. International Accreditation Service (2015) Guidelines for food testing laboratories
69. De Beer J (2011) Ensuring the quality of results from food control laboratories: laboratory accreditation, method validation and measurement uncertainty. In: Elsevier eBooks. Elsevier, Amsterdam, pp 194–222. https://doi.org/10.1533/9780857090973.2.194
70. Odumeru JA, Steele M, Fruhner L, Larkin C, Jiang J, Mann E, McNab WB (1999) Evaluation of accuracy and repeatability of identification of foodborne pathogens by automated bacterial identification systems. J Clin Microbiol 37(4):944–949. https://doi.org/10.1128/JCM.37.4.944-949.1999Va

Deep Eutectic Solvents for Extraction and Detection of Adulterants in Food Samples

Mussab Uddin, Zaharaddeen Muhammad, Abdulsalam Kabiru Saidu, Nur Nadhirah Mohamad Zain, and Waleed Alahmad

Contents

Mussab Uddin, Zaharaddeen Muhammad and Abdulsalam Kabiru Saidu contributed equally with all other contributors.

M. Uddin · A. K. Saidu · W. Alahmad (✉)
Faculty of Science, Department of Chemistry, Chulalongkorn University, Bangkok, Thailand
e-mail: waleed.al@chula.ac.th

Z. Muhammad
Faculty of Science, Department of Chemistry, Chulalongkorn University, Bangkok, Thailand

Department of Chemistry, Faculty of Natural and Applied Sciences, Sule Lamido University, Kafin Hausa, Nigeria

N. N. M. Zain (✉)
Department of Toxicology, Advanced Medical and Dental Institute (AMDI), Universiti Sains Malaysia, Bertam, Kepala Batas, Pulau Pinang, Malaysia
e-mail: nurnadhirah@usm.my

W. Alahmad, M. Ashfaq (eds.), *Advanced Techniques against Food Adulteration*,
https://doi.org/10.1007/978-3-031-99454-8_11

Abstract Deep eutectic solvents (DESs) have emerged as innovative green solvents with a high potential for food adulteration analysis. This chapter explores the fundamental characteristics, formation, and classifications of DESs, highlighting their eco-friendly nature, biocompatibility, and tunable properties for effective extraction and detection applications. The unique hydrogen bonding networks in DESs facilitate superior solubility and selectivity for diverse analytes, including dyes, metal oxides, phytosterols, and synthetic drugs, from complex food matrices. Various analytical techniques, such as HPLC-UV, GC-MS, and electrochemical approaches, are detailed for DES-based food safety assessments. Moreover, the chapter addresses the environmental and economic advantages of DESs alongside key challenges in their stability, recyclability, and viscosity management. The findings underscore DESs as sustainable and efficient alternatives for enhancing food quality control and ensuring compliance with regulatory standards.

Keywords Deep eutectic solvents · Food adulteration detection · Green analytical chemistry · Selective extraction · Sustainable solvents

Introduction

Food safety is essential due to its significant consequences and effects on the public. Adulterants are chemical substances that should not be contained in a food or beverage sample [1, 2]. The health of a population depends on the nature of the food they consume and its nutritive value [3]. Food adulteration is a pressing issue. It refers to the deliberate or accidental contamination of food products with inferior, harmful, or unapproved substances by misleading consumers, mainly to profit from it [4]. Adulteration has affected the entire food industry, leading to significant public health implications [5, 6]. This practice poses considerable health risks to consumers, including exposure to toxic compounds, allergens, and diminished nutritional value.

Detecting adulterants in complex food matrices presents substantial challenges, primarily from the diverse nature of food constituents, variations in chemical interactions, the presence of interferent, and the complex nature of food matrices [7]. A significant challenge is that scientists do not have the knowledge and access to the adulterants used in practice; as a result, the way these samples are adulterated in the lab may not accurately reflect how adulteration occurs in real-world scenarios [8]. Food products often contain a wide range of natural compounds that can interfere with detection methods, making it challenging to isolate/extract and identify specifics [2, 9]. Another factor is the limitation of specific methods for different food samples [10]. Traditional analytical methods offer precise quantification and are ideal for laboratory analysis but struggle with extraction and sample preparation stages [11]. This prompts the need for more selective solvents to enable rapid screening of food adulteration and supplement the analytical techniques.

In addressing these challenges, the field of green chemistry plays a vital role by promoting environmentally friendly practices in developing analytical methods. Sample preparation remains an essential preliminary step in all analytical studies. Greenness metrics are necessary for creating sustainable analytical procedures [12]. A range of practical techniques exists for the pre-concentration and extraction of target analytes before analysis. Traditionally, liquid-liquid extraction (LLE) and solid-phase extraction (SPE) have stood out as the primary methods for isolating analytes from samples, but with the drawback of requiring significant amounts of hazardous organic solvents [13]. Green chemistry principles encourage the reduction of harmful solvents, energy consumption, and waste. Thus, it is a critical necessity for adopting greener methodologies in sample analysis processes [14, 15]. The novel developments and latest technological trends, such as the use of surfactants and ionic liquids, as well as deep eutectic solvents and their combinations with microextraction, and even the implementation of organic solvent-free extraction methods, have been successfully applied to remove contaminants and adulterants in several food samples [15, 16].

Deep eutectic solvents (DESs) have emerged as an innovative solution in food analysis. They are formed by combining two or more components that interact to produce a melting point lower than that of individual components. They are Lewis or Brønsted acids and bases described by the general formula $Cat^{+}X^{-}Zy$ [17]. DESs are biocompatible, non-toxic, and can be used for specific applications, making them an attractive alternative to traditional organic solvents. DESs were first introduced by Abbot et al. in 2004 and began to draw attention because of their fascinating properties, which include high biodegradability, low volatility, affordability, eco-friendliness, thermal stability, and, mostly, ease of synthesis, positioning them as promising agents for the extraction and detection of food adulterants. Since their introduction, DESs have become popular among researchers from various fields, and many food additives are extracted based on DES microextraction techniques [13, 18, 19]. Their extraction ability has also been confirmed through the utilization of DES as a functional monomer for molecularly imprinted polymers (MIPs) electrochemical food analysis [20].

DES: Fundamentals

Definition and Characteristics of DES

In an updated theory, Martins et al. define a deep eutectic solvent as "a mixture of two or more pure compounds with a eutectic point temperature lower than that of the ideal liquid mixture and with significant negative deviations from ideality ($\Delta T2>0$)", where the ideal and actual eutectic points are stated. Let $\Delta T2$ be the temperature drop [21]. Deep eutectic solvents were described by Abbott et al. as the type of ionic liquids obtained by the mixing of quaternary ammonium salt with metal salt, resulting in the formation of complex anions, for instance, $Al_2Cl_7^-$ and $Zn_2Cl_5^-$. Melting point depression was first observed in a mixture of choline chloride (Tm ≈ 302 °C) and crystalline urea (Tm ≈ 133 °C) when mixed in a 1:2 molar ratio. This mixture, known as reline, appeared to have a low melting point: $T_{eutectic}$ = 12 °C. The dip in the melting point lies in variable hydrogen bonding and Van der Waals interactions due to asymmetric ions resulting in low lattice energy, and this lattice energy damping gives DESs lower melting points than their parent compounds [22].

Formation of DES

The following methods can form deep eutectic solvents.

Heating and Stirring Method

DES can be prepared by conventional heating method while stirring with the help of a magnetic stirrer. The DES constituents, for instance, choline chloride 2.688 g and urea 2.312 g, are heated at 80 °C while stirring with constant temperature monitoring at intervals of every 5 seconds. This process continues until a precise homogenous mixture is obtained. This process involves no actual reaction, solvent use, or residue left; thus, it commits to being environmentally friendly in DES preparation [23].

Vacuum Evaporation

The vacuum evaporation method for preparing DESs uses an evaporation setup, usually a rotary oven. The first step involves the dissolution of DESs components into a solvent, i.e., water; after the complete dissolution of the salts, the solution goes for evaporation with a rotary oven at 323 K until most of the water evaporates. Then, the residual solution is kept inside the desiccator and the silica gel until it reaches the constant weight [24].

Grinding

Florindo et al. first used this method. They deduced a technique for the preparation of DESs without the use of any heat. Before grinding, HBA is subjected to drying for 2 days by keeping it in a vacuum oven at 40 °C, and then it is kept under high vacuum in a Schlenk for 4 days due to its highly hygroscopic nature. The components are ground with the help of a mortar and pestle under an inert environment or inside the glove box until a clear, homogeneous liquid is formed [25].

Freeze Drying

Gutierrez, in 2009, devised a freeze-drying method to prepare DES. Choline chloride and urea solution 5% by weight were prepared separately and then mixed in a ratio of 1:2. The aqueous solution was dried at 77 K and 253 K after freezing. Sublimation was done under low pressure and temperature to eventually sublimate the water directly from solid to vapors to obtain a clear, viscous liquid [16].

Types of Deep Eutectic Solvents

Deep eutectic solvents are classified into five types based on the type of solvent constituent [17]. Table 1 shows the types and composition of deep eutectic solvents.

Advantages of DESs for Food Adulterant Detection

Deep eutectic solvents (DESs) have two or more components that significantly lower the melting point when combined. DESs are increasingly utilized in food adulterant detection due to their distinctive physicochemical properties, such as high solvating capacity, low volatility, and exceptional tunability. In contrast to conventional organic solvents, deep eutectic solvents (DESs) are eco-friendly, biodegradable, and economically viable [26]. These characteristics render them an advantageous alternative for formulating sustainable, efficient, and precise solvents

Table 1 Types and composition of deep eutectic solvents

Type	General formula	Terms
Type I (a quaternary ammonium salt and metal halide)	$Cat^+ X^-{}_z MCl_x$	M=Zn[1,56], Sn[7], Fe, Al[8], Ga[9],In[10]
Type II (a quaternary ammonium salt and a metal halide hydrate)	$Cat^+ X^-{}_z MCl_x \cdot {}_y H_2O$	M=Cr[11], co, cu, Ni, Fe
Type III (a quaternary ammonium salt and HBA, which is an organic molecular moiety like amide, carboxylic acid, or a polyol)	$Cat^+ X^-{}_z RZ$	$Z=CONH_2$[12], COOH[13], OH[14]
Type IV (a metal halide hydrate and an HBD)	$MCl_x + RZ = MCl_{x-1} + \cdot RZ + MCl_{x+1}^-$	M=Al, Zn and $Z=CONH^2$, OH
Type V (completely non-ionic entities, both the HBA and HBD are molecular)	HBD + HBA	N/A

for analytical methods in food safety. The simplicity of customizing DES compositions enables researchers to focus on adulterants, thus enhancing selectivity. Furthermore, the minimal toxicity and diminished environmental impact of DESs correspond with green chemistry principles, rendering them suitable for regular food safety monitoring and regulatory adherence [27]. Below, we explore these advantages in detail under three key aspects.

High Selectivity and Solubility for Targeted Analytes

DESs demonstrate exceptional selectivity and solubility for analytes, which is mainly attributed to their adjustable physicochemical properties, which can be customized by modifying the molar ratios and varieties of hydrogen bond donors (HBDs) and hydrogen bond acceptors (HBAs) incorporated in their formulation [28].

High Solubility of DESs for Analyte

Factors that affect the solubility of analytes in DESs are:

1. Hydrogen bonding network.
2. Cocrystal formation.
3. Solvent polarity hydrophilic/hydrophobic balance.
4. Reduced nucleation and crystallization tendencies.

Hydrogen Bonding Network

The primary mechanism behind the enhanced solubility of analytes in DESs lies in their hydrogen bonding network. In a DES, the HBA and HBD, such as (choline chloride and urea or malonic acid, respectively) interact through strong hydrogen

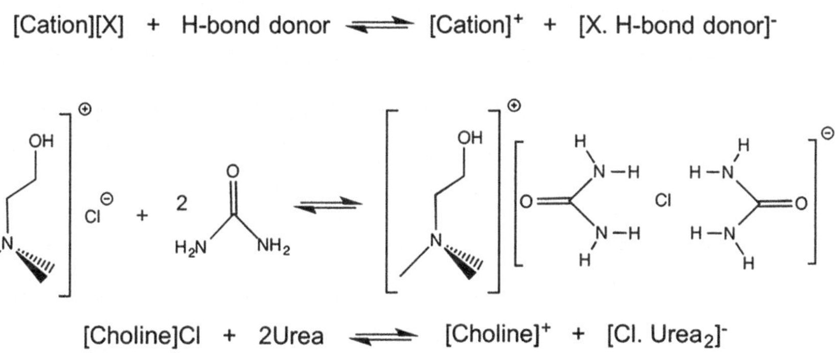

Fig. 1 Interaction of hydrogen bonding anions with choline chloride enhances the dissolved analyte's solubility. Reprinted from [30] with permission from the Royal Society of Chemistry

bonds, creating a highly stable and structured liquid environment. These hydrogen bonds disrupt the analyte's crystalline lattice structure, reducing the energy barrier for solubilization. Analytes with functional groups such as –OH, –COOH, or $–NH_2$ readily interact with the hydrogen-bonding sites within the DES, leading to increased solubility. Benzoic acid, griseofulvin, and danazol showed significant solubility enhancement in DESs compared to water [29]. Moreover, DESs have a distinct characteristic of complex formation via hydrogen bonding; anion binds to the organic molecule through the hydrogen bonding (Fig. 1), illustrating the complex formation within choline chloride urea (1:2) DES. Those DESs consisting of poorly complexing HBDs have increased proton activity; hence, the H^+ ability to accept oxygen increases the solubility in DES. An experimental work to determine the dissolution of metal oxide in an EG: ChCl was done by Pateli et al. The impressive attributes of low viscosity and high conductivity convinced the selection of EG: ChCl DES coupled with ICP-MS for detecting metal. Interestingly, the solubility of PbO in this DES was found to be 4453 ± 10 mol L^{-1} at room temperature, almost double that of the previous study using ICP-OES [31]. Besides the role of hydrogen bonding, there are other factors which influence the solubility of metal oxides in DESs, such as differences in physical properties like crystallinity water content of the DES (metal oxides are dependent on the water content for solubility; usually metal oxides show less solubility in aqueous solutions), and particle size.

Cocrystal Formation

Some analytes can form cocrystals with DES components, further enhancing solubility and stability. Cocrystallization in DESs is a phenomenon of crystal formation typically when a drug and a conformer combine in a DES medium. Cocrystal formation can assist in detecting antibiotics as food adulterants in baby food. For instance, in the urea–choline chloride DES system, a novel 2:1 cocrystal of urea and choline chloride was observed. This cocrystal has a lower melting point and higher disorder, facilitating the dissolution of poorly soluble compounds. The formation of cocrystals suggests that the enhanced solubility in DESs is not merely due to a eutectic melting phenomenon but rather a result of specific molecular interactions and structural rearrangements within the DES environment [32]. Figure 2 shows the cocrystal formation of antibiotics in DES. Another study by Ito et al. demonstrated that they found five kinds of cocrystals using DESs on active pharmaceutical ingredients. When mixed with water, DES consisting of malonic acid and choline chloride increases the solubility of poorly soluble compounds by 5–22,000 times. Moreover, olanzapine and acetaminophen showed crystal formation with volatile DESs, and drugs like betamethasone, meloxicam, and piroxicam showed enhanced water solubility in ChCl-based DESs [33].

Solvent Polarity Hydrophilic/Hydrophobic Balance

DESs have tunable polarity, which allows them to dissolve both polar and non-polar analytes. The hydrogen bonding and Van der Waals interactions allow DESs to create microenvironments suitable for the dissolution of analytes with diverse chemical properties. This characteristic enables DESs to outperform water or other single-component solvents in solubilizing certain compounds [34]. Hydrogen bond acidity is a significant factor influencing the solubilities of solutes in DESs [35]. DESs can inhibit the nucleation and crystallization of dissolved analytes. The viscous nature and strong molecular interactions within DESs prevent the formation of stable nuclei, keeping analytes in the dissolved state for more extended periods (Table 2). This property is beneficial for compounds with high crystallization tendencies, where maintaining them in solution is essential for bioavailability or analytical purposes [37, 38].

Solubility of Metal Oxides in DESs

Deep eutectic solvents are also very efficient in the solubilization of metal oxides. Factors which assist the solubilization of metal oxides in DESs are as follows.

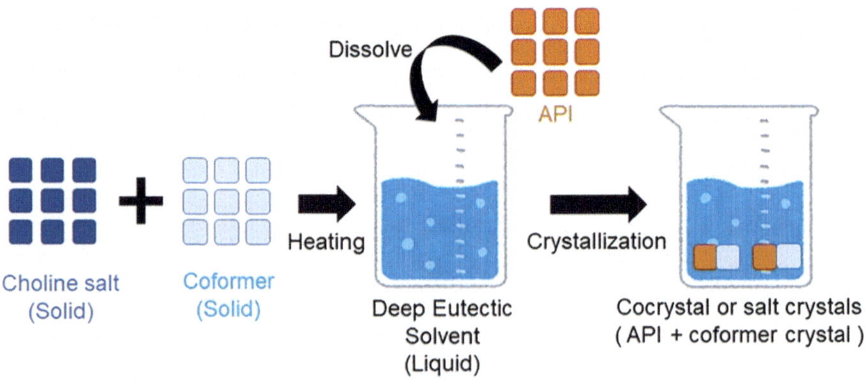

Fig. 2 Cocrystal formation of antibiotics in DES. Reprinted from [33] with permission from ACS publications

Table 2 Depiction of increased solubility of different antibiotics/drugs in DES vs. water

Drug	Solubility (mg/mL) in pure DES	Fold increase (vs water)	References
Benzoic acid	229	76×	[36]
Danazol	0.048	96×	[36]
Griseofulvin	0.034	5×	[36]
AMG517	0.010	22,000×	[36]
Itraconazole	<0.001	Significant (not quantified)	[36]

Formation of Metal Complexes in DESs

The formation of metal complexes plays a crucial role in stabilizing the dissolved state of the analytes and preventing precipitation. DESs facilitate the formation of soluble complexes between analytes (e.g., metal oxides) and solvent molecules. Zinc acetate dissolves in a DES formed by oleic acid and alkyl amine at specific conditions (286 °C, for 1 h). It produces a hexagonal micro-pyramid. Similarly, α-Fe_2O_3 forms when $FeCl_3$ is heated to 200 °C with choline chloride and urea DES. The reaction is completed through the intermediate stage of $Fe(NH3)_2Cl_3$ complex formation, which reacts with Fe_2O_3 by subsequent water addition and finally forms nanospindles of Fe_2O_3 [39]. The speciation and coordination environment in DESs is influenced by the nature of the metal oxide and the type of HBD present. Metal oxides with higher ionic character (e.g., ZnO, Cu_2O) exhibited superior solubility in DESs compared to more covalent oxides (e.g., TiO_2). The solubility patterns suggest that the DES matrix facilitates specific coordination interactions between the metal centers and the solvent components, resulting in enhanced stabilization of dissolved species. The coordination shell around the dissolved metal is influenced by the metal oxide's HBD and ionic character [40].

Role of Proton Donors (HBDs)

In certain DESs, for instance, malonic acid-based DESs, protons act as oxygen acceptors, enhancing the dissolution of oxides by modifying the chemical speciation of the analytes. The acidity of malonic acid contributes to enhanced solubility through protonation of analyte surface functional groups and stabilization of metal complexes in solution. The solubility of metal oxides in malonic acid-based DESs was significantly higher compared to urea or ethylene glycol-based systems, emphasizing the role of proton donors in solubility enhancement. An increased proton activity in DESs composed of poorly complexing HBDs enhances the solubility of metal oxides because of H^+ capacity to function as an oxygen acceptor. Nevertheless, using HBDs with enhanced complexing capabilities significantly enhances the solubility of the metal oxides through surface complexation reactions [31, 41].

High Solvation Potential of DESs

DESs provide a high solvation potential even for hardly soluble metal oxides under relatively mild conditions, showing their potential for processing metal oxides. In general, DESs containing an organic acid as a melt component are strongly acidic,

Table 3 Comparison of solubility of different metal oxides in malonic acid, urea, and ethylene glycol DES

Metal oxide	Solubility in malonic acid DES (ppm)	Solubility in urea DES (ppm)	Solubility in ethylene glycol DES (ppm)	References
ZnO	16,217	1894	469	[40]
Cu_2O	18,337	219	394	[40]
MnO	6816	0	12	[40]
TiO_2	4	0.5	0.8	[40]

which gives them a high potential to dissolve metal oxides. The solubility can partly be correlated with the approximate pH value of the DESs (Table 3). In addition, notable differences in the solubilities of metal oxides were observed according to the nature of DESs, which could be applied for the separation and selective recovery of metals [41].

Malonic acid-based DES exhibited superior solubility performance due to proton donation and complexation.

Selectivity of DESs in Analyte Extraction

DESs have gained considerable attention as highly selective solvents for extracting and preconcentrating analytes from complex matrices. Their selectivity arises from unique molecular interactions, including hydrogen bonding, acid-base interactions, π–π, and hydrophobic forces. A study by Shishov et al. demonstrates the application of thymol-nonanoic acid-based DES for the selective extraction of melamine from dairy products, highlighting key factors influencing selectivity [42].

Mechanisms of Selectivity in DES Systems

Acid-Base Interactions

Acid-base interactions are the primary drivers of selectivity; the carboxylic group of a DES constituent shows susceptibility for oppositely charged species and aids in the formation of ionic bonds, for instance, the extraction of melamine in the thymol-nonanoic acid DES system. Melamine, a nitrogen-rich compound with amino groups, interacts with the carboxylic acid groups of nonanoic acid via acid-base interactions. During extraction, the amino groups of melamine form stable ionic bonds with the carboxylic acid groups, promoting efficient partitioning into the DES phase. FTIR analysis confirmed these interactions, showing significant shifts in the C=O stretching frequency (from 1713 cm^{-1} to 1730 cm^{-1}) and NH_2 bending frequencies (from 1535 cm^{-1} to 1585 cm^{-1}) [42].

Hydrogen Bonding and π–π Bonding Effect on Selectivity

The mechanism of hydrogen bonding in enhancing the selectivity of DESs toward specific analytes, particularly for separating aromatic hydrocarbons from aliphatic mixtures, is studied by Wang et al. Both HBDs and HBAs are required to tune the activity and selectivity of DESs. Hydrogen bonding influences both the solubility and the strength of interactions of DESs with target analytes. In this study, shorter alkyl side chains, a smaller central atom in the cation and larger anions in HBAs led to improved selectivity of DESs to extract toluene from n-heptane mixtures. Nonetheless, the π–π interactions of aromatic molecules with DESs were strengthened when the HBD contained a carbonyl (–C=O) or hydroxyl (–OH) functional group, which forms hydrogen bonds with toluene, thereby enhancing the solubility of the aromatic molecule. The research indicated that elevating extraction temperature enhanced selectivity by diminishing hydrogen bond interactions within the DES matrix, rendering the solvent more proficient in extracting aromatics than aliphatics. Excessive temperature increases reduced efficiency due to solvent degradation [43]. Hydrogen bonding is essential in modulating the selectivity of DESs, positioning them as viable alternatives to traditional solvents for separating aromatic and aliphatic hydrocarbons. The results offer insights into developing extraordinarily selective and efficient deep eutectic solvents for applications in green chemistry and industrial separations. Figure 3 depicts the molecular interaction between salt cation, anion, HBD, and nitrogen compounds.

π–π interactions act as an auxiliary mechanism, enhancing the overall selectivity of DES toward melamine. Melamine possesses conjugated bonds capable of interacting with the aromatic ring of thymol through π–π stacking. This interaction enhances the retention of melamine within the DES phase, especially in systems containing aromatic components like thymol. The spectral evidence is the shift in the aromatic group absorption band from 2937 cm^{-1} to 2945 cm^{-1}, which supports the role of π–π interactions in improving selectivity [42].

Synergistic Effect Between DES Components

The selectivity and efficiency of DESs result from the combined action of both components, not their isolated performance. A comparison between the extraction efficiencies of thymol, nonanoic acid, and the DES (thymol-nonanoic acid) individually done by Shishov et al. where DES outperformed its components, demonstrating a synergistic effect arising from the combined contributions of acid-base interactions (nonanoic acid), hydrogen bonding (thymol), and π–π interactions (thymol). The DES system extraction recovery reached 95%, significantly higher than thymol or nonanoic acid alone [42].

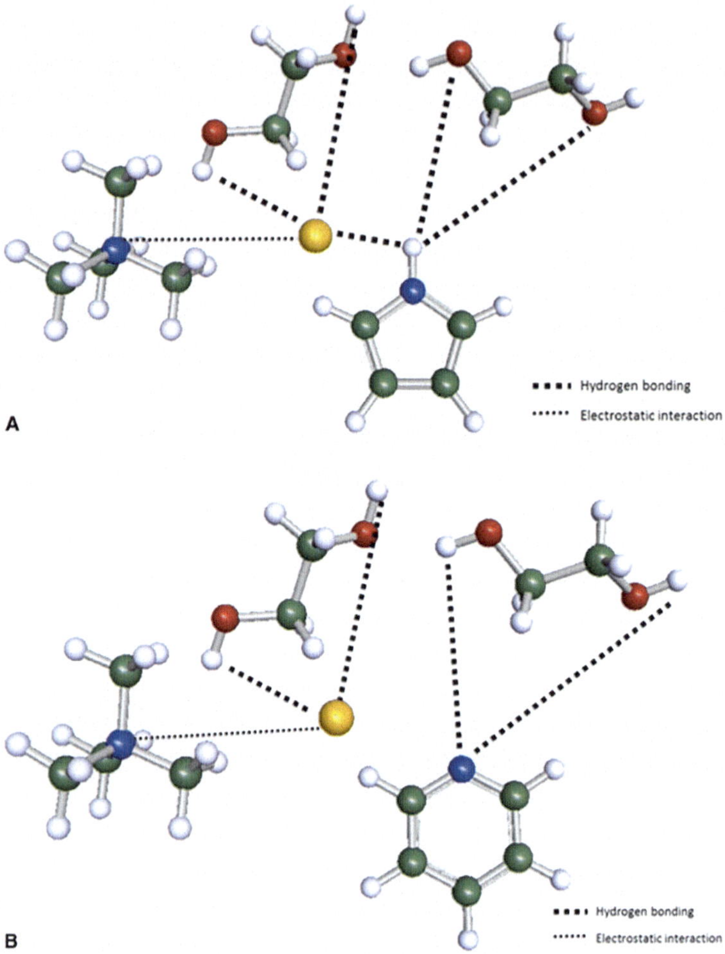

Fig. 3 Hydrogen bonding (thick dashes) and electrostatic interactions (small dots): A represents the interaction of TMAC: EG (1:2) DES with pyrrole (five-membered), and B represents the TMAC: EG interaction with six-membered pyridines. Reprinted from [44] with permission from Wiley Analytical Science

Compatibility with Various Food Matrices

In recent years, deep eutectic solvents (DESs) have gained substantial attention as crucial and environmentally benign alternatives to conventional solvents, especially in the food industry. They are highly compatible with various food matrices due to their distinctive physicochemical properties, which include low volatility, high solubility for a wide range of compounds [45]. DESs are conventionally produced by combining a hydrogen bond donor (HBD) and a hydrogen bond acceptor (HBA) in precise ratios, resulting in a eutectic composite with a melting point lower than that of its components. This adaptability enables DESs to be adapted to a specific applications, improving their compatibility with various food systems [17].

DESs are exceptional for extracting bioactive compounds, including polyphenols, carotenoids, and tocopherols, from lipid-rich food matrices, such as oils and lipids [45]. For instance, choline chloride and organic acid antioxidants from olive oil and other plant-based oils have been extracted using DESs composed of choline chloride and organic acids (e.g., citric acid or lactic acid). Their ability to dissolve non-polar compounds while maintaining stability in lipid environments renders them optimal for these applications. Furthermore, DESs can be engineered to minimize the interference with the sensory properties of the food, thereby guaranteeing that the extracted compounds maintained their quality and functionality [46].

DESs have demonstrated potential in extracting, purifying, and modifying proteins in food matrices which are high in protein, such as dairy products and livestock [17]. Certain DESs, particularly those based on choline chloride and urea or glycerol, have been employed to solubilize and stabilize proteins without denaturing them. This is very crucial for preserving proteins' nutritional and functional properties in food applications. Additionally, DESs can remove allergens or antinutritional factors from protein isolates, enhancing their suitability for human use. Their biocompatibility and low toxicity further increase their appeal for use in protein-rich diets [47]. Sugars, polysaccharides, and dietary fibers from carbohydrate-abundant matrices such as cereals, fruits, and vegetables can be extracted using DESs [48]. Combining choline chloride and carbohydrates like glucose or fructose creates DESs that selectively extract pectin from fruit peels through an environmentally friendly approach instead of acidic extraction methods. DESs serve the dual purpose of offering structural property modifications to carbohydrates for culinary purposes, including starch. DES systems demonstrate high compatibility with carbohydrate-containing systems because their hydrophilic compound binding ability preserves stability within aqueous conditions [47].

The adjustable characteristics and nature of DESs allow practical application with different food components. It also helps to simplify purification processes along with extraction operations and modifications because they contain features such as minimal toxicity and environmental effects and can dissolve polar and non-polar compounds [46, 48]. Research expansion reveals that DESs present an emerging potential to enhance food processing effectiveness while ensuring high-quality and safe preserved products.

Environmental and Economic Benefits

DESs Are Environmentally Benign

DESs are developing as a viable eco-friendly alternative to conventional organic solvents, providing substantial environmental advantages. A significant benefit is their low toxicity and biodegradability. In contrast to traditional solvents, frequently sourced from petrochemicals and environmentally persistent, deep eutectic solvents (DESs) are generally formulated from natural substances, including organic acids, sugars, and choline derivatives. The components that make up DESs help both the environment and quick decomposition happen, leading to less extended periods of environmental contamination [17]. Industrial facilities view DESs as a sustainable substitute for reducing their environmental impact.

The additional environmental advantage of DESs lies in their energy efficiency and minimal carbon emissions during production and use. Traditional solvents seldom need high-energy processes for manufacturing and cleansing, which increases greenhouse gas emissions. In contrast, DESs can be produced by simply amalgamating readily available, cost-effective precursors at moderate temperatures, significantly decreasing energy use [49]. Furthermore, DESs show high solvation ability, and their stability promotes efficient extraction and separation processes, limiting the need for several applications and resulting in plummeting energy usage. The combination of low-energy preparation and high efficiency makes DESs an attractive option for reducing the total environmental impact of chemical processes. Deep eutectic solvents act as tools for minimizing waste outputs and protecting resources. The reusability of deep eutectic solvents (DESs) keep their performance stable while regular solvents degrade during their first usage. Reusability is the quality of DESs, which reduces waste output and minimizes the requirements for raw materials [50]. The synthesis of deep eutectic solvents can be channelized to create a selective DES that extracts important compounds from waste products, thus aiding in collecting recovered materials that previously went to waste. DESs enable businesses to implement a circular economy model to embrace sustainable practices because they support global pollution and resource depletion mitigation programs [49]. DESs function as crucial tools to lead sustainable chemistry and industrial practice development because of their environmental advantages.

Economic Benefits of DESs

DESs present economic viability by using inexpensive choline chloride, urea, and organic acids as raw materials instead of traditional solvents. Lowering production costs happens through inexpensive raw materials, but speciality materials such as menthol alongside high-purity organic acids increase the overall expenses. The expense model for DESs depends on their HBD and HBA selection; however, using renewable starting materials helps achieve budget optimization. Mild mixing temperatures maintain low production costs by creating simple synthesis routes that resist energy costs. To achieve large-scale industrial applications, researchers must resolve viscosity and energy consumption problems when recovering and reusing DES solvents. The costs of running DES solutions fall below conventional solvents because DESs exhibit less volatility and have lower flammability. The high viscosity levels of specific DES formulations increase operational energy requirements for mixing and extraction procedures, affecting total cost-effectiveness [17, 45].

The long-term operational potential of DESs significantly improves economic viability because they show stable performance through multiple extraction procedures. The economic value increases through efficient solvent recovery systems, yet some DES formulations need replacement due to their time-based degradation. High recovery rates and repeated use capabilities exist in DES-based food adulterant detection through LLE and SPE methods, thus reducing the amount of waste produced. The expanding market need for eco-friendly technological solutions has increased commercial interest in DESs, which target food safety, pharmaceuticals, and bioenergy industries. The appeal of DESs increases because of their regulatory benefits, which include less toxicity exposure and reduced environmental harm, which justifies initial capital expenditures while delivering enduring economic gains [47, 48].

Analytical Techniques Coupled with DESs

Foods and ingredients commonly associated with food fraud include oil, fish, honey, dairy products, meat, grain-based items, fruit juices, wine and alcoholic beverages, organic foods, spices, coffee, tea, and specific highly processed meals. The magnitude of global food fraud is uncertain. Previous assessments mainly focused on the analytical techniques for detecting adulteration in particular foods and products. Many businesses involved with food production are following malpractices to get the most benefit out of their businesses. Nonetheless, a paucity of comprehensive evaluations of the analytical methods utilized for various categories of contaminated food exists. Diverse methodologies have been utilized to detect adulteration and fraud in food, encompassing chromatographic techniques (thin layer chromatography, high-performance liquid chromatography, gas chromatography) and mass spectrometry-based methods.

Microextraction of Melamine from Milk by HPLC-UV

To extract and determine the concentration of fraudulent substances from milk, Shishov et al. [42] put forward a study where a DES was employed by combining thymol and nonanoic acid in a 1:2 molar ratio, utilizing their synergistic extraction properties. The procedure entails protein precipitation from the milk matrix utilizing sodium sulfate, succeeded by DES-based liquid-phase microextraction of melamine. Following the separation of the DES phase, melamine was quantified via HPLC-UV under optimized chromatographic conditions. Figure 4 shows the graphical description of the microextraction of melamine.

Essential parameters for this process include pH, DES composition, solvent volume, extraction duration, and temperature to achieve high recovery and sensitivity. The optimal extraction occurred at pH 7, utilizing a DES volume of 100 μL, an

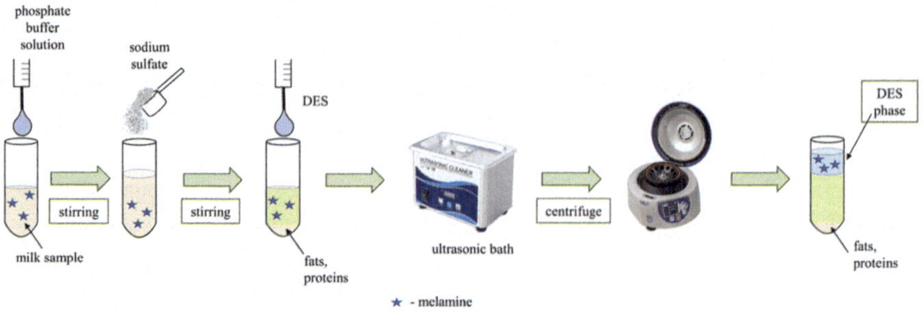

Fig. 4 A flow diagram of microextraction of melamine followed by HPLC-UV analysis, reprinted from [42] with permission from Elsevier

extraction duration of 5 minutes, and a temperature of 40 °C. The study examined the adulteration of milk with the nitrogen-containing compound melamine, which falsely increases the protein content. The method had 95% extraction recovery, 0.03 mg/L detection limit, and 0.1 mg/L quantification limit. The process was tested with powdered and whole milk, yielding 90%–110% recoveries.

Method Development for Extraction of Tryptophan from Nuts and Seeds Oil by HPLC-DAD

Tryptophan (Trp) occurs naturally in nuts and seeds; Trp can be induced artificially in any low-standard oil to fake it as an original oil from nuts and seeds. To curb this problem, a sensitive detection method for Trp was necessary. Research by Ražić et al. [51] employed DES-based reversed-phase dispersive liquid-liquid microextraction (RP-DLLME) to extract free Trp from cold-pressed oils, subsequently quantified via HPLC-DAD. This method can be a precedent in determining the adulteration of edible oil. A hydrophilic deep eutectic solvent composed of choline chloride and urea was the most effective extractant due to its capacity to establish robust hydrogen bonds with Trp. The optimal conditions for the RP-DLLME procedure comprised a dilution ratio of 1:9, 450 μL DES, vortex agitation, and a centrifugation duration of 4 minutes at 6000 rpm. Under these conditions, the method attained a detection limit of 11 mg/kg, with recovery rates ranging from 91% to 94% and relative standard deviations under 9%. The HPLC-DAD analysis employed isocratic elution with methanol: water ratio of 40:60 (v/v) and detection at 280 nm, guaranteeing high sensitivity and selectivity.

The study validated DES-based RP-DLLME in conjunction with HPLC-DAD to detect and quantify food adulterants in complex matrices using cold-pressed oils from nuts and seeds, with Trp concentrations between 11 and 38 mg/100 g, with sunflower oil having the highest Trp. Figure 5 illustrates the extraction of tryptophan from oil.

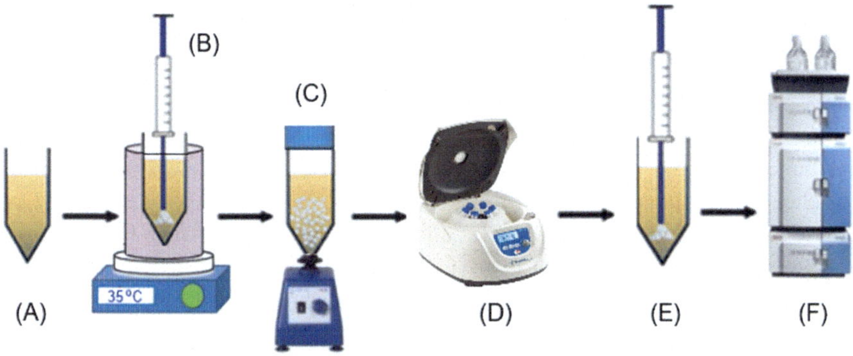

Fig. 5 Step-by-step illustration of extraction of tryptophan from oil. (**a**) Diluted oil sample; (**b**) injection of DES; (**c**) vortexing; (**d**) centrifugation; (**e**) retrieval of the DES phase; (**f**) HPLC measurement. Reprinted from [51] with permission under common creative licence

Extraction of Synthetic Dyes from Jellies and Drinks by HPLC-DAD/PDA

The food industry consumes several synthetic dyes to enhance product appearance; however, dyes are associated with potential toxicity and adverse health impacts, such as carcinogenicity and attention deficit hyperactivity disorder (ADHD), which necessitate strict regulation and monitoring. The method used to determine dyes from food samples presents vortex-assisted dispersive liquid-liquid microextraction (VA-DLLME) using hydrophobic DESs as a more sustainable and effective alternative.

The VA-DLLME method was applied for the first time to extract and quantify eight synthetic dyes—Tartrazine, Quinoline Yellow, Sunset Yellow, Brilliant Blue, Ponceau 4R, Indigo Carmine, Allura Red, and Carmoisine—in jellies and drinks [52]. The DES phase was introduced into the aqueous sample via vortex mixing, resulting in fine droplets that enhanced dye extraction efficiency. After centrifugation, the DES phase containing the extracted dyes was separated, cleaned, and analyzed using high-performance liquid chromatography (HPLC) with a photodiode array (PDA) detector. The optimized conditions for the significant extraction are benzyl triethylammonium chloride (BTEAC) and thymol with a 1:4 molar ratio, 100 μL as an optimal DES volume, with a neutral sample pH of 6 and 5% w/v sodium chloride identified as ideal for maximizing dye recoveries via the salting-out effect. The proposed method demonstrated superior analytical performance, with a linear range of 5–1000 μg/L for all dyes, limits of detection (LOD) between 0.02 and 0.05 μg/L, and limits of quantification (LOQ) ranging from 0.04 to 0.15 μg/L. The relative recoveries for the eight dyes were between 87.2% and 107.4% in jelly samples and between 89.1% and 104% in drink samples, demonstrating high accuracy and precision.

Development of a Green Analytical Method for Detecting Phytosterols in Cow Butter Using Deep Eutectic Solvent-Based Microextraction Coupled with GC-FID

Cow butter is widely consumed as an edible fat and is susceptible to adulteration with cheaper vegetable oils to reduce production costs. The methods discussed in this section involve a combination of ultrasound and heat-assisted liquid-liquid extraction (USHA-LLE) and dispersive liquid-liquid microextraction (DLLME) with deep eutectic solvents (DESs) and air-assisted liquid-liquid microextraction followed by gas chromatography with a flame ionization detector (GC-FID). Both methods were developed by the same research group with variable extraction methods. The USHA-LLE extraction process involves dissolving butter samples in ethanol, ultrasonic treatment at 50 °C for 7 min, and centrifugation to separate the analytes [53]. The DES phase is then subjected to DLLME, where a hydrophobic DES (comprising ethyl(methyl)ammonium chloride and pivalic acid) acts as the extractant. The DES is dispersed into deionized water to enhance analyte recovery, and a syringe-based dispersion technique ensures thorough mixing. Fatty acid interferences are removed through a saponification step using 3 M sodium hydroxide. GC-FID analyzes the final DES extract without requiring derivatization.

The method demonstrated excellent analytical performance, with linear detection range (LR) for four target phytosterols (campesterol, brassicasterol, stigmasterol, and β-sitosterol) spanning 1.7 to 5000 ng/g, coefficients of determination (R^2) greater than 0.993, limits of detection (LOD) from 0.51 to 1.3 ng/g, and limits of quantification (LOQ) from 1.7 to 4.3 ng/g. Enrichment factors varied between 340 and 425, while extraction recoveries ranged from 68% to 85%. Intra-day and interday precision tests yielded relative standard deviations (RSD) below 8.4%, confirming the method's reliability and reproducibility. Similarly, air-assisted liquid-liquid microextraction (AALLME) with gas chromatography-flame ionization detection (GC-FID) for the sensitive and efficient analysis of phytosterols in butter and oil samples was developed by [54]. The two-step extraction process involves an LLE step to dissolve butter samples in ethanol at elevated temperatures to facilitate the release of phytosterols into the ethanol phase. In the second step, an 80 μL aliquot of DES is introduced to the solution, followed by repeated air-assisted microextraction cycles to improve dispersion and optimize analyte extraction. The phase of DES containing enriched phytosterols is isolated through centrifugation and saponified with 4 M NaOH to eliminate interfering fatty acids. The method exhibited significant sensitivity and reproducibility, with limits of detection (LOD) for four target phytosterols ranging from 0.73 to 1.5 ng/g and limits of quantification (LOQ) ranging from 2.6 to 5.2 ng/g. The linear detection range spanned from 2.6 to 5000 ng/g, with a correlation coefficient exceeding 0.991, demonstrating superior linearity. These studies present an innovative and sustainable analytical method for detecting adulteration in cow butter by identifying phytosterols as markers. Integrating DES-based microextraction and GC-FID offers a rapid, sensitive solution for food quality assurance with widespread application in regulatory and industrial settings.

Determination of Illegally Added Amphetamines in Snacks and Beverages by GC-MS

The illicit incorporation of amphetamines into food and beverages, frequently marketed as dietary supplements, presents a considerable public health threat. Improper use of these potent stimulants can lead to substantial cardiovascular, neurological, and psychological consequences. Insufficient regulatory oversight in numerous countries results in the undisclosed incorporation of substances in products marketed for energy enhancement, cognitive improvement, or athletic performance. A new method for detecting amphetamines in snacks and beverages is proposed by Hashemi et al. [55] using metal-organic framework (MOF)-based deep eutectic solvents and molecularly imprinted polymer fibers for solid-phase microextraction. They used GC-MS to analyze the amphetamines, combining the high surface area and selectivity of MOFs with the tailored binding capacity of MIPs for efficient extraction and quantification.

MIL-Al-53 synthesized the MOF-DES/MIP fibers as the metal-organic framework. Then, the metal-organic framework was incorporated into a deep eutectic solvent matrix consisting of choline chloride and glycerol, followed by polymerization with methacrylic acid. After polymerization, porous fibers with cross-linked binding sites and enhanced thermal stability were produced, facilitating their application in GC-MS. The fibers were analyzed through FT-IR, SEM, and XRD to verify their chemical and physical properties. Samples of snacks and beverages were diluted and extracted utilizing MOF-DES/MIP fibers under optimized parameters: pH 3.5, an extraction duration of 35 min, and a stirring rate of 600 rpm. The analytes were desorbed using 1-octanol as the solvent, then evaporated and reconstituted for GC-MS analysis. The parameters for GC-MS, including injection temperature, column gradient, and flow rate, were optimized to improve the separation and quantification of amphetamines.

The MOF-DES/MIP SPME-GC-MS method demonstrated notable sensitivity and selectivity for amphetamines. The method exhibited a linear detection range of 0.1–400 μg/L for dextroamphetamine, methamphetamine, methylphenidate, and modafinil, with limits of detection (LOD) ranging from 0.023 to 0.033 μg/L and limits of quantification (LOQ) between 0.076 and 0.115 μg/L. The enrichment factors varied between 158 and 163, indicating effective analyte preconcentration.

The method underwent validation via recovery studies conducted on actual samples, resulting in recoveries ranging from 94% to 110% for snacks and beverages, accompanied by relative standard deviations (RSD) of less than 6.7%. Repeatability and reproducibility were validated across various fibers, with intra-day relative standard deviations (RSDs) ranging from 1.1% to 4.6% and inter-day RSDs between 2.2% and 6.7%.

Extraction of Phytosterols from Cow Cream Samples and Detection by GC-MS

Phytosterols, as the name suggests, are sterols derived from plants with bioactive properties that significantly benefit human health by plummeting cholesterol absorption and imparting anti-cancer, anti-inflammatory, and antioxidant effects. Compounds such as β-sitosterol, campesterol, stigmasterol, brassicasterol, and lupeol are naturally found in various oils and fats. Nevertheless, analyzing milk-based products for phytosterols is necessary to identify adulteration of milk-based products with vegetable oils to ensure product quality and authenticity. A study by Nemati et al. explores a novel method combining sodium hydroxide-induced homogeneous liquid-liquid extraction (HLLE) with effervescent-assisted dispersive liquid-liquid microextraction (EA-DLLME) using deep eutectic solvents (DESs) as an extraction agent [56].

Their approach integrates sodium hydroxide-induced homogeneous liquid-liquid extraction (HLLE) with effervescent-assisted dispersive liquid-liquid microextraction (EA-DLLME), utilizing deep eutectic solvents (DESs) as extraction agents. They used this method to extract phytosterols from cow milk cream; after extraction, the samples were analyzed through gas chromatography-mass spectrometry (GC-MS). The technique utilized two deep eutectic solvents (DESs): tetrabutylammonium bromide (TBAB) and ethylene glycol (EG) for the high-level liquid extraction (HLLE) step and TBAB in conjunction with dichloroacetic acid and octanoic acid for the enhanced dispersive liquid-liquid microextraction (EA-DLLME) step. Deep eutectic solvents (DESs) were selected due to their low toxicity, straightforward preparation, and effectiveness in extracting analytes from complex matrices. The method demonstrated excellent analytical performance, with limits of detection (LOD) for phytosterols ranging from 0.06 to 0.26 μg/kg and limits of quantification (LOQ) between 0.19 and 0.86 μg/kg. Depending on the phytosterol, the linear range (LR) was wide, from 0.19 to 10,000 μg/kg. The enrichment factors (EF) were between 215 and 232, indicating significant analyte concentration during extraction. Recovery rates ranged from 86% to 93%, and intra-day and inter-day precision, expressed as relative standard deviations (RSD), were below 8.8%, highlighting the method's reproducibility and reliability. The process was utilized on 15 samples of cow cream, comprising both pasteurized (Tetra Pak and polyethylene-packed) and non-pasteurized varieties. Phytosterols were absent in Tetra Pak-packed pasteurized creams, whereas they were found in polyethylene-packed pasteurized and non-pasteurized creams. The differences in phytosterol content among samples were linked to factors including the animal's diet, season, and breed, underscoring the impact of production conditions on phytosterol levels.

Determination of Antioxidants in EVOO by Electroanalytical Method

Extra virgin olive oil (EVVO) frauds can be intercepted by determining the antioxidant composition of the oils. A usual malpractice is mixing substandard oils, commonly virgin olive oil, lampante oil, and several other seed oils, and then marketing them as counterfeit extra virgin olive oil. A step forward in counterfeiting is adulterating seed oils with green-yellow additives like chlorophyll and carotenoids. These fraudulent practices can be very troublesome for those with allergies to specific kinds of oils. A study by Shabani et al. explored the electroanalytical approach employing DES [57]. Polar antioxidants like caffeic acid (CA) and vanillic acid (VA) can be detected by electroanalytical methods, but conventional organic solvents can be limited.

Process optimization includes the ratio adjustment of EVOO to DES, which resulted in a 1:2 molar ratio, giving the best outcomes. Moreover, the sonication effect showed the best transfer of antioxidants from oil to DES between 2 and 20 min of the sonication period. Electrochemical analysis for antioxidants was performed through a screen-printed electrode modified with multi-walled carbon nanotubes (MWCNTs) and titanium oxide (TiO_2). This enhanced the sensitivity and detection threshold for CA and VA. The electrochemical detection involves square wave voltammetry (SWV), revealing unique oxidation peaks for CA and VA. Figure 6 depicts the differences in electrochemical responses of antioxidants between the original EVOO and antioxidants added to the other seed oils. The result demonstrated more than 97% recovery percentages compared to the approved

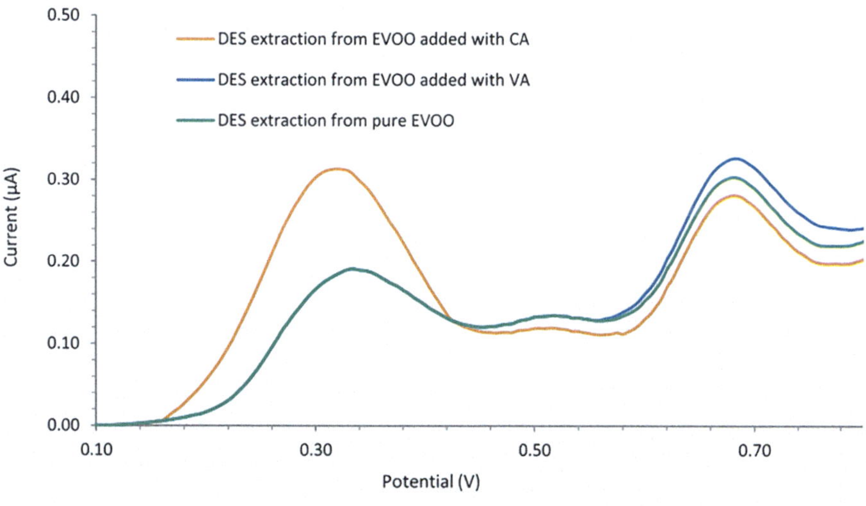

Fig. 6 The detection of caffeic acid (CA) and vanillic acid (VA) from the pure EVOO vs. the oils with externally added CA and VA. Reprinted from [57] with permission from Elsevier

spectrophotometric method. This comparison validates the electrochemical process, while commercial seed oils showed no significant peaks for antioxidants, proving this technology's authenticity in detecting counterfeit EVOOs.

Electrical Impedance Spectroscopy for Food Adulterant Detection

The method of electrochemical impedance spectroscopy (EIS) employs time-dependent current application between two fundamental electrodes alongside a measurement of independent differential voltage. Computer simulation evaluates the impedance spectrum for material electrical property analysis. Electronic impedance spectroscopy effectively detects food adulteration across various products, from meat to fruits to vegetables, milk to honey [58]. Chakraborty et al. studied electrical properties like capacitance, conductance, and impedance, which displayed direct relationships to adulterant amounts with a minimum measurable threshold of 10% adulteration [59]. The FT-MIR method delivered equivalent analytical results, validating its precision capability. EIS effectively detected adulterants like water and hydrogen peroxide, achieving a classification accuracy of 94.6% when combined with artificial neural networks [60]. However, temperature variations can influence impedance measurements, necessitating standardized conditions.

This method offers a cost-effective, rapid, and user-friendly approach to detecting food adulteration. It has the potential to replace traditional methods and reduce the need for highly specialized personnel, making it a practical solution for ensuring food safety. It can also detect infections and advantageous bacteria. Nonetheless, obstacles persist in data interpretation, system calibration, and measurement standardization owing to variables such as temperature, electrode characteristics, and sample architecture. Although effective in laboratory environments, additional study is required to enhance EIS for extensive industrial applications in food safety.

Sudan IV Preconcentration Using DES and Spectrophotometric Determination

Sudan dyes, especially Sudan IV, are synthetic lipophilic azo dyes illicitly employed in food items such as chili, ketchup, spices, and snacks to augment color. These dyes provide significant health hazards, such as thyroid tumors, allergic reactions, and carcinogenic effects, resulting in their designation as banned additives by the FDA, EU, and the International Agency for Research on Cancer (IARC). Sudan IV is designated as a category three carcinogen. This study sought to establish a straightforward, quick, and selective technique for the preconcentration and detection of Sudan IV in water and food, utilizing a green, environmentally benign deep eutectic

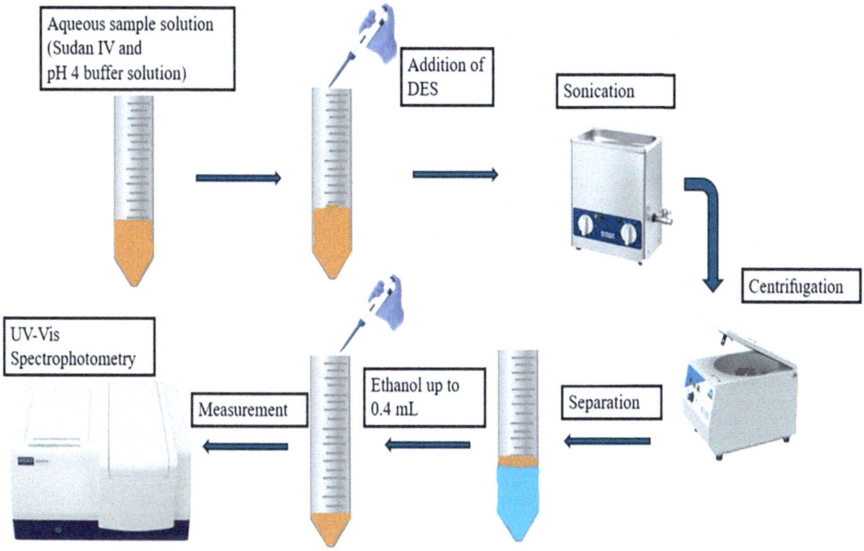

Fig. 7 A schematic diagram of the spectrophotometric method for determining Sudan IV as a food adulterant. Reprinted from [61] with permission from Taylor & Francis

solvent (DES) and spectrophotometry. The DES was formulated by combining tetrabutylammonium bromide and 1-octanol in different molar ratios (1:1 to 1:4). The (1:2) molar ratio was selected as ideal owing to its transparency and simplicity of separation.

The extraction procedure entailed combining a 4 mg/mL Sudan IV solution with a buffer and 200 μL of DES. The solution was treated in an ultrasonic bath for 5 min to achieve dispersion and centrifuged at 4000 rpm for 5 min to facilitate phase separation. The DES-rich phase containing Sudan IV was isolated, diluted with 0.4 mL of ethanol, and subjected to spectrophotometric analysis at 515 nm. The method exhibited a preconcentration factor (PF) of 62.5, with limits of detection (LOD) and quantitation (LOQ) of 7.5 mg/L and 25 mg/L, respectively. This method offers a cost-efficient substitute for intricate and labor-intensive chromatographic procedures for detecting Sudan IV in aqueous and food matrices (Fig. 7).

Detection of Undeclared Pharmaceuticals in Dietary Supplements and Traditional Medicines Using NMR

The extensive consumption of dietary supplements creates uncertainties about pharmaceutical compounds, which may harm human health because they are incorporated without disclosure. Manufacturers seldom include artificial compounds because they seek product enhancement without following necessary regulatory

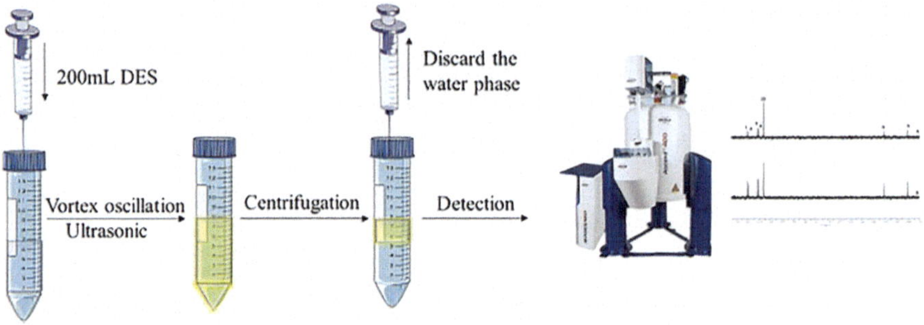

Fig. 8 A process showing UADLLME ultimately analyzed by NMR. Reprinted from [62] with permission from Elsevier

controls. Infrequent adulteration findings in modern products reveal fluoroquinolone antibiotics and phenethylamines (PEAs) as major adulterants that can create health dangers. The identification and quantification of compounds in these complex formulations now rely on nuclear magnetic resonance (NMR) spectroscopy, a helpful method that performs compound identification without extensive chromatographic separation methods. A study by Xue et al. examined the occurrence of fluoroquinolones in typical prostate medications, utilizing 19F quantitative NMR (qNMR) alongside DES-based ultrasonic-assisted dispersive liquid-liquid microextraction (DLLME). Researchers produced a hydrophobic deep eutectic solvent from methyltrioctyl-ammonium bromide and n-capric acid, which facilitated the successful extraction of fluoroquinolones like ciprofloxacin, levofloxacin, sparfloxacin, gatifloxacin, and enrofloxacin [62]. Figure 8 showing UAD-LLME. After optimizing conditions including DES volume, salt concentration, and ultrasonication time, the method achieved a detection limit of 0.1 μg/mL and a recovery range of 92.4% to 103.22%. The approach proved highly effective, detecting sparfloxacin at a 2.04 μg/mL concentration in one prostate medicine sample. This study demonstrated that NMR, in combination with environmentally friendly solvents like DES, provides a practical method for screening pharmaceutical adulterants in traditional medicines.

A separate experiment concentrated on identifying PEAs in sports dietary supplements conducted by Zhao et al., a category of chemicals recognized for their stimulating characteristics. The typical ingredients in weight-loss supplements alongside performance-enhancing products include phenethylamine with synephrine and oxilofrine, as well as hordenine and β-methylphenethylamine with N-methyltyramine overlaying octopamine and deterenol [63]. Due to their performance-enhancing capabilities, numerous substances are now restricted by the World Anti-Doping Agency (WADA). In the discussed study, the researchers studied 32 commercial nutritional supplements using 1D and 2D NMR analyses and disclosed massive inaccuracies and fraudulent modifications in their content.

Several products contained undeclared PEAs, while others listed them on labels despite containing none. The qNMR method developed in this research achieved detection limits between 2.7 and 4.3 μg/mL and recovery rates from 96.8% to 104.1%. Some supplements contained excessively high levels of PEAs, with estimated daily intakes exceeding regulatory safety thresholds, for instance, oxilofrine (up to 87 mg/serving), β-methylphenethylamine (up to 106.5 mg/day), and octopamine (up to 90.3 mg/day).

Detection of Vegetable Oils as Adulterants from Essential Oils by NMR Spectroscopy

Essential oils (EOs) are valuable products used in the food sector and studied intensively as potential bio preservatives to extend food shelf-life. Unfortunately, the EOs may be fake to boost industrial profits. Vegetable oils (VOs) should be considered among the potential adulterants due to their properties and low prices. Truzzi et al. reported that nuclear magnetic resonance (NMR) spectroscopy was applied to detect and identify VOs in mixtures with EOs. This approach compares the peak area ratio matrices of specific VO ^{13}C NMR fatty acid signals to those from the adulterated EOs. Through matrix similarity calculations at multiple confidence levels, the VOs were diagnosed. The strategy showed the potential to identify adulteration in mixtures efficiently. Lavender, citronella, rosemary, and orange EOs were combined with almond, corn, peanut, sunflower, soybean, colza seed oils, and wheat germ oil to get binary mixtures. Other dilutions were prepared from a stock solution with adulterant concentrations of 50% w/w (25, 12.5, 6.25, 3.12, 1.5, and 0.8% w/w) by increasing the quantity of EOs. Based on the results from the EO–VO binary mixture experiments, two dilutions were chosen for each EO–VO combination for 56 binary mixtures evaluated using NMR experiments. The current study successfully detected and identified adulterating volatile organic compounds in counterfeiting essential oils utilizing high-resolution nuclear magnetic resonance. Proton and carbon NMR spectra were used to identify VOs by analyzing the glycerol backbone signals of triglycerides. In ^{1}H NMR spectra, glycerol displayed two signals at 4.12 and 4.27 ppm, corresponding to 2H at the sn-1′ and sn-3′ locations, respectively, whereas the glycerol carbons were observed at 62.4 and 69.2 ppm for C1/C3 and C2, respectively.

Challenges in DES Applications for Food Adulterant Detection

DESs have shown great promise as substitutes for traditional organic solvents in detecting food adulterants because of their distinct physicochemical characteristics and possible environmental advantages. However, several issues with their use in this field demand careful consideration. Here, we have examines essential factors that affect the effectiveness and viability of DESs in food analysis, with particular

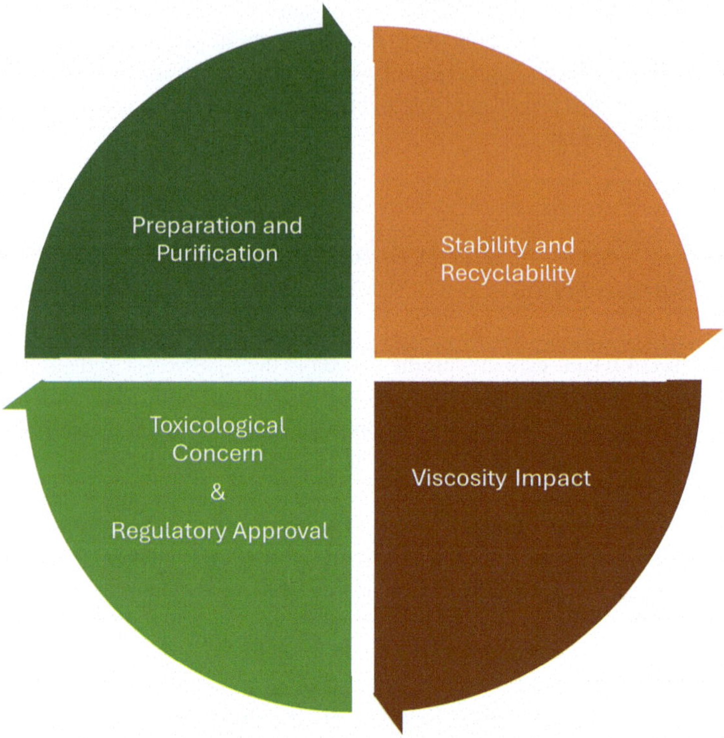

Fig. 9 Challenges associated with DESs in food adulterant detection

attention to solvent preparation and purification, the stability of the system and recyclability for sustainable practice, the viscosity impacts, as well as the toxicological effects and regulatory compliance of the DESs. The significant challenges associated are shown in Fig. 9.

Solvent Preparation and Purification

Despite the simplicity of making DESs, the preparation and purification bring some notable difficulties, especially in sensitive applications like the food industry. The physicochemical properties of DESs are primarily determined by the selection of the hydrogen bond acceptor (HBA) and the hydrogen bond donor (HBD) [64]. According to Florindo and his co-workers [25], the extensive hydrogen bond network in DES underpins their unique properties, including low melting points, minimal volatility, non-flammability, low vapor pressure, dipolar nature, chemical and thermal stability, excellent solubility, tunable characteristics, compatibility, presumed low toxicity, and high biodegradability. These factors are very critical in addressing the selectivity of the analyte and the overall efficiency of the process. In preparing DESs for food adulterant detection, the primary challenges emanate from the selection of HBAs and HBDs, as well as the technique employed. For example, effective detection of dietary adulterants requires careful consideration of their chemical properties and solubility in solvents with appropriate polarity. Generally, polar adulterants, such as specific dyes, are more effectively detected using polar DESs, while non-polar contaminants are better addressed with less polar solvents. This polarity-based approach ensures a targeted and efficient separation process tailored to the specific chemical characteristics of the adulterants. For example, The results of DES made of organic acids and ChCl were more polar than those made using sugars [65]. In a study by Abbott et al., components containing electronegative groups interact with polar liquids; they can produce dipole-dipole interactions, explaining their hydrophilic qualities [22]. Similarly, determining the optimal molar ratio in both HBA and HBD is a critical factor in addressing the physicochemical properties of the DES, which are responsible for the selectivity and efficiency of the detection process, for example, in a study conducted by Omar et al. It was discovered that the polarity values for ChCl: glycerol-based DES at different molar ratios of 1:1, 1:2, and 1:3 were 58.49 kcal/mol, 58.00 kcal/mol, and 57.96 kcal/mol, respectively. Generally, DES is prepared by either heating, grinding, freeze-drying, vacuum evaporation, microwave, or ultrasound-assisted method [25, 66]. The application of concentrated solar radiation is another technique that appears intriguing from a sustainability perspective [67]. In a review conducted by Zhou et al., it was reported that different preparation methods significantly influence the properties of DES, including variations in their moisture content, which can arise from the specific technique used [68]. For example, controlling the moisture during preparation might be difficult because of the hygroscopic nature of some DES precursors. In some cases, the excess water content can alter the properties, especially the polarity of the DES [69]. Accurate temperature regulation during the preparation is also necessary to prevent component thermal deterioration [70]. These characteristics directly affect the extraction process's efficiency and selectivity, essential to the mass transfer phenomenon [71]. DES purification is another crucial step that significantly impacts the precision and dependability of adulterant detection techniques. One of the primary challenges is the existence of leftover DES components in the isolated analytes [72]. Water can be classified as an impurity in some situations, particularly when its presence is excessive or unintentional and adversely affects the desirable attributes of a DES [73].

Stability and Recyclability of DESs

Thermal stability is one of the key challenges in DES formulation. The greatest operating temperature at which DESs can be effective is limited by their thermal stability, making it a crucial characteristic [74]. Because of their temperature sensitivity, DESs decompose at high temperatures. Research has indicated that the thermal stability of various DES formulations varies, most likely due to their thermodynamic characteristics. Variables like glass transition temperature, melting temperature, and thermal decomposition are essential to address this stability. Understanding their uses in various systems requires knowledge of specific heat capacity, total enthalpy, and phase transition entropy [75]. It has been reported that several DESs remain stable in liquid form at temperatures above 150 °C [76]. However, some DESs experience weight loss at a very high temperature. Choline chloride-based DESs, including citric acid monohydrate or oxalic acid, can lose much weight at higher temperatures due to component breakdown and water evaporation [75]. Another disturbing factor is the hydrolytic stability of the DES. This characteristic describes DES's capacity to maintain structural and operational integrity when water is present. Although DESs are renowned for being environmentally friendly and having adjustable characteristics, water can drastically change their behavior, which presents both potential and difficulties. As pointed out earlier, many DES can absorb moisture, which alters their viscosity and polarity and consequently affects the process of detecting or extracting the analyte. For example, if too much water is incorporated into DES, it may lead to the breakdown of the hydrogen bond network in the DES and hence result in an aqueous solution of the DES components [77]. Another research also demonstrated that adding trace amounts of water to DESs improves enzyme stability by preserving the network of hydrogen bonds between HBDs and HBAs [78]. Instability may arise due to the disruption of hydrogen bonds by water molecules. This impact is more noticeable in hygroscopic DESs like those based on choline chloride. Chemical degradation can also occur in certain situations, particularly when some DES components are susceptible to hydrolysis by water. For example, DES formulations containing esters or other reactive groups may undergo hydrolysis, rendering them ineffective [79]. This may often lead to phase separation which would ruin the solvent homogeneity and lessen its usefulness in the detection or medium [80]. These effects may lead to inconsistent analytical results, especially in food diaries and powdered products, and pose limited application in wet food samples such as milk, honey, etc. The recyclability of DESs presents a significant challenge, impacting their cost-efficiency, environmental advantages, and overall analytical effectiveness. Contamination, deterioration during reuse, and energy-intensive recovery techniques are the main problems with DES recycling. DES has been purified and recovered using a variety of techniques, including crystallization, antisolvent addition, solid-liquid extraction, membrane filtration, short path distillation, liquid-liquid extraction, supercritical fluid extraction, and separation based on density differences [81]. However, several issues must be resolved for DESs to be practical for long-term, large-scale applications. In some instances, water is frequently added during the recovery process to dissolve the DES and separate it from the recovered analytes. Cui et al. reported that after six recycling cycles, the yields remained unchanged, indicating the potential of DESs for use in industrial applications [82]. Nevertheless, partial loss of DES components due to evaporation or filtration processes may lower efficiency over time [82]. Another challenge is that impurities from the materials that DESs interact with, including food matrices in adulterant detection, can contaminate them during recycling. In different research conducted by Xu et al. and Falcini and Gonzalo also discusses the reusability of DESs, pointing out that although some DESs can be used repeatedly, their effectiveness frequently decreases due to component deterioration or impurity buildup [83, 84]. Adulterants such as azo dyes (e.g., Allura Red AC), melamine, and urea demand specific extraction conditions for accurate detection. Variations in solvent composition across reuse cycles can adversely affect the selectivity and sensitivity of the detection process [85].

Impact of DES Viscosity on Extraction Efficiency

A critical characteristic of DESs that significantly impacts their extraction effectiveness is viscosity. Understanding how viscosity impacts mass transfer, solute-solvent interactions, and overall performance is crucial for maximizing the usage of DESs in food adulterant detection. Three main factors make it challenging to deal with viscosities in deep eutectic solvents (DESs): the lengthy molecular chains of HBDs, the substantial effect of temperature on viscosity, and the effect of water content. Viscosity directly affects DES fluidity, which controls the mass transfer rate between the solvent and the target analyte. High viscosity can reduce the detection effectiveness by preventing solutes from diffusing into the solvent. On the other hand, reduced viscosity improves mass transfer and solute-solvent interaction, which raises the detection performance. Although temperature can significantly decrease viscosity, it is essential to consider how natural substances' thermolability may be mediated [73]. Water can also be added to adjust the viscosity. However, excess water can inevitably disrupt the hydrogen bond and affect the performance of the DESs. According to Zhekonov et al. 2019, the system should only have a maximum of 50% molar water fraction [86]. The molar ratio of HBDs and HBAs and the type of DESs (hydrophilic or hydrophobic) can significantly influence the viscosity of DESs. Hydrophilic DESs have a high viscosity, which makes it challenging to transfer mass during extraction procedures effectively, so water has to be added to adjust the viscosity, as discussed earlier [87–89]. The effect of the molar ratio on the viscosity of Chcl; glycerol and ChCl; fructose was studied by Fuad et al., and the results show an increase in viscosity from 1:1 to 1:2 in both DESs [90]. The study by Ghaedi et al. also found that increasing the mole fraction of HBDs in allyl triphenyl phosphonium bromide (ATPPB) with diethylene glycol (DEG) or trimethylene glycol (TEG) DESs leads to a significant decrease in viscosity. For instance, at 20 °C, the viscosity of ATPPB-DEG DES decreased from 213.18 mPa·s (1:4 ratio) to 57.19 mPa·s (1:16 ratio), while ATPPB-TEG DES showed a similar trend, dropping from 233.75 mPa·s (1:4 ratio) to 76.631 mPa·s (1:16 ratio). The team also reported that an increase in molecular weight, especially HBDs, with long chains increases the viscosity of DESs [91]. Vazquez et al. also reported that higher viscosities are typically found in HBDs with longer chains, such as glycerol or polyols, as opposed to those with shorter chains, such as ethylene glycol [92]. Table 4 shows

Table 4 Effect temperature, water content, and molar ratio on the viscosities of DESs

DES precursor	Molar ratio	Water content (wt %)	Temperature (°C)	Viscosity (mPa s)	References
ChCl: Eth-glycol	1:1	0.0790	20, 40, 60	60.00, 27.77, 15.31	[93]
	1:2	0.0651	20, 40, 60	37.35, 18.27, 10.47	
	1:3	0.0389	20, 40, 60	31.80, 15.52, 9.01	
	1:4	0.0579	20, 40, 60	28.49, 13.72, 7.50	
	1:5	0.0465	20, 40, 60	25.56, 12.30, 6.91	
ChCl: 1,2-propanediol	1:1	0.0642	20, 40, 60	94.05, 35.89, 17.11	
	1:4	0.0613	20, 40, 60	80.19, 30.40, 14.43	
	1:5	0.0608	20, 40, 60	70.57, 26.42, 12.21	
ChCl: 1,3-propanediol	1:3	0.0355	20, 40, 60	69.74, 31.35, 16.88	
	1:4	0.0340	20, 40, 60	61.31, 27.31, 14.66	
	1:5	0.0446	20, 40, 60	53.95, 24.01, 12.83	
ChCl: 1,4-butanediol	1:3	0.0402	20, 40, 60	112.88, 47.89, 23.60	
	1:5	0.0383	20, 40, 60	93.60, 39.41, 16.64	
	1:6	0.0505	20, 40, 60	91.44, 38.31, 16.41	
ChCl: phenol	1:2	0.256	20, 30, 40	120.77, 68.41, and 42.42	[94]
	1:3	0.358	20, 30, 40	57.84, 35.17, and 23.08	
	1:4	0.130	20, 30, 40	40.23, 25.20, and 16.71	
	1:5	0.186	20, 30, 40	31.96, 19.75, and 13.44	
	1:6	0.094	20, 30, 40	27.03, 16.82, and 11.45	
ChCl: Eth-glycol	1:2	–	25, 45, 65, and 85	24.7, 12.6, 7.6, and 5.6	[95]
	1:2.5	–	25, 45, 65, and 85	26.1, 16.0, 8.5, and 5.8	
	1:3	–	25, 45, 65, and 85	41.2, 19.0, 11.0, and 6.9	
ChCl: Tri-Eth-glycol	1:1	–	25, 45, 65, and 85	44.9, 21.6, 10.9, and 6.5	
	1:2	–	25, 45, 65, and 85	61.9, 30, 15.1, and 8.1	
	1:2.5	–	25, 45, 65, and 85	68.0, 33.8, 16.8, and 9.0	
ChCl: glycerol	1:0.5	–	20, 30, 40, and 50	80, 314, 163, and 120	
	1:1	–	20, 30, 40, and 50	700, 349, 180, and 98	
	1.15	–	20, 30, 40, and 50	600, 270,140, and 80	
	1:2		20, 30, 40, and 50	503, 246, 132, and 86	
	1:3	–	20, 30, 40, and 50	401, 213, 120, and 71	

how temperature, water content, molar ratio, and long-chain variations affect the viscosities of various DESs.

Toxicological Concerns and Regulatory Approval

Although DESs have many benefits, regulatory licensing and toxicological issues still pose major obstacles to their widespread use in food analysis. Their safety, biodegradability, and adherence to food safety laws are all unknown, which leads to these problems. DESs frequently have biological effects that differ significantly from those of their constituent parts. Despite DES's constituent parts being identified as harmless, the final mixture often shows possible toxicity to many environmental biological species because of synergistic effects [96]. Research on the toxic effects of DESs primarily focuses on assessing cytotoxicity and cell viability using human cell lines and evaluating their toxicity and antimicrobial properties through various biological assays [97]. Knowing that the combination of HBDs and HBAs determines how poisonous DESs are is paramount. ChCl: glycerol-based DESs, for instance, are typically considered harmless, but those containing ammonium or oxalic acid salts may be more cytotoxic [98, 99]. Additionally, DES containing organic acids may pose a significant risk to human cells, as revealed by Hayyan et al. [100]. Interestingly, all studies using carboxylic acids show no inhibiting effects on human cells, even at the highest concentration (2000 mg/L) [101]. Another factor that affects the toxicity in DES is the variation of the molar ratio of HBDs and HBAs. In a study, Ahmadi et al. proved that the toxicity in ChCl: 1,2-propanediol increased with an increase in the molar ratio (1:1 = LC_{50} 75.46 and 1:2 = LC_{50} 48.18) [102]. pH is another critical property to consider when designing a DES for food application. The pH of DES is determined by the relative acidity of its anionic and cationic constituents [99]. Studies reported that the acidic nature of HBDs is the primary factor influencing the variation in pH and the effects of HBAs [103, 104]. Although no human toxicity studies are currently in the literature, various studies have proven the possible toxicity. For example, a study based on the common HBDs conducted by Radosevic et al. found that ChCl: oxalate when applied to human tumors and fish cells leads to the formation of calcium oxalate crystals which can damage the cells [105]. Temperature plays a crucial role in modulating pH levels, as evidenced by studies on ChCl: D-glucose mixtures. When measured between 25 °C and 45 °C, the pH remained nearly neutral. However, as the temperature increased to 85 °C, a noticeable decline in pH was observed, exhibiting a slight linear correlation with rising temperature. This trend underscores the thermal sensitivity of DES systems and highlights the need for precise temperature control in their applications. Viscosity is another critical factor while evaluating the toxicity of DESs. High viscosity may contribute to elevated toxicity levels, potentially influencing the substance's biological interactions and safety profile [100]. The findings indicated that ChCl: glucose and ChCl: fructose were less toxic than traditional DESs in in vitro studies. However, in vivo tests revealed higher toxicity levels for these natural deep eutectic solvents (NADESs), suggesting that viscosity influences toxicity differently in more complex biological systems [46]. Another key factor in applying DESs in food adulterant detection is their biodegradability. Numerous research has looked into the biodegradability of DESs, offering information about their sustainability and environmental effects. A large number of DESs were reported to exhibit good biodegradability. The majority of evaluated DESs were easily biodegradable, with biodegradability levels surpassing 80% after 28 days, according to a study on cholinium-based DESs [106]. Chemicals are considered rapidly biodegradable if they can attain a biodegradation level greater than 60% during a 28-day period, and the biodegradation test is often regarded as genuine if a 60% ThOD is reached for the reference molecule after 14 days [107]. However, in most of the DESs reported, it was found that the lower the toxicity the higher the biodegradability [105]. Some DESs are hazardous to different organisms, which can prevent them from biodegrading. Based on the studies evaluating the toxicity and biodegradability of organic acids DESs, it was found that some DESs may harm the environmental microbes responsible for the biodegradation processes that impede their degradation [108]. The chemical structure and the elemental composition of HBAs and HBDs also significantly affect their biodegradability. For example, Martinez et al. highlighted that DESs made of particular HBDs and quaternary ammonium salt may be less biodegradable due to their resistance to microbial degradation [97]. The persistence of DESs that are not easily biodegradable may lead to long-term exposure to environmental organisms, potentially disrupting the system. Another big setback for DESs' applicability in food adulterant detection is regulatory compliance. The major challenge in this regard is the lack of standardized testing methods, safety evaluation techniques, and values for safe, effective application in the food industry. Conventional solvents are generally covered by established guidelines for safety evaluation, unlike DESs which do not have toxicity data, specified purity criteria, and long-term stability [109]. Because of this, DES-based techniques' method validation and broad adoption fall short of their scientific potential, highlighting the need for more precise safety standards and focused risk assessments to close the gap between innovation and regulatory acceptance.

Conclusion and Future Perspectives

DESs represent a groundbreaking advancement in food adulteration analysis, offering green, sustainable, and efficient alternatives to traditional organic solvents. Their customizable physicochemical properties, such as high solubility and selectivity, tunability, and low toxicity, make DESs suitable for the extraction and detection of a wide range of food adulterants, including synthetic dyes, phytosterols, and hazardous contaminants like amphetamines. Integrating DESs with analytical techniques, including HPLC, GC-MS, and electrochemical methods, has demonstrated their superior performance in improving detection sensitivity and reducing environmental impact. Despite their remarkable potential, challenges remain in the widespread adoption of DESs. Issues such as viscosity management, stability under varying environmental conditions, recyclability, and selecting optimal hydrogen bond donors and acceptors for specific applications require continued research. Furthermore, standardization of DES formulations and extraction procedures is necessary to enhance reproducibility and enable industrial applications.

Looking ahead, future research should focus on:

1. Designing task-specific DESs: Developing novel DES formulations tailored for complex food matrices to enhance selectivity and efficiency.
2. Improving viscosity control: Exploration of strategies such as solvent additives and temperature optimization to mitigate the effects of high DES viscosity on extraction processes.
3. Enhancing recyclability and sustainability: Innovations in DES recovery and purification methods to reduce waste and operational costs.
4. Integration with emerging technologies: Coupling DESs with advanced analytical tools like sensor-based platforms and artificial intelligence (AI) for real-time, high-throughput food analysis.
5. Toxicological and regulatory evaluations: Comprehensive studies on the safety of DES residues in food products and their environmental impact to support regulatory approval.

By addressing these challenges, DESs have the potential to revolutionize food safety practices, providing a greener and more efficient approach to combating food adulteration and ensuring consumer protection.

Acknowledgments This work is supported by the Second Century Fund (C2F), Chulalongkorn University, Thailand. Waleed Alahmad gratefully acknowledges the Department of Chemistry, Faculty of Science, Chulalongkorn University, for the provision of research facilities, resources, and technical support.

References

1. Tamanna IM (2024) Food adulteration and inadequate hygiene practices endangering public health in Bangladesh. Discov Food 4:109
2. Misgana B (2020) Food adulteration and some methods of detection, review. Int J Nutr Food Sci 9:86–94
3. Anagaw YK, Ayenew W, Limenh LW, Geremew DT, Worku MC, Tessema TA, Simegn W, Mitku ML (2024) Food adulteration: causes, risks, and detection techniques-review. SAGE Open Med 12:20503121241250184
4. Das P, Altemimi AB, Nath PC, Katyal M, Kesavan RK, Rustagi S, Panda J, Avula SK, Nayak PK, Mohanta YK (2025) Recent advances on artificial intelligence-based approaches for food adulteration and fraud detection in the food industry: challenges and opportunities. Food Chem 468:142439
5. Song D, Dong K, Liu S, Fu S, Zhao F, Man C, Jiang Y, Zhao K, Qu B, Yang X (2024) Research advances in detection of food adulteration and application of MALDI-TOF MS: a review. Food Chem 456:140070

6. Elumalai A, Natarajan V (2025) Advancements in analytical technologies for ensuring food quality and authentication: a comprehensive review. J Food Compos Anal 139:107075
7. Srivastava S, Pandey VK, Singh A, Dash KK, Dar AH, Rustagi S (2025) Recent insights on microfluidics applications for food quality and safety analysis: a comprehensive review. Food Control 168:110869
8. Erasmus S, Ruth S (2021) Key challenges and developments in non-targeted methods or systems to identify food adulteration. Wageningen University, Wageningen, pp 175–204
9. Momtaz M, Bubli SY, Khan MS (2023) Mechanisms and health aspects of food adulteration: a comprehensive review. Foods 12:199
10. Agrawal U, Bawane N, Alsubaie N, Alqahtani MS, Abbas M, Soufiene BO (2024) Design & development of adulteration detection system by fumigation method & machine learning techniques. Sci Rep 14:25366
11. Balaji A, Bathinapatla A, Manuel M, Mulpuri RK, Kanchi S (2025) Recent developments in melamine detection: applications of gold and silver nanostructures in colorimetric and fluorometric assays. J Hazard Mater Adv 17:100553
12. Hashim NM, Waras MN, Yahaya N, Raoov M, Kabir A, Mohamad Zain NN (2024) Green metrics for analytical methodologies: uncovering sustainable chromatography approaches for detecting emerging contaminants in food and environmental water. TrAC Trends Anal Chem 172:117598
13. Khan WA, Varanusupakul P, Ul Haq H, Arain MB, Marchel M, Boczkaj G (2025) Diverse roles, advantages and importance of deep eutectic solvents application in solid and liquid-phase microextraction techniques—a review. Sep Purif Technol 358:130362
14. Peng J, Hassan FA, Wu J, Xiong C (2024) Determination of fifteen illegal colorants in traditional Chinese medicines by two hydrophobic DES-based microextraction methods coupled with an HPLC-DAD. Talanta 277:126236
15. Llompart M, Celeiro M, Dagnac T (2019) Microwave-assisted extraction of pharmaceuticals, personal care products and industrial contaminants in the environment. TrAC Trends Anal Chem 116:136–150
16. Brito TA, Costa FS, Oliveira RC, Amaral CDB, Labuto G, Gonzalez MH (2025) Green extraction using natural deep eutectic solvents for determination of as, cd, and Pb in plant and food matrices by ICP-MS. Food Chem 464:141922
17. Smith EL, Abbott AP, Ryder KS (2014) Deep eutectic solvents (DESs) and their applications. Chem Rev 114:11060–11082
18. Lanjwani MF, Tuzen M, Khuhawar MY, Afshar Mogaddam MR, Farajzadeh MA (2024) Deep eutectic solvents for extraction and preconcentration of organic and inorganic species in water and food samples: a review. Crit Rev Anal Chem 54:1290–1303
19. Abbott AP, Capper G, Davies DL, Rasheed RK, Tambyrajah V (2003) Novel solvent properties of choline chloride/urea mixtures. Chem Commun 2003:70–71
20. Deng W, Fan C, Zhang R, Jin M (2024) Deep-eutectic-solvent-decorated metal–organic framework for food and environmental sample preparation. Foods 13:3614
21. Martins MAR, Pinho SP, Coutinho JAP (2019) Insights into the nature of eutectic and deep eutectic mixtures. J Solut Chem 48:962–982
22. Abbott AP, Boothby D, Capper G, Davies DL, Rasheed RK (2004) Deep eutectic solvents formed between choline chloride and carboxylic acids: versatile alternatives to ionic liquids. J Am Chem Soc 126:9142–9147
23. Długosz O, Banach M (2024) Green methods for obtaining deep eutectic solvents (DES). J Clean Prod 434:139914
24. Taghizadeh M, Taghizadeh A, Vatanpour V, Ganjali MR, Saeb MR (2021) Deep eutectic solvents in membrane science and technology: fundamental, preparation, application, and future perspective. Sep Purif Technol 258:118015
25. Florindo C, Oliveira FS, Rebelo LPN, Fernandes AM, Marrucho IM (2014) Insights into the synthesis and properties of deep eutectic solvents based on cholinium chloride and carboxylic acids. ACS Sustain Chem Eng 2:2416–2425

26. Boateng ID (2022) A critical review of emerging hydrophobic deep eutectic solvents' applications in food chemistry: trends and opportunities. J Agric Food Chem 70:11860–11879
27. Wang S, Hui J, Si Y, Geng T, Yang C, Kang Y, Li D, Ye M, Yan T, Yue C (2025) Deep-eutectic solvent coated paper: a reusable multi-purpose device for rapid visualization, adulteration identification, efficient removal and recycling of dyes in both food and environmental samples. Microchem J 209:112625
28. Lu Y, Zhang Z, Zhu D, Lu C (2025) Magnetic nanoparticles-based dispersive solid phase extraction and deep eutectic solvent for the extraction and preconcentration of phenethylamines in sports dietary supplements followed by HPLC–FL analysis. Microchem J 208:112605
29. Pandey A, Bhawna D, Dhingra S (2017) Pandey, hydrogen bond donor/acceptor cosolvent-modified choline chloride-based deep eutectic solvents. J Phys Chem B 121:4202–4212
30. Ashworth CR, Matthews RP, Welton T, Hunt PA (2016) Doubly ionic hydrogen bond interactions within the choline chloride–urea deep eutectic solvent. Phys Chem Chem Phys 18:18145–18160
31. Pateli IM, Thompson D, Alabdullah SSM, Abbott AP, Jenkin GRT, Hartley JM (2020) The effect of pH and hydrogen bond donor on the dissolution of metal oxides in deep eutectic solvents. Green Chem 22:5476–5486
32. Alhadid A, Jandl C, Mokrushina L, Minceva M (2022) Cocrystal formation in choline chloride deep eutectic solvents. Cryst Growth Des 22:1933–1942
33. Ito M, Iwata Y, Iwasaki H, Hayashi K, Nakayama T, Hikawa H, Azumaya I, Suzuki H, Noguchi S (2023) Cocrystal or salt crystallization for active pharmaceutical ingredients by using deep eutectic solvents. Cryst Growth Des 23:6298–6307
34. Farooq MQ, Abbasi NM, Anderson JL (2020) Deep eutectic solvents in separations: methods of preparation, polarity, and applications in extractions and capillary electrochromatography. J Chromatogr A 1633:461613
35. Abbasi NM, Farooq MQ, Anderson JL (2021) Investigating the variation in solvation interactions of choline chloride-based deep eutectic solvents formed using different hydrogen bond donors. ACS Sustain Chem Eng 9:11970–11980
36. Morrison HG, Sun CC, Neervannan S (2009) Characterization of thermal behavior of deep eutectic solvents and their potential as drug solubilization vehicles. Int J Pharm 378:136–139
37. Su M, Wei Y, Huang M, Zhang J, Gao Y, Qian S (2024) Sustainable deep eutectic solvents induced the polymorph selectivity and high purification efficiency. Sep Purif Technol 349:127856
38. Chu K, Yuan C, Li H, Yang K, Wang Y, Cheng X, Zhu X, Wang Z, Su L (2017) Pressure-induced amorphization and crystallization of choline chloride/ethylene glycol deep eutectic solvent. J Mol Liq 242:109–114
39. Richter J, Ruck M (2020) Synthesis and dissolution of metal oxides in ionic liquids and deep eutectic solvents. Molecules 25(1):78
40. Abbott AP, Capper G, Davies DL, McKenzie KJ, Obi SU (2006) Solubility of metal oxides in deep eutectic solvents based on choline chloride. J Chem Eng Data 51:1280–1282
41. Chen W, Jiang J, Lan X, Zhao X, Mou H, Mu T (2019) A strategy for the dissolution and separation of rare earth oxides by novel Brønsted acidic deep eutectic solvents. Green Chem 21:4748–4756
42. Shishov A, Nizov E, Bulatov A (2023) Microextraction of melamine from dairy products by thymol-nonanoic acid deep eutectic solvent for high-performance liquid chromatography-ultraviolet determination. J Food Compos Anal 116:105083
43. Wang Y, Hou Y, Wu W, Liu D, Ji Y, Ren S (2016) Roles of a hydrogen bond donor and a hydrogen bond acceptor in the extraction of toluene from n-heptane using deep eutectic solvents. Green Chem 18:3089–3097
44. Li X, Row KH (2016) Development of deep eutectic solvents applied in extraction and separation. J Sep Sci 39:3505–3520
45. Dai Y, van Spronsen J, Witkamp G-J, Verpoorte R, Choi YH (2013) Natural deep eutectic solvents as new potential media for green technology. Anal Chim Acta 766:61–68

46. Mbous YP, Hayyan M, Wong WF, Looi CY, Hashim MA (2017) Unraveling the cytotoxicity and metabolic pathways of binary natural deep eutectic solvent systems. Sci Rep 7:41257
47. Liu Y, Friesen JB, McAlpine JB, Lankin DC, Chen S-N, Pauli GF (2018) Natural deep eutectic solvents: properties, applications, and perspectives. J Nat Prod 81:679–690
48. Zdanowicz M, Wilpiszewska K, Spychaj T (2018) Deep eutectic solvents for polysaccharides processing. A review. Carbohydr Polym 200:361–380
49. Abbott AP, Harris RC, Ryder KS, D'Agostino C, Gladden LF, Mantle MD (2011) Glycerol eutectics as sustainable solvent systems. Green Chem 13:82–90
50. Zhang Q, Vigier KDO, Royer S, Jérôme F (2012) Deep eutectic solvents: syntheses, properties and applications. Chem Soc Rev 41:7108–7146
51. Ražić S, Bakić T, Topić A, Lukić J, Onjia A (2023) Deep eutectic solvent based reversed-phase dispersive liquid–liquid microextraction and high-performance liquid chromatography for the determination of free tryptophan in cold-pressed oils. Molecules 28(5):2395
52. Gholami Z, Marhamatizadeh MH, Yousefinejad S, Rashedinia M, Mazloomi SM (2021) Vortex-assisted dispersive liquid-liquid microextraction based on hydrophobic deep eutectic solvent for the simultaneous identification of eight synthetic dyes in jellies and drinks using HPLC-PDA. Microchem J 170:106671
53. Ebadnezhad H, Afshar Mogaddam MR, Farajzadeh MA, Mohebbi A, Torbati M (2021) Development of an ultrasonic and heat-assisted liquid–liquid extraction method combined with deep eutectic solvent-based dispersive liquid–liquid microextraction for the extraction of some phytosterols from cow milk butter samples. J Iran Chem Soc 18:2483–2491
54. Ebadnezhad H, Afshar Mogaddam MR, Farajzadeh MA, Mohebbi A, Nemati M, Torbati M (2021) Combining a liquid-liquid extraction with successive air assisted liquid-liquid microextraction for the analysis of phytosterols present in animal based butter and oil samples. J Chromatogr A 1642:462025
55. Hashemi M, Rashedinia M, Khezeli T, Shariati S, Mahdavinia M, Noori SMA (2024) Application of deep eutectic solvent@ metal-organic framework molecularly imprinted polymer fiber for solid phase microextraction: determination of illegally added amphetamines in snacks and beverages and quantitation by Gc-Ms. J Pharm Biomed Anal 242:116005
56. Nemati M, Tuzen M, Altunay N, Farajzdeh MA, Abdi F, Afshar Mogaddam MR (2022) Development of sodium hydroxide-induced homogenous liquid-liquid extraction-effervescent assisted dispersive liquid-liquid microextraction based on deep eutectic solvents; application in the extraction of phytosterols from cow cream samples. J Food Compos Anal 106:104291
57. Shabani E, Zappi D, Berisha L, Dini D, Antonelli ML, Sadun C (2020) Deep eutectic solvents (DES) as green extraction media for antioxidants electrochemical quantification in extra-virgin olive oils. Talanta 215:120880
58. Das C, Chakraborty S, Acharya K, Bera NK, Chattopadhyay D, Karmakar A, Chattopadhyay S (2017) FT-MIR supported electrical impedance spectroscopy based study of sugar adulterated honeys from different floral origin. Talanta 171:327–334
59. Chakraborty M, Biswas K (2018) Limit of detection for five common adulterants in milk: a study with different fat percent. IEEE Sensors J 18:2395–2403
60. Bertemes-Filho P, Negri LH, Paterno AS (2012) Detection of bovine milk adulterants using bioimpedance measurements and artificial neural network. In: Jobbágy Á (ed) 5th European conference of the international federation for medical and biological engineering. Springer, Berlin, pp 1275–1278
61. Erbas Z, Soylak M (2025) Simple deep eutectic solvent-based liquid phase microextraction for Sudan IV from food, water, and cosmetics. Anal Lett. https://doi.org/10.1080/00032719.2024.2447940
62. Xue Q, Wang C, Lin Y, Jiang T-F, Lv Z (2021) Determination of fluoroquinolones illegally added in traditional prostate medicines by ultrasonic-assisted dispersive liquid-liquid microextraction based on deep eutectic solvent combined with quantitative 19F nuclear magnetic resonance method. Microchem J 170:106725

63. Zhao J, Wang M, Avula B, Khan IA (2018) Detection and quantification of phenethylamines in sports dietary supplements by NMR approach. J Pharm Biomed Anal 151:347–355
64. Mbous YP, Hayyan M, Hayyan A, Wong WF, Hashim MA, Looi CY (2017) Applications of deep eutectic solvents in biotechnology and bioengineering—promises and challenges. Biotechnol Adv 35:105–134
65. Craveiro R, Aroso I, Flammia V, Carvalho T, Viciosa M, Dionísio M, Barreiros S, Reis R, Duarte A, Paiva A (2016) Properties and thermal behavior of natural deep eutectic solvents. J Mol Liq 215:534–540
66. Zhang X, Zhu P, Li Q, Xia H (2022) Recent advances in the catalytic conversion of biomass to furfural in deep eutectic solvents. Front Chem 10:911674
67. Ghorpade P, Gadilohar B, Pinjari D, Shinde Y, Shankarling G (2015) Concentrated solar radiation enhanced one pot synthesis of DES and N-phenyl phthalimide. Sol Energy 122:1354–1361
68. Zhou Y, Wu W, Zhang N, Soladoye OP, Zhang Y, Fu Y (2022) Deep eutectic solvents as new media for green extraction of food proteins: opportunity and challenges. Food Res Int 161:111842
69. Chen Y, Yu D, Chen W, Fu L, Mu T (2019) Water absorption by deep eutectic solvents. Phys Chem Chem Phys 21:2601–2610
70. Ortega-Zamora C, González-Sálamo J, Hernández-Borges J (2021) Deep eutectic solvents application in food analysis. Molecules 26:6846
71. Brusas JR, Dela Pena EMB (2021) Hygroscopicity of 1: 2 choline chloride: ethylene glycol deep eutectic solvent: a hindrance to its electroplating industry adoption, journal of electrochemical. Sci Technol 12:387–397
72. Bowen H, Durrani R, Delavault A, Durand E, Chenyu J, Yiyang L, Lili S, Jian S, Weiwei H, Fei G (2022) Application of deep eutectic solvents in protein extraction and purification. Front Chem 10:912411
73. Cannavacciuolo C, Pagliari S, Frigerio J, Giustra CM, Labra M, Campone L (2022) Natural deep eutectic solvents (NADESs) combined with sustainable extraction techniques: a review of the green chemistry approach in food analysis. Foods 12:56
74. Ferreira C, Sarraguça M (2024) A comprehensive review on deep eutectic solvents and its use to extract bioactive compounds of pharmaceutical interest. Pharmaceuticals 17:124
75. Jablonský M, Škulcová A, Ház A, Šima J, Majová V (2018) Long-term isothermal stability of deep eutectic solvents. Bioresources 13:7545–7559
76. Chen W, Xue Z, Wang J, Jiang J, Zhao X, Mu T (2018) Investigation on the thermal stability of deep eutectic solvents. Acta Phys Chim Sin 34:904–911
77. Dai Y, Witkamp G-J, Verpoorte R, Choi YH (2015) Tailoring properties of natural deep eutectic solvents with water to facilitate their applications. Food Chem 187:14–19
78. Guo J, Zhao N, Zhao Y, Jin H, Sun G, Yu J, Zhang H, Yu M, Yang D, Liang Z (2023) Effect of deep eutectic solvents on the activity and stability of cellulases and pectinases. ACS Omega 8:45678–45686
79. Ferreira RS, Ferreira AM, Coutinho JA, Batista EAC (2024) Stability of cholinium chloride-based deep eutectic solvents with carboxylic acids: a study on ester formation. ACS Sustain Chem Eng 12:15893–15900
80. Campodónico PR, Alarcón-Espósito J, Alcázar JJ, Olivares B, Suárez-Rozas C (2024) Analysis of the behavior of deep eutectic solvents upon addition of water: its effects over a catalytic reaction. Molecules 29:3296
81. Isci A, Kaltschmitt M (2022) Recovery and recycling of deep eutectic solvents in biomass conversions: a review. Biomass Convers Biorefinery 12:197–226
82. Cui Y, Li C, Bao M (2019) Deep eutectic solvents (DESs) as powerful and recyclable catalysts and solvents for the synthesis of 3, 4-dihydropyrimidin-2 (1 H)-ones/thiones. Green Process Synth 8:568–576

83. Xu L, Liaqat F, Khazi MI, Sun J, Zhu D (2024) Natural deep eutectic solvents-based green extraction of vanillin: optimization, purification, and bioactivity assessment. Front Nutr 10:1279552
84. Falcini C, Gonzalo GD (2024) Deep eutectic solvents as catalysts in the synthesis of active pharmaceutical ingredients and precursors. Catalysts 14:120
85. Elangovan T, Nor MH, Nurina IMH, Noorfatimah Y, Farhanini Y, Nur NMZ (2024) Exploring deep eutectic solvents as a potential solvent for the extraction of food additive allura red in food technology application. Malays J Anal Sci 28:645–663
86. Zhekenov T, Toksanbayev N, Kazakbayeva Z, Shah D, Mjalli FS (2017) Formation of type III deep eutectic solvents and effect of water on their intermolecular interactions. Fluid Phase Equilib 441:43–48
87. Huang Y, Feng F, Chen Z-G, Wu T, Wang Z-H (2018) Green and efficient removal of cadmium from rice flour using natural deep eutectic solvents. Food Chem 244:260–265
88. Bajkacz S, Rusin K, Wolny A, Adamek J, Erfurt K, Chrobok A (2020) Highly efficient extraction procedures based on natural deep eutectic solvents or ionic liquids for determination of 20-hydroxyecdysone in spinach. Molecules 25:4736
89. López R, D'Amato R, Trabalza-Marinucci M, Regni L, Proietti P, Maratta A, Cerutti S, Pacheco P (2020) Green and simple extraction of free seleno-amino acids from powdered and lyophilized milk samples with natural deep eutectic solvents. Food Chem 326:126965
90. Fuad FM, Nadzir MM, Harun A (2021) Hydrophilic natural deep eutectic solvent: a review on physicochemical properties and extractability of bioactive compounds. J Mol Liq 339:116923
91. Ghaedi H, Ayoub M, Sufian S, Shariff AM, Lal B (2017) The study on temperature dependence of viscosity and surface tension of several Phosphonium-based deep eutectic solvents. J Mol Liq 241:500–510
92. Serna-Vázquez J, Ahmad MZ, Boczkaj G, Castro-Muñoz R (2021) Latest insights on novel deep eutectic solvents (DES) for sustainable extraction of phenolic compounds from natural sources. Molecules 26:5037
93. Gajardo-Parra NF, Cotroneo-Figueroa VP, Aravena P, Vesovic V, Canales RI (2020) Viscosity of choline chloride-based deep eutectic solvents: experiments and modeling. J Chem Eng Data 65:5581–5592
94. Guo W, Hou Y, Ren S, Tian S, Wu W (2013) Formation of deep eutectic solvents by phenols and choline chloride and their physical properties. J Chem Eng Data 58:866–872
95. Mjalli FS, Naser J (2015) Viscosity model for choline chloride-based deep eutectic solvents. Asia Pac J Chem Eng 10:273–281
96. Juneidi I, Hayyan M, Mohd Ali O (2016) Toxicity profile of choline chloride-based deep eutectic solvents for fungi and Cyprinus carpio fish. Environ Sci Pollut Res 23:7648–7659
97. Martínez GM, Townley GG, Martínez-Espinosa RM (2022) Controversy on the toxic nature of deep eutectic solvents and their potential contribution to environmental pollution. Heliyon 8:e12567
98. Boateng ID (2023) Evaluating the status quo of deep eutectic solvent in food chemistry. Food Chem 406:135079
99. Skulcova A, Russ A, Jablonsky M, Sima J (2018) The pH behavior of seventeen deep eutectic solvents. Bioresources 13:5042–5051
100. Hayyan M, Mbous YP, Looi CY, Wong WF, Hayyan A, Salleh Z, Mohd-Ali O (2016) Natural deep eutectic solvents: cytotoxic profile. Springerplus 5:1–12
101. Mitar A, Panić M, Prlić Kardum J, Halambek J, Sander A, Zagajski Kučan K, Radojčić Redovniković I, Radošević K (2019) Physicochemical properties, cytotoxicity, and antioxidative activity of natural deep eutectic solvents containing organic acid. Chem Biochem Eng Q 33:1–18
102. Ahmadi R, Hemmateenejad B, Safavi A, Shojaeifard Z, Mohabbati M, Firuzi O (2018) Assessment of cytotoxicity of choline chloride-based natural deep eutectic solvents against human HEK-293 cells: a QSAR analysis. Chemosphere 209:831–838

103. Taysun MB, Sert E, Atalay FS (2017) Effect of hydrogen bond donor on the physical properties of benzyltriethylammonium chloride based deep eutectic solvents and their usage in 2-ethyl-hexyl acetate synthesis as a catalyst. J Chem Eng Data 62:1173–1181
104. Hayyan M, Aissaoui T, Hashim MA, AlSaadi MA, Hayyan A (2015) Triethylene glycol based deep eutectic solvents and their physical properties. J Taiwan Inst Chem Eng 50:24–30
105. Radošević K, Bubalo MC, Srček VG, Grgas D, Dragičević TL, Redovniković IR (2015) Evaluation of toxicity and biodegradability of choline chloride based deep eutectic solvents. Ecotoxicol Environ Saf 112:46–53
106. Wen Q, Chen J-X, Tang Y-L, Wang J, Yang Z (2015) Assessing the toxicity and biodegradability of deep eutectic solvents. Chemosphere 132:63–69
107. Lomba L, Ribate MP, Sangüesa E, Concha J, Garralaga MP, Errazquin D, García CB, Giner B (2021) Deep eutectic solvents: are they safe? Appl Sci 11:10061
108. Hayyan A, Zainal-Abidin MH, Putra SSS, Alanazi YM, Saleh J, Nor MRM, Hashim MA, Gupta BS (2024) Evaluation of biodegradability, toxicity and ecotoxicity of organic acid-based deep eutectic solvents. Sci Total Environ 948:174758
109. U.S. Food and Drug Administration (2016) Economically motivated adulteration (food fraud). FDA, Silver Spring

Emerging Trends in Portable Analytical Devices (Miniaturized Systems) for Food Adulteration Analysis

Mkliwa Koumbia, Koffi Aquilas Agbevohia, Agoura Diantom, and Mohamed Aly Saad Aly

Contents

M. Koumbia (✉)
Graduate School of Engineering, University of Tokyo, Tokyo, Japan

K. A. Agbevohia
Institute of Natural and Applied Science, University of Çukurova, Adana, Turkey

A. Diantom
Laboratoire des Sciences Biomédicales, Alimentaires et de Santé Environnementale (LaSBASE), Ecole Supérieure des Techniques Biologiques et Alimentaires (ESTBA), Université de Lomé, Lomé, Togo

M. Aly Saad Aly (✉)
School of Electrical and Computer Engineering, Georgia Institute of Technology, Atlanta, GA, USA

Department of Electrical and Computer Engineering, Georgia Tech Shenzhen Institute (GTSI), Shenzhen, Guangdong, China
e-mail: mohamed.alysaadaly@ece.gatech.edu

W. Alahmad, M. Ashfaq (eds.), *Advanced Techniques against Food Adulteration*,
https://doi.org/10.1007/978-3-031-99454-8_12

Abstract The rapid transition from traditional diagnostic systems to portable analytical devices (PADs) over the past few decades has led to significant progress in food safety, offering innovative solutions to combat the pervasive issue of food adulteration. Adulterants, encompassing chemical contaminants like melamine and pesticides and biological hazards such as *Salmonella* and mycotoxins, threaten public health and economic stability. In this regard, PADs have emerged as an effective real-time and on-site detection tool. By addressing the limitations of conventional laboratory methods, such as lengthy analysis times and the dependence on extensive infrastructure, PADs empower stakeholders across the food supply chain to mitigate the risks associated with foodborne illnesses and financial loss. Research efforts are directed toward improving the accuracy, sensitivity, specificity, and multiplexing features of portable analytical devices (PADs) while ensuring their effective deployment in real-world applications. This initiative leverages state-of-the-art technologies, including nanomaterials, biorecognition elements with extended shelf life, and the incorporation of wearable/portable technology with artificial intelligence. The aim is to facilitate these tools' rapid and efficient application in combating food adulteration.

Keywords Portable analytical devices · Wearable device · Biosensor · Food Adulteration Analysis · Artificial intelligence

Introduction to Portable Analytical Devices

The advancement of analytical diagnostics has experienced a significant transformation in recent decades, shifting from traditional laboratory-centric approaches to more decentralized, field-ready systems. Established diagnostic methods, including high-performance liquid chromatography (HPLC), gas chromatography-mass spectrometry (GC-MS), enzyme-linked immunosorbent assays (ELISA), spectrophotometry, and polymerase chain reaction (PCR) technologies, have historically been regarded as benchmarks due to their accuracy and sensitivity. Nonetheless, these techniques are limited by their dependence on advanced infrastructure, extended analysis times resulting from sample transport, pre-processing requirements, intricate analytical procedures, and the necessity for specialized personnel. Such constraints have prompted considerable research to create alternative, decentralized analytical solutions that can deliver swift, cost-effective, and user-friendly diagnostics, culminating in the emergence of portable analytical devices [1, 2].

Early developments in portable analytical devices (PADs) technology include the introduction of handheld glucometers in 1970–1976, illustrating the potential of miniaturized electrochemical sensing in biomedical contexts [3, 4]. In the late twentieth and early twenty-first centuries, remarkable advancements in microfluidic technologies transformed paper-based analytical devices into sophisticated lab-on-a-chip (LOC) systems. The integration of microfluidics enabled the miniaturization of extensive analytical processes, allowing for precise control over fluid dynamics at the microscale, which enhanced the sensitivity and specificity of diagnostic tests. A notable breakthrough occurred in the early 2000s with the introduction of paper-based analytical devices with the first reported immunochromatographic assay for detecting melamine in milk in 2008, which utilized patterned nitrocellulose matrices for colorimetric detection of analytes while reducing reagent consumption [5]. Subsequent innovations, such as smartphone-integrated optical sensors and wearable biosensors, have expanded the functionality and accessibility of PADs. Today, these devices exemplify cutting-edge engineering and applied science convergence, driven by the demand for rapid, portable, and accessible diagnostic solutions.

PADs are revolutionizing food quality monitoring by providing rapid, on-site detection of contaminants, adulterants, and spoilage indicators, ensuring the safety and integrity of food products throughout the supply chain. Unlike traditional laboratory methods, which are often time-consuming and require specialized infrastructure, PADs enable real-time analysis at critical points such as production facilities, distribution centers, and retail outlets [6, 7]. This capability is vital for detecting economically motivated adulteration (EMA), where substances like melamine, Sudan dyes, or starch are added to food products to increase volume or mimic quality. By actively detecting fraudulent activities in the food industry through portable analytical devices (PADs), potential public health dangers can be avoided, economic losses minimized, and consumer confidence in the food supply is upheld. Chemical contaminants, such as pesticides, heavy metals, and allergens, are another primary focus of PAD-based food quality monitoring. Devices employing colorimetric, electrochemical, or spectroscopic principles can screen for organophosphates, mercury, or gluten with high sensitivity and specificity. For instance, paper-based microfluidic PADs have been designed to detect pesticide residues in fruits and vegetables through simple color changes, making them accessible even to non-technical users. Additionally, PADs are being employed to monitor the presence of

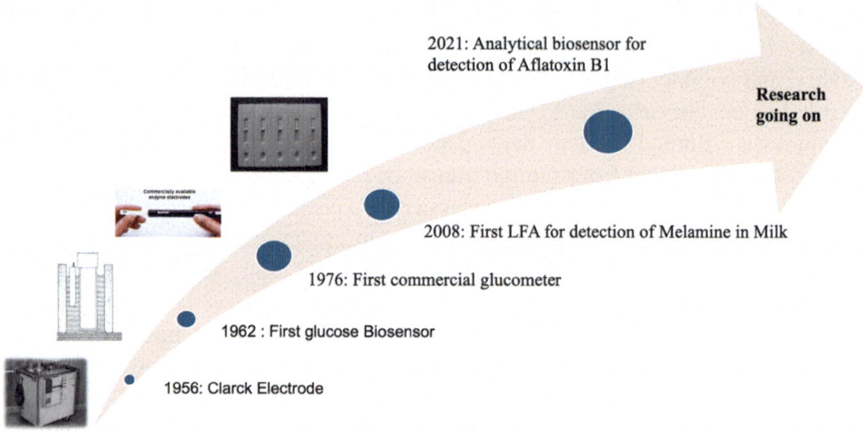

Fig. 1 Key progress in the research of PADs for monitoring food adulteration

allergens like peanuts or shellfish in processed foods, ensuring compliance with labeling regulations and protecting vulnerable consumers [7, 8]. Figure 1 a graphical representation of the key progress in the research of PADs for monitoring food adulteration.

This chapter provides an in-depth examination of the latest developments and ongoing research in the design of PADs aimed at detecting food adulterants, with a focus on next-generation innovations such as multiplexed detection systems, smartphone-compatible optical sensors, and nanotechnology-enhanced biosensors. The discussion also emphasizes ongoing research employing microfluidics, molecular imprinting, and machine-learning techniques to improve sensitivity, specificity, and user-friendliness. It critically addresses the challenges that impede widespread adoption, including scalability issues, cross-reactivity in complex food matrices, and the necessity for regulatory harmonization. Furthermore, the chapter outlines potential pathways to accelerate progress in this area. By facilitating the shift from laboratory to field applications and fostering interdisciplinary collaboration, this synthesis underscores the essential role of PADs in preserving food integrity, ensuring compliance with international safety standards, and ultimately boosting consumer confidence in an increasingly interconnected agri-food environment.

Definition of PADs

PADs are compact, self-contained systems capable of performing qualitative or quantitative analyses of chemical, biological, or environmental samples without conventional laboratory environments [9, 10]. Unlike traditional benchtop instruments, PADs are characterized by their low energy consumption, miniaturized sample and reagent volumes, usually in the microliter to nanoliter range, and ability to

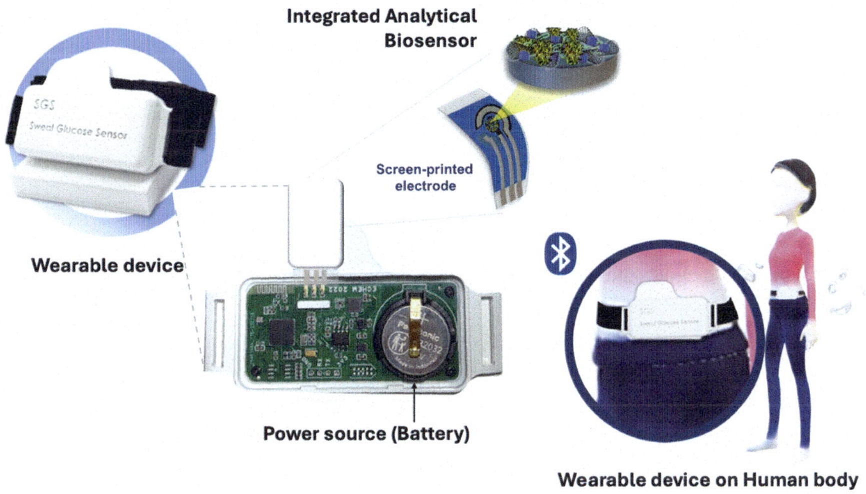

Fig. 2 Schematic illustration of a wearable PAD along with its constituents (adapted from [11])

integrate sample preparation, detection, and data analysis into a single platform. Typically, PADs are designed with a sophisticated architecture that combines sample collection, biosensing, signal transduction, and data processing into a streamlined, user-friendly device, as shown in Fig. 2. The sample collection component is where the food sample is introduced, often using a simple loading mechanism that relies on passive diffusion, active pumping, or capillary action to transport the sample to the biosensor. The biosensor is the core element of the device, hosting biorecognition components such as antibodies, enzymes, or aptamers that selectively bind to or react with the target adulterant. This binding event generates a biological signal, which is then amplified and converted into an optical or electrical signal by a transducer. The transducer, which is often functionalized with the same biorecognition components, plays a crucial role in transforming the biological interaction into a measurable output [11, 12].

Classification of PADs

As depicted in Fig. 3, PADs can be categorized into three primary groups: qualitative, semi-quantitative, and quantitative PADs. The quantitative PADs can be further distinguished based on their detection mechanisms, encompassing optical, electrochemical, thermal, gravimetric, electronic, and acoustic sensors. Optical PADs function through detection methods such as fluorescence, colorimetric analysis, or chemiluminescence, while electrochemical PADs rely on potentiometric, amperometric, voltammetry, conductometry, or impedimetric techniques to evaluate the concentration of analytes. Additionally, incorporating smartphone-based readout systems and wireless connectivity significantly enhances the functionality of PADs

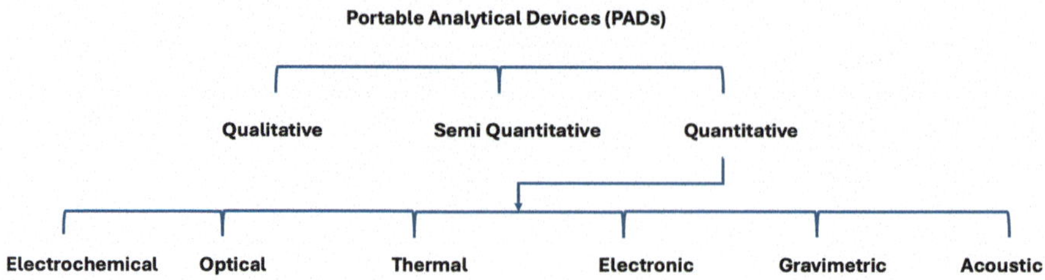

Fig. 3 Graphical diagram showing different groups of PADs

in areas such as telemedicine, remote diagnostics, and the transmission of real-time data [13]. The versatility inherent in PADs has led to their implementation across diverse sectors (healthcare, agriculture, food quality, environment, air quality, forensic science, sports and defense, etc.), emphasizing their transformative capacity in tackling global issues through decentralized data-informed decision-making [14]. As analytical science progresses, portable analytical devices are emerging as pivotal players in technological innovation, fundamentally altering the frameworks of diagnostics and environmental stewardship.

PADs Transforming Diagnostics Across Fields

Apart from food adulteration, the emergence of PADs has transformed diagnostic practices in numerous disciplines, delivering swift, effective, and accessible alternatives to traditional laboratory methods. Due to their compact design, minimal energy requirements, and portability, PADs are simple to use in various settings, particularly remote or limited resources. PADs have drastically revolutionized conventional diagnostic procedures by allowing for real-time, on-site analysis, encouraging innovation and widening the horizons of analytical research [1, 2, 6, 9].

PADs in Healthcare

PADs have significantly transformed the healthcare landscape by facilitating swift, precise, and decentralized diagnostic processes, effectively bridging the gaps inherent in conventional laboratory systems. Their impact is particularly pronounced in resource-constrained environments, where access to sophisticated medical facilities is limited. PADs enable point-of-care (POC) testing, empowering healthcare professionals to diagnose illnesses, oversee chronic conditions, and respond to outbreaks in real time [15]. During the COVID-19 pandemic and flu outbreaks, PADs were crucial for mass testing through rapid antigen tests, highlighting their adaptability

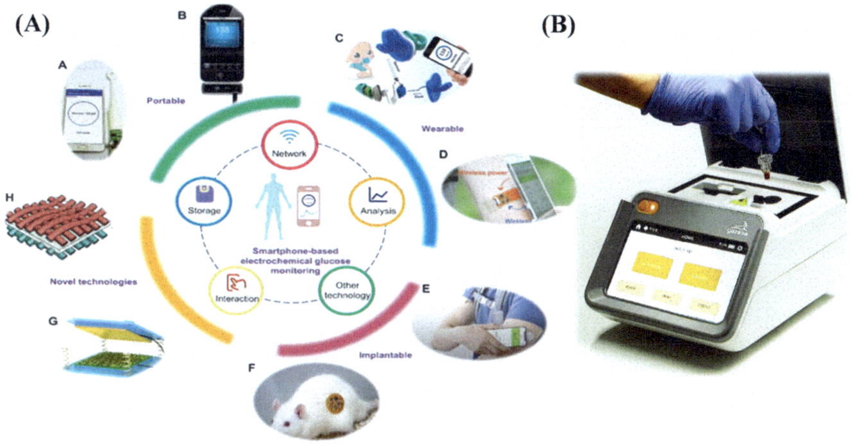

Fig. 4 (**a**) Smartphone-integrated PADs for glucose monitoring [16]. (**b**) Example of PAD for malaria [17]

and scalability in critical situations. Figure 4 shows the smartphone-integrated PADs for glucose monitoring and PAD for malaria.

PADs in Agricultural

Increasingly, PADs are acknowledged as transformative assets in the agricultural landscape, offering innovative strategies to enhance crop productivity, ensure food safety, and promote sustainable farming practices. PADs help farmers and agronomists make informed, data-driven decisions that optimize resource use while reducing environmental impact by allowing on-site, real-time soil, water, and plant health assessments [18, 19]. For example, handheld nutrient analyzers can swiftly evaluate the concentrations of critical nutrients such as nitrogen, phosphorus, and potassium, thereby guiding accurate fertilizer application and minimizing the likelihood of over-fertilization, which can contribute to soil degradation and water contamination [20]. Likewise, PADs that monitor pH and moisture levels assist farmers in maintaining ideal growing conditions, thereby enhancing crop productivity and resilience against climate fluctuations. In the area of pest and disease management, PADs are transforming early detection and intervention methodologies. Devices equipped with biosensors can detect plant pathogens, including fungi, bacteria, and viruses, from leaf or soil samples, allowing for the timely application of targeted treatments and reducing crop losses [21]. For instance, advancements in paper-based microfluidic devices have enabled the detection of specific plant diseases through observable colorimetric changes, offering a cost-effective and user-friendly solution for smallholder farmers.

PADs in Food Quality Monitoring

As they are revolutionizing the landscape of food quality monitoring systems, PADs have been emphasized for their importance in the food industry. These devices facilitate swift, on-site identification of contaminants, adulterants, and spoilage indicators, thereby protecting the safety and integrity of food products throughout the supply chain [6, 8]. A prominent application of PADs in this context is the detection of pathogens such as *Salmonella*, *Listeria*, and *E. coli*, which represent substantial risks to food safety; their prompt identification is essential for preventing outbreaks. Additionally, PADs focus on identifying chemical contaminants, including pesticides, as shown in Fig. 5, heavy metals, and allergens, which are critical to food quality monitoring. By assessing spoilage and freshness indicators in food, these devices provide essential data for retailers and consumers, thus enabling informed choices regarding food safety and quality. Moreover, their capacity to detect heavy metals, nitrates, and microbial contaminants in water sources allows farmers to take corrective actions before these pollutants compromise crop quality. They can also identify mycotoxins in stored grains and assess fruit ripeness, ensuring quality control and minimizing post-harvest losses [23, 24].

PADs in Environmental Monitoring

In environmental monitoring, PADs make possible rapid, on-site identification of pollutants across various mediums, including water, air, and soil. This capability provides immediate insights into environmental conditions, rendering them

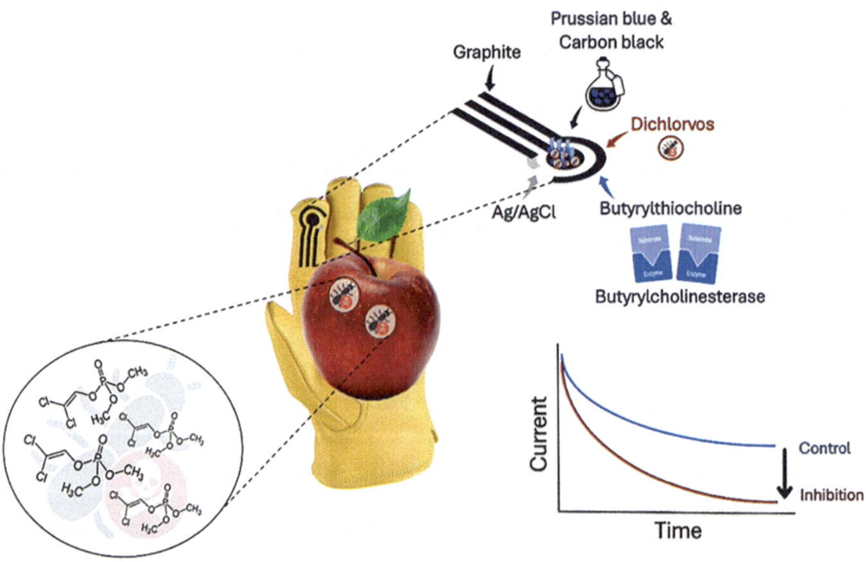

Fig. 5 Schematic representation of the detection of dichlorvos directly on apple fruit peels using an on-glove electrochemical strip, nanoengineered with the electrocatalyst [22]

particularly suitable for field applications. Such real-time data acquisition supports prompt decision-making to safeguard ecosystems and public health [25]. The key application of PADs is in water quality monitoring, allowing for rapid, on-site detection of contaminants like heavy metals, nitrates, phosphates, and microbial pathogens, which is particularly valuable for rapid assessments and early warning systems in various water sources. Beyond pollutant detection, PADs are also utilized to assess ecological parameters, including pH, temperature, and dissolved oxygen levels in natural environments. These measurements are crucial for evaluating aquatic ecosystems' health, understanding climate change's effects and guiding conservation initiatives [25, 26].

PADs for Air Quality Monitoring

PADs have reshaped the landscape of air quality assessment by enabling fast, on-site identification of various pollutants, including particulate matter (PM), volatile organic compounds (VOCs), nitrogen oxides (NO_x), sulfur dioxide (SO_2), carbon monoxide (CO), and ozone (O_3). These devices employ sophisticated electrochemical, optical, and nanomaterial-based sensors, which offer high sensitivity and real-time analytical capabilities. Electrochemical sensors detect gaseous pollutants through redox reactions, while optical sensors utilize laser-based methodologies for measuring particulate matter. In addition to their applications in urban air quality monitoring, PADs play a crucial role in enhancing occupational safety within sectors such as manufacturing and mining by identifying hazardous gases and ensuring adherence to exposure regulations. Furthermore, they contribute to evaluating indoor air quality in both residential and commercial environments by detecting pollutants like formaldehyde and airborne allergens, thereby promoting healthier living and working conditions [27].

PADs in Forensic Science, Sports, and Defense

Timely detection and analysis conducted on-site are essential in forensic science, athletics, and defense, where prompt decision-making significantly influences security, performance, and investigative outcomes. PADs have transformed these sectors by providing accuracy, dependability, and real-time data acquisition. The appraisal of evidence is sped up in forensic science by tools such as handheld spectrometers and microfluidic sensors, which enable the real-time detection of substances, including drugs, explosives, biological fluids, and gunshot residue [28]. In sports, wearable biosensors monitor physiological indicators such as lactate concentrations and hydration levels, enhancing training regimens and ensuring adherence to anti-doping standards through instantaneous substance analysis [29]. In defense applications, PADs allow to identify chemical, biological, radiological, and nuclear (CBRN) threats, utilizing portable electrochemical and optical sensors to detect explosives, nerve agents, and pathogens. Furthermore, these devices contribute to environmental surveillance in conflict areas and enable health monitoring for military personnel [30].

The Critical Need for PADs in Food Adulteration Analysis

It is essential to recognize that food adulteration involves intentionally adding inferior or harmful substances to food products, which can significantly jeopardize public health, economic stability, and consumer trust. Traditional methods for detecting adulterants, such as chromatography and spectroscopy, are often characterized by their time-consuming, costly nature and reliance on centralized laboratories, limiting their practicality for real-time monitoring throughout the food supply chain. PADs solve these issues by enabling rapid, on-site detection of contaminants, including melamine, synthetic dyes, and microbial pathogens. The capacity for real-time monitoring is vital to prevent the distribution of adulterated products, reduce the risk of foodborne illness outbreaks, and minimize economic losses. By enabling immediate analysis at critical junctures such as production facilities, markets, and border checkpoints, PADs ensure timely interventions, uphold regulatory standards, and protect the integrity of the global food system [31, 32].

Food Adulteration Has a Significant Impact on Human Health

Food adulteration is a significant public health concern, potentially causing a variety of adverse health impacts, including digestive issues like diarrhea and vomiting, allergic reactions, chronic diseases such as cancer and kidney failure, and even nutritional deficiencies due to the replacement of valuable nutrients with cheaper substitutes. Essentially, consuming adulterated food can lead to a range of mild to severe health problems depending on the type and amount of contaminant present [33] Furthermore, there are specific conditions linked to food adulteration including dizziness, joint pain, liver complications, dropsy, respiratory challenges, edema, cardiac arrest, glaucoma, and carcinogenic risks [34].

The primary sources of contamination stem from poor supply chain management and unsanitary market conditions, where food storage and handling practices are inadequately regulated. Nationally, approximately 13% of both packaged and loose food items have been identified as contaminated, with state-level discrepancies observed in Chandigarh (40%), Uttarakhand (34%), Uttar Pradesh (29%), among others. A separate investigation revealed that 70% of milk samples did not conform to established standards, with 46% showing signs of water dilution and 8% containing detergent residues [35]. Some past studies have identified hazardous chemicals and artificial colors in fruits and vegetables, which pose serious health risks. For instance, calcium carbide, commonly used for ripening mangoes and bananas, and copper sulfate, utilized to hasten fruit ripening, are toxic substances.

Oxytocin, a hormone that promotes rapid growth in various vegetables, and wax coatings applied to apples and pears for aesthetic purposes are also harmful. Furthermore, toxic substances such as metallic lead enhance the color of bitter gourds and leafy greens. The excessive use of pesticides and herbicides compounds these risks. Consumption of these chemically treated fruits and vegetables can lead to severe health issues, including digestive disorders, liver damage, and ocular problems, with children particularly vulnerable to vomiting, diarrhea, and potential kidney failure. The neurotoxic effects of oxytocin also raise concerns regarding brain health [36, 37].

The 2008 melamine crisis in China is a well-known example of food contamination where infant formula was deliberately adulterated with melamine, leading to widespread kidney stone formation and renal failure in over 300,000 infants, tragically resulting in the deaths of six babies; this incident alerted global attention to the great harm food safety practices can cause [38]. Melamine, a nitrogen-rich compound utilized to enhance protein content in milk artificially, is harmful to renal function and may result in long-term health issues for those who survive. This event underscored the severe repercussions of food adulteration, particularly for vulnerable groups such as children.

Another critical health concern stems from the adulteration of edible oils and fats. The Spanish toxic oil syndrome outbreak of 1981 is a significant example of food poisoning where industrial-grade rapeseed oil, contaminated with aniline and intended for non-food use, was falsely sold as olive oil, causing a large-scale epidemic affecting over 20,000 people with an estimated 1000 deaths due to the toxic oil consumption; this incident is often cited as a prime example of the dangers of food adulteration and improper labeling [39]. Affected individuals experienced severe symptoms, including respiratory issues, neurological impairment, and chronic autoimmune conditions, with many suffering from long-term disabilities. Additionally, the incorporation of Sudan dyes in spices and sauces, known carcinogens, has been associated with heightened risks of liver and bladder cancers. Research conducted in India revealed that about 60% of chili powder samples tested positive for Sudan dyes, representing a significant public health concern [40].

Microbial contamination resulting from food adulteration represents a significant public health issue. A notable instance is the 2011 outbreak of *E. coli* O104:H4 in Germany, which was associated with contaminated fenugreek sprouts. This outbreak led to more than 4000 reported illnesses, including 855 cases of hemolytic uremic syndrome (HUS), a serious kidney disorder, and resulted in 50 fatalities [41]. The strain involved was particularly virulent, producing Shiga toxins that inflicted extensive damage on blood vessels and various organs. Such food safety incidents, like outbreaks of foodborne illnesses linked to contaminated produce or ready-to-eat foods, emphasize the importance of implementing strict procedures to stop food contamination with harmful microbes, particularly when dealing with fresh fruits and vegetables and foods that require no further cooking before consumption.

Moreover, chronic exposure to low concentrations of adulterants can result in enduring health complications. For example, the contamination of spices, seafood, or rice with heavy metals such as lead and cadmium can lead to neurological impairments, developmental issues in children, and renal dysfunction. Recent research revealed that turmeric samples collected from India, Pakistan, Sri Lanka, and Nepal were tainted with lead chromate (more than 500 times higher than the regulated level), a hazardous substance used to enhance color, raising blood lead levels among consumers [42]. Additionally, adulterating milk with substances like hydrogen peroxide or formalin, used as preservatives, can indeed lead to gastrointestinal issues, respiratory complications, and potentially carcinogenic effects if consumed over a prolonged period due to their toxic nature, essentially harming the body through direct irritation of the digestive tract and potential long-term cellular damage. The ramifications of food adulteration extend beyond physical health issues, encompassing psychological and social dimensions. Individuals affected by adulteration scandals frequently experience diminished trust in food systems, which can lead to anxiety and alterations in dietary habits. The 2013 "horsemeat scandal" in Europe, where horse meat was falsely labeled as beef, serves as a prime example of food fraud that caused significant public outcry and eroded consumer trust, even though there were no immediate health concerns linked to the horse meat itself; this incident highlighted the vulnerabilities within the food supply chain and the need for stricter quality control measures [43].

Food Adulteration Has a Significant Impact on the Global Economy

Addressing food adulteration is not only a public health imperative but also an economic necessity. In fact, food adulteration has far-reaching economic consequences, affecting industries, governments, and consumers alike. The deliberate addition of inferior or harmful substances to food products undermines market integrity, leading to significant financial losses for legitimate producers and suppliers. One notable example is the 2008 melamine contamination scandal in China, as previously reported, where milk and infant formula were tainted with melamine; it stands as a significant example of a food safety crisis, causing widespread illness among infants, several deaths, and a major loss of consumer trust in Chinese dairy products globally, leading to massive product recalls and substantial economic damage to the industry, with estimates of over $3 billion in lost sales and compensation claims [38, 44]. Similarly, the 2013 horsemeat scandal in Europe, where horsemeat was fraudulently labeled as beef, caused a significant decline in consumer confidence, leading to a 43% drop in frozen burger sales in the UK alone and costing the European food industry approximately €1 billion in lost revenue [43].

Adulterated foods can cause immediate and future health issues due to toxic additives. Increased medical costs and loss of workforce productivity impact the economy. Erosion of trust in food quality and safety leads to decreased market confidence. Countries involved in adulteration disputes often face export bans or strict import restrictions that reduce their market access and lead to lost trade opportunities. For example, India has experienced multiple bans on its seafood exports to the European Union due to antibiotic residues and other pollutants, significantly affecting numerous fishermen's and exporters' livelihoods [45]. Likewise, adulterating honey with sugar syrups can prompt trade barriers and economic setbacks for beekeepers in countries where honey exports constitute a vital income source. Furthermore, the economic ramifications extend to healthcare systems, as foodborne diseases stemming from adulterated products escalate medical costs and diminish workforce efficiency. As previously reported, the 2011 *E. coli* outbreak in Germany, associated with contaminated fenugreek sprouts, led to an estimated economic repercussion surpassing €1 billion due to healthcare costs, lost productivity, and trade interruptions [41].

Example Practices in Food Adulteration

Water Adulteration in Drinks

One prevalent method of adulteration involves increasing the water content of beverages such as juices and milk. While water itself is not harmful, its excessive addition dilutes the nutritional value of the drinks and may lead to microbial growth if untreated or contaminated water is used [46]. Regulatory agencies, such as the Food and Drug Administration (FDA) and the European Food Safety Authority (EFSA), mandate strict guidelines for water quality and the permissible limits for addition in food and beverages.

Color Additives and Texture Enhancers

Incorporating synthetic colorants and various additives to enhance food items' visual appeal and texture is prevalent in the food industry. These substances may create a false impression regarding the quality and freshness of the products [47]. A typical example is the addition of Sudan dyes in chili powder and sauces, industrial colorants prohibited for food applications. The ingestion of these colorants poses potential health risks, including the possibility of carcinogenic effects.

Adulteration of Edible Oils

The practice of combining low-quality oils with high-quality oils is a misleading strategy designed to enhance profit margins. This may include mixing premium oils like olive oil with less expensive options like palm or soybean oil [48]. The presence of adulterated oils not only deceives consumers but may also lead to health complications.

Weight Enhancers in Cereals

Incorporating non-consumable substances, such as stones or husks, into cereals to augment their weight constitutes an unethical practice that jeopardizes food safety. This form of adulteration poses potential physical risks to consumers and diminishes the nutritional integrity of cereals [49].

Non-edible Waxes on Fruits

Applying non-edible waxes to fruits improves their visual appeal and prolongs their shelf life. Nonetheless, these waxes may present health hazards due to the potential presence of harmful chemicals. In contrast, edible waxes like beeswax and carnauba wax are sanctioned for use on fruits, while non-edible waxes are subject to stringent regulations [50].

Use of Ripening Agents

Artificial ripening agents, such as calcium carbide, are frequently employed to accelerate the ripening of fruits. Nonetheless, applying these substances may result in the buildup of harmful residues, which can present significant health risks [51]. Ethylene gas, a naturally occurring plant hormone, is a safer option for fruit ripening and has received approval from numerous food safety regulatory bodies.

Pesticide Residue

Over-application of pesticides in agricultural practices may result in food products containing residues that surpass established safety thresholds. Prolonged exposure to elevated levels of these pesticide residues has the potential to cause significant health problems, including the development of cancer and disruptions to the endocrine system. Implementing integrated pest management (IPM) strategies and adopting organic farming techniques can diminish dependence on synthetic pesticides [52].

Adulteration of Alcoholic Beverages

The incorporation of methanol or industrial alcohol into alcoholic drinks represents a perilous type of adulteration that may result in fatal outcomes. Methanol is extremely toxic, with the potential to induce blindness or death, even when ingested in minimal quantities [53].

Tackling the Issue of Food Adulteration with PADs

When it comes to analyzing food authenticity and verifying traceability using current analytical methods, the significant challenges include high cost per sample due to expensive equipment and reagents, the need for skilled personnel to operate and interpret results, lengthy analysis times often involving complex extraction and incubation steps, limited portability for on-site testing, and intricate sample preparation procedures that may require hazardous solvents and potentially damage the sample, making it challenging to implement widespread and efficient food authenticity checks across the supply chain [54, 55]. Therefore, considerable research has been conducted in this field, resulting in the commercialization of portable analytical devices (PADs). The potential of PADs to address food adulteration is significant; however, a comprehensive strategy that integrates policy reforms, technological advancements, regulatory enforcement, and public involvement is essential to expedite progress.

Governments and international bodies must focus on creating and executing policies that deter food adulteration and foster ethical practices within the food sector. Public awareness campaigns are crucial to informing consumers about the risks of adulterated food, allowing them to recognize potentially unsafe products and make informed purchasing decisions, thereby protecting their health and promoting a safer food supply. For instance, the "Detect Adulteration with Rapid Test (DART)" program established by the Food Safety and Standards Authority of India's (FSSAI) empower consumers by providing straightforward, at-home techniques for detecting prevalent food adulterants [56]. Furthermore, policies should encourage high-quality production by offering financial incentives, such as subsidies or grants, to farmers and food producers who comply with safety regulations. By reducing the financial burden associated with high-quality food through subsidies or tax incentives, the economic motivations for adulteration can be alleviated.

Establishing a strong regulatory framework is essential to combat food adulteration, requiring governments to actively enforce strict quality control standards across all stages of food production, from farm to table, to protect consumer health and ensure food safety. For example, the European Union's Rapid Alert System for Food and Feed (RASFF) allows for prompt responses to adulterated products by enhancing information exchange among member states [57]. Regular inspections, mandatory certifications, and severe penalties for violations can serve as effective deterrents against fraudulent activities. Moreover, international cooperation is essential to align food safety standards and combat the cross-border trade of adulterated products. Technological innovations, especially in PADs, present a formidable means for immediately detecting food adulterants. This topic will be elaborated upon in the subsequent section.

Importance of PADs for Tracking Food Adulteration

PADs present a unique opportunity to address the critical challenge of food adulteration by allowing rapid, on-site detection of contaminants within minutes, thereby obviating the necessity for lengthy laboratory testing. In contrast to qualitative approaches, PADs can yield quantitative data, allowing for accurate assessment of adulterant concentrations, essential for adherence to regulatory standards and informed decision-making. Their portable nature permits utilization in various environments, including customs checkpoints, manufacturing sites, and import/export locations, enabling continuous monitoring throughout the supply chain. PADs are designed to be simple and easy to operate, so even people without extensive technical knowledge or training can use them effectively, allowing a wide range of individuals to access their analytical capabilities. Furthermore, their ability to operate independently of external power sources ensures dependable functionality in remote or resource-constrained settings, making them vital for maintaining food safety and public health. Incorporating PADs into the food supply chain, from agricultural production to retail distribution, can significantly improve surveillance, lessen dependence on centralized laboratory facilities, and facilitate prompt interventions.

Current Design of PADs

As briefly introduced earlier, PADs used for food authenticity verification can be categorized into three main types based on the level of information they provide: qualitative PADs (simply indicating the presence or absence of a specific component), semi-quantitative PADs (providing a rough estimate of the amount of an element), and quantitative PADs (giving precise measurements of the analyte concentration).

Qualitative PADs (LFA and some Colorimetric Sensor)

Qualitative PADs, commonly called lateral flow assays (LFA), have gained significant attention for their application in detecting various adulterants in food products [58]. Their popularity stems from their straightforward design, affordability, and the rapidity of the results they provide. These assays yield qualitative outcomes, indicating either a negative (−) or positive (+) result, which is determined by a color change in the reactive zone or the emergence of distinct colored lines, typically known as the "test line." LFAs are structured as paper-based strips that utilize capillary action to move liquid samples through designated functional zones, where specific biorecognition elements interact with the target analytes. A standard LFA comprises several essential components: a sample pad, a conjugate pad, a nitrocellulose membrane, and an absorbent pad. The sample pad serves as the initial point for the liquid sample, which travels to the conjugate pad that houses labeled detection reagents, such as gold nanoparticles or fluorescent dyes, which are conjugated with antibodies specific to the adulterant of interest. As the sample progresses through the nitrocellulose membrane, it encounters test and control lines. The test line features immobilized capture molecules that specifically bind to the target analyte, while the control line verifies the proper functioning of the assay. The absorbent pad plays a crucial role by absorbing excess fluid, maintaining a continuous flow and preventing backflow [59]. Figure 6 depicts a schematic illustration of an LFA showing its different parts and how results can be visualized.

The operational principle of LFAs is fundamentally based on the specific binding interactions between the target analyte and the biorecognition elements. When a food sample containing an adulterant is applied to the sample pad, the labeled detection reagents in the conjugate pad become rehydrated. As the sample progresses laterally, these labeled reagents attach to the target analyte, forming a complex. This complex subsequently interacts with the capture molecules at the test line, generating a visible signal, such as a colored line, indicating the adulterant's presence (positive test result). The control line, which captures any unbound labeled reagents, validates the assay. The entire procedure generally requires 5–15 min, rendering LFAs suitable for rapid, on-site testing. Commonly utilized materials in LFAs include nitrocellulose membranes, which offer a stable matrix for the immobilization of capture molecules, and gold nanoparticles or latex beads, which act as signal-generating labels. Antibodies are predominantly employed as biorecognition elements due to their exceptional specificity and affinity for target analytes. For instance, Kuo-Hui Wu and colleagues have developed an LFA designed to detect melamine in milk, utilizing a nitrocellulose membrane, silver nanoparticles (AgNPs), and anti-MEL monoclonal antibodies (mAb) [61]. Figure 7 shows the LFA for melamine detection in milk. Similarly, Tong Bu and associates have created an LFA capable of identifying *Salmonella* bacteria in food samples, employing ampicillin-coated magnetic nanoparticles as a substitute for traditional antibodies [62]. Several qualitative PADs have been developed and are commercially available for monitoring various adulterants, including meat, salmonellas, milk, nuts, etc.

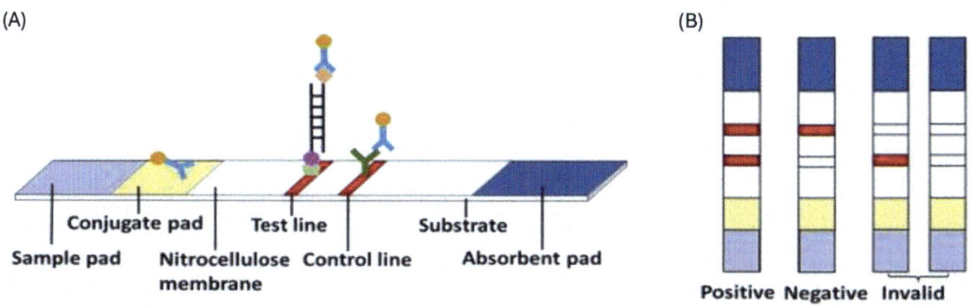

Fig. 6 Schematic illustration of a LFA showing its different parts (**a**) and how results can be visualized (**b**) [60]

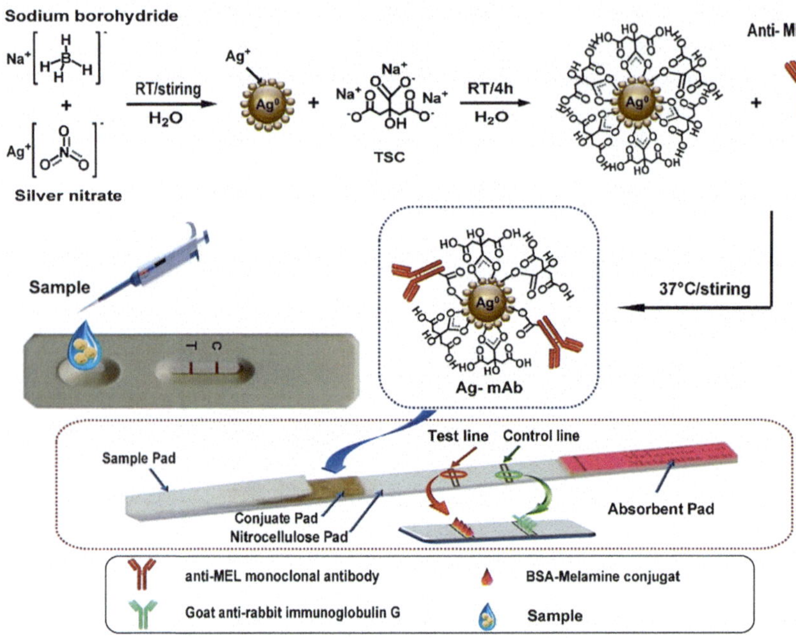

Fig. 7 A LFA for melamine detection in milk. Adapted with permission [61]

Companies like Meizheng provide commercially available qualitative PADs for monitoring food adulteration, offering a diverse array of LFAs that can detect adulterants across various food products, including honey, milk, meat, nuts, grains, and poultry. Unisensor also offers a broad selection of LFAs for detecting aflatoxins, melamine, beta-lactams, and other contaminants in food. Creative Diagnostics and other firms contribute to this market as well.

Semi-Quantitative PADs

Semi-quantitative PADs are an advanced variant of LFAs that approximate the target adulterant's concentration in food samples, bridging the gap between qualitative detection and precise quantitative analysis. These devices are designed to offer a more informative assessment of adulterant levels, which is crucial for determining the severity of contamination and ensuring compliance with regulatory thresholds. Like traditional LFAs, semi-quantitative PADs typically consist of a sample pad, conjugate pad, nitrocellulose membrane, and absorbent pad. However, they incorporate additional features to enable semi-quantitative analysis, such as multiple test lines with varying capture molecule concentrations or integrated signal intensity readers like the camera [62]. Figure 8 shows a semi-quantitative LFA that allows qualitative and quantitative detection of *Salmonella* in food samples; a camera is

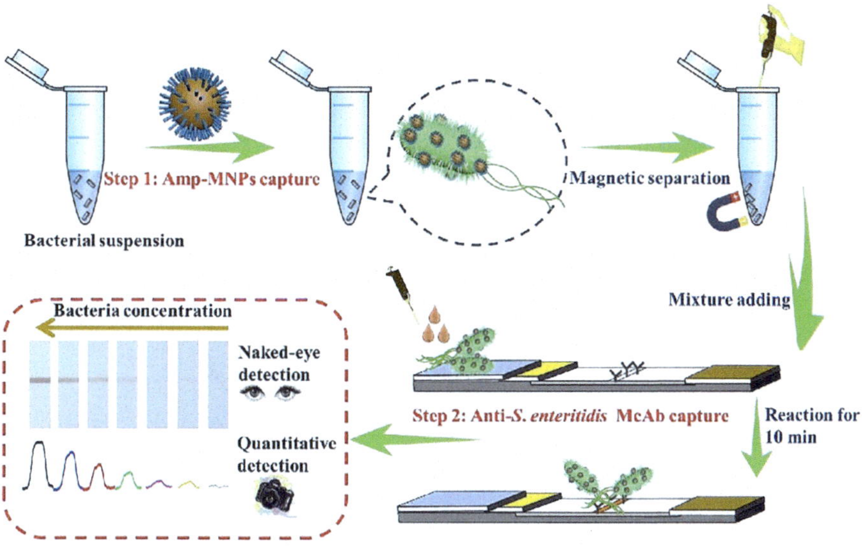

Fig. 8 A semi-quantitative LFA that allows qualitative and quantitative detection of *Salmonella* in food samples; a camera is required for quantitative detection, adapted with permission [62]

required for quantitative detection. For example, a device designed to detect melamine in milk might include several test lines, each corresponding to different concentration ranges, allowing users to estimate the adulterant level based on the number of lines that produce a visible signal.

The working mechanism of semi-quantitative PADs builds on the principles of LFAs but enhances their functionality through innovative design elements. When a food sample is applied to the device, it flows through the conjugate pad, where labeled detection reagents (e.g., gold nanoparticles or fluorescent dyes) bind to the target adulterant. As the sample migrates across the nitrocellulose membrane, it interacts with multiple test lines, each containing a different density of capture molecules. The intensity of the signal generated at each test line correlates with the adulterant concentration, providing a visual or digital readout that indicates the approximate level of contamination. For instance, a stronger signal at a higher concentration test line suggests a more severe adulteration issue. Some semi-quantitative PADs also integrate smartphone-based readers or handheld scanners to measure signal intensity more accurately, enhancing their utility. Materials commonly used in semi-quantitative PADs include nitrocellulose membranes for capturing molecules, gold nanoparticles or quantum dots as signal-generating labels, and antibodies or aptamers as biorecognition elements. More advanced versions may incorporate nanomaterials like graphene oxide or carbon nanotubes to improve sensitivity and enable the simultaneous detection of multiple adulterants. For example, a semi-quantitative PAD designed to detect pesticide residues in fruits and vegetables might use enzyme-linked antibodies to produce a colorimetric signal, with the intensity of the color indicating the pesticide concentration. These devices are particularly effective for monitoring contaminants such as heavy metals, allergens, and microbial pathogens, as even rough quantification can provide valuable insights into food safety issues [58, 59].

Semi-quantitative PADs are highly versatile and can be employed in various settings, including production facilities, customs checkpoints, and retail environments. Their ability to provide quick, on-site evaluations of adulterant levels makes them crucial for ensuring food safety, preventing economic fraud, and protecting public health. By combining the ease and portability of lateral flow assays (LFAs) with enhanced analytical capabilities, semi-quantitative PADs represent a significant advancement in food adulteration monitoring, offering a practical solution for stakeholders across the food supply chain.

Quantitative PADs

Quantitative PADs demonstrate a notable progression in food safety technology, facilitating the accurate identification and quantification of adulterant levels within food samples, even at minimal concentrations. The operational mechanism of quantitative PADs encompasses a sequence of meticulously coordinated steps designed to guarantee both accuracy and precision. Upon introducing the sample to the biosensor, the target adulterant interacts with the immobilized biorecognition element, instigating a biochemical reaction or a physical transformation. This transformation is subsequently detected by a transducer, which may be optical, electrochemical, or piezoelectric, contingent upon the specific design employed. For instance, optical transducers assess variations in light absorption, fluorescence, or surface plasmon resonance, whereas electrochemical transducers monitor changes in electrical characteristics such as current, voltage, or impedance. The transducer amplifies the resultant signal and translates it into an electrical output, which is then analyzed by an electronic component. This component utilizes a pre-established calibration curve to convert the electrical signal into quantitative value, thereby accurately measuring adulterant concentration [63, 64].

Materials commonly utilized in quantitative PADs include high-affinity biorecognition elements such as monoclonal antibodies or molecularly imprinted polymers, which ensure specificity toward the target adulterant. Transducers are frequently constructed from advanced materials like gold nanoparticles, carbon nanotubes, or graphene, enhancing sensitivity and signal stability. For example, electrochemical PADs designed for detecting heavy metals, such as lead or cadmium, in food samples may incorporate screen-printed electrodes coated with conductive nanomaterials to improve detection thresholds. Likewise, optical PADs that identify microbial pathogens may amplify the signal with quantum dots or fluorescent dyes. Integrating these materials with microfluidic designs allows for the miniaturization and portability of quantitative PADs without sacrificing performance [64].

Problems with Current Analytical Devices

Several recent reports have highlighted the limitations, accuracy, and sensitivity of current PADs utilized for monitoring food adulteration. One concern is the employment of non-biodegradable polymers in the fabrication and packaging of LFAs, which raises significant environmental waste issues associated with these devices [65–67]. Additionally, antibodies' stability and production costs have prompted researchers to explore alternative solutions.

Despite notable progress in the development of PADs for both qualitative and quantitative analysis, several obstacles impede their broader implementation and efficacy. These challenges encompass technical issues, such as biofouling, reliance on specific reagents, and extensive concerns, including environmental impact and limited access in low- and middle-income countries (LMICs) [68]. It is essential to address these challenges to facilitate the ongoing advancement and refinement of PADs for diverse applications.

Surface Limitations and Biofouling in Electrochemical PADs

Electrochemical-based sensors, particularly those utilizing flat electrodes, face inherent limitations due to their flat design, which restricts the available surface area for analyte capture. This constraint can affect the sensitivity and detection thresholds, complicating the identification of biomolecules present in low concentrations. Furthermore, biofouling, characterized by the accumulation of proteins, cells, or other biological entities on the electrode surface, can lead to a decline in sensor efficacy over time. A pertinent example is the glucose sensors employed in diabetes management, which frequently experience performance degradation due to protein adsorption, resulting in diminished signal precision and sensor lifespan. In response to these challenges, researchers are investigating the potential of nanostructured electrodes and anti-fouling coatings [68]; however, issues related to scalability and cost-effectiveness continue to pose significant obstacles.

Limited Multiplexing Capability: Single Analyte Detection

Most paper-based analytical devices are engineered to detect a single analyte, which restricts their ability to measure only one biomarker or contaminant at any time. This constraint poses significant challenges in intricate sample matrices, such as those encountered in food adulteration assessments or disease diagnostics, where the simultaneous evaluation of multiple analytes is essential. For example, LFAs, commonly employed in COVID-19 and pregnancy testing, exhibit high specificity but are inherently limited in their multiplexing capabilities [69].

Reagent Dependency: Challenges in Acquisition and Storage

Numerous PADs depend on liquid reagents for conducting chemical reactions, including enzymatic and colorimetric assays. Adequate storage and handling of these reagents are critical, as they are susceptible to degradation when exposed to extreme temperatures or prolonged air contact. In settings with limited resources, ensuring a consistent cold chain for reagent storage presents a considerable obstacle, thereby restricting the application of some PADs in LMICs. For instance, nitrate-detecting PADs utilized for monitoring water quality rely on colorimetric reagents that deteriorate over time, which can compromise the accuracy of the tests.

Short Shelf Life of Enzyme- and Antibody-Based PADs

Most current PADs utilize biomolecular recognition elements, including enzymes and antibodies, to achieve high specificity in detecting analytes. Nevertheless, these biological entities exhibit limited stability and are prone to degradation over time, which compromises the reliability and shelf life of PADs. For example, glucose sensors that rely on enzymes experience a decline in performance due to enzyme denaturation, necessitating regular recalibration or replacement. Similarly, antibody-based PADs, particularly those employed for pathogen detection in food safety, face significant stability challenges, complicating long-term storage and transportation [59].

Environmental Concerns: Non-Eco-Friendly Materials and Waste Generation

The increasing prevalence of PADs has sparked significant apprehension regarding their material sustainability and ecological footprint. Many PADs are constructed from plastic substrates, synthetic polymers, and heavy metals, contributing to the generation of non-biodegradable waste. For example, lateral flow tests produce millions of single-use plastic strips yearly, culminating in substantial biomedical waste. The pursuit of environmentally friendly PADs utilizing biodegradable materials, such as cellulose-based paper and bioplastics derived from plants, is an emerging focus; however, large-scale adoption faces obstacles related to cost and regulatory challenges [65].

Limited Accessibility in LMICs

Although PADs hold promises for decentralized testing, their development and commercialization predominantly occur in high-income nations, thereby restricting access in LMICs where they are critically needed. High production costs, intellectual property limitations, and insufficient distribution networks impede their broad availability in resource-limited environments. For instance, rapid diagnostic tests for diseases like malaria and HIV remain prohibitively expensive in numerous African and South Asian nations despite their essential role in public health initiatives. Initiatives to create affordable, open-source PAD technologies, including 3D-printed biosensors and paper-based microfluidics, seek to address this disparity; however, ongoing investment and supportive policies are crucial for scaling production and ensuring fair distribution.

Design Concept for Next-Generation PADs

The constraints associated with contemporary PADs have catalyzed significant research efforts to develop novel materials, sensing technologies, and design methodologies to improve their efficacy, durability, and accessibility. Specifically, research efforts focus on enhancing electrode surface characteristics, expanding multi-analyte detection capabilities, reducing reagent dependence, developing more stable biomolecular recognition components, employing environmentally friendly materials, and making these devices available in LMICs. The subsequent sections will provide some of the most promising strategies to address these challenges.

Advancing Electrochemical PADs: Nanostructured and Anti-Fouling Electrode Surfaces

In response to the challenges posed by limited surface area and biofouling in electrochemical PADs, researchers are innovating with nanostructured electrode materials. By incorporating nanomaterials like gold nanoparticles (AuNPs), graphene, and carbon nanotubes (CNTs) onto electrode surfaces, researchers can significantly increase the active area available for interaction, leading to improved sensitivity and signal stability in electrochemical detection due to their high surface-to-volume ratio and excellent electrical conductivity [13, 70–74]. Figure 9 shows the graphical representation of the improved sensitivity of LFA by using magneto-Au-nanomaterials. Research has demonstrated that electrodes modified with graphene can significantly accelerate electron transfer rates and mitigate biofouling by decreasing nonspecific adsorption of proteins and other biomolecules. Furthermore, exploring anti-fouling coatings, such as polyethylene glycol (PEG)-based films, is underway to sustain sensor performance over prolonged periods [75].

Multi-Analyte Detection: The Convergence of Microfluidics and Sensor Arrays

To address the limitations associated with single-analyte detection in PADs, researchers are exploring integrating microfluidic systems with sensor arrays capable of concurrently identifying multiple analytes. Lab-on-a-chip (LOC) technologies facilitate accurate fluid manipulation, enabling the establishment of various detection zones within a single PAD framework [8, 14, 76, 77]. For example, recent

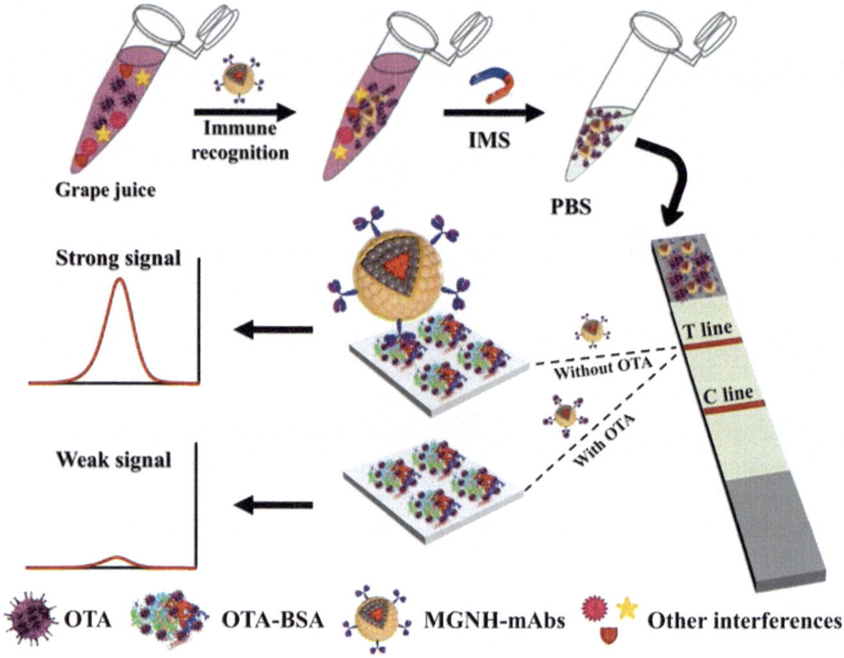

Fig. 9 Improving the sensitivity of LFA by using magneto-gold nanomaterials. (Adapted with permission [70])

developments in multiplexed electrochemical biosensors have made it possible to simultaneously detect various foodborne pathogens, thereby enhancing comprehensive metabolic assessment.

Strategies for Reagent-Free and Long-Term Stable Detection

Minimizing reliance on reagents is essential for enhancing the practicality of PADs in real-world applications. To this end, research focuses on dry-reagent PADs that preserve reagents in stable formats, such as lyophilized powders or hydrogel matrices. Furthermore, developing reagent-free detection methods, including sensors based on surface-enhanced Raman spectroscopy (SERS), aims to eliminate the necessity for chemical reagents [78–80].

Replacing Enzymes and Antibodies with Synthetic and Aptamer-Based Recognition Elements

The limited shelf life of enzyme- and antibody-based paper analytical devices has prompted investigations into synthetic alternatives, including molecularly imprinted polymers (MIPs) and aptamers. MIPs serve as artificial receptors that replicate biological binding sites, offering enhanced specificity while exhibiting excellent stability and resistance to environmental factors. Similarly, aptamers are short, single-stranded DNA or RNA sequences that present another viable substitute for antibodies. They are characterized by remarkable stability, ease of synthesis, and reduced susceptibility to denaturation. For example, sensors utilizing aptamers for food safety have shown significant sensitivity in identifying aflatoxins in contaminated grains, thus providing a reliable and cost-effective alternative to traditional antibody-based assays [81–83].

Eco-Friendly and Biodegradable Materials for Sustainable PADs

In response to the environmental challenges posed by disposable PADs, researchers are increasingly exploring biodegradable and recyclable materials for constructing these devices. Paper-based microfluidic devices (μPADs) emerge as a sustainable and economically viable option. Recent developments in cellulose-based nanocomposites have enhanced the mechanical and chemical durability of paper-based PADs, rendering them more suitable for practical applications in the field. Furthermore, the creation of biodegradable plastics from polylactic acid (PLA) and other plant-derived polymers is being pursued to replace conventional plastic substrates. These initiatives aim to reduce environmental waste while maintaining the operational efficacy of the devices.

Enhancing Accessibility of PADs in LMICs

To ensure the availability of PADs in LMICs, it is essential to tackle issues related to cost, production scalability, and regulatory hurdles. Various initiatives focus on creating affordable PADs through straightforward manufacturing methods, including wax, inkjet, and screen printing. Exploring open-source designs and decentralized production methods aims to enable local fabrication of PADs in settings with limited resources. Additionally, partnerships among academic institutions, nongovernmental organizations (NGOs), and policymakers are instrumental in developing regulatory frameworks that support the implementation of PADs in LMICs.

Challenges in Developing PADs and Future Perspectives

Despite the significant advancements in PADs for qualitative and quantitative detection, several limitations hinder their widespread adoption and effectiveness [3, 21, 84]. These challenges range from technical constraints, such as biofouling and reagent dependency, to broader issues like environmental concerns and limited accessibility in LMICs. Addressing these limitations is crucial for developing and optimizing PADs for various applications.

Many PADs, particularly electrochemical-based sensors, rely on flat electrodes for signal transduction. However, these flat electrodes have a limited surface area, restricting their ability to capture analytes efficiently. This limitation reduces sensitivity and detection limits, making it challenging to detect low-concentration biomolecules. Additionally, biofouling, where proteins, cells, or other biological materials accumulate on the electrode surface, can degrade sensor performance over time. For example, glucose sensors used in diabetic monitoring often suffer from biofouling due to protein adsorption, which decreases signal accuracy and longevity. Although there is ongoing research to explore nanostructured electrodes and anti-fouling coatings to overcome this challenge, scalability and cost remain concerns. Many PADs employ biomolecular recognition elements, such as enzymes and antibodies, for high specificity in analyte detection. However, these biological molecules have limited stability and tend to degrade over time, reducing the reliability and shelf life of PADs. For instance, enzyme-based glucose sensors lose efficiency due to enzyme denaturation, requiring frequent recalibration or replacement. Antibody-based PADs, such as those used for pathogen detection in food safety, also suffer from stability issues, making long-term storage and transport challenging. Researchers are investigating synthetic alternatives, such as molecularly imprinted polymers (MIPs) and aptamers, to replace fragile biomolecules and enhance durability.

Most PADs are designed for single-analyte detection, meaning they can only measure one biomarker or contaminant at a time. This limitation is particularly problematic in complex sample matrices, such as food adulteration analysis or disease diagnostics, where multiple analytes must be assessed simultaneously. For instance, lateral flow assays (LFAs), widely used for COVID-19 and pregnancy tests, are highly specific but lack multiplexing capability. Researchers are working on integrating microfluidic designs and multi-electrode systems to enable multi-analyte detection, yet these advancements often increase production costs and complexity. Many PADs rely on liquid reagents for chemical reactions, such as enzymatic or colorimetric assays. These reagents often require careful storage and handling, as they can degrade under extreme temperatures or prolonged exposure to air. In resource-limited settings, maintaining a cold chain for reagent storage is a significant challenge, limiting the usability of PADs in LMICs. For example, nitrate-detecting PADs for water quality monitoring depend on colorimetric reagents that degrade over time, affecting test accuracy. Alternative reagent-free detection methods, such as electrochemical and surface-enhanced Raman spectroscopy (SERS)-based PADs, are being explored but are still in the developmental phase. The widespread adoption of PADs has raised concerns regarding material sustainability and environmental impact. Many PADs are fabricated using plastic substrates, synthetic polymers, and heavy metals, contributing to non-biodegradable waste. For instance, lateral flow tests generate millions of single-use plastic-based strips annually, accumulating significant biomedical waste. Developing eco-friendly PADs using biodegradable materials, such as cellulose-based paper and plant-derived bioplastics, is a growing area of interest. Still, large-scale implementation remains a challenge due to cost and regulatory barriers.

Despite their potential for decentralized testing, PADs are primarily developed and commercialized in high-income countries, limiting access to LMICs where they are most needed. High production costs, intellectual property restrictions, and inadequate distribution networks hinder their widespread availability in resource-poor settings. For example, rapid diagnostic tests for malaria and HIV remain expensive in many African and South Asian countries despite their necessity for public health interventions. Efforts to develop low-cost, open-source PAD technologies, such as 3D-printed biosensors and paper-based microfluidics, aim to bridge this gap. Still, sustained investment and policy support are needed to scale production and ensure equitable distribution.

Conclusion

Portable analytical devices (PADs) offer a promising solution for efficiently detecting food adulterants. These devices can be further developed by leveraging techniques, such as optical and electrochemical sensing, to overcome current limitations, including limited surface area, low sensitivity, and single-analyte detection. Ongoing research is focused on enhancing sensitivity by incorporating nanomaterials and improving stability using aptamers and molecularly imprinted polymers as biorecognition elements. Additionally, the integration of microfluidic chips is being explored to enable multiplex detection, allowing simultaneous identification of multiple biomarkers. Furthermore, the sustainability of next-generation PADs is a key research priority, addressing environmental concerns associated with electronic waste and climate change. While these technological advancements are crucial, they must be complemented by strong regulatory frameworks and policies to ensure the effective and standardized implementation of PADs in food safety monitoring.

References

1. Li XJ, Yang C (2021) Analytical chemistry for infectious disease detection and prevention. Anal Bioanal Chem 413(18):4561–4562. https://doi.org/10.1007/S00216-021-03441-1
2. Bharti A, Jain U, Chauhan N (2024) Progressive analytical techniques utilized for the detection of contaminants attributed to food safety and security. Talanta Open 10:100368. https://doi.org/10.1016/J.TALO.2024.100368
3. Tonyushkina K, Nichols JH (2009) Glucose meters: a review of technical challenges to obtaining accurate results. J Diabetes Sci Technol 3(4):971–980. https://doi.org/10.1177/193229680900300446
4. Newman JD, Turner APF (2005) Home blood glucose biosensors: a commercial perspective. Biosens Bioelectron 20(12):2435–2453. https://doi.org/10.1016/J.BIOS.2004.11.012

5. Li X, Luo P, Tang S, Beier RC, Wu X, Yang L, Li Y, Xiao X (2011) Development of an immunochromatographic strip test for rapid detection of melamine in raw milk, milk products and animal feed. J Agric Food Chem 59(11):6064–6070. https://doi.org/10.1021/JF2008327
6. Kapse S, Kausley S, Rai B (2022) Portable food diagnostic devices and methods: a review. J Food Process Eng 45(11):e14159. https://doi.org/10.1111/JFPE.14159
7. Gopal J, Muthu M (2024) Handheld portable analytics for food fraud detection, the evolution of next-generation smartphone-based food sensors: the journey, the milestones, the challenges debarring the destination. TrAC Trends Anal Chem 171:117504. https://doi.org/10.1016/J.TRAC.2023.117504
8. Alahmad W, Cetinkaya A, Kaya SI, Ozkan SA (2024) Innovative and cutting-edge approaches in microfluidic paper-based analytical devices for detection of food adulteration. TrAC Trends Anal Chem 181:118012. https://doi.org/10.1016/J.TRAC.2024.118012
9. Zarei M (2017) Portable biosensing devices for point-of-care diagnostics: recent developments and applications. TrAC Trends Anal Chem 91:26–41. https://doi.org/10.1016/J.TRAC.2017.04.001
10. Park S, Zhang Y, Lin S, Wang TH, Yang S (2011) Advances in microfluidic PCR for point-of-care infectious disease diagnostics. Biotechnol Adv 29(6):830–839. https://doi.org/10.1016/J.BIOTECHADV.2011.06.017
11. Promphet N, Thanawattano C, Buekban C, Laochai T, Lormaneenopparat P, Sukmas W, Rattanawaleedirojn P, Puthongkham P, Potiyaraj P, Leewattanakit W, Rodthongkum N (2024) Smartphone based wearable sweat glucose sensing device correlated with machine learning for real-time diabetes screening. Anal Chim Acta 1312:342761. https://doi.org/10.1016/J.ACA.2024.342761
12. Sher M, Zhuang R, Demirci U, Asghar W (2017) Paper-based analytical devices for clinical diagnosis: recent advances in the fabrication techniques and sensing mechanisms. Expert Rev Mol Diagn 17(4):351–366. https://doi.org/10.1080/14737159.2017.1285228
13. Naresh V, Lee N (2021) A review on biosensors and recent development of nanostructured materials-enabled biosensors. Sensors 21(4):1109. https://doi.org/10.3390/S21041109
14. Koumbia M, Ngoepe M, Holman JB, Mufamadi MS, Takai M (2024) Microfluidic electrochemical biosensors: tools for advancing the sustainable development goals. Trends Biotechnol 42(10):1207–1210. https://doi.org/10.1016/J.TIBTECH.2024.04.005
15. Tsutsui H, Lillehoj PB (2020) Flexible analytical devices for point-of-care testing. SLAS Technol 25(1):6–8. https://doi.org/10.1177/2472630319896762
16. Xu J, Yan Z, Liu Q, Xu J, Yan Z, Liu Q (2022) Smartphone-based electrochemical systems for glucose monitoring in biofluids: a review. Sensors 22(15):5670. https://doi.org/10.3390/S22155670
17. Kumar R, Verma AK, Shrivas S, Thota P, Singh MP, Rajasubramaniam S, Das A, Bharti PK (2020) First successful field evaluation of new, one-minute haemozoin-based malaria diagnostic device. EClinicalMedicine 22:100347. https://doi.org/10.1016/j.eclinm.2020.100347
18. Wang X, Luo Y, Huang K, Cheng N (2022) Biosensor for agriculture and food safety: recent advances and future perspectives. Adv Agrochem 1(1):3–6. https://doi.org/10.1016/J.AAC.2022.08.002
19. Kim MY, Lee KH (2022) Electrochemical sensors for sustainable precision agriculture—a review. Front Chem 10:848320. https://doi.org/10.3389/FCHEM.2022.848320/BIBTEX
20. Hossain MI, Khaleque MA, Ali MR, Bacchu MS, Hossain MS, Shahed SMF, Saad Aly MA, Khan MZH (2024) Development of electrochemical sensors for quick detection of environmental (soil and water) NPK ions. RSC Adv 14(13):9137–9158. https://doi.org/10.1039/D4RA00034J
21. Chen M, Yazdani M, Murugappan K (2024) Non-destructive pest detection: innovations and challenges in sensing airborne semiochemicals. ACS Sens. https://doi.org/10.1021/ACSSENSORS.4C02049
22. Miglione A, Raucci A, Mancini M, Gioia V, Frugis A, Cinti S (2025) An electrochemical biosensor for on-glove application: organophosphorus pesticide detection directly on fruit peels. Talanta 283:127093. https://doi.org/10.1016/J.TALANTA.2024.127093

23. Ling Z, Yang L, Zhang W, Yao T, Xu H (2024) Detection of food contaminants: a review of established rapid analytical techniques and their applications and limitations. Food Saf Health 2(1):72–95. https://doi.org/10.1002/FSH3.12032
24. Choi JR, Yong KW, Choi JY, Cowie AC (2019) Emerging point-of-care technologies for food safety analysis. Sensors 19(4):817. https://doi.org/10.3390/S19040817
25. Meredith NA, Quinn C, Cate DM, Reilly TH, Volckens J, Henry CS (2016) Paper-based analytical devices for environmental analysis. Analyst 141(6):1874. https://doi.org/10.1039/C5AN02572A
26. Trinh TND, Tran NKS, Nguyen HA, Chon NM, Trinh KTL, Lee NY (2024) Recent advances in portable devices for environmental monitoring applications. Biomicrofluidics 18(5):051501. https://doi.org/10.1063/5.0224217
27. Singh D, Dahiya M, Kumar R, Nanda C (2021) Sensors and systems for air quality assessment monitoring and management: a review. J Environ Manag 289:112510. https://doi.org/10.1016/J.JENVMAN.2021.112510
28. Tejashwini DM, Naik R, Sunagar V, Nagaswarupa HP, Naik YV (2024) An introduction to forensic electrochemistry. ACS Publication, Washington, pp 1–23. https://doi.org/10.1021/BK-2024-1481.CH001
29. Seshadri DR, Li RT, Voos JE, Rowbottom JR, Alfes CM, Zorman CA, Drummond CK (2019) Wearable sensors for monitoring the physiological and biochemical profile of the athlete. Npj Dig Med 2(1):1–16. https://doi.org/10.1038/s41746-019-0150-9
30. Velayutham J, Mariappan SA, Manickam P (2024) Emerging (bio)sensor technologies for monitoring vital markers of military, mining, and defense healthcare. In: Health and environmental applications of biosensing technologies: clinical and allied health science perspective. Elsevier, Amsterdam, pp 393–412. https://doi.org/10.1016/B978-0-443-19039-1.00018-3
31. Bissonnette L, Bergeron MG (2010) Diagnosing infections—current and anticipated technologies for point-of-care diagnostics and home-based testing. Clin Microbiol Infect 16(8):1044–1053. https://doi.org/10.1111/J.1469-0691.2010.03282.X
32. Huckle D (2008) Point-of-care diagnostics: an advancing sector with non-technical issues. Expert Rev Mol Diagn 8(6):679–688. https://doi.org/10.1586/14737159.8.6.679
33. Haji A, Desalegn K, Hassen H (2023) Selected food items adulteration, their impacts on public health, and detection methods: a review. Food Sci Nutr 11(12):7534–7545. https://doi.org/10.1002/FSN3.3732
34. Bansal S, Singh A, Mangal M, Mangal AK, Kumar S (2017) Food adulteration: sources, health risks, and detection methods. Crit Rev Food Sci Nutr 57(6):1174–1189. https://doi.org/10.1080/10408398.2014.967834
35. Savvy Soumya M (2011). Pesticide-rich food. Accessed https://www.downtoearth.org.in/environment/pesticiderich-food-32964
36. Anita G, Neetu S, Scholar R, Professor A (2025) Hazards of new technology in promoting food adulteration. IOSR J Environ Sci 5(1):8–10
37. Faraz A, Lateef M, Mustafa M, Akhtar P, Yaqoob M, Rehman S (2013) Detection of adulteration, chemical composition and hygienic status of milk supplied to various canteens of educational institutes and public places in Faisalabad. J Anim Plant Sci 23:119–124
38. Chen JS (2009) What can we learn from the 2008 melamine crisis in China? Biomed Environ Sci 22(2):109–111. https://doi.org/10.1016/S0895-3988(09)60031-3
39. Gelpí E, Paz MP, Terracini B, Abaitua I, Cámara AG, Kilbourne EM, Lahoz C, Nemery B, Philen RM, Soldevilla L, Tarkowski S, Tóxico WSC (2002) The Spanish toxic oil syndrome 20 years after its onset: a multidisciplinary review of scientific knowledge. Environ Health Perspect 110(5):457–464. https://doi.org/10.1289/EHP.110-1240833
40. Mishra KK, Dixit S, Purshottam SK, Pandey RC, Das M, Khanna SK (2007) Exposure assessment to Sudan dyes through consumption of artificially coloured chilli powders in India. Int J Food Sci Technol 42(11):1363–1366. https://doi.org/10.1111/J.1365-2621.2006.01356.X
41. Mellmann A, Harmsen D, Cummings CA, Zentz EB, Leopold SR, Rico A, Prior K, Szczepanowski R, Ji Y, Zhang W, McLaughlin SF, Henkhaus JK, Leopold B, Bielaszewska M, Prager R, Brzoska PM, Moore RL, Guenther S, Rothberg JM, Karch H (2012) EHEC

O104:H4 in Germany 2011: large outbreak of bloody diarrhea and haemolytic uraemic syndrome by Shiga toxin–producing *E. coli* via contaminated food. PLoS One 6(7):22751. https://doi.org/10.1371/JOURNAL.PONE.0022751
42. Forsyth JE, Mistree D, Nash E, Angrish M, Luby SP (2024) Evidence of turmeric adulteration with lead chromate across South Asia. Sci Total Environ 949:175003. https://doi.org/10.1016/J.SCITOTENV.2024.175003
43. Risch PK (2017) The horsemeat scandal: the unknown victims of economically motivated crime. J Victimol 5:63–84
44. Gossner CME, Schlundt J, Embarek PB, Hird S, Lo-Fo-Wong D, Beltran JJO, Teoh KN, Tritscher A (2009) The melamine incident: implications for international food and feed safety. Environ Health Perspect 117(12):1803–1808. https://doi.org/10.1289/EHP.0900949
45. Nicki H (2017) Non-compliance on EU antibiotics order may cost India shrimp industry - responsible seafood advocate. Global Seafood Alliance, Portsmouth. Accessed https://www.globalseafood.org/advocate/eu-antibiotics-india-shrimp/
46. Shaikh N, Soomro AH, Sheikh SA, Khaskheli M (2013) Extent of water adulteration and its influence on physical characteristics of market milk. Pak J Nutr 12(2):178–181. https://doi.org/10.3923/PJN.2013.178.181
47. Ramesh M, Muthuraman A (2018) Flavoring and coloring agents: health risks and potential problems. In: Natural and artificial flavoring agents and food dyes. Elsevier, Amsterdam, pp 1–28. https://doi.org/10.1016/B978-0-12-811518-3.00001-6
48. Nazrim Marikkar JM (2022) Adulteration in oils and fats industry. In: Recent advances in edible fats and oils technology: processing, health implications, economic and environmental impact. Springer, Cham, pp 463–480. https://doi.org/10.1007/978-981-16-5113-7_17
49. Faller AC, Kesanakurti P, Arunachalam T (2021) Fraud in grains and cereals. In: Food fraud: a global threat with public health and economic consequences. Elsevier, Amsterdam, pp 281–308. https://doi.org/10.1016/B978-0-12-817242-1.00007-5
50. Priya K, Thirunavookarasu N, Chidanand DV (2023) Recent advances in edible coating of food products and its legislations: a review. J Agric Food Res 12:100623. https://doi.org/10.1016/J.JAFR.2023.100623
51. Nasir U, Ismail A, Riaz M, Razzaq K, Ali S, Hussain A, Ameen M, Saif A, Aslam F, Fernandes de Oliveira CA (2024) Exploring fruit ripening methods: conventional, artificial, and novel approaches for quality and health. Food Control 165:110626. https://doi.org/10.1016/J.FOODCONT.2024.110626
52. Gerage JM, Meira APG, da Silva MV (2017) Food and nutrition security: pesticide residues in food. Nutrire 42(1):1–9. https://doi.org/10.1186/S41110-016-0028-4/TABLES/2
53. Magnúsdóttir K, Kristinsson J, Jóhannesson P (2010) Adulterated alcoholic beverages. Laeknabladid 96(10):626–628. https://doi.org/10.17992/LBL.2010.10.322
54. McVey C, Elliott CT, Cannavan A, Kelly SD, Petchkongkaew A, Haughey SA (2021) Portable spectroscopy for high throughput food authenticity screening: advancements in technology and integration into digital traceability systems. Trends Food Sci Technol 118:777–790. https://doi.org/10.1016/j.tifs.2021.11.003
55. He H-J, da Silva Ferreira MV, Wu Q, Karami H, Kamruzzaman M (2024) Portable and miniature sensors in supply chain for food authentication: a review. Crit Rev Food Sci Nutr. https://doi.org/10.1080/10408398.2024.2380837
56. Pawan A (2017) Detect adulteration with rapid test Food Safety and Standards Authority of India inspiring trust, assuring safe & nutritious food. Accessed https://fssai.gov.in/book-details.php?bkid=201
57. Pigłowski M (2020) Food hazards on the European Union market: the data analysis of the rapid alert system for food and feed. Food Sci Nutr 8(3):1603–1627. https://doi.org/10.1002/FSN3.1448
58. Kalogianni DP (2021) Lateral flow assays for food authentication. Springer, Cham, pp 343–363. https://doi.org/10.1007/978-3-030-66165-6_16

59. Bahadır EB, Sezgintürk MK (2016) Lateral flow assays: principles, designs and labels. TrAC Trends Anal Chem 82:286–306. https://doi.org/10.1016/J.TRAC.2016.06.006
60. Kumar Y, Narsaiah K (2021) Rapid point-of-care testing methods/devices for meat species identification: a review. Compr Rev Food Sci Food Saf 20(1):900–923. https://doi.org/10.1111/1541-4337.12674
61. Wu KH, Huang WC, Chang SC, Kao CH, Shyu RH (2019) Colloidal silver-based lateral flow immunoassay for rapid detection of melamine in milk and animal feed. Mater Chem Phys 231:121–130. https://doi.org/10.1016/J.MATCHEMPHYS.2019.04.035
62. Bu T, Yao X, Huang L, Dou L, Zhao B, Yang B, Li T, Wang J, Zhang D (2020) Dual recognition strategy and magnetic enrichment based lateral flow assay toward salmonella enteritidis detection. Talanta 206:120204. https://doi.org/10.1016/J.TALANTA.2019.120204
63. Kurita R, Niwa O, Ueno Y (2019) Development of portable immunoassay device for future internet of things applications. In: Chemical, gas, and biosensors for internet of things and related applications. Elsevier, Cham, pp 87–103. https://doi.org/10.1016/B978-0-12-815409-0.00006-1
64. Li F, You M, Li S, Hu J, Liu C, Gong Y, Yang H, Xu F (2020) Paper-based point-of-care immunoassays: recent advances and emerging trends. Biotechnol Adv 39:107442. https://doi.org/10.1016/J.BIOTECHADV.2019.107442
65. Abi M (2022) How to make lateral flow tests more sustainable? Medical Device Network, New York. Accessed https://www.medicaldevice-network.com/features/lfts-plastic-waste-sustainability/
66. De La Guardia M, Garrigues S (2020) Past, present and future of green analytical chemistry. RSC Green Chem 66:1–18. https://doi.org/10.1039/9781788016148-00001
67. Cook E, Woolridge A, Stapp P, Edmondson S, Velis CA (2023) Medical and healthcare waste generation, storage, treatment and disposal: a systematic scoping review of risks to occupational and public health. Crit Rev Environ Sci Technol 53(15):1452–1477. https://doi.org/10.1080/10643389.2022.2150495
68. Wisniewski N, Reichert M (2000) Methods for reducing biosensor membrane biofouling. Colloids Surf B Biointerfaces 18(3–4):197–219. https://doi.org/10.1016/S0927-7765(99)00148-4
69. Sultana S, Azlan A, Desa MNM, Mahyudin NA (2023) Multiplex platforms in biosensor based analytical approaches: opportunities and challenges for the speciation of animal species in food chain. Food Control 149:109727. https://doi.org/10.1016/J.FOODCONT.2023.109727
70. Hao L, Chen J, Chen X, Ma T, Cai X, Duan H, Leng Y, Huang X, Xiong Y (2021) A novel magneto-gold nanohybrid-enhanced lateral flow immunoassay for ultrasensitive and rapid detection of ochratoxin a in grape juice. Food Chem 336:127710. https://doi.org/10.1016/J.FOODCHEM.2020.127710
71. Bier FF (2012) Nanobiotechnology for integration in bioanalysis. New Biotechnol 29:S33. https://doi.org/10.1016/J.NBT.2012.08.084
72. Lim SA, Ahmed MU (2015) A carbon nanofiber-based label free immunosensor for high sensitive detection of recombinant bovine somatotropin. Biosens Bioelectron 70:48–53. https://doi.org/10.1016/J.BIOS.2015.03.022
73. Lim SA, Yoshikawa H, Tamiya E, Yasin HM, Ahmed MU (2014) A highly sensitive gold nanoparticle bioprobe based electrochemical immunosensor using screen printed graphene biochip. Mater Sci Chem Eng RSC Adv 4(102):58460–58466. https://doi.org/10.1039/C4RA11066H
74. Althomali RH, Uinarni H, Gandla K, Mayet AM, Romero-Parra RM, Cahalib I, Oudaha KH, Almulla AF, Bisht YS (2023) Applications of magnetic nanomaterials in the fabrication of lateral flow assays toward increasing performance of food safety analysis: recent advances. Food Biosci 56:103149. https://doi.org/10.1016/J.FBIO.2023.103149
75. Hamami M, Mars A, Raouafi N (2021) Biosensor based on anti-fouling PEG/gold nanoparticles composite for sensitive detection of aflatoxin M1 in milk. Microchem J 165:106102. https://doi.org/10.1016/J.MICROC.2021.106102

76. Wu G, Li H, Chen J, Chiu Lai KW, Xiong L, Yang R (2024) Microfluidics enhanced biosensor technology for advancing food safety monitoring. Trends Food Sci Technol 149:104556. https://doi.org/10.1016/J.TIFS.2024.104556
77. Jung W, Han J, Choi JW, Ahn CH (2015) Point-of-care testing (POCT) diagnostic systems using microfluidic lab-on-a-chip technologies. Microelectron Eng 132:46–57. https://doi.org/10.1016/J.MEE.2014.09.024
78. Srivastava SK, Shalabney A, Khalaila I, Grüner C, Rauschenbach B, Abdulhalim I (2014) SERS biosensor using metallic nano-sculptured thin films for the detection of endocrine disrupting compound biomarker vitellogenin. Small 10(17):3579–3587. https://doi.org/10.1002/SMLL.201303218
79. Hu C, Wei G, Zhu F, Wu A, Luo L, Shen S, Jia N, Zhang J (2023) Label-free electrochemical biosensing architecture based on a protein-decorated Pt@BSA nanocomposite for sensitive immunoassay of pathogenic Escherichia coli O157:H7. ACS Sustainable Chem Eng 11(20):7894–7907. https://doi.org/10.1021/ACSSUSCHEMENG.3C01241
80. Khansili N, Rattu G, Krishna PM (2018) Label-free optical biosensors for food and biological sensor applications. Sensors Actuators B Chem 265:35–49. https://doi.org/10.1016/J.SNB.2018.03.004
81. Ayerdurai V, Cieplak M, Kutner W (2023) Molecularly imprinted polymer-based electrochemical sensors for food contaminants determination. TrAC Trends Anal Chem 158:116830. https://doi.org/10.1016/J.TRAC.2022.116830
82. Castillo G, Lamberti I, Mosiello L, Hianik T (2012) Impedimetric DNA aptasensor for sensitive detection of Ochratoxin a in food. Electroanalysis 24(3):512–520. https://doi.org/10.1002/ELAN.201100485
83. Castillo G, Spinella K, Poturnayová A, Šnejdárková M, Mosiello L, Hianik T (2015) Detection of aflatoxin B1 by aptamer-based biosensor using PAMAM dendrimers as immobilization platform. Food Control 52:9–18. https://doi.org/10.1016/J.FOODCONT.2014.12.008
84. Krska R, Becalski A, Braekevelt E, Koerner T, Cao XL, Dabeka R, Godefroy S, Lau B, Moisey J, Rawn DFK, Scott PM, Wang Z, Forsyth D (2011) Challenges and trends in the determination of selected chemical contaminants and allergens in food. Anal Bioanal Chem 402(1):139–162. https://doi.org/10.1007/S00216-011-5237-3

Electronic Tongue and Nose Technologies for Flavor and Adulteration Analysis

Kashif Ameer, Isam A. Mohamed Ahmed, Faima Atta Khan, Muhammad Abubakar, and Afaq Ahmad

Contents

K. Ameer (✉) · M. Abubakar · A. Ahmad
Institute of Food Science and Nutrition, University of Sargodha, Sargodha, Pakistan
e-mail: kashif.ameer@uos.edu.pk

I. A. Mohamed Ahmed
Department of Food Science and Nutrition, College of Food and Agricultural Sciences, King Saud University, Riyadh, Saudi Arabia

F. A. Khan
The University Institute of Diet and Nutritional Sciences, University of Lahore, Sargodha, Pakistan

W. Alahmad, M. Ashfaq (eds.), *Advanced Techniques against Food Adulteration*,
https://doi.org/10.1007/978-3-031-99454-8_13

Abstract Electronic sensing has emerged as powerful detection techniques for detection of adulteration and profiling of the flavoring compounds. E-sensing techniques include electronic nose (E-nose) and electronic tongue (E-tongue). These sensor-based instrumental techniques mimic the human olfactory sense and sensory perception owing to possible detection of the chemical compound which impart the characteristic flavor and taste to the food products. This chapter provides the recent advances electronic tongue and nose technologies for flavor and adulteration analysis. Both E-tongue and E-nose employs the usage of the array-based sensors and involve the pattern recognition for analysis of the liquid samples and volatile compounds through use of the gas sensors, respectively. The E-nose and E-tongue applications are diverse spanning agricultural, pharmaceuticals, and food industries for verification of authenticity, safety assessment, and quality control. These instrumental techniques are versatile in terms of their rapidity, objective analysis, and nondestructive nature as compared to those of traditional counterparts involving conventional sensory evaluation and instrumental methods, such as spectroscopy and chromatography. E-tongue is specifically employed for characterization of states profile of beverages, pharmaceutical formulations, and dairy products. Moreover, E-nose is widely utilized for detecting food contamination, off-flavors, and food spoilage. Recent advances in the fields of machine learning and artificial intelligence have led to increased accuracy of the instrumental methods and helped to achieve the real-time monitoring of food samples and predictive analysis. The miniaturization of these instruments have expanded their field-specific uses and industrial exploitation. There are several challenges regarding the E-nose and E-tongue analysis of the food commodities, such as durability and sensor selectivity in conjunction with multimodal screening. These instrumental methods adoption may lead to improved food safety, enhanced consumer trusts, and standardization of regularity compliance. This chapter provides the recent advances electronic tongue and nose technologies for flavor and adulteration analysis.

Keywords Electronic nose · Nose technology · Adulteration technology · Electronic tongue · Sensor

Introduction

Electronic tongue (E-tongue) together with electronic nose (E-nose) represent groundbreaking instruments which duplicate human sensory abilities to sense taste and smell. These sensing technologies work across various industries, particularly in food and beverage production, and pharmaceutical development, environmental testing, and healthcare institutions. Sensor technology designs the electronic tongue to simulate how humans perceive taste through its device array. The set of chemical sensors in the device can detect and measure sweet, salty, bitter, sour, and umami taste stimuli. Ameer et al. [1] state that such sensors track electrical alterations which occur after exposure to chemical compounds linked to taste elements in a tested solution.

Acceptance of electronic tongues originates from a union of sensor arrays along with data processing algorithms. Detecting devices made of conductive or electrochemical materials create reactions to particular ions and molecules during measurement process [1]. Electronic tongue devices measure voltage changes and resistance fluctuations and current fluctuations applied to the sensors so they can generate flavor fingerprint data from the analyzed sample [2]. After collecting data from the sample the chemometric techniques execute pattern recognition or machine learning algorithms for sample classification and taste quantification. An electronic tongue consists of sensor array units along with signal processing units and data analysis algorithms and different sensor types which respond to specific chemical compounds [3].

Electronic noses (E-noses) function as sensor arrays which perform odor detections and identifier processes according to the principles of human scent recognition. The sensor array detects volatile compounds that exist in the air through its set of sensing technologies [4]. The sensors which make up E-nose operate using MOS devices together with conducting polymers and quartz crystal microbalances (QCMs) [5]. An electronic nose has the same operating mechanism as an electronic tongue by deploying sensor arrays to analyze airborne odors. Each VOC-sensitive sensor in the array changes its electrical function when exposed to specific volatile organic compounds. Sensors respond collectively to air compounds which generates a distinctive odor fingerprint that represents the active substances [6]. A fundamental electronic nose consists of three major components such as sensor array (multiple detectors that analyze different odor molecules), detection system (measures sensor interactions with target volatile compounds), and pattern recognition (algorithms analyze sensor responses to classify odor profiles) [7].

Importance of Flavor Profiling and Adulteration Detection in Food

The electronic nose functions as a sensory system that uses sensors to measure volatile organic compounds (VOCs) similar to human olfaction. Food aromas and flavors depend on VOCs, while their chemical makeup determines food quality aspects including freshness along with ripeness rates and preservation quality [8]. The primary tasks of an electronic nose include identification of food odors through aroma profiling in addition to quality monitoring and authenticity assessment. The E-nose technology serves four distinct functions for food monitoring: (1) sensory analysis of aromas and flavors in food processing and packaging, (2) perishable item spoilage detection, (3) aroma-based product batch consistency testing, and (4) safety enhancement through mold and volatile pathogen detection [9, 10]. The comparative features of E-nose and E-tongue systems appear in Table 1.

The assessment of food flavors together with examination for possible adulteration acts as an essential mechanism to protect food quality and consumer acceptance of both food and beverages and pharmaceutical products [11]. Consumer preferences along with marketplace triumph depend heavily on sensory perception of flavor as a vital attribute. Manufacturers need exact flavor evaluations for designing products that match consumer expectations by maintaining consistent high-quality outputs. The iterative process of flavor profiling enables beverage industries to enhance their product quality during optimization through standardized taste profiles [12, 13].

Consumer health along with trust relies heavily upon detection methods for adulterated products. The food and beverage industry regularly introduces adulterants comprising synthetic colors artificial sweeteners and low-grade oils to lower production expenses which harms product safety and quality [14]. Detection of adulterants protects both consumers' health and regulatory compliance and preserves brand integrity. Stakeholders depend on the detection of melamine in dairy products combined with synthetic additives in honey because this helps prevent industry scandals while securing public health [15].

The process of flavor profiling improves the tastiness of oral drugs designed for pediatric and geriatric patients in pharmaceutical products. Flavor optimization techniques provide masking functions to reduce the detectable bitterness found in active pharmaceutical ingredients thus improving acceptance rates from patients. Govari et al. [16] explain that adulteration detection methods protect pharmaceutical authenticity by identifying counterfeit drugs which might either be ineffective or dangerous to health. Product integrity along with public health and regulatory compliance require flavor profiling and adulteration detection instruments as essential tools. Advanced electronic detection methods such as tongues and noses augment the precision and productivity of these procedures which boost market innovation as well as consumer trust among quality-demanding markets [17]. The electronic tongue exists as a sensor array which functions to mimic human taste capabilities so it can evaluate liquid and semi-liquid food measurements. The device interprets

Table 1 Electronic tongue and nose technologies for flavor and adulteration analysis [1–3]

Aspect	Electronic tongue (E-tongue)	Electronic nose (E-nose)
	A sensor-based analytical device that mimics the human gustatory system to detect and quantify flavors and tastes	A sensor-based analytical device designed to mimic the human olfactory system to detect and identify volatile compounds
Key components	Sensor array (e.g., potentiometric, voltammetric, or impedance-based) Data acquisition system Chemometric software	Sensor array (e.g., metal oxide sensors, polymer-based sensors, or quartz crystal microbalance) Data acquisition system Chemometric software
Working principle	Measures liquid-phase properties like pH, ion concentration, and electrochemical responses to differentiate tastes	Detects and analyzes volatile organic compounds (VOCs) in the gas phase to identify aroma profiles or adulterants
Analytes detected	Sweet, salty, sour, bitter, umami Organic acids, sugars, salts, and peptides Liquid-based contaminants	Aromatic compounds VOCs such as esters, ketones, aldehydes, and alcohols Gaseous adulterants or contaminants
Applications	Beverage quality analysis (e.g., wine, tea, coffee) Detection of adulteration in honey, milk, and oils Pharmaceutical taste masking	Aroma profiling in food and beverages Detection of spoilage or rancidity in dairy, meat, or oils Adulteration detection (e.g., ethanol in perfumes)
Advantages	Rapid and consistent flavor profiling High sensitivity to taste variations Reduced reliance on human panels	Noninvasive and fast analysis Effective in detecting low concentrations of volatile adulterants Versatility in analyzing a wide range of products
Challenges	Difficulty in detecting nonionic compounds Limited applicability to gas-phase analytes Sensor fouling in complex matrices	Limited by cross-sensitivity of sensors Drift in sensor performance over time Challenges in distinguishing similar VOCs
Data analysis techniques	Principal component analysis (PCA) Partial least squares regression (PLSR) Artificial neural networks (ANN)	PCA Linear discriminant analysis (LDA) ANN Fuzzy logic
Recent advances	Nanomaterial-based sensors for improved sensitivity Integration with IoT for real-time monitoring Bioelectronic tongue systems	Development of miniaturized portable E-noses Enhanced machine learning algorithms for odor fingerprinting Hybrid sensor systems
Limitations	Inability to detect volatile flavors or odors Relatively high cost of development	Sensitivity to environmental conditions (e.g., humidity and temperature) Sensor degradation over time
Industrial usage	Food and beverage quality control Pharmaceutical industry for taste masking evaluationfermentation monitoring	Food and beverage quality control (e.g., coffee aroma, wine bouquet) Detection of spoilage in packaging Cosmetics and perfumery

food chemical ingredients which correspond to sweet, salty, sour, bitter, and umami taste compounds [18]. Food taste evaluation through E-tongues provides an objective method which acts as an efficient replacement for human taste panels and subjective assessment methodologies [19]. Quality control together with nutrient analysis and taste profiling represent the main functions of the system. This technology provides its applications through (1) flavor production optimization in product development and quality control and (2) nutritional assessment of foods for dietary requirements such as low-sodium and low-sugar products combined with solution-based testing that substitutes human taste panels and generates dependable objective answers [20].

Detection Mechanism: Chemical and Biochemical Interactions

E-tongue and E-nose represent modern analytical instruments which enable artificial sensing of human taste and olfactory functions [21]. Sensors arranged in arrays serve to identify and evaluate compounds that exist in foods alongside beverages and atmospheric elements. The main purpose of these devices is to taste products and detect adulteration within products [22].

Detection Mechanism in E-Tongue

Chemical Interaction

E-tongues use several sensors which specifically interact with different taste compounds. The sensors monitor electrical modifications including conductivity and impedance and voltage variations when they come in contact with sample substances [22]. The changes in electrical conductivity serve as a measureable detection method for conductometric sensors when these sensors interact with bitter food components. The detection of specific ions which generate salty or sour sensations occurs through ion-selective electrodes. Taste-related chemicals expose the impedance-based sensor to alterations of electrical impedance in test samples [23].

Biochemical Interactions

The design of some E-tongues includes biochemical reactions performed through biological receptors combined with enzymes. The combination of enzyme-coated sensors enables identification of both glucose and amino acids because these chemical compounds contribute to food sweetness and umami taste [24]. The sensors initiate electrical signals through the process of chemical and biochemical reactions yet produce different results such as resistance modifications or voltage shifts. The data processing system operates on recorded signals to interpret test results [25].

The system identifies specific flavors through analysis of sensor data patterns against predetermined taste profiles. The e-tongue proves useful for determining the flavor attributes of both food and beverages [26]. The sensor system detects chemical substances which indicate food adulteration when they produce unexpected or unwanted taste compounds including synthetic sweeteners and excess salt. The monitoring system maintains product quality by detecting deviations from standard taste profiles through its technology [27].

Detection Mechanism in E-Nose

Chemical Interaction

Standard E-nose sensors utilize materials which include metal-oxide semiconductors, conducting polymers, and piezoelectric crystals as their components. The sensor surface interacts with volatile compounds leading to physical or chemical sensor changes which measurement systems detect [28].

Metal-Oxide Semiconductor (MOS) Sensors

The sensors transform their electrical resistance when encountering gases. The sensor detects modifications to resistance when it interacts with alcohols or aldehydes which enables the system to analyze the smell [29].

Quartz Crystal Microbalance (QCM)

The sensor's surface takes up volatile compounds that result in a change of quartz crystal resonant frequency which gets detected to determine particular smells [30].

Conductometric Sensors

The sensors monitor electrical conductivity alterations which occur after they encounter gases from ammonia or acidic materials. Different volatile compounds found in spoiled food and fermented products are detected through sensor arrays [31].

Biochemical Interactions

Modern E-nose technology includes biochemical reaction sensors for selective compound detection through the use of enzymatic and immunological selective compounds. Enzymes integrated into E-noses serve as detection elements for volatile organic compounds including ethanol which enhances their sensitivity and specificity [32].

Signal Output

The sensor-stimulant chemical or biochemical binding results in electrical signals that the system transforms into analysis. The system uses pattern recognition algorithms which include principal component analysis (PCA) and artificial neural networks (ANNs) to evaluate sensor-generated signals for odor identification purposes [33].

Flavor and Adulteration Analysis

An E-nose system functions as a vital tool for flavor analysis because it detects the odor-creating volatile compounds. The device identifies adulteration through comparison of sample odor profiles against profiles of known pure products. The device performs two actions including flagging artificial flavors while detecting spoilage through gas analysis of food decomposition [34].

Machine Learning in E-Tongue and E-Nose Systems

Analyzing Data and Making Predictions

The fundamental work of E-tongues and E-noses consists of receiving sensor array measurements about interactions between compounds in liquids and gases. The acquired raw data needs sophisticated processing to provide beneficial insights. The systems use machine learning to find patterns in sensor data which assists in making predictions about sample characteristics as well as the classification of products [35].

Classification

The main role which machine learning performs in E-tongue and E-nose systems involves determining the categories of different smells and flavors. Support vector machines (SVMs) together with random forests and k-nearest neighbors (k-NN) analyze sensor outputs to provide data categorization. The system successfully identifies between sweet liquids and bitter beverages as well as detects floral and fruity odors [36].

Regression: Software engineers implement machine learning for regression-based forecasting which requires systems to predict continuous results such as sugar content or alcohol level in beverages and wine. Linear regression algorithms and neural networks serve to predict values based on sensor [37].

Clustering: Under some circumstances the data lacks labels so the system must organize similar samples through their defining characteristics. The system uses unsupervised learning approaches that include k-means clustering and hierarchical clustering to discover patterns inside unlabeled data. The process of clustering allows identification of homogenous sample groups based on chemical compositions and detects outlier samples likely to indicate adulteration [1].

Pattern Recognition in E-Tongue and E-Nose Systems

The identification of established patterns within data stream leads to category matching during the pattern recognition process. Pattern recognition plays a critical role in E-tongue and E-nose systems to analyze sensor signals for resulting actionable information [1].

Applications in Flavor Analysis

The analysis of flavors stands as a vital knowledge field in food science because it enables quality control programs and consumer expectation fulfillment together with product development initiatives [2].

Detection of Taste Profiles

Taste detection requires both identification and precise measurement of sweet, sour, bitter, salty and umami tastes. These taste components play a fundamental role in creating the sensory reactions to food. Electronic tongues together with high-performance liquid chromatography (HPLC) enable this process through advanced technology [10]. Modern electronic tongues combine sensor arrays with pattern recognition algorithms to simulate human taste functions. The evaluation system enables manufacturers to maintain product consistency by providing them with clear and impartial assessment tools that fulfill consumer requirements [12].

Analysis of Aroma Compounds and Volatile Organic Compounds (VOCs)

The volatile compounds together with VOCs serve as essential elements for determining what food tastes like. Researchers use GC-MS and SPME as techniques to evaluate these compounds. Research methods analyzing VOCs allow scientists to understand the complex effects which generate food sensory attributes. The assessment of aroma compounds through this process requires immediate attention for quality management and detecting undesirable off-flavors that occur from food spoilage or chemical reactions in processing and storage [14].

Evaluation of Food Quality and Consistency

Food manufacturers require consistent management of product quality as an essential target. The monitoring tools from flavor analysis enable manufacturers to inspect chemical and sensory qualities of products for quality fulfillment. The evaluations heavily depend on spectrophotometry and nuclear magnetic resonance (NMR) spectroscopy techniques. Instrumental methods in food production settings receive additional assessment from sensory evaluation panels which together provide complete information about flavor consistency for consumer satisfaction [15].

Development of New Food Products Based on Flavor Profiling

The assessment of flavors through profiling methods supports businesses in creating new food products. The evaluation of consumer choices together with market dynamics helps manufacturers to develop products featuring designed flavor combinations for their intended customer base. Computational tools such as flavoromics together with machine learning algorithms allow manufacturers to predict flavor interactions while modifying their product formulations for improved flavor outcomes. Technological advances enable producers to introduce new food products which strengthen both the competitiveness and dish variety in the food industry [16].

Case Study and Real-World Applications

E-Sensing in Wine and Beverage Industries for Flavor Profiling

Customer satisfaction primarily depends on flavor quality and aroma standards within the wine and beverage manufacturing sector. The use of traditional sensory tests requires extensive time which drives manufacturers to adopt the advanced E-tongue and E-nose analytical systems. A touchscreen device reproduces human taste detection abilities by identifying chemical substances in liquids [16]. The E-tongue sensors monitor taste attributes in wine to detect flavors. The E-tongue system measures sensory properties including bitterness and the perception of astringency with sweetness detection and it assists quality control through task profiling among wine products across different batches [17].

Detection of Rancidity in Edible Oils Using Electronic Tongue and Nose Technologies

The process known as rancidity deteriorates both taste quality and nutritional characteristics of foods. The detection of rancidity uses chemical analysis along with these other technologies which provide immediate results. The E-tongue features sensors that function as taste sensors to detect specified properties in the same manner as human taste receptors [19]. E-tongue systems provide the best performance for monitoring oxidative rancidity in oils. In a study, an E-tongue technique was used for monitoring of the oxidative rancidity of olive oil. The instrument proved successful for detecting bitterness and astringency. Rancid olive oil showed distinct taste attributes alongside elevated acidity which separated it from fresh oil according to the E-tongue system. This research evaluated the use of E-nose technology for measuring sunflower oil rancidity. The E-nose device monitored shifting volatile concentrations of the substance throughout the research duration [20].

Monitoring Freshness in Perishable Goods (Meat, Fish, and Fruits)

Sensor-based devices working in combination with these technologies produce detection outcomes for flavor and aroma modifications. The E-nose system assessed chicken meat freshness while it was stored in different conditions. E-nose technology demonstrated ability to identify volatile organic compounds associated with spoilage during the research study. The analysis of apple ripening and spoilage was achieved using an E-tongue system. The E-tongue device system monitors changes in apple acidity together with sweetness levels [20].

Adulteration Analysis

E-tongue and E-nose systems are powerful tools for detecting adulteration in food products:

Milk

Detection methods for water, starch, and detergent adulterants in milk play an important role to verify milk safety and quality. The identification of these adulterants in milk can be performed efficiently through E-tongue and E-nose systems which eliminate the requirement for complicated chemical analysis [21]. Table 2 showcases how E-tongue and E-nose systems measure various milk adulterants through combined taste and aromatic profile assessment.

The E-tongue evaluates milk chemical elements through its detection of taste variations between samples. Milk consists of a precise combination of dissolved substances in a specific condition that changes when water is added since the water affects conductivity and adds detectable sweetness while viscous properties change. The sensor array of an E-tongue detects bitter together with soapy taste signals that detergents generate. The E-nose detects volatile compounds which release from milk samples. The specific aromatic pattern from detergents helps distinguish these chemicals from milk's natural profile indicating that either water or starch adulteration occurred [22]. The combination of pattern recognition algorithms permits these systems to both identify and measure adulterants in quick succession. Both E-tongue and E-nose systems work as a single Bluetooth instrument which performs nondestructive automated sensitive milk adulteration analysis to protect food safety regulations and consumer health. These devices find substantial use for quality control operations in dairy industry establishments [20].

Table 2 Detection of adulterants in milk using E-tongue and E-nose, adapted with permission [1, 8, 10, 15]

Adulterant	Detection method	E-tongue response	E-nose response	Key advantages
Water	Conductivity and taste	Reduced conductivity, altered taste profile (diluted flavor)	No direct response, but change in volatile compounds may be detected indirectly	Easy detection in real-time, nondestructive testing
Starch	Viscosity and sweetness	Increased sweetness or altered aftertaste	May not produce significant volatile changes	Quick and reliable for common adulterants like starch
Detergents	Bitter or soapy taste	Bitter or soapy taste signals	Detection of unusual aromatic compounds (e.g., chemical scents)	Highly sensitive to chemical adulterants like detergents
Urea	pH, conductivity	Shift in pH, slight bitterness or off-flavor	Not directly detectable, but certain VOC changes may occur	Cost-effective and nondestructive analysis
Formalin	Chemical reaction	Possible chemical reactions causing changes in the taste profile	Detects formaldehyde's distinct odor	Rapid identification of harmful chemicals in milk

Honey

The quality of honey gets compromised when producers add glucose syrups and sugar or enhance their products by adding additional volume. Testing such adulterated samples matters because it safeguards both consumer health and quality of products. E-tongue instruments couple with E-nose instruments serve as accurate tools which detect honey adulterants through noninvasive methods. The E-tongue conducts adulteration analysis through measuring the taste characteristics of honey samples. The sugar profile of raw honey contains multiple tastes that include sweetness combined with different intensities of acidity and bitterness recognized in published literature [21]. The addition of glucose or syrups and sugar leads to changes in sweetness and acidity which generate unique signals. An E-tongue utilizes its sensors to record changes in compounds that enable the device to spot pure honey from contaminated samples. The electrochemical response changes when using glucose as an additive because it intensifies sweetness and syrup additives result in changes to viscosity [25].

Honey can be evaluated by an E-nose that senses volatile organic compounds emitted from honey to identify various compounds. Glucose as well as syrups introduce new emission compounds that modify the original odor profile of honey. Pattern recognition through the E-nose detects chemical changes in samples that allow it to identify pure honey from adulterated products [25].

Oils

Edible oils acquire their overall quality along with nutritional properties from their purity assessment. The practice of using mixed or substandard oils for adulteration affects consumer confidence and health security levels. The rapid analysis of oil purity relies on two modern analytical instruments known as E-tongue and E-nose according to Chung et al. [10]. The E-tongue monitors oil purity through taste profiling that depends on natural fatty acids alongside solvent compounds. Both pure oils exhibit unique electrochemical fingerprints but these signals change when inferior oils are mixed together. Olive oil adulteration through addition of soybean or palm oil causes detectable changes in bitterness and acidity as well as aftertaste properties [35]. The E-nose identifies that oils lack authenticity through analysis of volatile compounds released from the oils. Each oil type possesses an individual scent profile which is fully defined by its volatile organic compounds (VOCs). The E-nose detects sensor patterns from VOC mixtures produced by using substandard or mixed oils. Advanced machine learning technologies perform both classification of adulterated elements and quantify the adulteration process. The edible oil industry can perform rapid and accurate quality checks through these systems which do not damage product samples. The E-tongue measurement systems assist both regulatory compliance and consumer health maintenance while ensuring product authenticity for producers [35]. The detection of adulterants in edible oils using E-tongue and E-nose is explained in Table 3.

Table 3 The detection of adulterants in edible oils using E-tongue and E-nose, adapted with permission [1, 8, 10, 15]

Adulterant	Detection method	E-tongue response	E-nose response	Key advantages
Soybean oil	Taste and viscosity	Altered bitterness, sweetness, and aftertaste (compared to pure olive oil, for example)	Changes in volatile compounds (aroma differences between oils)	Detects adulteration in high-value oils like olive oil
Palm oil	Flavor profile and acidity	Increased bitterness, altered acidity and sweetness	Detects difference in VOCs between palm oil and other oils	Effective for differentiating expensive oils from cheaper alternatives
Sunflower oil	Taste and texture analysis	Slightly altered sweetness or a decrease in natural flavor intensity	Distinct changes in aroma due to mixing with lower-quality oils	Can detect low-quality or mixed oils in premium oil types
Mustard oil	Bitter compounds and viscosity	Altered bitterness and pungency	Change in VOCs and aroma profile	Ideal for identifying cheaper fillers in mustard oil
Rapeseed oil	Sweetness, acidity, and aftertaste	Changes in sweetness and acidity, altered aftertaste	Different VOC signatures due to adulteration with cheaper oils	Provides fast, cost-effective quality control for oils
Industrial/used oils	Viscosity and chemical composition	Increased bitterness and off-flavors	Detection of rancidity and oxidative by-products	Identifies harmful, reused oils that may be harmful to health

Beverages

The verification of wine along with juices and spirits requires special attention because it protects both quality standards and consumer trust and safety measures. Labeling beverages with cheaper elements together with flavor additives or water content presents substantial difficulties to the industry [30]. The authenticity verification process of alcoholic and nonalcoholic beverages can be achieved through the use of E-tongue and E-nose systems which deliver efficient and dependable evaluation methods. Beverages pass through E-tongue analysis where this device identifies taste alterations stemming from adulteration [34]. Water or sugar additives in wine create dilution that affects its typical combination of sweetness alongside bitterness and acidity which an E-tongue device can detect without difficulty. Detection systems can identify specific taste variations when synthetic flavorings exist in juices and spirits. The sensors based on E-tongue technology produce electrical output signals from dissolved compounds to classify distinct authentic and adulterated samples. The E-nose combines with the E-tongue to analyze the volatile organic compounds (VOCs) that influence beverage aromas. All drinks including wine together with juice and spirit possess their own particular aroma characteristic patterns [35]. E-nose technology detects adulterated products through changes in their volatile organic compounds which occur when both artificial flavorings and

Table 4 The detection of adulterants in beverages (e.g., wine, juices, spirits) using E-tongue and E-nose, adapted with permission [1, 8, 10, 15]

Adulterant	Detection method	E-tongue response	E-nose response	Key advantages
Water (in wine/ spirits)	Conductivity and taste	Reduced acidity, dilution of sweetness, and altered flavor	No direct response but can affect aromatic compounds slightly	Quick and nondestructive; ideal for verifying wine or spirit purity
Sugar (in juices)	Sweetness and acidity	Increased sweetness or altered balance of acidity	No direct response, but VOC profile may change slightly	Effective for detecting hidden sugars or syrups in juices
Alcohol (in nonalcoholic beverages)	Alcohol content and taste	Altered bitterness or sweetness, depending on the beverage	Detection of alcohol-based volatiles in the VOC profile	Nondestructive, cost-effective analysis for nonalcoholic beverage verification
Synthetic flavorings	Taste and aftertaste	Altered or synthetic sweetness, bitterness, or aftertaste	Detection of artificial or off-aromas in the VOC profile	Detects artificial flavor additives without destroying the sample
High-fructose corn syrup (HFCS)	Sweetness and texture	Unnatural sweetness profile with more pronounced sugar notes	Detects changes in VOC patterns from added sugars or syrups	Can quickly detect common sweetener adulteration in juices and soft drinks
Flavor enhancers (e.g., citric acid)	Acidity and sourness	Altered acidity or sourness, possible bitter aftertaste	Changes in volatile acid profiles, detectable sourness	Ideal for verifying juice authenticity and detecting artificial souring agents
Cheaper alcohol (in spirits)	Flavor and bitterness	Altered bitterness or loss of characteristic flavor notes	Different VOC patterns due to lower-quality alcohol	Rapid detection of counterfeit or low-quality spirits

low-quality alcohol ingredients are added. The pattern recognition algorithms operated by the system enable the sample identification process before authenticating it [36]. Table 4 presents the methods of detection of adulterants in beverages (e.g., wine, juices, spirits) using E-tongue and E-nose.

Spices

Food safety together with quality and authenticity depend heavily on proper identification of synthetic additives and low-quality fillers that are found in spices. The addition of artificial colorants and starches and filler materials in spices leads to diminishing flavor and nutritional value of the products [36]. E-tongue and E-nose prove to be highly capable in detecting adulterants. The E-tongue performs analysis on the chemical makeup and taste property waveform of the spice. Each spice type including cumin or paprika displays unique taste characteristics derived from its natural chemical elements [30]. Changes in bitterness and acidity and sweetness profiles from synthetic colorants or starch add-ons serve as detection indicators for E-tongue systems. The system detects false adoption through its sensors by measuring alterations in taste caused by synthetic colorant introduction [28]. Table 5 presents details about how E-tongue and E-nose detect adulterants that appear in spices and herbs.

E-nose performs VOC analysis to detect organic volatile compounds emitted from spices which helps completion of the identification process. Spices can only emit distinct odors naturally until they receive additional fillers or artificial components. The E-nose detects how adulterants modify volatile organic compounds which permits the detection of contaminants. The system performs spice classification based on patterns executed by pattern recognition algorithms which guard against adulteration or purity determination [28].

Table 5 The detection of adulterants in spices and herbs using E-tongue and E-nose, adapted with permission [1, 8, 10, 15]

Adulterant	Detection method	E-tongue response	E-nose response	Key advantages
Starch (in chili powder)	Sweetness and texture	Increased sweetness, altered bitterness, or aftertaste	May show minor changes in VOCs due to added fillers	Quick detection of common filler adulterants in chili powder
Synthetic colorants (e.g., in turmeric)	Flavor and acidity	Altered or diminished bitterness or slight increase in sweetness	No direct flavor change, but VOC profile may be altered	Effective for detecting artificial colorants without damaging the sample
Wood or sawdust (in paprika)	Texture and mouthfeel	Altered mouthfeel and slight sweetness change	No distinct VOC change, but the odor profile may be altered	Nondestructive, quick, and reliable for detecting nonedible fillers
Salt (in spices like pepper)	Salinity and bitterness	Increased salinity, altered bitterness	Minor changes in volatile compounds related to salt	Detects adulteration in spice powders and ensures purity
Flour (in black pepper)	Taste and aftertaste	Altered bitterness or sweetness, or an increase in texture	Change in aroma profile, may show earthy or flour-like odors	Identifies starch-based fillers or adulterants
Chemical additives (preservatives)	Taste and acidity	Altered acidity or bitterness, potential aftertaste	Change in VOCs due to preservatives (e.g., synthetic aromas)	Detects harmful preservatives and ensures safety
Low-quality fillers (e.g., cheaper herbs)	Flavor profile and texture	Altered or reduced intensity of natural flavor	Identifiable difference in the VOC profile from pure spices	Quickly identifies low-quality herbs mixed with fillers

Conclusions

Detection of adulteration and profiling of the flavoring compounds in foodstuffs is tedious in recent era and electronic sensing-based methods have provided the rebut analysis for flavor characterization and adulteration detection. E-sensing techniques include E-nose and E-tongue. These sensor-based instrumental techniques mimic the human olfactory sense and sensory perception owing to possible detection of the chemical compound which impart the characteristic flavor and taste to the food products. This chapter provides the recent advances electronic tongue and nose technologies for flavor and adulteration analysis. Both E-tongue and E-nose employs the usage of the array-based sensors and involve the pattern recognition for analysis of the liquid samples and volatile compounds through use of the gas sensors, respectively. The E-nose and E-tongue applications are diverse spanning agricultural, pharmaceuticals, and food industries for verification of authenticity, safety assessment, and quality control. These instrumental techniques are versatile in terms of their rapidity, objective analysis, and nondestructive nature as compared to those of traditional counterparts involving conventional sensory evaluation and instrumental methods, such as spectroscopy and chromatography. E-tongue is specifically employed for characterization of states profile of beverages, pharmaceutical formulations, and dairy products. Moreover, E-nose is widely utilized for detecting food contamination, off-flavors, and food spoilage. Recent advances in the fields of machine learning and artificial intelligence have led to increased accuracy of the instrumental methods and consequently helped to achieve the real-time monitoring of food samples and predictive analysis. The miniaturization of these instruments have expanded their field-specific uses and industrial exploitation. There are several challenges regarding the E-nose and E-tongue analysis of the food commodities, such as durability and sensor selectivity in conjunction with multimodal screening. These instrumental methods adoption may lead to improved food safety, enhanced consumer trust, and standardization of regularity compliance. This chapter provides the recent advances of electronic tongue and nose technologies for flavor and adulteration analysis.

References

1. Ameer K, Jo Y, Amir RM, Shahbaz HM, Kwon J-H (2020) Screening and identification of electron-beam irradiated dried spice-mixture products by electronic sensing and standard analytical methods through dose estimation. LWT 125:108957
2. Ameer K, Jo Y, Kim G-R, Chung N, Kwon J-H (2017) Profiles of phenolic acids and amino acids from stevia extracts obtained by optimized heat-reflux method. Korean Soc Food Sci Nutr 10:228–229
3. Ameer K, Shahbaz HM, Kwon JH (2017) Green extraction methods for polyphenols from plant matrices and their byproducts: a review. Compr Rev Food Sci Food Saf 16(2):295–315
4. Anwar H, Anwar T, Murtaza S (2023) Review on food quality assessment using machine learning and electronic nose system. Biosens Bioelectron X 14:100365

5. Arshad MS, Kwon JH, Ahmad RS, Ameer K, Ahmad S, Jo Y (2020) Influence of E-beam irradiation on microbiological and physicochemical properties and fatty acid profile of frozen duck meat. Food Sci Nutr 8(2):1020–1029
6. Arshak K, Moore E, Lyons GM, Harris J, Clifford S (2004) A review of gas sensors employed in electronic nose applications. Sens Rev 24(2):181–198
7. Bonah E, Huang X, Aheto JH, Osae R (2020) Application of electronic nose as a non-invasive technique for odor fingerprinting and detection of bacterial foodborne pathogens: a review. J Food Sci Technol 57:1977–1990
8. Calvini R, Pigani L (2022) Toward the development of combined artificial sensing systems for food quality evaluation: a review on the application of data fusion of electronic noses, electronic tongues and electronic eyes. Sensors 22(2):577
9. Chilo J, Pelegri-Sebastia J, Cupane M, Sogorb T (2016) E-nose application to food industry production. IEEE Instrum Measure Magaz 19(1):27–33
10. Chung N, Ameer K, Jo Y, Kwon J-H (2019) Comparison of electronic sensing techniques for screening dried shrimps irradiated using three types of approved radiation with standard analytical methods. Food Chem 286:395–404
11. del Mar Contreras M, Arroyo-Manzanares N, Arce C, Arce L (2019) HS-GC-IMS and chemometric data treatment for food authenticity assessment: olive oil mapping and classification through two different devices as an example. Food Control 98:82–93
12. Furizal F, Ma'arif A, Firdaus AA, Rahmaniar W (2023) Future potential of E-nose technology: a review. Int J Robot Control Syst 3(3):449–469
13. Galli BD, Baptista DP, Cavalheiro FG, Negrão F, Eberlin MN, Gigante ML (2019) Peptide profile of camembert-type cheese: effect of heat treatment and adjunct culture *Lactobacillus rhamnosus* GG. Food Res Int 123:393–402
14. Gliszczyńska-Świgło A, Chmielewski J (2017) Electronic nose as a tool for monitoring the authenticity of food. A review. Food Anal Methods 10:1800–1816
15. Gonçalves WB, Teixeira WSR, Cervantes EP, Mioni MSR, Sampaio ANCE, Martins OA et al (2023) Application of an electronic nose as a new technology for rapid detection of adulteration in honey. Appl Sci 13(8):4881
16. Govari M, Tryfinopoulou P, Panagou EZ, Nychas G-JE (2022) Application of Fourier transform infrared (FT-IR) spectroscopy, multispectral imaging (MSI) and electronic nose (E-nose) for the rapid evaluation of the microbiological quality of gilthead sea bream fillets. Foods 11(15):2356
17. Han F, Zhang D, Aheto JH, Feng F, Duan T (2020) Integration of a low-cost electronic nose and a voltammetric electronic tongue for red wines identification. Food Sci Nutr 8(8):4330–4339
18. Huang L, Liu H, Zhang B, Wu D (2015) Application of electronic nose with multivariate analysis and sensor selection for botanical origin identification and quality determination of honey. Food Bioprocess Technol 8:359–370
19. Illahi AAC, Dadios EP, Bandala AA, Vicerra RRP (2021) Electronic nose technology and application: a review. In: 2021 IEEE 13th international conference on humanoid, nanotechnology, information technology, communication and control, environment, and management (HNICEM). IEEE, Piscataway, pp 1–5
20. Jia W, Liang G, Jiang Z, Wang J (2019) Advances in electronic nose development for application to agricultural products. Food Anal Methods 12:2226–2240
21. Jo Y, Ameer K, Chung N, Kang Y-H, Ahn DU, Kwon J-H (2020) E-sensing, calibrated PSL, and improved ESR techniques discriminate irradiated fresh grapefruits and lemons. J Food Sci Technol 57:364–374
22. Jo Y, Ameer K, Chung N, Kang Y-H, Kwon J-H (2019) Quality differentiation of low-dose irradiated navel oranges by electronic sensing techniques during storage. Food Anal Methods 12:1041–1054
23. Kiani S, Minaei S, Ghasemi-Varnamkhasti M (2016) Application of electronic nose systems for assessing quality of medicinal and aromatic plant products: a review. J Appl Res Med Aromatic Plants 3(1):1–9

24. Lozano J, Santos JP, Horrillo MC (2016) Wine applications with electronic noses. In: Electronic noses and tongues in food science. Elsevier, Amsterdam, pp 137–148
25. Lu L, Hu Z, Hu X, Li D, Tian S (2022) Electronic tongue and electronic nose for food quality and safety. Food Res Int 162:112214
26. Mahanti NK, Shivashankar S, Chhetri KB, Kumar A, Rao BB, Aravind J et al (2024) Enhancing food authentication through E-nose and E-tongue technologies: current trends and future directions. Trends Food Sci Technol 150:104574
27. Malik T, Sharma R, Ameer K, Bashir O, Amin T, Manzoor S et al (2023) Potential of high-pressure homogenization (HPH) in the development of functional foods. Int J Food Prop 26(1):2509–2531
28. Munekata PE, Finardi S, de Souza CK, Meinert C, Pateiro M, Hoffmann TG et al (2023) Applications of electronic nose, electronic eye and electronic tongue in quality, safety and shelf life of meat and meat products: a review. Sensors 23(2):672
29. Rasekh M, Karami H (2021) Application of electronic nose with chemometrics methods to the detection of juices fraud. J Food Process Preservation 45(5):e15432
30. Roy M, Yadav B (2022) Electronic nose for detection of food adulteration: a review. J Food Sci Technol 59(3):846–858
31. Schaller E, Bosset JO, Escher F (1998) 'Electronic noses' and their application to food. LWT - Food Sci Technol 31(4):305–316
32. Seesaard T, Wongchoosuk C (2022) Recent progress in electronic noses for fermented foods and beverages applications. Fermentation 8(7):302
33. Śliwińska M, Wiśniewska P, Dymerski T, Wardencki W, Namieśnik J (2016) Advances in electronic noses and tongues for food authenticity testing. In: Advances in food authenticity testing. Elsevier, Amsterdam, pp 201–225
34. Vanaraj R, Mayakrishnan G, Kim IS, Kim S-C (2025) A systematic review of the applications of electronic nose and electronic tongue in food quality assessment and safety. Chem 13(5):161
35. Wang M, Chen Y (2024) Electronic nose and its application in the food industry: a review. Eur Food Res Technol 250(1):21–67
36. Wang Q, Chen X, Zhang C, Li X, Yue N, Shao H et al (2023) Discrimination and characterization of volatile flavor compounds in fresh oriental melon after forchlorfenuron application using electronic nose (e-nose) and headspace-gas chromatography-ion mobility spectrometry (HS-GC-IMS). Foods 12(6):1272
37. Wang W, Zhang S, Li Z, Li J, Yang X, Wang C et al (2020) Construction of covalent triazine-based frameworks and application to solid phase microextraction of polycyclic aromatic hydrocarbons from honey samples. Food Chem 322:126770

Advances in Surface-Enhanced Raman Spectroscopy for the Detection of Synthetic Dyes in the Food Matrices

Harish Farale and Suvardhan Kanchi

Contents

H. Farale · S. Kanchi (✉)
Department of Chemistry, CHRIST University, Bengaluru, India

Centre for Renewable Energy and Environmental Sustainability, CHRIST University, Bengaluru, India

W. Alahmad, M. Ashfaq (eds.), *Advanced Techniques against Food Adulteration*,
https://doi.org/10.1007/978-3-031-99454-8_14

Abstract Food adulteration is a global concern that is a significant public health risk. Such issues demand sophisticated analytical devices with high-speed sensitivity toward detecting adulterants. The surface enhanced Raman spectroscopic (SERS) technique is an advanced tool that gains wide appreciation due to its very high sensitivity and selectivity. Being an excellent ultra-trace contaminant-detecting tool even within complex food matrices, this resource becomes fundamental in food safety protection. SERS is found to detect trace concentrations of synthetic dyes. Recent advancements have been made on new substrates that enhance signal stability among noble metals, hybrid nanostructures, and metal-organic frameworks (MOFs). These substrates are further functionalized with specific chemical moieties to improve the selectivity and reduce the matrix interference. SERS, combined with advanced computing tools like machine learning and chemometric algorithms, has completely changed the data analysis scenario; nowadays, high-throughput simultaneous detection and accurate adulterant quantification are possible. With rapid outputs and less sample preparation, portable SERS devices revealed promise in on-site food safety monitoring. Subsequent developments that address several fundamental concerns regarding cost, reusability, and scalability ultimately make SERS devices more widely adaptable. This chapter covers the emerging aspects of SERS substrates, enhancement strategies, and computational improvements concerning synthetic dye detection. It speaks of the potential of SERS as a flexible tool for ensuring public health and food safety worldwide while discussing the challenges and opportunities in developing inexpensive, reusable substrates.

Keywords Surface-enhanced Raman spectroscopy (SERS) · Synthetic dyes · Food safety · Nanostructured substrates · Trace detection

Introduction

Food adulteration is a change in food quality for various reasons. Frequent adulterants include chemical additives, artificial colors, and fillers that have harmful effects on health, such as allergic reactions, chronic illnesses, and even carcinogenic effects. Furthermore, adulteration poses risks not only to health but also undermines consumer confidence and burdens healthcare systems; it disrupts global trade. An incident such as China's milk scandal in 2008 exemplifies the consequences of adulteration for public health and confidence [1, 2]. One of the most common adulterants is food colorants, which are used to improve the appearance of food products and make their color vibrant. These are classified based on several criteria: first, based on their origin in nature, they are nature-identical, or, if synthetic, whether they are organic or inorganic. Another classification could be based on their solubility (e.g., soluble or insoluble) or covering ability (e.g., transparent or opaque). However, an overlap may exist among one or more of these classifications. The most common and widely accepted classification is based on the difference between soluble and insoluble color additives (colorants or pigments), which can be divided into natural or synthetic [3].

SERS is thus an emerging analytical technique for trace-level detection of food adulterants. The technique is significantly powerful due to its high sensitivity, specificity, and speed. This has become one of the most significant methods used to identify contaminants at trace concentrations and plays a key role in evaluating food safety and quality. Enhancing Raman signals by as much as 10^{14} makes it possible to trace contaminants even at parts-per-billion levels. This high sensitivity is particularly good for identifying harmful substances, such as pesticides, antibiotics, and illegal additives, in trace amounts but with enormous health risks [4].

Recent developments in the engineering of SERS substrates have further improved their applications in food safety. For example, nanostructures such as gold and silver nanoparticles have been engineered to provide uniform and reproducible enhancement, which was previously observed to be problematic due to variability in signals. For example, the association of metal nanoparticles with adsorbent materials has dramatically improved the detection of traces of lean meat powder residues within meat products, thus enhancing accurate and sensitive analyses [5]. The integration of SERS with advanced separation techniques has also expanded its capabilities. Combining SERS with liquid chromatography enables the separation and identification of complex mixtures of contaminants, providing a comprehensive analysis of food samples. This combination improves detection accuracy and decreases the possibility of false positives, which is necessary for regulatory compliance and consumer safety [6]. Furthermore, the SERS technique effectively detects food adulterants, mainly chemical, biological, or harmful metals. Its utilization is wide, from identifying foodborne pathogens and allergens to identifying genetically modified organisms. Rapid detection by SERS would be helpful during public health crises when food safety must be ascertained, such as during the COVID-19 pandemic [4].

Overview of Synthetic Dyes

Synthetic dyes in the food industry have been a common practice for decades. These dyes are added to food products to enhance their appearance, making them more appealing to consumers [7]. These dyes are used in various food products, including candies, baked goods, beverages, and cereals [8]. They are commonly used in the food industry, including Tartrazine, Sunset Yellow, and Brilliant Blue [9]. Synthetic dyes in foods are associated with various health risks. Some studies have reported that synthetic dyes may cause carcinogenic effects, whereas others associate synthetic dyes with hyperactivity among children [10, 11]. Some of these factors have also been reported to cause allergic reactions in some people [12]. However, synthetic dyes are still widely used in the food industry. This is because they are relatively inexpensive and easy to use, in addition to the fact that there are no effective

Table 1 Categories of synthetic dyes are based on their chemical nature

Category	Example dyes	Characteristic colors	Food products	References
Azo dyes	Tartrazine, allura red	Yellow, red	Cheese, wafers	[14]
Xanthene dyes	Erythrosin B, rhodamine B	Pink, red	Beverages	[15]
Triarylmethane dyes	Brilliant blue, fast green	Blue, green	Beverages, candies	[16]
Indigoid dyes	Indigo carmine	Deep blue	Dairy-based drinks and products	[17]

natural alternatives. However, recently, there has been an increasing trend of applying natural dyes to food products. In fact, consumer demand has been the driver behind the quest for more natural and sustainable food products [13].

Based on their chemical nature, synthetic dyes frequently used in the food industry can be classified into azo, xanthene, triaryl methane, and indigo dyes; some examples are tabulated in Table 1. Azo dyes account for most dyes, and they are compounds bearing one or more azo groups (–N=N–), which are linked with aromatic compounds. Azo dyes account for 65% of the total uptake and are consumed as food additives in soft drinks, jams, candies, and pickles [18]. Some of the most commonly used azo dyes include Tartrazine (Yellow 5), Sunset Yellow (Yellow 6), Amaranth (E123), Allura Red (FD&C Red 40), and Carmoisine (Red 3) [14]. Non-azo dyes, such as Brilliant Blue (E133), Erythrosine (E127), Xanthene dye, Fast Green FCF (E143), and Indigo Carmine (Blue 2), are less frequently used but still play a role in coloring foods [19]. Recent research highlights that azo dyes are associated with potential genotoxic and oxidative stress effects. The FDA and EFSA have set limits on its consumption and require it to be labeled explicitly to inform the customer [18]. Xanthene dyes, which are derived from the xanthene structure, produce bright fluorescence. These dyes have wide applications in beverages and brightly colored candies.

An example is fluorescein, whose derivatives are very conspicuous with some types of lighting. Such a synthesis is carried out via multistep reactions involving aromatic compounds, which results in some health risks if the process is not well regulated. These dyes are very effective in enhancing color but need rigorous safety testing because of their interaction with biological systems, especially in sensitive populations such as children [20, 21]. Triarylmethane dyes, including Brilliant Blue FCF (Blue No. 1) and Fast Green FCF, are widely applied in foods and beverages because of their stability and bright blue and green shades. They are triphenylmethane and thus highly color-implying but are surrounded by health concerns over allergenicity and cytotoxic effects. Brilliant Blue FCF is associated with hyperactivity in children and other behavioral problems, raising controversy about whether it should be allowed in children's formulations. However, the product has passed various regulatory reviews by agencies such as the EFSA and the FDA that declare its safety within specified limits. These deep blue dyes, such as Indigo Carmine, also called Blue No. 2, are used in food products, especially candies and desserts. These artificial dyes are synthesized to mimic natural dyes isolated from indigo plants. They are often considered safe, but even large doses have been suggested for

Synthetic Dye	Colour	Physical Properties	Chemical Structure	Food Products Used	Health Hazards
Erythrosine		Fluorescent; soluble in water; bright appearance	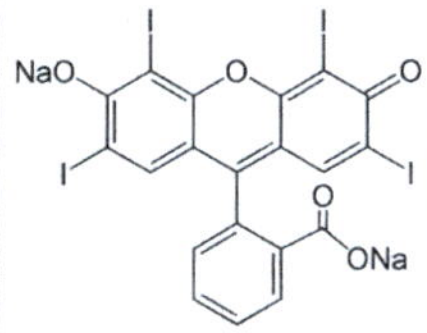		Potential thyroid toxicity; linked to carcinogenicity in animal studies
Brilliant Blue FCF		Water-soluble; stable in light and heat	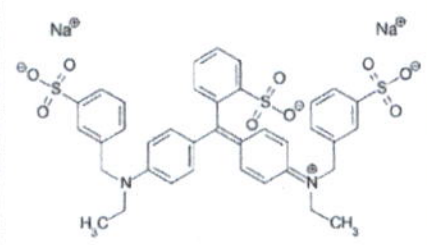		Linked to hyperactivity in children; potential allergen
Indigo Carmine		Water-soluble; stable under light	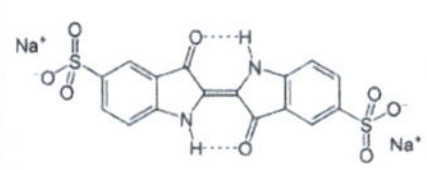		Gastrointestinal disturbances; potential genotoxicity
Tartrazine		Soluble in water; bright yellow; crystalline	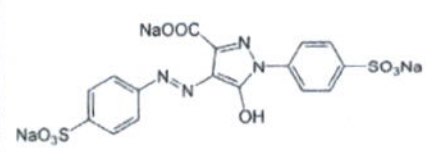		Allergies; hyperactivity in children (e.g., ADHD); asthma exacerbation
Allura Red AC		Soluble in water; bright and stable color	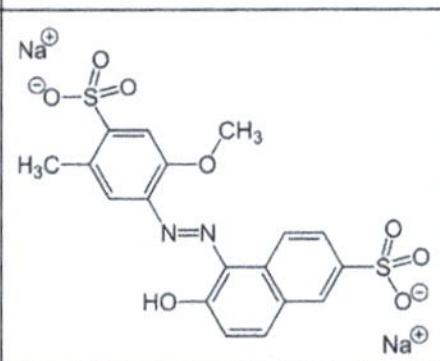		Potential allergen; suspected carcinogen in high doses
Rhodamine B		Soluble in water; fluorescent; stable under normal conditions	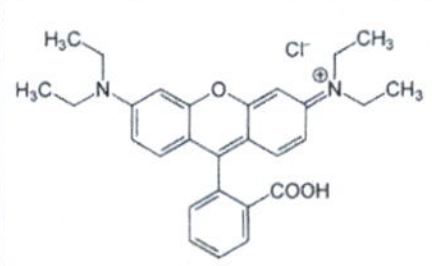		Carcinogenic potential; liver damage; toxic effects on reproductive health
Fast Green FCF		Water-soluble; stable under heat and light	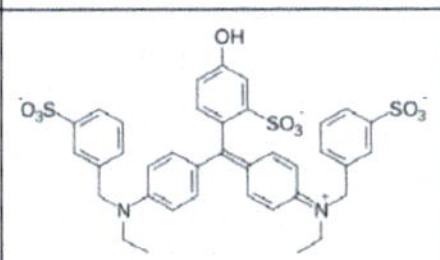		Possible allergen; mild gastrointestinal effects at high doses

Fig. 1 Overview of some commercially used synthetic dyes in food products, their properties, and health hazards

gastrointestinal effects with probable genotoxic effects. Indigo Carmine is approved in many countries and is subject to specific usage [22]. Figure 1 represents a comparison of various synthetic food dyes, highlighting their color, physical properties, chemical structures, food products where they are used, and associated health hazards.

Fundamentals of Surface-Enhanced Raman Spectroscopy (SERS)

Surface-enhanced Raman spectroscopy, commonly abbreviated as SERS, is an extremely sensitive analytical technique that enhances the Raman scattering of molecules adsorbed on metallic nanostructured surfaces. This enhancement is achieved via electromagnetic and chemical mechanisms; the Raman signals are amplified between 10^6 and 10^{10}. The primary mechanism in question is the electromagnetic field, through which localized surface plasmon resonance (LSPR) is responsible; this happens by exciting the conduction electrons in the metal nanoparticle with light, resulting in high-intensity localized electromagnetic fields. These fields are strongly confined to "hot spots", which generally occur at tiny gaps between nanoparticles or on the sharp edges of nanostructures. Chemical enhancement, though weaker, arises from the transfer of charge between the metal surface and the adsorbate molecules, which then modifies the molecule's polarizability and, hence, the molecule's Raman scattering efficiency [23, 24]. In any case, the performance of SERS critically depends on the design and fabrication of metallic nanostructures, typically comprising gold, silver, or copper, because of their excellent plasmonic properties. The most common approaches for fabricating SERS substrates include top-down techniques such as electron-beam lithography and focused ion beam milling, as well as bottom-up techniques such as chemical synthesis and self-assembly. These methods tend to enhance the substrate morphology to maximize the density of hot spots and increase signal reproducibility and sensitivity [25, 26]. The latest breakthroughs in nanotechnology have made hybrid nanostructures, such as core-shell nanoparticles and bimetallic systems, a reality. They provide easily tunable plasmonic properties and improved chemical stability concerning applications in the SERS field [27]. SERS is an effective technique with applications in biomedical diagnostics, environmental monitoring, and food safety. It enables highly sensitive detection of disease biomarkers in medicine, including cancer indicators, pathogens, and drug delivery markers [28]. Trace amounts of additives, pesticides, antibiotics, illegal food dyes, mycotoxins, and other small-molecule toxins can be detected in complex samples of food products via SERS. Similarly, food safety ensures that products meet safety standards by identifying contaminants and adulterants [29]. Another grand challenge within SERS lies in having reproducible and scalable substrates applicable to practical problems. Substrate nanoscale morphology variations give way to wide variability in the signal amplitude, thus posing severe constraints toward quantitative applications. Therefore, the application of standard procedures for fabrication with the inclusion of artificial intelligence algorithms for pattern recognition for spectral analysis also represents potential solutions [30]. Furthermore, integrating SERS with complementary techniques such as chromatography and microfluidics has improved its potential in the analysis of complex mixtures by separating components and detecting them with very high accuracy [31]. In addition, major theoretical progress has been made in better understanding fundamental processes that govern SERS processes. Computational modeling, especially density functional theory (DFT) and finite-difference time-domain (FDTD) simulations, has greatly facilitated an understanding of the interactions between molecules and plasmonic nanostructures, allowing more efficient substrates to be designed [32]. In addition, incorporating artificial intelligence into data processing has made spectral analysis faster and more accurate; thus, it is possible to identify target molecules in noisy environments. Despite its many developments, SERS still has several drawbacks, such as interference from non-target species and the inability to detect non-adsorbing analytes. Scientists are developing functionalized substrates with selective binding sites for target molecules. Portable SERS devices and fiber-optic probes also allow real-time, onsite analysis, increasing the practicality and versatility of SERS [27, 33].

Application and Recent Advancements of SERS in the Detection of Synthetic Dyes

Recent advancements in SERS have shown its potential for detecting illegal food additives and synthetic dyes with high sensitivity, selectivity, and reusability. The key findings and methodologies from the reviewed studies are summarized below:

Advances in SERS Substrates for Enhanced Sensitivity and Specificity

Recent advances in SERS substrates have emphasized their sensitivity and specificity for environmental monitoring, food safety, and biomedical diagnostics applications. Metallic nanoparticles, especially those based on gold (Au) and silver (Ag), are the cornerstone of SERS substrates because of their superior plasmonic properties. Hybrid substrates of plasmonic nanomaterials with semiconductors (e.g., TiO_2, ZnO) or graphene have increased signal amplification via charge transfer mechanisms and high adsorption affinity [34, 35]. MOFs have emerged as promising hybrid materials, combining the high surface area and tunable porosity of MOFs with the plasmonic properties of metals such as gold (Au) and silver (Ag). The use of MOF-based substrates permits the generation of hotspots uniformly and enables the selective adsorption of analytes, thus further enhancing signal enhancement and molecular specificity [36]. Functionalized substrates tailored with specific receptors or molecular recognition elements also increase specificity by selectively capturing target molecules. These developments open avenues for precise and robust SERS-based sensing in complex matrices [37].

Hybrid Nanostructures

A group of researchers constructed semiconductor/noble-metal heterostructures via a facile photo-reduction method to deposit AgNPs on flower-like ZnO microcrystals. High SERS activity for detecting Rhodamine 6G, Sunset Yellow, and Tartrazine can be achieved via flower-like ZnO@Ag nanostructures with ultralow detection limits of 10^{-12} M, 10^{-10} M, and 10^{-11} M, respectively. These flower-like ZnO@Ag nanostructures have excellent analytical enhancement factors, sensitivities, and uniformities for sensing multiple food colorants [38]. Another study detected low concentrations of Methyl Orange and Rhodamine B in the presence of silver, a noble metal, and zinc oxide, ZnO. It was synthesized through chemical procedures. The nanoparticles and nanocomposites were subsequently investigated via field emission scanning electron microscopy, which revealed spherical AgNPs and hexagonal nanorods of ZnO. The nanocomposites indicate the attachment of clusters of Ag nanoparticles to the hexagonal ZnO nanorods. The same was further confirmed via transmission electron microscopy. The changes in the optical absorption of Ag attached to the ZnO nanorods were monitored via UV–visible spectroscopy. Raman spectroscopy was used to monitor the presence of Methyl Orange dye at concentrations of 10^{-3} M and 10^{-6} M and RhB dye at concentrations ranging from 10^{-6} and even to as low as 10^{-9} M in the presence of Ag nanoparticles and Ag–ZnO nanocomposites [39]. According to SERS testing, the food items yellow croaker, braised pork, red pepper, green dumpling, and red wine contained Acid Orange II and Brilliant Blue. The preparation of Fe_3O_4@Au core shells for fast sensitivity, convenient detection, and preservation of integrity from the sample tested due to a prominent SERS impact was shown. DFT studies were performed, and Raman spectrum investigations of the solid samples revealed characteristic peaks for the two dyes. The average recovery was approximately 92.95–99.36% [40]. In one of these studies, researchers developed a multifunctional Fe_3O_4@SiO_2−Au core-satellite SERS substrate for the selective detection of food dyes. A very large Raman signal enhancement was achieved by adjusting the gap of the gold nanoparticle to sub 10 nm. Functionalization with charged polyelectrolytes enabled the selective adsorption of oppositely charged molecules. MDS also showed high affinity, with near-approach values of ~20 Å for the substrates and analytes; however, DFT analysis confirmed that charge transfer was the primary mechanism of this type of interaction. For Sunset Yellow, the limit of detection (LOD) value reached 1×10^{-7} M. Its magnetic nature helped separate compounds, thus making it suitable for complex chemical mixture analysis [41]. A TiO_2 nanorod array @PDA/Ag (TNR@PDA/Ag) SERS sensor was developed for sensitive duplex detection of illicit food dyes. Polydopamine (PDA) has dual functions, which improve the efficiency of electron transfer and enrich analytes such as Rhodamine B and crystal violet through hydrogen bonding, π–π stacking, and electrostatic interactions. The LOD achieved by the sensor for Rhodamine B was 1 nM, and that for crystal violet was 5 nM. The bifunctional SERS platform has been proven to be feasible for duplex detection even when real samples of chili meals are used and would be highly applicable for monitoring food safety [42]. Another article reported a plasmonic AgNP/SnS_2 NH nanostructure as an extremely sensitive SERS substrate for detecting synthetic dyes where hexagonal-like tin disulfide (SnS_2 NHs) were grown hydrothermally. The AgNPs provide localized electromagnetic enhancement, whereas the SnS_2 NHs contribute to effective charge transfer, which is optimized through the photo-reduction time. The substrate achieves a detection limit of 4.12×10^{-10} M for Methylene Blue (MB) and 3.01×10^{-9} M for Tartrazine (TZ), with enhancement factors of 10^8 (MB) and 10^7 (TZ). It shows high homogeneity (RSD: 8.71% for MB, 7.09% for TZ) and reproducibility (RSD: ~8%). For effective detection in real samples, such as pond water and orange juice, 10^{-8} M is suitable for environmental and food safety monitoring [43]. Researchers developed a graphene oxide/Au@Ag nanobone membrane as a substrate for simultaneous enrichment and SERS detection of beverage colorants as shown in Fig. 2. The sensitivity enhancement is attributed to the enhanced Raman properties of the Au@Ag nanobones together with the enrichment capacity of GO. The microporous membrane allows uniform, high-density GO/Au@Ag NBs loading to enable efficient analyte enrichment during flow through the filtration unit of beverages. The membrane was able to reach a limit of detection of 1.12×10^{-9} M for Rhodamine 6G, with great uniformity (RSD: 7.87%), repeatability (RSD: 10.44%), and time stability for up to 60 days. Six colorants were detected accurately to monitor food safety [44]. An AgNP@alumina bead-based SERS sensor was developed for ultrasensitive point-of-care detection. The sensor achieves high sensitivity by chemically partitioning the bead surface via a hydrophilic–hydrophobic gradient and chloride etching, promoting aggregation-induced hot spots. This setup

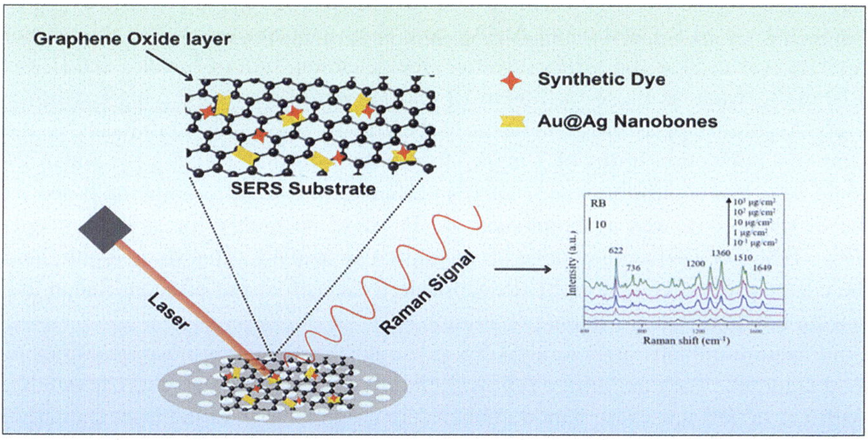

Fig. 2 SERS detection of dyes combined with GO/Au@Ag nanobones membrane. Graphene oxide (GO) layer: Serves as a chemical enhancer that captures analyte molecules through strong adsorption due to its high surface area and π-π interactions and offers a planar platform for securing and dispersing the nanobones; Au@Ag nanobones: Plasmonic nanostructures that enable electromagnetic amplification through localized surface plasmon resonance (LSPR) also produces so-called "hot spots" where Raman signals of captured analytes are highly amplified. Adapted with permission [44]

detected 4-aminophenyl disulfide (4-APDS) at an ultralow concentration of 10^{-15} M. The sensor was applied successfully to track the SERS fingerprints of five food colorants in candies, demonstrating the practicality of this sensor for real sample analysis. The sensor is easy to fabricate and portable and thus provides a robust platform for the analysis of food additives with high sensitivity and specificity in solid-state SERS applications [45].

Noble Metal Nanoparticles

Flower-shaped silver nanoparticles were produced by reducing a simple aqueous phase of silver nitrate using ascorbic acid with polyvinylpyrrolidone (PVP) as a surfactant. The diameters of the nanoparticles ranged from 450 to 1000 nm, with surface protrusions of up to 10–25 nm. The flower-shaped silver nanostructures obtained were used as stable SERS substrates with high SERS activity for detecting Rhodamine 6G at a concentration of only 10^{-9} mol/L, where the SERS signal is still clear. SERS spectra of four different food colorants (e.g., food blue, Tartrazine, Sunset Yellow, and Acid Red) were analyzed, and the characteristic bands were identified. This detection of four food colorants with concentrations as low as approximately 10^{-8} mol/L relies on an improved PCA principle. The LODs for blue in food, Tartrazine, Sunset Yellow, and Acid Red are 79.285 μg/L, 5.3436 μg/L, 45.238 μg/L, and 50.244 μg/L, respectively [46]. Sudan dyes are illegal color additives for food. Therefore, they represent a significant risk to the safety of food. For this purpose, an analysis of Sudan I–IV in standard solutions and chili flakes was conducted via SERS and using Au-Ag core-shell nanospheres. When 90 ± 5 nm Au@Ag was used as the SERS substrate, the detection limit of Sudan I and II was 0.10 mg/L, whereas those for Sudan III and IV were 0.08 and 0.2 mg/L, respectively. Acetonitrile was shown to be the best extractive solvent for chili flakes. At the same time, detection limits of 1 mg/kg for Sudan I–III and 2 mg/kg for Sudan IV could be achieved despite some matrix interferences. Dye differentiation based on the SERS spectra was accomplished via PCA, while highly linear PLS models ($R^2_{cv} = 0.869$–0.959) were obtained for quantitative analysis. This study highlighted the potential of Au@Ag as a SERS substrate for sensitive and efficient Sudan dye detection with minimal sample preparation [47].

Gold nano-dumbbells are useful SERS substrates for detecting trace quantities of food colorants, such as Sunset Yellow, Lemon Yellow, Orange II, and Chrysoidin. With λ_{max} at 785 nm, these nanodumbbells are compatible with the laser excitation wavelengths of portable Raman spectrometers employing diode lasers at 785 nm, which enable site-specific detection. This substrate's detection limit was 10^{-8} M for crystal violet in aqueous solutions. The reproducibility was excellent, with RSD values below 15% for 10^{-6} M crystal violet across 50 random points on the same substrate and 10 spots across different batches. The successful onsite detection of pigments in orange juice and coke indicates that this substrate may be applicable for practical food safety purposes [48]. SERS with gold nanorods was applied to detect Allura Red and Sunset Yellow in beverages. For Allura Red, nanorods with an aspect ratio (AR) of 2.4 showed the best performance, whereas the AR of 1.8 nanorods worked best for Sunset Yellow. The detection limit was 0.1 mg/L for both dyes. Adding $MgSO_4$ to the samples increased the SERS signal for Sunset Yellow but had no effect on that for Allura Red. The method provided good reproducibility, with RSD values less than 15%. The partial least squares models for quantitative analysis produced a high linear correlation from $R^2 = 0.925–0.982$ and reliable predictability, indicating that the technique may find rapid, sensitive applications for food safety [49]. A novel, low-cost Au nanorod-incorporated melamine foam (AuNR-MF) SERS substrate was fabricated to detect banned food dyes, such as Rhodamine B and Basic Orange II. The substrate showed powerful SERS activity because of the aggregation effect of the partially absent CTAB. The detection limits for the dyes were 0.1 μg/mL for Rhodamine B and 0.5 μg/mL for Basic Orange II. The substrate showed good reproducibility, with RSD values of 8.6% for Rhodamine B and 9.2% for Basic Orange II. The simultaneous detection of these dyes in food samples, such as chili products, was achieved, thereby demonstrating the substrate's potential in food safety applications [50]. Herein, a highly sensitive SERS substrate based on stable spherical hemp AgNPs was developed by employing a simple hydrothermal process to quickly detect adulterated colorants in black tea. The AgNPs exhibited excellent stability over 60 days, and the enhancement factor obtained was 10^8 when R6G was used as the probe molecule. The method successfully detected and quantified four colorants, Sunset Yellow, Lemon Yellow, Carmine, and Erythrosine, using a CARS-PLS model with correlation coefficients of 0.95, 0.97, 0.99, and 0.88, respectively. The mixed colorants were effectively differentiated, with recoveries ranging from 91.87% to 106.5% and RSD values less than 1.97%. This rapid and straightforward method requires no pretreatment and thus has the potential for precise and efficient colorant detection in tea products [51]. A paper-based SERS method for detecting Rhodamine 6G using grape skin-gold nanoparticles/graphene oxide (GE-AuNPs/GO) was developed. The obtained GE-AuNPs/GO showed a plasmon resonance at 539 nm, a particle size of 19.5 nm, and a zeta potential of −59.7 mV. The substrate achieved an enhancement factor of 1.92×10^9 and a low detection limit of 7.33×10^{-11} M in the linear range of 10^{-11}–10^{-5} M. The RSD values for repeatability and reproducibility were 9.6% and 12.6%, respectively. Recovery yields for Rh6G were achieved in tap water samples as high as 93.9–100.8% and in food samples as high as 91.0–95.4% for chili powder, 96.6–98.3% for glutinous rice balls, and 98.9–102.3% for tomato ketchup. The substrate maintained an efficiency of less than 20% loss after 60 days, promising for sensitive and stable onsite detection of organic toxins in food and water [52]. A silver nanocube (Ag NC) thin-film SERS substrate was developed for the sensitive detection of Rhodamine B (RhB) in food samples such as chili powder and oil. The dynamic linear range of the developed ultralow detection limit, 10^{-10} M, was between 10^{-4} and 10^{-10} M, with an excellent correlation coefficient of $R^2 = 0.99$, ensuring high reliability. The Ag NC thin film was stable over 60 days, indicating long-term usability. In this regard, an efficient, cost-effective SERS-based platform is a powerful tool capable of detecting harmful food dyes at trace levels by offering molecular fingerprinting ability for food safety [53].

Metal-Organic Frameworks (MOFs)

This method was developed with UiO-66(NH_2) @Au SERS for synthetic dyes such as new coccine and Orange II, which are targeted for their detection in food samples, where the adsorption properties of the MOF (UiO-66(NH_2)) combined with Au nanoparticles on the MOF surface provided enhanced Raman scattering. This method exhibited extreme sensitivity, with LODs of 0.4015 mg/L for NC and 0.0546 mg/L for Orange II, along with very strong linear correlations between the concentration and peak area: $R^2 = 0.9684$ for NC and $R^2 = 0.9912$ for OII. The SERS platform can quantify NC and OII in Mirinda soft drink and paprika samples, with recovery rates ranging from 82.92% to 109.63%. This versatile platform provides a reliable and efficient method for monitoring synthetic colorants in food and offers significant potential for food safety applications [54]. Two UiO-66 MOF-based SERS substrates constructed with gold nanoparticles for food safety analysis were developed. The UiO-66/AuNP suspension substrate allows for the selective detection of a known carcinogenic heterocyclic amine in barbecue meat. The UiO-66(NH_2)/AuNP/Nylon-66 membrane substrate, as shown in Fig. 3, facilitated the separation, enrichment and in situ analysis of Sudan Red 7B in chili products. Both substrates showed good sensitivity and recovery, with 82.3–110% for heterocyclic amines and 84.5–114% for Sudan dye. The RSDs were 3.1–11.0% and 1.9–5.6%, respectively. These substrates combine molecular enrichment with SERS

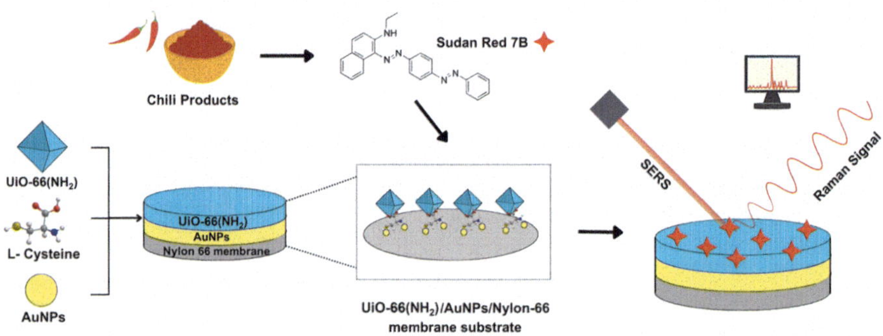

Fig. 3 SERS detection system for identifying Sudan Red 7B in chili products using a composite SERS substrate. UiO-66(NH_2): A metal-organic framework with functional amine groups that should increase the adsorption of analytes; AuNPs (gold nanoparticles): Provide necessary plasmonic activity for amplifying the signal of Raman; L-cysteine: a linker molecule used to functionalize and stabilize the interaction between MOFs and AuNPs; Nylon-66 membrane: It is used as a supporting platform for the composite; thus it provides stability and usability in practical applications. Adapted with permission [55]

activity and provide a reliable and versatile approach for the quantitative analysis of complex food samples, thereby advancing practical applications of SERS in food safety monitoring [55]. A self-cleaning SERS substrate, CC-L-MOF@F-Ag, was fabricated via in situ growth of a leaf-like metal-organic framework and one-step electrodeposition of flower-like Ag nanoparticles followed by chemical modification. This substrate shows excellent reusability (usable for six sensing cycles) and long-term stability for at least 40 days, with a low RSD of 9.5%. This technique allows the sensitive detection of azo dyes, including Congo Red, Chrysoidine, and Acid Orange II, in soft drinks. The limits of detection were 0.013 μM for CR, 0.017 μM for CYD, and 0.0067 μM for AO II, with linear ranges of 0.1–1000 μM for CR and CYD, and 0.01–100 μM for AO II. Satisfactory recoveries were obtained for azo dyes in six soft drinks, indicating the substrate's potential for cost-effective and efficient dye monitoring in beverages [56]. A high-performance SERS substrate, MOF(Zr)@Ag NPs, was synthesized via a hydrothermal two-step process to detect dyes such as crystal violet, Rhodamine 6G, and 4-mercaptobenzoic acid. The MOF's large surface area, porosity, and defect engineering enhance silver nanoparticle adsorption, thereby inducing an abundance of "hot spots" through localized surface plasmon resonance. This design achieved ultralow detection limits for 10^{-15} M CV and successfully detected CV residues in shrimp tissues at parts per billion concentrations. Amino and sulfhydryl groups in the structure of the MOF improved the adsorption efficiency and detection of the Raman signal. This MOF(Zr)@AgNP platform shows great sensitivity and stability, and this method offers a reliable approach to detecting residues of dyes in food and environmental samples [57]. A flexible NH_2-MIL-101(Cr)@Au SERS substrate was fabricated by uniformly decorating amino-functionalized Cr-based MOFs with gold nanoparticles via in situ reduction. The substrate combines adsorption and detection capabilities for Acid Orange II, a banned azo pigment. The 3D structure and $-NH_2$ interactions improve gold nanoparticle dispersion, resulting in many SERS-active "hot spots." The nanocomposite has a high adsorption capacity of 419.85 mg g^{-1} and a low detection limit of 0.05 mg L^{-1} AO II, with a wide linear range of 0.05–50 mg L^{-1}. The nanocomposite-based SERS assay was successfully applied for the detection of AO II in orange juice and chili powder and was highly sensitive and reliable. This research provides a fast, sensitive tool that is portable for use in monitoring illegal dyes in food samples [58]. A high-performance SERS sensor made from AuNPs and Zeolitic Imidazolate Framework-8 photonic crystals (ZIF-8 PCs) was prepared for trace organic dye and pesticide enrichment and detection. The sensor uses the band edge effects of ZIF-8 PCs and AuNP enrichment to realize an enhancement factor of 1.439×10^{11} and a detection limit as low as 1×10^{-12} M for 4-Mercaptobenzoic acid with an RSD of 5.0%. Detection limits as low as 1×10^{-12} M for Rhodamine 6G and Methylene Blue and 1×10^{-11} M for Thiabendazole and Parathion-methyl have also been obtained for the sensor. The proposed portable SERS platform does not require pretreatment in natural lake water, enabling efficient application in onsite monitoring of environmental and food contaminant contamination [59]. A novel SERS-imprinted composite material, Ag@UIO-66(NH_2)/PDA-MIPs, was designed to selectively detect organic dyes in river water. The substrate was synthesized by wrapping UIO-66(NH_2) with a polydopamine (PDA) layer and loading Ag nanoparticles on its rough surface. The MIP layer enhanced the selectivity of the substrate. Optimization of the mass ratio of PDA to $AgNO_3$ enhanced the SERS sensitivity up to a detection limit of less than 10^{-10} M Acid Orange II. A strong linear relationship $R^2 = 0.9985$ was shown between the Acid Orange II concentration (10^{-8} to 10^{-10} M) and the SERS intensity at 1596 cm^{-1}. This substrate provided excellent sensitivity and selectivity, enabling the expansion of SERS applications for practical water sample contaminant detection [60]. Table 2 presents a summary of materials used, analytes detected, merits/key figures, and applications in studies employing SERS-based materials for selective and ultra-trace detection of various synthetic dyes as the contaminants.

Table 2 Summary of materials used, analytes detected, merits/key figures, and applications in studies employing SERS-based materials for selective and ultra-trace detection of various synthetic dyes as the contaminants

Materials used	Analytes	Merits/key figures	Applications	References
ZnO@ag nanostructures	Rhodamine 6G, sunset yellow, Tartrazine	LODs: 10^{-12} M (Rh6G), 10^{-10} M (sunset yellow), 10^{-11} M (Tartrazine)	Detection of multiple food colorants	[38]
AgNPs@ZnO nanocomposites	Methyl Orange (MO), rhodamine B	LODs: 10^{-3} M (MO), 10^{-6} M (RhB), 10^{-9} M (RhB with Ag-Zno nanocomposites)	Dye detection in food colorants	[39]
Fe_3O_4@au core-shells	Acid Orange II, brilliant blue	Recovery rate: 92.95–99.36%; excellent SERS enhancement	Food dye detection and sample integrity preservation	[40]
Fe_3O_4@SiO_2–au core-satellite	Sunset yellow	LOD: 1×10^{-7} M; high signal enhancement from nanoparticle gap tuning	Food dye detection, chemical mixture analysis	[41]
TiO_2 nanorod arrays@PDA/ag	Rhodamine B, crystal violet (CV)	LODs: 1 nM (rhodamine B), 5 nM (crystal violet)	Duplex detection in real food samples	[42]
AgNPs/SnS_2 NHs nanostructures	Methylene blue (MB), Tartrazine (TZ)	Detection limits: 4.12×10^{-10} M (MB), 3.01×10^{-9} M (TZ); enhancement factors: 10^8 (MB), 10^7 (TZ)	Food safety, environmental monitoring	[43]
GO/au@ag nanobones membrane	Rhodamine 6G	LOD: 1.12×10^{-9} M; RSD: 7.87% (uniformity), 10.44% (repeatability); 60 days stability	Beverage colorant detection	[44]

(continued)

Table 2 (continued)

Materials used	Analytes	Merits/key figures	Applications	References
AgNPs@alumina bead-based SERS sensor	4-Aminophenyl disulfide (4-APDS)	LOD: 10^{-15} M; successful detection of food colorants in candies	Point-of-care detection of food additives	[45]
Flower-shaped ag NPs	Rhodamine 6G (R6G)	LOD: 10^{-9} Mol/L; ability to detect food colorants like Tartrazine, sunset yellow, acid red	Sensitive detection of food colorants using SERS	[46]
Au@ag core-shell nanospheres	Sudan I–IV dyes	LOD: 0.10 mg/L (Sudan I), 0.08 mg/L (Sudan III), 0.2 mg/L (Sudan IV)	Illegal Sudan dye detection in food	[47]
Gold nanodumbbells	Sunset yellow, lemon yellow, Orange II, Chrysoidin	LOD: 10^{-8} M for crystal violet; excellent reproducibility (RSD < 15%)	Food safety, on-site pigment detection	[48]
Gold nanorods	Allura red, sunset yellow	LOD: 0.1 mg/L for both dyes; good reproducibility (RSD < 15%)	Beverage colorant detection, rapid on-site application	[49]
Au nanorod-incorporated melamine foam	Rhodamine B, basic Orange II	LOD: 0.1 μg/mL (rhodamine B), 0.5 μg/mL (basic Orange II); reproducibility (RSD < 10%)	Detection of banned food dyes, particularly in chili products	[50]
Hemp spherical AgNPs	Sunset yellow, lemon yellow, carmine, erythrosine	Detection limit of 10^{-8} M; recovery rates between 91.87% and 106.5% with RSD < 1.97%	Rapid, efficient detection of colorants in tea	[51]
Grape skin-AuNPs/GO paper-based SERS sensor	Rhodamine 6G	LOD: 7.33×10^{-11} M; RSD for repeatability: 9.6%, reproducibility: 12.6%, reproducibility: 12.6%	Detection of organic toxins in water and food samples	[52]
Ag nanocube thin-film SERS substrate	Rhodamine B	LOD: 10^{-10} M; excellent dynamic range; stability over 60 days	Sensitive detection of harmful food dyes	[53]
UiO-66(NH_2)@ au SERS	New Coccine (NC), Orange II (OII)	LOD: 0.4015 mg/L (NC), 0.0546 mg/L (OII); $R^2 = 0.9684$ for NC, 0.9912 for OII	Synthetic dye detection in beverages like Mirinda, paprika	[54]

(continued)

Table 2 (continued)

Materials used	Analytes	Merits/key figures	Applications	References
UiO-66/AuNPs suspension	Carcinogenic heterocyclic amines	Good sensitivity, recovery: 82.3–110%; RSD: 3.1–11.0%	Meat safety analysis, SERS-based detection of foodborne toxins	[55]
CC-L-MOF@F-ag substrate	Congo red, Chrysoidine, acid Orange II	LOD: 0.013 μM (CR), 0.017 μM (CYD), 0.0067 μM (AO II); stable for 40 days	Detection of azo dyes in beverages	[56]
MOF(Zr)@ag NPs	Crystal violet, rhodamine 6G, 4-Mercaptobenzoic acid	LOD: 10^{-15} M (CV); good stability, high surface area for hotspot generation	Residue detection in food (shrimp) and environmental samples	[57]
NH2-MIL-101(Cr)@au SERS substrate	Acid Orange II	LOD: 0.05 mg/L; high adsorption capacity: 419.85 mg g^{-1}; linear range: 0.05–50 mg/L	Food and beverage monitoring for banned dyes	[58]
ZIF-8 photonic crystals@AuNPs SERS sensor	Rhodamine 6G, methylene blue, Thiabendazole	LOD: 1×10^{-12} M (rhodamine 6G); enhancement factor: 1.439×10^{11}	On-site monitoring of environmental and food contaminants	[59]
Ag@UIO-66(NH_2)/PDA-MIPs	Acid Orange II	Detection limit: $<10^{-10}$ M High linear relationship: $R^2 = 0.9985$ SERS peak at 1596 cm^{-1}	Selective detection of organic dyes in river water	[60]

Integration of SERS with Machine Learning and Chemometrics for Improved Detection Accuracy

The integration of SERS with chemometric methods and machine learning algorithms helps better analyze such complex food samples, increasing the accuracy of the detection results. Chemometrics is the mathematical and statistical technique used to interpret complex chemical data. When combined with SERS, chemometric methods such as principal component analysis (PCA) and partial least squares (PLS) regression can effectively handle the overlapping spectral features and matrix interferences that are generally common in food samples. This integration facilitates rapid detection and quantification of synthetic dyes, increasing the analysis's reliability. For example, a study demonstrated the use of SERS in combination with chemometrics for the determination of Sudan I–IV dyes. It demonstrated the possibility of this technique for the fast detection of synthetic dyes in food matrices [61]. Integrating machine learning algorithms into the processing of SERS has extended the ability to detect synthetic dyes in food samples. When learning from the complex spectral data of foods, ML models classify and accurately predict dye presence. Techniques include support vector machines (SVMs), artificial neural networks (ANNs), and random forests (RFs) as promising models for predictive modeling given the variability in food matrices. A critical review outlines the detailed procedure for using ML techniques to analyze SERS, noting that such integration is highly significant in improving detection sensitivity [62]. A combined method of SERS with chemometrics and ML has been used to detect several synthetic dyes added to food products. For example, by applying SERS-based sensors, the detection of Rhodamine dyes, which are widely utilized synthetic dyes, has been reported; thus, the applicability of this method for food safety monitoring and environmental pollution can be demonstrated [63]. Additionally, high-performance SERS substrates based on machine learning have enabled ultrasensitive and reusable probes to detect synthetic dyes in the food industry [64]. A portable Raman spectrometer with surface-enhanced Raman scattering (SERS) was used to measure Sudan-I in complex food matrices. The method successfully detected the analyte over a 10^{-3} to 10^{-4} mol/L concentration range. Quantitative analysis was performed via multivariate techniques, such as PCA, PLS regression, ANNs, and SVR, taking advantage of the full spectral information to simplify and accelerate the process. These techniques outperform conventional univariate methods and obviate the selection and integration of marker bands, thus simplifying food safety monitoring [65]. The SERS spectra of the dye Carmine were monitored, and the characteristic peaks were found at 1019, 1360, and 1573 cm^{-1}. Using an enhanced principal component analysis (PCA) method, Carmine was detected at concentrations as low as 10^{-8} M. The presence of these peaks at 10^{-8} M confirms that the detection limit of the AgNP-based substrate for Carmine is approximately 10^{-8} M. This reflects the high sensitivity of the substrate and its suitability for trace-level dye detection [66]. Another study applied an advanced principal component analysis (PCA) method to detect four food colorants, namely, food blue, Tartrazine, Sunset Yellow, and Acid Red, at concentrations ranging from 10^{-8} mol/L. This approach highlights the potential of PCA-enhanced SERS analysis for the sensitive detection of multiple food colorants, which could be crucial for ensuring effective monitoring in food safety applications [46]. A method was developed that combines SERS with machine learning algorithms to detect Auramine O in traditional Chinese medicines, where a sparrow search algorithm-backpropagation (SSA-BP) neural network model was developed that accurately predicted the Acid Orange content in *Dendrobium officinale*, with the coefficient of determination (R_p^2) of 0.962 and root mean square error of prediction of 0.080 mg/mL. Using *D. officinale* as a calibration model, the Auramine O in saffron and curcuma were analyzed. The secondary wavelength screening was improved: the support vector machine model yielded an R_p^2 of 0.780 ± 0.035 for saffron when key wavelengths such as 579, 781, and 1195 cm^{-1} were used, whereas partial least squares regression predicted an R_p^2 of 0.500 ± 0.035 for Curcuma. These results show that the integration of SERS with machine learning offers a rapid, robust approach to Auramine O detection across diverse traditional Chinese medicines [67]. A gold nanorod-based SERS sensor was developed for sensing hazardous azo dyes in natural products such as safflowers. Using the optimized sensor, all the tested dyes in the panel, including Lemon Yellow, Sunset Yellow, Golden Orange II, Acid Red 73, Coccine, and Azorubine, had a detection limit of 50 mg/kg. A high-definition SERS spectral dataset from 960 replicates derived from 64 dye combinations supported robust model training. The K-nearest neighbor (KNN) classification model achieved 91% accuracy in the detection of dye additives when validated via nine batches of safflower, thereby identifying three compounds that contained banned dyes confirmed via high-performance liquid chromatography (HPLC). Density functional theory (DFT) validations ensure the accuracy of characteristic SERS peaks [68].

Development of Portable SERS Devices for Onsite Analysis of Food Samples

The development of portable SERS devices has greatly facilitated the onsite analysis of synthetic dyes in food samples through rapid, sensitive, and nondestructive detection methods. However, the analytical techniques based on HPLC and MS are very accurate but time-consuming and require sophisticated sample preparation that is inappropriate for immediate, onsite analysis. In contrast, portable SERS devices offer a convenient real-time food safety monitoring approach.

Recent studies have shown that portable SERS systems can effectively detect synthetic dyes such as Allura Red and Sunset Yellow in different food matrices. For example, Atta et al. recently developed a solution-based SERS detection method using GNS with high-aspect-ratio spikes; this method achieved detection limits as low as 0.5 parts per billion (ppb) for AR and 1 ppb for SY without the need for aggregating agents, which enhances reproducibility and applicability in real-world samples [69]. Pilot et al. recently reviewed portable Raman instrumentation for detecting food contaminants, including synthetic dyes in foods, via SERS, and these dyes were sensitively and selectively detected for identification [70].

A portable, low-cost Ag nanoflowers @cotton swab (AgNF@CS) SERS substrate has been developed, and limits of detection of 10^{-10} M for Methylene Blue and 10^{-8} M for Carmine have been obtained. The sensor showed high reproducibility, a 1.4×10^7 enhancement factor, and stability over 30 days, allowing residue detection on irregular food surfaces [71]. Commercially available devices, such as the Metrohm Instant SERS Analyzer (MISA), were designed for fast onsite detection of contaminants in food matrices. It has a user-friendly interface with minimal sample preparation and analyte detection at parts per billion levels, which would be very helpful in monitoring food safety [72]. The portability and ease of use make these devices applicable in all settings, from field inspections to quality control in food processing facilities. Real-time analysis allows faster decision-making processes, which enable the identification and removal of contaminated products before they can reach consumers and endanger health. In addition to the detection of synthetic dyes, portable SERS devices also have great versatility. They have been used to detect a variety of food contaminants, including pesticides, mycotoxins, and bacterial pathogens. These applications prove that handheld SERS devices can be widely applied in food safety. For example, handheld SERS devices can be used for rapid onsite pesticide residue detection in rice [73]. The improvement of portable SERS devices has dramatically transformed the detection of synthetic dyes and other contaminants within food samples. They ensure very sensitive and rapid onsite analysis and have become tools for safe food quality at any institution. Further research in this area continues to improve the performance and usability of these devices, which might yield more efficient solutions for detecting contaminants in food in the near future.

Challenges and Limitations

SERS application is fraught with challenges such as sensitivity and reproducibility issues, matrix interference, and cost and scalability issues for its widespread adoption.

Sensitivity and Reproducibility for Complex Food Matrices

There is still a serious challenge in obtaining consistent sensitivity and reproducibility when analyzing complex food matrices via SERS. Food samples are quite heterogeneous, which can lead to variable signals that make quantitative analysis challenging. The distributions of target analytes, interfering substances, and other physical properties of the food matrix can affect SERS performance. For example, detecting limits and validating SERS methods using naturally contaminated samples have proven challenging. This indicates that real-world applications still have sensitivity and reproducibility problems [74]. Moreover, the synthesis of SERS substrates with uniform and dense hotspots is difficult, which may result in inconsistent signal enhancement and can impact reproducibility [75].

Matrix Interference and Mitigation Strategies

Matrix interference is a considerable challenge in SERS-based food analysis. Interference with food matrix components may generate background signals or interact with analytes of interest and their corresponding SERS substrates, which results in signal enhancement or suppression that affects the outcome of the analysis. Therefore, advanced sample purification methods have been used to minimize this interference. Techniques such as single-drop microextraction have been used to extract target analytes from complex matrices, thus reducing interference that leads to accurate detection [76]. Functionalizing SERS substrates with specific capture molecules or ligands can also enhance selectivity, enriching analytes of interest and minimizing matrix effects. In addition, correct sample preparation, which involves homogenization, filtration, extraction, concentration, and purification, is needed to reduce interference and enhance the accessibility of the analyte to the SERS-active substrates [77].

Cost and Scalability Issues

The cost and scalability of SERS technology are two major issues limiting its further application in food safety monitoring. The preparation of high-quality, reproducible SERS substrates is usually performed with complicated and costly nanofabrication techniques, which cannot be easily scaled up for mass production. Developing cost-effective and reproducible SERS substrates is the key issue that needs to be solved for practical applications. The most recent development has been the effort to make new SERS-active substrates sensitive and reproducible simultaneously while attempting to decrease production costs and improve scalability [78]. Another notable breakthrough is the integration of SERS with portable and user-friendly devices that now allow onsite and point-of-care applications. Miniaturizing and creating portable SERS systems have allowed for the real-time and rapid detection of analytes in diverse fields, including food safety [77].

Future Directions

Surface-enhanced Raman spectroscopy has recently emerged as a powerful analytical tool that can detect highly sensitive chemical and biological analytes. Improvements in nanomaterial engineering have proven to be instrumental in increasing the performance of SERS substrates. Traditional substrates are often made of rigid materials, such as glass or silicon, offering uniform and reproducible signals. For example, excellent sensitivity and biocompatibility were achieved by immobilizing silver nanoparticles onto organosilane-functionalized glass surfaces. The latest inventions of flexible and stretchable SERS substrates have further extended their applications to various fields. Since these substrates retain stability even upon mechanical deformation, they find applications in wearable sensors as well as in in vivo diagnostics [79]. SERS can also be used with different complementary analytical techniques. Analyzing through SERS accompanied by chromatography allows the isolation of complex mixtures so that all the components might be simultaneously identified, increasing the analysis accuracy [80]. Similarly, coupling with fluorescence spectroscopy provides a dual detection mode, enhancing both the sensitivity and specificity during analyte detection. With complex sample matrices such as food products, this multimode has been very helpful when trying to identify contaminants through detailed identification [81].

In the field of food safety, SERS has demonstrated great promise. Its ability to detect trace levels of chemical, biological, and metal contaminants makes it a valuable tool for ensuring food quality [4]. However, the widespread adoption of SERS for routine food safety checking would require developing new enhanced regulatory frameworks. Currently, standardized protocols are lacking, and variations in the performance of the substrates used for SERS cause regulatory hurdles in accepting them [82]. The standardized fabrication of SERS substrates and comprehensive databases of SERS spectra for different contaminants could be a step toward their integration into regulatory practices. In addition, portable, user-friendly SERS devices would allow for onsite testing, facilitating faster decision-making in food safety management.

Conclusion

The potential of surface-enhanced Raman spectroscopy (SERS) to detect synthetic dyes in food matrices is covered in this chapter. SERS has shown its potential in detecting adulterants at trace levels owing to its unparalleled sensitivity and specificity, beating the core challenges associated with food safety. Recent improvements, including the invention of new substrates, hybrid materials, and integration with machine learning, enhance applicability and boost reproducibility while gradually reducing matrix interference. Portable SERS devices also make it even more practical by allowing rapid, onsite analysis, crucial for real-time food quality assessment. Despite these advances, there are still prominent challenges, including high cost, lack of scalability, and a lack of standardized protocols to gain mass applications. All these limitations may be addressed using innovative substrate engineering, regulatory alignment, and multimodal analytical approaches so that they can be included in routine food safety practices. In summary, this chapter highlights SERS as an intelligent tool that promises a safer future by using precise and accessible detection methods for food safety.

References

1. Momtaz M, Bubli SY, Khan MS (2023) Mechanisms and health aspects of food adulteration: a comprehensive review. Foods 12(1):199
2. Sawyer WE, Izah SC (2024) Unmasking food adulteration: public health challenges, impacts and mitigation strategies. ES General 4:1091

3. Ntrallou K, Gika H, Tsochatzis E (2020) Food colorants analysis in foods. In Encyclopedia
4. Xu M-L, Gao Y, Han X-X, Zhao B (2022) Innovative application of SERS in food quality and safety: a brief review of recent trends. Foods 11(14):2097
5. Guo Q, Peng Y, Qin J, Chao K, Zhao X, Yin T (2023) Advance in detection technique of lean meat powder residues in meat using SERS: a review. Molecules 28(22):7504
6. Liao W, Lu X (2016) Determination of chemical hazards in foods using surface-enhanced Raman spectroscopy coupled with advanced separation techniques. Trends Food Sci Technol 54:103–113
7. Downham A, Collins P (2000) Colouring our foods in the last and next millennium. Int J Food Sci Technol 35(1):5–22
8. Amchova P, Kotolova H, Ruda-Kucerova J (2015) Health safety issues of synthetic food colorants. Regul Toxicol Pharmacol 73(3):914–922
9. Dilrukshi PGT, Munasinghe H, Silva ABG, De Silva PGSM (2019) Identification of synthetic food colours in selected confectioneries and beverages in Jaffna District, Sri Lanka. J Food Qual 2019(1):7453169
10. McCann D et al (2007) Food additives and hyperactive behaviour in 3-year-old and 8/9-year-old children in the community: a randomised, double-blinded, placebo-controlled trial. Lancet 370(9598):1560–1567
11. Potera C (2010) Diet and nutrition: the artificial food dye blues. Environ Health Perspect 118(10):A428
12. Asero R (2002) Food additives intolerance: a possible cause of perennial rhinitis. J Allergy Clin Immunol 110(6):937–938
13. Novais C et al (2022) Natural food colorants and preservatives: a review, a demand, and a challenge. J Agric Food Chem 70(9):2789–2805
14. Barciela P, Perez-Vazquez A, Prieto MA (2023) Azo dyes in the food industry: features, classification, toxicity, alternatives, and regulation. Food Chem Toxicol 178:113935
15. Niu P et al (2019) Preparation of magnetic carbonized polyaniline nanotube and its adsorption behaviors of xanthene colorants in beverage and fish samples. J Chromatogr A 1605:460369
16. Fiorito S, Epifano F, Palumbo L, Collevecchio C, Spogli R, Genovese S (2023) Separation and quantification of Tartrazine (E102) and brilliant blue FCF (E133) in green colored foods and beverages. Food Res Int 172:113094
17. Ristea M-E, Zarnescu O (2023) Indigo carmine: between necessity and concern. J Xenobiot 13(3):509–528
18. Dey S, Nagababu BH (2022) Applications of food color and bio-preservatives in the food and its effect on the human health. Food Chem Adv 1:100019
19. Vander Leek TK (2024) Food additives and reactions: antioxidants, benzoates, parabens, colorings, flavorings, natural protein-based additives. In: Sicherer SH (ed) Encyclopedia of food allergy, 1st edn. Elsevier, Oxford, pp 862–881
20. Kamino S, Uchiyama M (2023) Xanthene-based functional dyes: towards new molecules operating in the near-infrared region. Org Biomol Chem 21(12):2458–2471
21. Ghulam S, Aamer S, Pervaiz Ali C (2018) A review on the recent trends in synthetic strategies and applications of xanthene dyes. Mini-Rev Org Chem 15(3):166–197
22. Amchova P, Siska F, Ruda-Kucerova J (2024) Food safety and health concerns of synthetic food colors: an update. Toxics 12(7):466
23. Moskovits M (1985) Surface-enhanced spectroscopy. Rev Mod Phys 57(3):783–826
24. Kneipp K et al (1997) Single molecule detection using surface-enhanced Raman scattering (SERS). Phys Rev Lett 78(9):1667–1670
25. Jeanmaire DL, Van Duyne RP (1977) Surface Raman spectroelectrochemistry: part I. Heterocyclic, aromatic, and aliphatic amines adsorbed on the anodized silver electrode. J Electroanal Chem Interfacial Electrochem 84(1):1–20
26. Nie S, Emory SR (1997) Probing single molecules and single nanoparticles by surface-enhanced Raman scattering. Science 275(5303):1102–1106

27. Zong C et al (2018) Surface-enhanced Raman spectroscopy for bioanalysis: reliability and challenges. Chem Rev 118(10):4946–4980
28. Cialla-May D et al (2024) Recent advances of surface enhanced Raman spectroscopy (SERS) in optical biosensing. TrAC Trends Anal Chem 181:117990
29. Zheng J, He L (2014) Surface-enhanced Raman spectroscopy for the chemical analysis of food. Compr Rev Food Sci Food Saf 13(3):317–328
30. Lussier F, Thibault V, Charron B, Wallace GQ, Masson J-F (2020) Deep learning and artificial intelligence methods for Raman and surface-enhanced Raman scattering. TrAC Trends Anal Chem 124:115796
31. Willets KA, Van Duyne RP (2007) Localized surface Plasmon resonance spectroscopy and sensing. Annu Rev Phys Chem 58:267–297
32. Sahbafar H et al (2024) Prepared plasmonic glass substrates via electrodeposition for detecting trace glucose: SERS, DFT, and FDTD investigations. Plasmonics 19(4):2087–2096
33. Tran V, Walkenfort B, König M, Salehi M, Schlücker S (2019) Rapid, quantitative, and ultrasensitive point-of-care testing: a portable SERS reader for lateral flow assays in clinical chemistry. Angew Chem Int Ed 58(2):442–446
34. Mosier-Boss PA (2017) Review of SERS substrates for chemical sensing. Nano 7(6):142
35. Ding S-Y et al (2016) Nanostructure-based plasmon-enhanced Raman spectroscopy for surface analysis of materials. Nat Rev Mater 1(6):16021
36. Wang P, Sun Y, Li X, Wang L, Xu Y, Li G (2021) Recent advances in metal organic frameworks based surface enhanced Raman scattering substrates: synthesis and applications. Molecules 26(1):209
37. Guerrini L, Graham D (2012) Molecularly-mediated assemblies of plasmonic nanoparticles for surface-enhanced Raman spectroscopy applications. Chem Soc Rev 41(21):7085–7107
38. Barveen NR, Wang T-J, Chang Y-H, Yuan-Liu Z (2021) Ultrasensitive and reusable SERS probe for the detection of synthetic dyes in food industry through hybrid flower-shaped ZnO@ ag nanostructures. J Alloys Compd 861:157952
39. Motla A, Sharma D, Soma VR, Annapoorni S (2024) Coupling effect and enhanced detection of organic dyes using SERS with ag noble metal embellished hexagonal ZnO nanorods. Opt Mater 149:115110
40. Xie Y, Chen T, Guo Y, Cheng Y, Qian H, Yao W (2019) Rapid SERS detection of acid orange II and brilliant blue in food by using Fe3O4@au core–shell substrate. Food Chem 270:173–180
41. Sun Z, Du J, Yan L, Chen S, Yang Z, Jing C (2016) Multifunctional Fe_3O_4@SiO_2–Au satellite structured SERS probe for charge selective detection of food dyes. ACS Appl Mater Interfaces 8(5):3056–3062
42. Tan L et al (2024) TiO2 nanorod arrays@PDA/ag with biomimetic polydopamine as binary mediators for duplex SERS detection of illegal food dyes. Anal Chim Acta 1287:342047
43. Barveen NR, Xu J-L, Cheng Y-W (2024) Photoassisted decoration of ag-NPs onto the SnS2 nanohexagons for the ultrasensitive SERS detection and degradation of synthetic dyes. J Environ Chem Eng 12(2):112200
44. Kong L, Chen J, Huang M (2021) GO/au@ag nanobones decorated membrane for simultaneous enrichment and on-site SERS detection of colorants in beverages. Sensors Actuators B Chem 344:130163
45. Chien Y-S, Chang C-W, Huang C-C (2022) Differential surface partitioning for an ultrasensitive solid-state SERS sensor and its application to food colorant analysis. Food Chem 383:132415
46. Ai Y-J et al (2018) Rapid qualitative and quantitative determination of food colorants by both Raman spectra and surface-enhanced Raman scattering (SERS). Food Chem 241:427–433
47. Ou Y et al (2017) Rapid analysis of multiple Sudan dyes in chili flakes using surface-enhanced Raman spectroscopy coupled with Au–Ag core-shell nanospheres. Food Anal Methods 10(3):565–574
48. Meng J, Qin S, Zhang L, Yang L (2016) Designing of a novel gold nanodumbbells SERS substrate for detection of prohibited colorants in drinks. Appl Surf Sci 366:181–186

49. Ou Y, Wang X, Lai K, Huang Y, Rasco BA, Fan Y (2018) Gold nanorods as surface-enhanced Raman spectroscopy substrates for rapid and sensitive analysis of allura red and sunset yellow in beverages. J Agric Food Chem 66(11):2954–2961
50. Sun Y et al (2021) Simultaneous SERS detection of illegal food additives rhodamine B and basic orange II based on au nanorod-incorporated melamine foam. Food Chem 357:129741
51. Li L et al (2023) Rapid detection of multiple colorant adulteration in Keemun black tea based on hemp spherical AgNPs-SERS. Food Chem 398:133841
52. Sridhar K, Inbaraj BS, Chen B-H (2022) An improved surface enhanced Raman spectroscopic method using a paper-based grape skin-gold nanoparticles/graphene oxide substrate for detection of rhodamine 6G in water and food. Chemosphere 301:134702
53. Phuong NTT et al (2021) Rapid and sensitive detection of rhodamine B in food using the plasmonic silver nanocube-based sensor as SERS active substrate. Spectrochim Acta A Mol Biomol Spectrosc 263:120179
54. Wu L, Pu H, Huang L, Sun D-W (2020) Plasmonic nanoparticles on metal-organic framework: a versatile SERS platform for adsorptive detection of new coccine and orange II dyes in food. Food Chem 328:127105
55. Fu J, Lai H, Zhang Z, Li G (2021) UiO-66 metal-organic frameworks/gold nanoparticles based substrates for SERS analysis of food samples. Anal Chim Acta 1161:338464
56. Wang Q et al (2021) Bio-inspired self-cleaning carbon cloth based on flower-like ag nanoparticles and leaf-like MOF: a high-performance and reusable substrate for SERS detection of azo dyes in soft drinks. Sensors Actuators B Chem 329:129080
57. Fan X, Wu P, Qu C, Gao Y, Yin P (2024) Functionalized ag/MOF nanocomposites based on defect engineering for highly sensitive SERS detection of organic dyes. J Environ Chem Eng 12(2):112240
58. Wang Q et al (2020) Rapid simultaneous adsorption and SERS detection of acid orange II using versatile gold nanoparticles decorated NH2-MIL-101(Cr). Anal Chim Acta 1129:126–135
59. Zheng W, Teng D, Kalsoom I, Murtaza G, Meng Z, Qiu L (2025) Au-modified metal-organic framework photonic heterostructure as a SERS sensor for enrichment and detection of trace organic dyes and pesticides. Sensors Actuators B Chem 422:136684
60. Xue Y, Shao J, Sui G, Ma Y, Li H (2021) Rapid detection of orange II dyes in water with SERS imprinted sensor based on PDA-modified MOFs@ag. J Environ Chem Eng 9(6):106317
61. Alomar TS et al (2022) Simultaneous multiplexed quantification of banned Sudan dyes using surface enhanced Raman scattering and chemometrics. Sensors 22(20):7832
62. Srivastava S, Wang W, Zhou W, Jin M, Vikesland PJ (2024) Machine learning-assisted surface-enhanced Raman spectroscopy detection for environmental applications: a review. Environ Sci Technol 58(47):20830–20848
63. Phuong NTT et al (2022) Sensors for detection of the synthetic dye rhodamine in environmental monitoring based on SERS. Micromachines 13(11):1840
64. Zhou W et al (2024) High-performance grating-like SERS substrate based on machine learning for ultrasensitive detection of Zexie-Baizhu decoction. Heliyon 10(9):e30499
65. Cheung W, Shadi IT, Xu Y, Goodacre R (2010) Quantitative analysis of the banned food dye Sudan-1 using surface enhanced Raman scattering with multivariate chemometrics. J Phys Chem C 114(16):7285–7290
66. Wu Y-X et al (2017) Design of a silver nanoparticle for sensitive surface enhanced Raman spectroscopy detection of carmine dye. Food Chem 237:974–980
67. Zhang L et al (2023) Rapid indentification of Auramine O dyeing adulteration in Dendrobium officinale, saffron and curcuma by SERS Raman spectroscopy combined with SSA-BP neural networks model. Foods 12(22):4124
68. Lin C, Zheng C, Fan B, Wang C, Zhao X, Wang Y (2025) Machine learning-assisted SERS sensor for fast and ultrasensitive analysis of multiplex hazardous dyes in natural products. J Hazard Mater 482:136584

69. Atta S, Watcharawittayakul T, Vo-Dinh T (2022) Ultra-high SERS detection of consumable coloring agents using plasmonic gold nanostars with high aspect-ratio spikes. Analyst 147(14):3340–3349
70. Pilot R (2018) SERS detection of food contaminants by means of portable Raman instruments. J Raman Spectrosc 49(6):954–981
71. Zhang S et al (2024) Cotton swabs wrapped with three-dimensional silver nanoflowers as SERS substrates for the determination of food colorant carmine on irregular surfaces. Microchim Acta 191(4):222
72. Metrohm (2024) MISA - metrohm instant SERS analyzer. Available https://www.metrohm.com/en_in/products/raman-spectroscopy/misa.html?utm
73. Logan N et al (2022) Handheld SERS coupled with QuEChERs for the sensitive analysis of multiple pesticides in basmati rice. npj Sci Food 6(1):3
74. Logan N, Cao C, Freitag S, Haughey SA, Krska R, Elliott CT (2024) Advancing mycotoxin detection in food and feed: novel insights from surface-enhanced Raman spectroscopy (SERS). Adv Mater 36(15):e2309625
75. Zhou Z-R et al (2024) Uniform and dense hotspots in au rough-nanocube monolayer for sensitive and reproducible SERS detection. ACS Appl Nano Mater 7(14):17009–17016
76. Terry LR, Sanders S, Potoff RH, Kruel JW, Jain M, Guo H (2022) Applications of surface-enhanced Raman spectroscopy in environmental detection. Anal Sci Adv 3(3–4):113–145
77. Bahlol HS, Li J, Deng J, Foda MF, Han H (2024) Recent progress in nanomaterial-based surface-enhanced Raman spectroscopy for food safety detection. Nano 14(21):1750
78. Chakraborty S et al (2024) PlQC based highly sensitive and reproducible novel SERS active substrate for biomolecule detection with high specificity. Sci Rep 14(1):29186
79. Peng R, Zhang T, Yan S, Song Y, Liu X, Wang J (2023) Recent development and applications of stretchable SERS substrates. Nano 13(22):2968
80. Wei Q, Dong Q, Pu H (2023) Multiplex surface-enhanced Raman scattering: an emerging tool for multicomponent detection of food contaminants. Biosensors 13(2):296
81. He H, Sun D-W, Pu H, Wu Z (2023) A SERS-fluorescence dual-signal aptasensor for sensitive and robust determination of AFB1 in nut samples based on apt-Cy5 and MNP@ag-PEI. Talanta 253:123962
82. Albini B, Galinetto P, Schiavi S, Giulotto E (2023) Food safety issues in the Oltrepò Pavese area: a SERS sensing perspective. Sensors 23(22):9015

Capillary Electrophoresis (CE) for High Resolution Food Component Separation

Aduloju Emmanuel Ibukun, Omachi Bosede Alice, Olawale Salaudeen Abdulwasiu, and Semail Nadhiratul-Farihin

Contents

A. Emmanuel Ibukun (✉)
Department of Toxicology, Advanced Medical and Dental Institute (AMDI), Universiti Sains Malaysia, Penang, Malaysia

Department of Biochemical and Chemical Sciences, The Federal Polytechnic Offa, Offa, Kwara State, Nigeria

O. Bosede Alice
Department of Dietetics and Human Nutrition, University of KwaZulu-Natal, Pietermaritzburg, KwaZulu-Natal, South Africa

O. Salaudeen Abdulwasiu
Chemistry Unit of Applied Mathematics Programme, National Mathematical Centre, Abuja, Nigeria

S. Nadhiratul-Farihin
Faculty of Science and Technology, Department of Chemical Sciences, Universiti Kebangsaan Malaysia, Bangi, Selangor, Malaysia

W. Alahmad, M. Ashfaq (eds.), *Advanced Techniques against Food Adulteration*,
https://doi.org/10.1007/978-3-031-99454-8_15

Abstract Advances in food technology demand efficient and precise analytical methods to ensure safety, quality, and regulatory compliance. Capillary electrophoresis (CE) has emerged as a powerful tool for food analysis, offering high-resolution separation of diverse components such as carbohydrates, proteins, lipids, phenolics, and contaminants within complex matrices. When coupled with mass spectrometry (CE-MS), it combines exceptional separation efficiency, minimal sample requirements, and rapid and efficient analysis of diverse food components such as carbohydrates, proteins, lipids, phenolics, and contaminants, outperforming traditional techniques like HPLC in resolving multicomponent mixtures with varying chemistries. CE also reduces the need for extensive sample cleanup while maintaining automation capabilities. Despite these advantages, its adoption in food analytics remains limited. This chapter explores CE's fundamental principles, methodologies, and applications in food science, covering nutritional profiling, contaminant detection, and authenticity assessment. It critically evaluates CE's performance against conventional methods, emphasizing its integration with advanced sample preparation, including online stacking and detection systems (such as MS LIF) to enhance sensitivity and specificity. It also discusses emerging innovations such as microchip CE and nanomaterial-enhanced capillaries, which are highlighted for their potential to revolutionize portable and high-throughput analysis. The discussion extends to future trends, including green chemistry approaches, automation, and compact analytical devices, positioning CE as a versatile solution for modern food industry challenges. It also compares separation modes such as zone electrophoresis, isoelectric, and micellar electrokinetic chromatography). The chapter guides practitioners in selecting optimal methods for specific needs. CE's adaptability and efficiency underscore its potential to advance food safety, quality control, and traceability, bridging research and industrial application gaps.

Keywords Capillary electrophoresis (CE) · High-resolution separation · Advanced analytical techniques · Green chemistry · CE-MS · CE-MS

Introduction

Capillary electrophoresis (CE) is a powerful analytical technique that separates ions based on their size-to-charge ratios in an applied electric field. This electrokinetic separation technique uses a narrow capillary filled with a buffer solution to facilitate the migration of charged species. As the electric field is applied, molecules move at different rates, depending on their charge, size, and shape, leading to high-efficiency separation. These separations enabled by high voltages may generate an electrophoretic and electroosmotic flow of ionic species and buffer solutions within the sub-millimeter diameter capillary. CE and high-performance capillary electromigration (HPCE) are advanced electroseparation techniques that have been widely adopted across numerous fields, particularly in proteomic [1–4] and food analysis to examine complex mixtures of components, such as proteins, carbohydrates, amino acids, lipids, food additives, and contaminants, due to its sensitivity, high resolution, and ability to analyze small sample volumes [1, 3, 5, 6]. CE has also been instrumental in the development of single-cell proteomics [7, 8] and analyzing a wide range of biological molecules [9–13]. Due to these, CE can be integrated with various detection techniques such as fluorescence and UV. It is also compatible with mass spectrometry (MS) detection, most commonly through electrospray ionization (ESI), though matrix-assisted laser desorption ionization (MALDI) is also used [14, 15].

Importance of High-Resolution Separation of Food Components

High-resolution separation is crucial in food analysis because food products are complex matrices containing many substances, many of which may have very similar chemical properties. The ability to effectively and precisely separate these components is essential for accurate identification and quantification. High-resolution separation allows for the precise measurement of individual food components, even when present in trace amounts. It also helps to distinguish between naturally occurring components and foreign substances, such as pesticides, preservatives, or food additives. It helps to accurately detect harmful contaminants or allergens that may pose a consumer risk. Finally, it ensures that food products meet the required ingredients, nutritional content, and safety specifications, helping maintain product consistency.

The food industry is increasingly focused on ensuring the quality and safety of food products. One of the critical aspects of food analysis is the separation and identification of various components within complex matrices. CE has emerged as a powerful technique for high-resolution separation of food components. This chapter explores the principles, applications, methodological advancements, and prospects of capillary electrolysis in food component separation. CE offers several advantages over traditional separation techniques, including high efficiency, rapid analysis times, and minimal sample requirements. These features make it particularly suitable for analyzing complex food matrices where multiple components may be present simultaneously. This chapter aims to provide a comprehensive overview of capillary electrolysis methods and their applications in food analysis.

Principles of CE

Overview of CE

Electrophoresis is the movement of electrically charged particles within an electrolyte solution when an electric current is applied. CE was introduced by Jorgenson and Lukacs in 1981 as a high-efficiency separation technique that separates charged species (sample ions) based on their size-to-charge ratio under the influence of an electric field. The method utilizes narrow-bore (25–100 μm) 30–100 cm capillaries filled with a buffer solution or background electrolyte (BGE), allowing for the efficient separation of analytes as they migrate through the capillary. The capillary is initially filled with the desired buffer solution. Its ends are then immersed in reservoirs or vials containing the inlet and outlet BGE solutions, along with high-voltage electrodes. It is crucial to ensure that both ends of the capillary remain in contact with the reservoirs throughout the separation process to preserve electrical neutrality. The sample ions will migrate toward the corresponding electrode and pass through the detector, as shown in Fig. 1. A solution ion's migration rate or mobility is primarily influenced by its size and the magnitude of its ionic charge. As the sample moves through the capillary, its components separate and exit the column at different times. Components that migrate more slowly are retained longer, resulting in a longer retention time. Conversely, faster-migrating components have shorter retention times.

The capillary's high electrical resistance allows the application of extremely high electric fields (100–500 V/cm) with minimal heat generation. Additionally, the capillary's large surface area-to-volume ratio efficiently dissipates any generated heat. These high electric fields enable rapid analysis with exceptional efficiency and resolution. Peak efficiency, often exceeding 10^5 theoretical plates, is partly attributed to the plug flow velocity profile of electroosmotic flow, an electrophoretic phenomenon responsible for the bulk movement of the solution within the capillary. This flow facilitates the simultaneous analysis of all solutes, regardless of their charge. Furthermore, CE offers diverse separation modes with unique mechanisms and selectivity, requires minimal sample volumes (1–50 nL), enables on-capillary detection, and supports quantitative analysis and automation.

Principles of CE

A typical CE system consists of a high-voltage power supply, a sample introduction system, a capillary tube, and a detector. The fundamental concepts of electrophoresis and CE, including ionic or electrophoretic mobility (μ) and electroosmotic flow (EOF), as shown in Fig. 2. Electrophoretic mobility depends on the analyte's

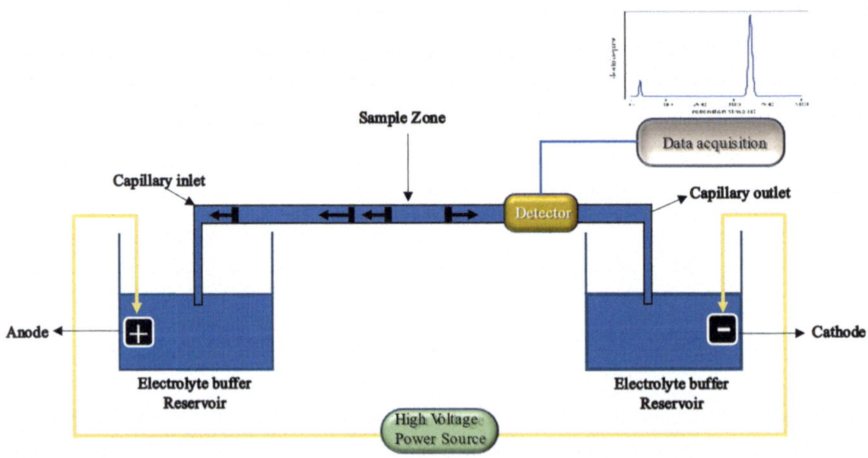

Fig. 1 Schematic representation of a typical CE system

charge-to-size ratio (q/r), being directly proportional to the analyte's charge (q) and inversely proportional to its hydrodynamic radius (r) and the medium's viscosity (η). This relationship is mathematically expressed in the Eq. (1)

$$\mu = \frac{q}{6\pi\eta r} \quad (1)$$

The migration of analytes is influenced by the combined effects of ionic or electrophoretic mobility and electroosmotic flow (EOF). While electrophoretic mobility is specific to each analyte, the EOF remains consistent for all analytes within a sample during a single separation, as shown in Fig. 2.

Stokes' Law gives the frictional coefficient of the molecule (f). This value depends on the molecule's size and shape and the solvent's viscosity (η).

The Principles of CE and its Mechanisms

The principles of separation in the CE are achieved in the narrow-bore capillary, usually ≥25 μm in internal diameter (ID) and ≥30 cm in total length, depending on the application. The capillary's exterior is coated with polyimide or Teflon to increase its flexibility and protection, except for the detection window, where the polyimide is removed for ease of optical detection. The capillary interior is coated with bare fused silica with an ionizable surface modified with silanol functional groups. The narrow capillary is filled with a background electrolyte (BGE) by submerging it into a BGE reservoir, and electrodes in all reservoirs connect to a voltage source. Both capillary ends must always contact the reservoirs for successful separation to maintain electrical neutrality.

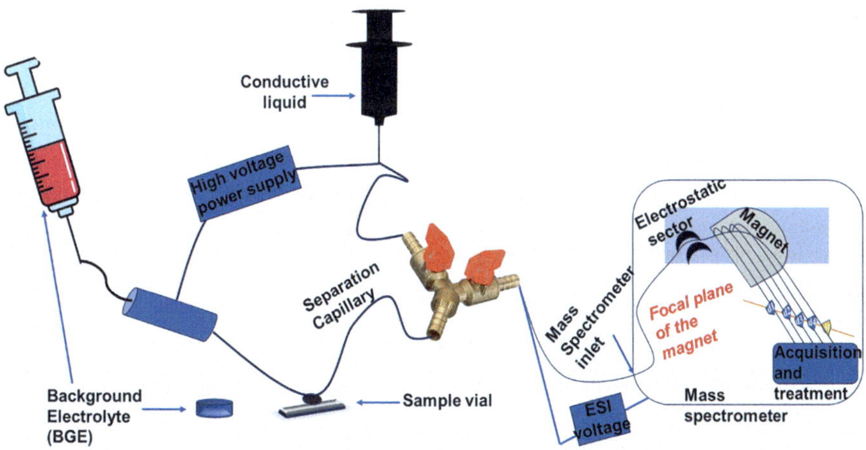

Fig. 2 A schematic representation of surface sampling CE coupled with mass spectrometry. The variable E is the applied electric field. The velocity describes the rate at which the components move through the capillary

The sample is injected by pressure or voltage through a capillary from the sample vial into the system (sample inlet) and plugged. The capillary is returned to the BGE reservoir for the separation process. In the detection area, the targeted analytes are detected as they flow via capillary through a transparent detection window via UV absorption or fluorescence (FLD) [16–18] such as laser-induced fluorescence (LIF) [19, 20] in conjunction with electrochemical detection such as conductivity detection (CD), amperometric detection (AD), and mass spectrometry (MS) [21] as shown in Fig. 2. The migration of the analytes from the sample injection point through the system is influenced by the combination of electrophoretic mobility (μ) and electroosmotic flow. While the electrophoretic mobility (μ) is unique to each analyte, the electroosmotic flow is constant. The results are displayed with signal on the y-axis against the migration time on the x-axis on the electropherogram.

Sample Injection Methods

Sample injection methods include hydrodynamic and electrokinetic injection. The injection processes used in the CE are hydrodynamic injection (pressure-based) and electrokinetic injection (electrophoresis-based). The hydrodynamic injection uses pressure on the capillary inlet or outlet achieved by siphoning in the traditional CE, while in the improved pneumatic CE systems, it is achieved by either applying positive pressure on the capillary inlet or a negative pressure on the outlet. The volume of the sample (V) injected into the CE system via hydrodynamic injection is given by Eq. (2).

$$V = \frac{\Delta P D^4 \pi t}{128 \eta L} \tag{2}$$

V is the sample volume, D is the capillary, and t is the injection time. The injected V from Eq. (2) is directly proportional to the capillary's inner diameter (D), a pressure difference between the two ends of the capillary (ΔP), but inversely proportional to both viscosity (η) and capillary length (L).

For electrokinetic injection, the migration of the targeted analytes relies on voltage based on both electroosmotic flow and /or their mobility. This injection method is not influenced by capillary volume like the hydrodynamic injection. It is more selective than pressure loading, though it can lead to biases if there are significant differences in the mobility of the analytes. For high-salt sample content, hydrodynamic injection is preferred because the number of anions, cations, and neutral ions injected into the systems differs, which might lead to variation in the injection of the analytes. Moreover, the hydrodynamic injection delivers a defined volume, but capillary volume is a significant limiting factor that can alter the sample loading; thus, the electrokinetic injection is preferred due to increased sensitivity [22]. Therefore, the number of targeted analytes injected for each sample must be calculated separately for each solute (Q), as given in Eq. (3).

$$Q = \left(\mu_{eo} + \mu_{ep}\right) \pi r^2 E C t \tag{3}$$

where Q is the amount of individual solute injected, $\mu_{eo}$is known as the electroosmotic mobility, μ_{ep} is the electrophoretic mobility, E is the strength of the electric field, r is the capilarry radius, C is the concentration of the particular solute, and t injection time.

Electric Fields

Electric fields in CE play significant roles in analyte separation. The electric field in the sample varies from that in the rest of the capillary because of the differences in ionic composition between the sample and the buffer. Generally, the sample exhibits lower conductivity due to reduced ionic strength. The electric field applies a force to charged analytes (ions) within the capillary.

The effective electric field is given in Eq. (4):

$$E' = E \times \frac{X_{\text{buffer}}}{X_{\text{sample}}} \tag{4}$$

This force allows the analytes to migrate according to their electrophoretic mobility under the influence of their charge, shape, and size. EFs also influence the electroosmotic flow (EOF), resulting in separation based on influences and electrophoretic mobility [23]. The electric field helps focus the targeted analytes into sharp/narrow bands, enhancing separation resolution in approaches such as isotachophoresis and isoelectric focusing. The strength of the electric field affects both the separation speed and resolution, with higher fields typically leading to faster separations. Still, excessive fields can cause Joule heating, reducing efficiency and resolution. Electric fields are essential in CE, driving analyte movement, enabling separation, and optimizing resolution and efficiency. Sample loading in the CE is affected by factors including the nature of the sample, the conductivity of the background electrolyte (BGE), the type of analysis to be performed and the ionic strength of the sample matrix.

Stacking Techniques

Stacking techniques such as field-amplified sample stacking (FASS) and field-amplified sample injection (FASI) enhance sensitivity by concentrating analytes into narrow bands. The FASS uses hydrodynamic injection, while the FASI uses electrokinetic injection to insert the plug. These techniques work by inserting a sample plug with a lower conductivity than the BGE into the capillary, focusing analytes into sharper and taller peaks as they migrate, thus improving detection sensitivity. Other developed stacking methods, including large volume sample stacking/injection (LVSS/LVSI), as shown in Fig. 3, sweeping [24, 25], electrokinetic supercharging (EKS), pH junction [26, 27], and transient isotachophoresis (tITP) [28], and are chosen based on various factors.

Large Volume Sample Stacking/Injection (LVSS/LVSI)

Large volume sample stacking (LVSS) is a technique that enhances sensitivity by compressing a large sample volume into a very narrow sample zone, improving detection sensitivity by 50–100 times. Chien and Burgi introduced the LVSS in 1992 for protein analysis. The sample was diluted in an aqueous solution and filled into a portion (1/2 or 1/3) of the capillary. The sample is introduced by gravity before applying reverse polarity voltage at the detection end of the capillary, which pushes the sample plug/ sample matrix from the capillary via the reversed EOF. The targeted analytes stack at the boundary between the sample plug and background electrolyte (BGE) zone and are then further separated by capillary zone electrophoresis. LVSS exhibits superior efficiency and sensitivity over the traditional CE techniques, as shown in Fig. 3. It allows large sample volumes to be concentrated online without affecting the peak shape.

Pressure Electrokinetic Supercharging (EKS)

Pressure-assisted electrokinetic injection (PAEKI) is an advanced online sample enrichment technique that enhances sample injection and sensitivity without compromising separation efficiency or altering separation efficiency. It combines field-amplified sample injection (FASI) and transient isotachophoresis (tITP) principles to achieve significant enrichment by applying external pressure while balancing the

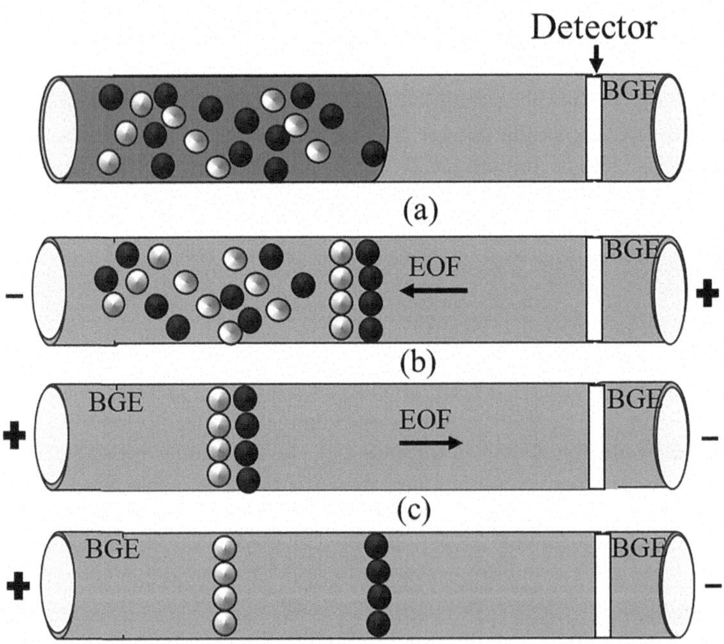

Fig. 3 Schematic diagram illustrating LVSS mechanisms

electroosmotic flow (EOF) in the opposite direction to stabilize the EOF, as shown in Fig. 4. This allows for longer injection times and more excellent sample volume introduction. The background electrolyte (BGE) is the leading electrolyte in this technique, followed by a long sample plug introduced under reversed polarity and external pressure. The sample solution is prepared in a high-resistivity matrix, while the BGE has low resistivity, facilitating analyte stacking at the boundary between the two zones. Applying external pressure is crucial for suppressing EOF and preventing velocity mismatches that could lead to band broadening. PAEKI enables three injection modes, especially for anionic analytes such as conventional electrokinetic injection (EKI) and high-density injection (HDI) and a novel pressure-assisted method that combines both, as shown in Fig. 4. The diagram illustrates three different sample injection and stacking mechanisms used before capillary electrophoresis (CE) separation, including hydrodynamic injection (HDI), electrokinetic injection (EKI), and pressure-assisted electrokinetic injection (PAEKI). Hydrodynamic injection (HDI) involves introducing a low-conductivity sample solution into a capillary prefilled with a high-conductivity background electrolyte (BGE) by applying pressure or vacuum (Fig. 4). This conductivity difference causes minimal analyte stacking at the interface of the sample and BGE, enhancing sensitivity. The effectiveness of the stacking depends on the injection time, pressure applied, and conductivity contrast. Electrokinetic injection (EKI) introduces ionic analytes using an electric field to preconcentrate analyte to migrate into the capillary, enhancing their concentration when sample conductivity is lower than the background electrolyte (Fig. 4). However, EKI is biased toward highly charged species. This method enhances stacking efficiency, particularly for weakly charged or hydrophobic analytes, making it highly effective for trace analysis before capillary electrophoresis (CE) separation. Unlike HDI and EKI, which limit the sample volume due to capillary constraints, PAEKI utilizes external pressure to stabilize EOF and allow the introduction of larger sample volumes without affecting the separation process, as shown in Fig. 4. However, there are limitations in this technique including excessive injection volume or incorrect optimization can disrupt the stacked zone, leading to poor resolution, incomplete focusing, and potential reproducibility issues. Therefore, careful optimization of injection parameters, such as voltage and pressure, is crucial for achieving optimal separation efficiency. The optimization of capillary length, external pressure, and voltage is critical to achieving effective preconcentration and maximizing sample injection without altering separation efficiency.

Sweeping

Sweeping is an online stacking technique that enhances the detection of neutral and hydrophobic analytes, improving sensitivity (SEF) by ≥700-fold, as shown in Fig. 5. It involves the aggregation of micelles that concentrate analytes during injection. The technique relies on introducing a high-conductivity sample solution and injecting a micellar BGE solution, accumulating and concentrating analytes through electrophoresis. When voltage is applied, the analytes that were initially stacked between the non-micellar sample and the BGE zones are rearranged and migrate toward the

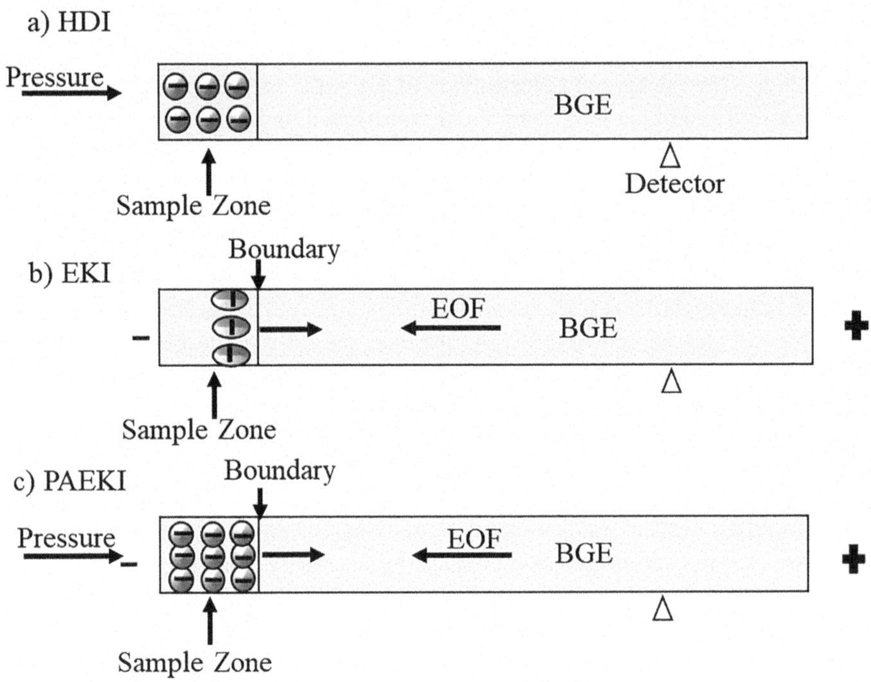

Fig. 4 The diagram illustrates three different sample injection and stacking mechanisms used before capillary electrophoresis (CE) separation, including hydrodynamic injection (HDI), electrokinetic injection (EKI), and pressure-assisted electrokinetic injection (PAEKI). Hydrodynamic injection (HDI) involves introducing a low-conductivity sample solution into a capillary pre-filled with a high-conductivity background electrolyte (BGE) by applying pressure or vacuum (Fig. 4). This conductivity difference causes minimal analyte stacking at the interface of the sample and BGE, enhancing sensitivity. The effectiveness of the stacking depends on the injection time, pressure applied, and conductivity contrast. Electrokinetic injection (EKI) introduces ionic analytes using an electric field to preconcentrate analyte to migrate into the capillary, enhancing their concentration when sample conductivity is lower than the background electrolyte (Fig. 4). However, EKI is biased toward highly charged species. This method enhances stacking efficiency, particularly for weakly charged or hydrophobic analytes, making it highly effective for trace analysis before capillary electrophoresis (CE) separation

other cathodic end, as shown in Fig. 5. This results in different electrophoretic velocities across the two zones, creating untacking and broadening of the target analytes peak bands. The inclusion of micellar electrokinetic chromatography into the sweeping results into the Sweeping-MEKC method [25, 29] further improves sensitivity by removing the sample matrix and eliminating diffusion effects, thus achieving sensitivity enhancement factors (SEF) ≥600-fold. The performance is optimized when the sample conductivity matches the BGE conductivity. This minimizes diffusion and enhances focus and separation efficiency, especially for hydrophobic analytes, as shown in Fig. 5.

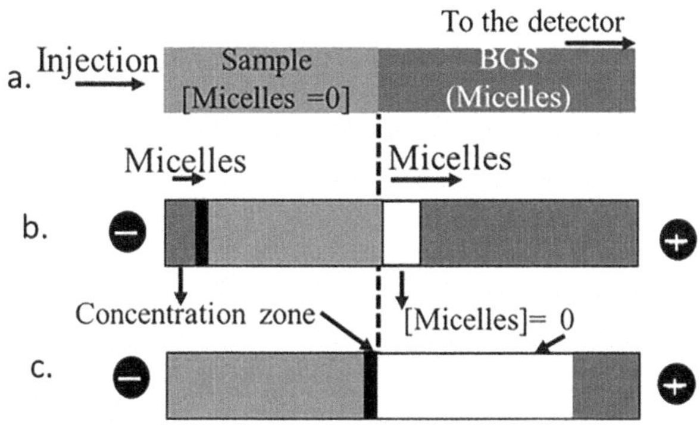

Fig. 5 Schematic diagram of sweeping mechanism. Sweeping in electrokinetic chromatography (EKC) involves the accumulation of analytes by a charged pseudostationary phase (e.g., micelles, CDs). Unlike stacking, sweeping does not depend on sample conductivity but rather on the presence of micelles in the background electrolyte (BGE) and their absence in the sample. Matrix effects must be considered. The diagram illustrates the mechanism of analyte separation in capillary electrophoresis using a pseudostationary (PS) phase, such as anionic micelles in micellar electrokinetic chromatography (MEKC). Sample components are injected into a capillary filled with background solution (BGS). As the PS zone (where micelle=0) sweeps through, hydrophobic analytes partition into the micelles and migrate at micellar velocity, while hydrophilic or charged analytes migrate differently based on electrophoretic mobility. PS vacancies indicate regions where micelles are absent, possibly showing unbound analytes. Differential interactions with the PS phase lead to separation, and resolved analytes are detected as distinct peaks

Isotachophoresis (ITP)

Isotachophoresis (ITP) was described by Kendall and Crittenden in 1923 as a versatile electrophoretic method that separates charged components based on differences in their electrophoretic mobilities and is effective for preconcentration, purification, and mixing, especially in trace analysis. The process moves charged electrolytes toward the equilibrium zone between the leading (LE) and trailing electrolytes (LE) with lower conductivity. When a voltage is applied across the device, the targeted analytes (from the injected sample matrix) are introduced between two electrolytes, migrating, separating and concentrating toward the equilibrium zone, forming a sharp band [30]. The movement of charged species in an electric field is governed by the eq. $V = nE$, where n represents the effective electrophoretic mobility and E denotes the electric field strength. In ITP, the trailing electrolyte (TE) and leading electrolyte (LE) are chosen such that the TE has a lower effective mobility than the LE, fulfilling the condition $n1_{\mathrm{e}} < n1_{\mathrm{e}}$ [30]. ITP is highly efficient for preconcentration and purification of analytes, relying on zone focusing and stacking principles. When ITP is under the influence of an electric field, ions travel through the medium at uniform velocities, creating separate zones according to their electrophoretic mobilities [31].Isotachophoresis (ITP) involves introducing a sample containing multiple ionic species into a buffer system composed of a leading electrolyte with higher ion mobility and a trailing electrolyte with lower ion mobility.Transient ITP

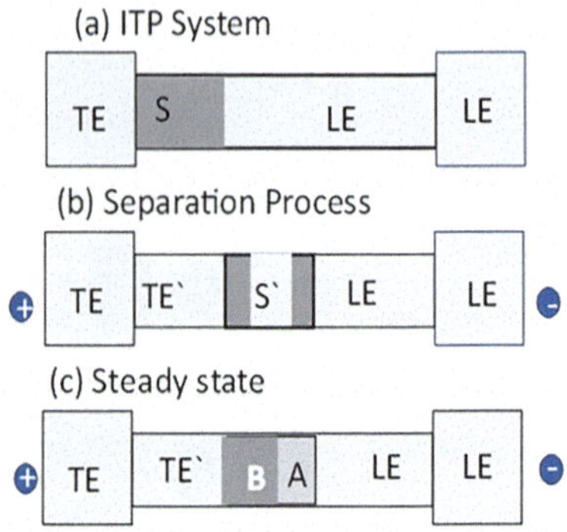

Fig. 6 Schematic mechanism of t-ITP in a CE

(t-ITP) uses a discontinuous electrolyte system with leading and terminating electrolytes to concentrate analytes by adjusting their migration velocities (Fig. 6) [32]. The leading electrolytes exhibit higher conductivity and greater mobility than the trailing electrolytes, while the targeted analyte ions possess intermediate mobilities. This process enables sample concentration in a smaller area in the capillary or paper-based systems. Sensitivity can be enhanced 10–1000 times using ITP, which can be combined with capillary zone electrophoresis (CZE) for further analysis.

Electrokinetic supercharging (EKS), a variation of ITP, combines FASI and t-ITP for improved sensitivity and allows higher sample volumes without peak broadening, achieving sensitivity enhancements of up to 1800-fold and limits of detection as low as 100 ng/L. The advantages of this method include the lack of sample preparation, speed, and applicability to a broad range of samples. It also makes the detection of low-concentration ions and analytes possible even with enhanced sensitivity, offering fast detection, quantification, and high-throughput analysis [30, 33]. However, it can only be used simultaneously for cations or anions, not both, such as small ions, proteins, nucleic acids, and peptides [34]. Other limitations include limited sample volume, ion compatibility, and small fluctuations in voltage, which can affect the reproducibility and resolution of separation.

Micelle-Based Stacking (MSS)

Joselito P. Quirino initially introduced micelle to solvent stacking (MSS) in 2009 [35–38]. It involves reversing the direction of the effective electrophoretic mobility of organic cations, using micelles in the sample solution and an organic solvent in the background electrolytes solution (BGS) (buffer solution), as shown in Fig. 7. This reversal leads to analyte focusing, which occur at the boundary zone between micellar solution without organic solvent and the BGS modified by an organic solvent known as the micelle to solvent stacking boundary, (MSSB). Recently, cyclodextrin (CD)-assisted stacking or sweeping techniques in CE have been developed to enhance sensitivity, enabling better detection and concentration of analytes across diverse and complex sample matrices [25, 39–43]. The effectiveness of analyte

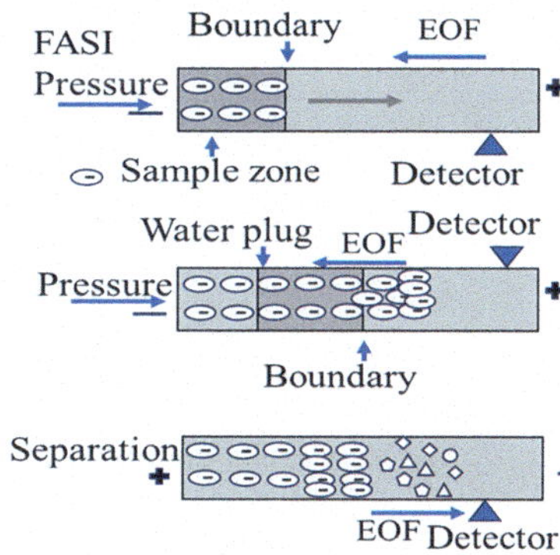

Fig. 7 Schematic mechanism of MSS in a CE

enrichment largely depends on the sample composition, which includes a micelle-to-trapping solution, the injection method, and the applied voltage.

To enhance the detection sensitivity of the MSS, a micelle-to-trapping solution stacking was created by introducing a section of trapping solution containing high concentrations of organic solvent, which induces micelle collapse. The organic solvent applied to induce micelle collapse and influence micelle interaction includes ethanol, acetonitrile, and methanol. The targeted analytes alter their effective electrophoretic directions in the trapping solution after being released by the collapse of the micelle. Nevertheless, these techniques have not yet been incorporated in non-aqueous capillary electrophoresis (NACE). Other recent innovations focus on micelle types, the location of micelles and organic solvents, and the suitable separation method to apply. Other researchers have focused on cationic surfactants to separate and detect anionic targeted analytes in the micelle to a solvent stacking boundary (MSSB). Other models used in the detection of the wide range of compounds include three-step stacking, where the three organic solvents are in the analytes and micelles are injected, respectively, and two-step stacking, where the organic solvent is in the BGS and the samples and micelles are injected separately, one-step stacking where the organic solvent is in the BGS and samples is mixed with micelles and other variations. The selection of the injection models can also lead to significantly enhanced performance, especially in achieving very low LOD. Advantages and limitations of stacking method are presented in Table 1.

Types of CE Methods

There are several types of CE such as capillary zone electrophoresis (CZE), capillary gel electrophoresis (CGE), electrokinetic capillary chromatography (EKC), which includes various methods such as micellar electrokinetic chromatography (MEKC), microemulsion electrokinetic chromatography (MEEKC), etc. Others include capillary electrochromatography (CEC), capillary isoelectric focusing (CIEF), capillary isotachophoresis (CITP), and non-aqueous capillary

Table 1 Advantages and limitations of stacking method

S. No.	Methods	Features	Advantages	Limitations
1.	t-ITP/EKS	Stacking techniques use two BGEs with different EOF rates, including LE with higher EOF and a TE with lower EOF. This focuses the sample as analytes migrate based on electrophoretic mobility into the LE/TE using EFs reaching a steady state	Effective for focusing a wide range of analytes – High sensitivity, SEF ≤10,000 folds – Simple to implement – 30–50% sample solution can be introduced into the capillary without altering separation efficiency	– Limited to low ionic strength samples – Varying analyte mobilities affects the steady state – Sensitivity can be affected by sample matrix – Maintaining the pH between BGE system and sample while identifying suitable LE and TE differences is challenging
2.	PAEKI	It involves the integration of both FASI and t-ITP stacking principles by applying external pressure and adjusting the EOF in the opposite direction, which extends the injection time until a steady state is reached	– Elimination of sample matrix using the EOF migrating at the opposite direction – Extended injection time ≤3600 s – Offers selective preconcentration for specific analytes – Enhances resolution and sensitivity – Allow the use of sample plug	– Complex operation due to pH control, requiring careful optimization of pH conditions – Peak overlap due to increased sample concentration altered by injection voltage – Maintaining the balance between EOF and counterbalance pressure sample is challenging – Extended ramping time
3.	Sweeping	Utilizes additives or surfactants as a pseudo-stationary phase (PSP), causing both analytes and micelle aggregates to accumulate and concentrate within the BGE zone	High preconcentration efficiency for both anions and cations Suitable for complex samples and neutral and zwitterionic analytes	Limited to hydrophobic analytes and amphoteric compounds Requires precise control of the sample and background solution May cause broadening of peaks

(continued)

Table 1 (continued)

S. No.	Methods	Features	Advantages	Limitations
4.	Dynamic pH junction	In this approach, a discontinuous basic BGE system causes band narrowing. The analyte's migration velocity is influenced by a reversal in its electrophoretic direction, moving from the acidic sample zone to the basic BGS zone or vice versa	A simple technique for analyte focusing – Effective for ionic species – Minimal sample loss – ≤95% targeted analyte focusing is achievable – Offers selective focusing of specific target analytes by significantly altering their ionization state through changes in pH	Limited to analytes with different pH behaviors – Requires careful optimization of pH conditions – The precise value in pH difference between BGE and sample zone system is very complicated to induce stacking
5.	Sequential stacking/ hybrid	It combines multiple stacking methods to enhance analytes preconcentration, detection and separation efficiency in a single analysis Sequential application of different conditions or strategies to focus analytes, improving sensitivity and resolution	Combines multiple techniques for enhanced sensitivity – Versatile, applicable to various analytes – Enhanced SEF ≤100,000 folds – Highly selective – Ideal for quantifying trace analytes in complex sample matrices	– Complex setup and operation – Potential for method interferences or inconsistencies in results – It is challenging to incorporate 2–3 models in a single analysis due to the use of co-solvents, multiple BGEs, and analyte mobility systems – Sample overloading and Joule heating due to sequential FASI injection negatively impact separation efficiency

electrophoresis (NACE), as shown in Fig. 8. The choice of technique depends on the physicochemical properties of the analyte and the specific purposes of the analysis.

(a) CITP: Separates ions/analytes based on their electrophoretic mobilities using a two-phase electrolyte system (leading (LE) and terminating electrolytes (TE)). In CITP, ions are separated into distinct bands based on their migration velocities in a discontinuous electrolyte system, allowing high-resolution separation.

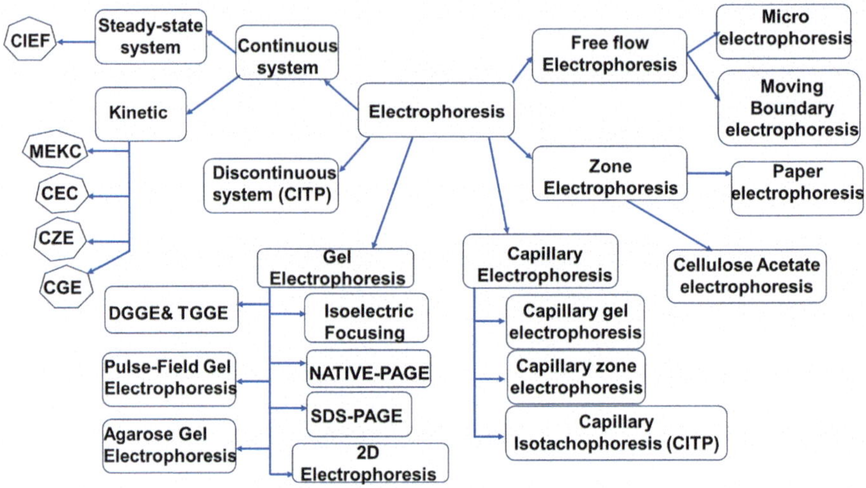

Fig. 8 Mind maps of types and modes of electrophoresis

This method is ideal for analyzing ions with similar mobilities and offers high sensitivity and resolution.

(b) CE: This is the most basic form of capillary electroseparation. It involves the separation of charged analytes in a continuous buffer-filled capillary tube under the influence of an electric field. The analytes migrate based on their electrophoretic mobility, which is determined by their charge and size ratio [44].

(c) Capillary gel electrophoresis (CGE): CGE is similar to traditional CE but uses a gel-filled matrix inside the capillary to separate analytes. The gel provides electrophoretic effects and additional size-based separation (sieving effects), allowing it to resolve larger molecules like proteins or nucleic acids based on charge and size. The background electrolyte (BGE) is substituted with a sieving gel matrix to facilitate separation. Before analysis, the protein sample is denatured by heating in the presence of sodium dodecyl sulfate (SDS), which uniformly coats the proteins at a ratio of approximately 1.4 g SDS per gram of protein [45], giving the proteins a uniform charge-to-size or q/r ratio. As a result, the inherent charges of individual proteins become insignificant once they are fully coated with SDS.

(d) CIEF: CIEF is applied for separating amphoteric substances, such as proteins, hemoglobin variants, immunoglobulins, and post-translational modifications of recombinant proteins, based on their isoelectric points (pI). In CIEF, the pH varies across the capillary, with the pH decreasing at the anodic section (toward the anode) and increasing toward the cathodic section. This creates a pH gradient within the capillary, and analytes migrate to the point where their net charge is zero (at their pI), allowing high-resolution separation. This pH gradient is created by a mixture of ampholyte molecules with both acidic and basic functional groups, which, when subjected to an electric field, generate a continuous pH gradient. In IEF, the migration of ampholytes stops when they reach their isoelectric point (pI), at which point they become neutral and no longer move.

The resolution of the separation improves as the number of ampholytes in the solution increases, creating a more uniform pH gradient. The high-resolution separation of complex mixtures in CIEF, particularly in protein analysis, relies on three fundamental processes: loading, focusing, and mobilization. To enhance the effectiveness of IEF, it is important to suppress electroosmotic flow (EOF) and other convective forces, which can be done by coating the capillary walls with materials like methylcellulose or polyacrylamide. This also helps to reduce protein adsorption to the capillary walls [46]. Therefore, the ΔpI of IEF also referred to as the resolving power, is quantified using Eqs. (5) and (6), which assess the effectiveness of separating different amphoteric substances according to their pI values.

$$\Delta pI = 3\sqrt{\frac{D(d\mathrm{pH}/dx)}{\mathrm{E}(d\mu/d\mathrm{pH})}} \quad (5)$$

or

$$3.D\left(\frac{d(pH)}{dx}\right) \div E\left(\frac{d\mu}{d(pH)}\right) \quad (6)$$

where D is the diffusion coefficient, E is the electric field strength, $\left(\frac{d(pH)}{dx}\right)$ is the rate of change of pH within the capillary, μ is the electrophoretic mobility of the protein, and $\frac{d\mu}{d(pH)}$ is the mobility gradient.

(e) CE Ionization mass spectrometry (CE-ESI-MS) combines CE with electrospray ionization mass spectrometry (ESI-MS) to separate and analyze complex mixtures. After separation in the capillary, analytes are ionized and analyzed by mass spectrometry, offering separation and structural information. The crucial components of the advanced ESI sources include a high-voltage (HV) power supply, a sampling tube with a spray tip, liquid chromatography and syringe pump for liquid injection, and auxiliary gas.

CEC: CEC combines CE and high-performance liquid chromatography (HPLC) principles. It utilizes a stationary phase inside the capillary and an electric field to enhance the separation of analytes. This technique is useful for separating neutral and charged compounds alike. These techniques, including chiral CEC [47, 48], chiral ligand exchange CE (CLE-CE) [49, 50], and chiral CE [51, 52], have demonstrated high-resolution enantiomers separation, underscoring their utility in analytical applications. The CLE-CE method generates ternary complexes exhibiting distinct stability differences involving enantiomers, metal ions, and chiral ligands [53–57]. Integrating CLE-CE-CEC [58–60], CLE-CEC-UV [61], and stimuli-responsive polymer [62] offers a promising strategy.

(f) Micellar electrokinetic chromatography (MEKC): MEKC is a modified form of CE that incorporates micelles (surfactant molecules aggregates) in the background electrolyte (BGE) [63, 64]. These micelles act as a pseudo-stationary phase, allowing for the enhanced separation of charged and neutral analytes based on their interactions with the micelles and their charge and size [41, 65, 66]. It separates charged and neutral solutes in natural products, biological samples, medical compounds, and the surrounding environment [64, 67–69]. To enhance the selectivity and sensitivity of micellar electrokinetic chromatography (MEKC), the selection of an appropriate surfactant is critical. Both the charge type (non-ionic, cationic, anionic, or zwitterionic) and chemical structure of the surfactant significantly influence separation efficiency. Commonly used surfactants include decyl glucoside, lauryl glucoside, and N-octyl-β-D-glucopyranoside. They exhibit outstanding stability and are recognized for being natural, environmentally friendly, and having low electrical conductivity [75, 76]. Examples are In-Cap-D-NS-MEKC-FL [77–81]. The modes of CE techniques are illustrated in Table 2.

These techniques can be classified into two systems based on the arrangement of the electrolytes:

(a) Continuous system (homogeneous BGE): In this system, a single background electrolyte (BGE) is present throughout the capillary, serving as a buffer to enhance separation, which is primarily based on the electrophoretic mobility of the analytes. This system can be further classified into kinetic and steady-state systems. There is a constant electrolyte system in the former, and the buffer remains unabated during the process. At the same time, there is a varying electrolyte composition system in the latter, where the buffer composition is altered during the process. Techniques such as CE and micellar electrokinetic chromatography (MEKC) typically operate continuously, where analytes are separated as they migrate through the homogeneous electrolyte, as shown in Fig. 8.
(b) Discontinuous system (two-phase BGE): This system uses two or more BGEs with different compositions, usually creating a pH or ionic strength gradient. This system divides the sample into unique bands, each separated by a different BGE, allowing for targeted and specific separations. Techniques like CIEF and capillary gel electrophoresis (CGE) operate in a discontinuous system, where analytes are separated based on different properties such as isoelectric point (pI) or size, influenced by varying conditions within the capillary, as shown in Fig. 8.

CE Optimization Parameters

Optimization in capillary electrophoresis involves fine-tuning multiple parameters to achieve the best separation efficiency, resolution, and reproducibility. To achieve optimal performance, several key parameters such as pH, buffer composition, voltage, capillary coating, and temperature must be carefully optimized to the specific characteristics of the analytes to achieve the best detection and separation of target analytes. Additionally, there is a need for systematic method development and optimization studies to improve the performance and reproductivity of CE methods.

Table 2 Summarized modes of CE techniques

S. No	Technique	Strenght	Limitations
1.	CE	• High resolution for small ions and molecules • Fast analysis • Requires minimal sample	• Limited to charged analytes • Sensitivity can be affected by matrix effects
2.	Capillary zone electrophoresis (CZE)	• Rapid, minimal preparation • Simple, widely used technique • Suitable for a wide range of analytes • High resolution • Suitable for ions and small molecules	• Limited to charged analytes • Lower resolution for closely related analytes • Complex samples may interfere • Limited resolution for complex molecules
3.	Capillary Gel Electrophoresis (CGE)	• Ideal for separating large molecules (proteins, Nucleic acid such as DNA). • Good resolution for biomolecules	• Requires gel matrix, which can make the process slow or complicated • Lower sensitivity than CE for small analytes
4.	CIEF	• Excellent for separating amphoteric substances (including peptides and proteins) based on their pI • High resolution	• Limited to amphoteric substances • pH gradient can be hard to maintain • Time-consuming • Challenging for complex matrices
5.	Capillary electrospray ionization mass spectrometry (CE-ESI-MS)	• Combines separation with mass spectrometric analysis. • Provides structural information • Highly sensitive	• Requires mass spectrometer • Complex setup • Expensive equipment
6.	CEC	• Suitable for separating neutral and charged analytes • High separation efficiency • Versatile	• Complex setup due to stationary phase • Requires optimization of conditions for each analyte
7.	Micellar electrokinetic chromatography (MEKC)	• It can separate both charged and neutral compounds • Good for complex mixtures • Simple and quick	• Limited resolution for closely related compounds • Micelle composition must be optimized, because its stability affects them methods reproducibility
8.	CITP	• Efficient for complex mixture • Excellent for high-resolution separation of ions with similar mobilities. High sensitivity • Effective for trace analysis	• Concentration effects • Limited to ionic analytes • Requires careful optimization of electrolyte systems • Complex sample preparation

Thorough validation is necessary to confirm these CE methods' accuracy, linearity, precision, and robustness, ensuring they offer reliable and reproducible results across various laboratories.

Effect of pH

The pH of the background electrolyte (BGE) plays a critical role in determining the ionic strength, directly impacting the analyte's charge, EOF, and the overall separation process. The pH of the BGE has a significant impact on the separation process. In CE, the buffer's pH must be carefully adjusted according to the analytes' pI (isoelectric point) to ensure adequate separation. The changes in the pH of the buffer affect the concentration of ions in the solution, which also alters the ionic strength (the total concentration of ions in the buffer). This altercation in ionic strength impacts the conductivity of the solution and influences the migration behavior of analytes, ultimately affecting the resolution of the separation. A suitable pH range must be selected to ensure that the analytes possess sufficient charge for effective migration and to control the EOF, facilitating efficient separation. If the pH is too high or too low, it can cause issues such as reduced resolution, poor separation, or inconsistency in reproducibility of the separation process. For example, at high pH, the EOF is stronger, causing the peptides such as zwitterions, which are negatively charged [82], to migrate toward the positive electrode (anode) due to the dominant EOF. This occurs due to the increased electrokinetic potential at the interface between the layer of cation (H^+) and the wall of the capillary (SiO^-). Conversely, peptides become positively charged at low pH, and with weaker EOF they migrate toward the negative electrode. In silica capillaries, solutes generally migrate toward the negative electrode at pH > 7.0.

Effect of BGE Type

The BGE plays a crucial role in the separation process because it serves as the medium through analyte migration under the influence of an applied electric field. In some experiments, a higher buffer concentration enhances the more considerable conductivity difference between BGE and the sample, which can be beneficial for improving the enrichment effect. However, an increased BGE concentration can lead to poor signal sensitivity and limit the analytes' enrichment. The types of BGE include Simple Buffers, which are commonly used in CE consisting of single ionizable species or a combination of weak acids, and their conjugate bases examples are phosphate buffer, borate, and acetate buffer, as listed in Table 1. They are usually applied in capillary zone electrophoresis (CZE), where the analytes are separated primarily based on their electrophoretic mobilities. They are suitable for separating small ions or small organic compounds. A unique background electrolyte (BGE) system free of inorganic species composed of Tris/CHES at pH 8.4 with an ionic strength of 50 mM was utilized for capillary zone electrophoresis (CZE) separation and laser-induced fluorescence (LIF) detection of fluorescently labelled peptides

Table 3 Buffers commonly used in CE

Buffer	Useful pH range
Phosphate buffer	2.5–8.0
Tris buffer (Tris(hydroxymethyl) aminomethane)	7.0–9.0
Acetate buffer	3.76–5.76
Borate buffer	8.14–10.5
Citric acid buffer	2.0–6.5
Ammonium acetate buffer	4.5–7.5
Tricine buffer	7.0–8.5
Phosphate	1.14–3.14
Phosphate	6.20–8.20
Zwitterionic buffer	Useful pH range
MES (2-(N-Morpholino) ethanesulfonic acid)	5.5–7.5
MOPS (3-(N-Morpholino) propanesulfonic acid)	6.5–7.9
HEPES (N-2-Hydroxyethylpiperazine-N′-2-ethanesulfonic acid)	6.8–8.2
PIPES (Piperazine-N,N′-bis(2-ethanesulfonic acid))	6.1–7.5
CHES (2-(N-Cyclohexylamino) ethanesulfonic acid)	7.0–9.0
TES (N-Tris(hydroxymethyl) methyl-2-aminoethanesulfonic acid)	6.8–8.8
TAPS (N-Tris(2-hydroxypropyl) methylamine)	7.5–9.0

and proteins [83]. Secondly, complex buffers such as salts, acids, bases, and co-solvents, such as a mixture of tris-hydrochloride and boric acid, are listed in Table 3. It can be applied in techniques such as CIEF or capillary gel electrophoresis (CGE). Other buffers such as MEKC (contain surfactants, usually sodium dodecyl sulfate (SDS)), CIEF (consists of ampholytes, used to separate proteins, peptides, and other amphoteric compounds based on their isoelectric points), and CEC buffers (used in high-resolution separation of a wide range of organic compounds). Finally, the mixed solvent buffers combine organic solvents (methanol and acetonitrile) and water, utilized when dealing with non-ionic or hydrophobic compounds. Adding methanol to the BGE must also be optimized for better peak resolution, usually >30%.

This table summarizes the useful pH ranges of various buffers commonly used in CE for different types of separations.

Effect of Sample Concentration

Sample concentration must be optimized to fall within the linear detection range without overloading the capillary. Although higher analyte concentrations can lead to better signal detection, it may result in sample overloading and poor resolution. It could also lead to peak broadening and potential migration (tailing) issues due to electroosmotic flow disturbance phenomena, analyte-analyte interactions, and overloading of the capillary, thus affecting the separation efficiency. Low concentrations may lead to poorly defined peaks, resulting in lower signal-to-noise. Factors affecting sample concentration include the volume of the injected sample, ionic strength of the BGE, and charge and size of the analyte. This can be managed by optimizing the injection time, voltage, sample dilution, and stacking techniques such as electrokinetic injection [84].

Effect of Electric Field Strength (Voltage)

An increase in electric field strength (voltage) in capillary electrophoresis directly accelerates the migration of charged analytes, resulting in quicker separations, analysis times, and improved resolution. This stronger electric field can also affect the movement of the buffer solution (electroosmotic flow). However, higher voltage can generate heat, potentially leading to band broadening. The limiting factor here is Joule heating. It is worthy of note that different analytes and experimental conditions require specific voltage ranges to achieve optimal separation. Also, the running buffer must be optimized and capillary heating should be carefully monitored.

The electroosmotic flow is given in Eq. (7) as:

$$v_{eo} = \frac{\in \zeta}{4\pi\eta} \mathrm{E} \tag{7}$$

where v_{eo} is the electroosmotic flow velocity (), $\in$ is the dielectric constant (permittivity of the buffer solution), η is the viscosity of the buffer solution, ζ is the zeta potential (the electrical potential at the surface (solid-liquid interface) of the capillary), and E is the applied electric field (V/cm) (voltage gradient).

The voltage gradient is defined in Eq. (8) as:

$$E = \mu_{ep} \frac{V}{L} \tag{8}$$

where V is the voltage, and L is the capillary length.

This equation shows that the electroosmotic flow velocity is proportional to the applied electric field, the zeta potential, and the permittivity of the buffer while inversely proportional to the viscosity of the solution. The electric field strength, E

(V/cm), is also related to electrophoretic mobility, μ_{ep} (cm²/Vs), and the electrophoretic velocity, v_{ep}(cm/s) as given in Eq. (9):

$$\mu_{ep} = \frac{v_{ep}}{E} = \frac{L_d / t_m}{V / L_t} \tag{9}$$

The velocities of the analytes in capillary electrophoresis are estimated by dividing the migration time t by the length of the capillary to the detector, L_d. The total length of the capillary from the inlet to the detector is given as L_t. To mitigate these challenges, various thermal control strategies can be employed, such as Joule heating, CE cooling systems, or temperature regulation mechanisms used for separation is estimated by dividing the applied voltage by the total length L_t. Although a higher voltage and a longer capillary length can result in greater separation efficiency, they can also increase the time it takes for the sample to reach the detector, potentially affecting the analysis speed. The excess capillary length, $L_t - L_d$ is needed to connect to the buffer reservoirs at both capillary ends. However, $L_t - L_d$ does not directly contribute to the separation process itself, although it is crucial for the flow of buffer and the overall operation of the system.

Migration time, t is given in Eq. (10) as:

$$t = \frac{L}{v_{ep}} = \frac{L^2}{\mu_{ep} V} \tag{10}$$

The migration of the analytes through the capillary results in molecular diffusion, leading to peak dispersion, denoted as σ^2 and can be estimated as given in Eq. (11):

$$\sigma^2 = 2D_m t = \frac{2D_m L^2}{\mu_{ep} V} \tag{11}$$

where D_m is given as the solute's diffusion coefficient (cm²/s). The number of theoretical plates (N) is given in Eq. (12):

$$N = \frac{L^2}{\sigma^2} = \frac{\mu_{ep} V}{2D_m} \tag{12}$$

The Effect of Temperature

Temperature plays a crucial role in influencing the electrophoretic mobility of analytes and the efficiency of the separation process in CE. As the temperature increases, the viscosity of the buffer solution decreases, which can enhance the mobility of charged analytes, potentially speeding up their migration times. However, this faster migration may negatively affect the resolution, as analytes could elute too quickly and overlap. Also, higher temperatures reduce the viscosity of the electrolyte solution, which lowers the resistance of the capillary and, as a result, increases the current. While this can result in faster separations, it may also lead to issues such as Joule heating (an excessive heat buildup due to high current) if not properly managed. Additionally, increased temperature can enhance separation efficiency by enhancing the ionic conductivity of the buffer, allowing ions to migrate more easily. However, excessive heating can cause peak broadening due to the reduced differential migration between analytes. Different thermal control techniques can be applied to mitigate these challenges, such as Joule heating, CE cooling systems, or temperature regulation mechanisms. These methods help to maintain a stable temperature during the separation process, ensuring optimal resolution and minimizing the risk of sample degradation.

Effect of Viscosity

Viscosity is influenced by temperature, so maintaining precise temperature control is essential. As the temperature rises, the diffusion of analyte molecules rises within the capillary, leading to increased electrophoretic mobility. However, this also reduces viscosity and separation resolution, broader peaks, and poorer separation quality. Therefore, separations are typically conducted at 20–25 °C (close to room temperature) to ensure optimal CE performance. Using liquid cooling for the capillary helps to maintain effective temperature control, even with high-concentration buffers and large-bore capillaries. When temperature control becomes challenging, common solutions include using smaller-bore capillaries (which reduce current and heat generation) or longer capillaries (which offer more surface area for heat dissipation). Alternatively, lowering the buffer concentration can also help, though this may reduce peak efficiency by weakening the focusing effect. Insufficient temperature control is the primary reason for using low-concentration buffers (e.g., 20 mM) or operating at higher temperatures.

Effect of SDS Concentration

SDS comprises a hydrophobic tail of a long hydrocarbon chain and a negatively charged ionic functional group at the head. It contains a negatively charged sulfate group ($-SO_4^-$) attached to a long hydrophobic carbon chain ($C_{12}H_{25}$), as shown in Fig. 9. When the concentration of SDS reaches a critical level, it forms a micelle. A micelle is a spherical cluster of 40–100 surfactant molecules, where the hydrophobic tails are oriented inward, while the negatively charged heads are directed outward. In capillary electrophoresis, increasing the concentration of SDS improves protein separation by promoting a more uniform negative charge distribution on the proteins. This enhances the size-based separation, where larger proteins migrate more slowly, based on higher resistance to the matrix (gel), but ultimately improving resolution based on molecular weight. Excessively high SDS concentrations can

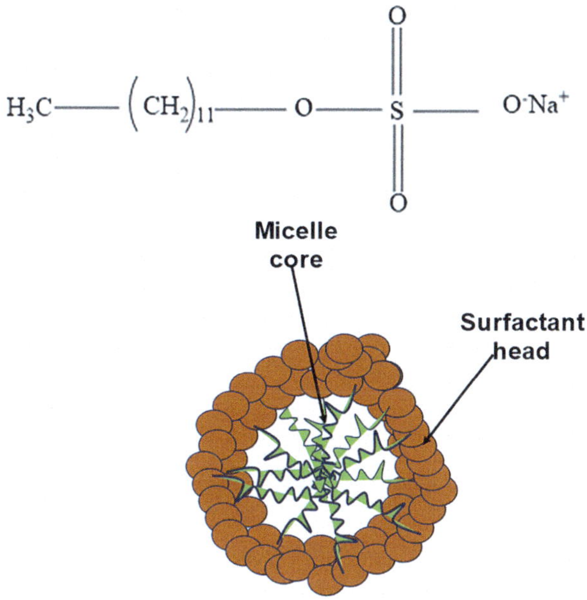

Fig. 9 Schematic illustration of the structure of the micelle

also lead to more significant capacity factors that enhance separation efficiency, thereby reducing separation times. The increased concentration of SDS also increases the retention factor in the separation buffer due to the strong interaction between micelle and analytes. However, it can lead to poor peak shape, suppression of signal sensitivity, and decreased migration efficiency (effective electrophoretic mobility of the targeted analytes). SDS acts as a denaturing agent, used in unfolding proteins and masking their native charge, enabling them to have a comparable charge-to-mass ratio, allowing for size-based separation. It is essential to optimize the SDS concentration for optimal separation. If the SDS concentration is too low, the proteins will not fully denature, while if it is too high, it can cause excessive protein-SDS complex formation, leading to migration issues. Additionally, the ideal SDS concentration can vary depending on the different protein types, and the capillary's characteristics, such as coating and inner diameter, can affect the ideal SDS concentration for optimal resolution.

Several key variables influence the efficiency and sensitivity of SIM-MEKC and need to be thoroughly examined, including injection voltage, applied pressure, and injection time during sample introduction.

Application of CE

CE is a versatile analytical technique commonly used in various fields, including molecular biology, biochemistry, pharmaceutical research, and environmental monitoring and analysis, to detect the residual levels of heavy metals and organic and inorganic ions. Others applications include the following:

- DNA/RNA analysis: It allows for rapid DNA sequencing, DNA fragments analysis, and gene expression studies, offering a cost-effective alternative with high throughput, accuracy, and efficiency.
- Food analysis: CE contributes to detecting allergens, additives, contaminants, and nutritional components to detect food product adulterations and monitor food safety and quality control. It allows the quantification of fatty acids, vitamins, and amino acids.
- Environmental monitoring: Testing for pollutants, organic compounds, and sulfonamide antibiotics in water and soil samples. It also contributes to detecting residual levels of pesticides and pharmaceutical pollutants and quantifying relevant anions like nitrates and sulfates, helping to monitor water quality. It can also preconcentrate and determine heavy metals, nonsteroidal anti-inflammatory drugs, and anticancer agents in biological, food, and environmental samples.
- Pharmaceutical analysis: CE contributes to drug purity, enantiomeric separation, and drug development. In pharmaceutical research, ACE is employed to understand the interactions between proteins and ligands, offering a quick, precise method for detecting binding constants, especially in protein-protein and protein-DNA interactions.
- Diagnostics: Biomarker detection, protein profiling, and disease diagnosis. It can also detect and quantify electrolytes, hormones, and enzymes in body fluids. Researchers have also employed CE to simultaneously detect ions Ca^{2+}, Na^{+}, NH^{4+}, K^{+}, and Mg^{2+} in saliva. It also offers rapid and precise protein quantification.
- Proteomics: CE is employed in studying post-translational modifications and protein interactions, which contributes to developing disease biomarkers and potential drug targets.
- Forensic science: CE has also been used by scientists for STR typing in biological materials, generating profiles based on highly polymorphic genetic markers unique to individuals. This technique is employed in DNA analysis for criminal investigations. It is also used to identify specific mRNA fragments, which helps to determine the tissue origin or biological fluid of forensic samples. It is also used in ink analysis to detect document counterfeiting, with techniques such as micellar electrophoretic capillary chromatography (MECC) distinguishing inks based on their chemical composition.
- Biotechnology: CE plays a crucial role in ensuring the quality of bioproducts, including recombinant proteins and antibodies. It is commonly utilized for precise, rapid separation and analysis of a broad range of molecules and for assessing protein folding and glycosylation, essential for biotherapeutics efficacy and safety. Table 4 summarizes the advantages and disadvantages of CE across different factors.

Table 4 Advantages and disadvantages of CE across different factors

S. No.	Factors	Advantages	Disadvantages
1.	Sensitivity and resolution	High sensitivity for low concentrations – Excellent resolution due to high separation power – Broad detection range such as UV, DAD, C4D and MS, enhanced with simple operating system for both ionic and non-ionic compounds	Detection limits may not match other methods (e.g., mass spectrometry) Possible interference or overlapping peaks Optical detectors have a short path length
2.	Speed	Fast analysis due to high electric field strength and small capillary size	Speed can vary depending on sample complexity Optimization may be needed for balancing speed and resolution
3.	Efficiency	High separation efficiency with sharp, defined peaks	Decreased efficiency with complex or large-volume samples due to peak broadening
4.	Accuracy	Precise and reliable results for determining migration times	Potential for variability from factors like temperature fluctuations, capillary conditions, or sample composition
5.	Cost and complexity	– Lower operational and maintenance costs (less solvent, fused silica capillary (column and smaller sample volumes) – Compact and simple instrumentation depends on voltage, temperature for analyte separation, and pH of the buffer solution	High initial equipment cost Requires specialized technical expertise, increasing operational complexity
6.	Basic principle	– Separation technique based on differential migration of ions (mass to charge ratio) in an electric field, flowing through capillary by plug-flow or capillary electro-osmotically	Primarily limited to charged species, neutral compounds require modifications or additional methods

This table provides a clear overview of the advantages and disadvantages of CE across different factors

Conclusion and Future Directions

CE is a powerful analytical tool that significantly contributes to food analysis, and its potential for enhancing food safety and quality remains promising. The advantages of this method include excellent resolution, high sensitivity, and cost-effectiveness. It is a rapid, high-resolution, highly sensitive analytical technique suitable for various applications. These capabilities can be further enhanced by integrating various offline and/or online enrichment techniques, facilitating ultra-trace analysis in complex sample matrices. The development of compact and real-time capillary electrophoresis (CE) devices has shown great potential, offering more efficient, faster, and user-friendly solutions. These innovations are poised to revolutionize various analytical applications, and with ongoing technological improvements, the capabilities of CE are expected to expand even further in the coming years. CE can be paired with a variety of detection methods, such as ultraviolet (UV) and fluorescence (FLD) detection, as well as electrochemical techniques, such as conductivity detection (CD) and amperometric detection (AD), which are commonly used. Additionally, more advanced and powerful detection options, such as mass spectrometry (MS), are increasingly popular. A miniaturized version of CE, known as microchip capillary electrophoresis (MCE), utilizes a microchip with a small separation channel to perform analyte separation. Advancements in this field also support the development of in-line extraction systems and stacking methods combined with contactless conductivity detection (C4D) or microchip CE. Such integration paves the way for portable systems that are ideal for on-site analysis, particularly for monitoring environmental samples. This technique is considered environmentally friendly due to its minimal use of reagents. As these devices become more accessible and sophisticated, they will undoubtedly play a crucial role in advancing the accuracy and speed of analytical processes, particularly in fields like food analysis, clinical diagnostics, and environmental monitoring. Nevertheless, when applying CE in specific scenarios, limitations in detection sensitivity and operational challenges must be considered. Overcoming these issues requires continued innovation in microfabrication technologies and materials science to enhance the technique's reliability and effectiveness. As new developments in CE technology emerge, its applications will likely expand, offering even greater precision and reliability for food analysis in the future.

Acknowledgments The Ministry of Higher Education Malaysia supported this study through the Fundamental Research Grant Scheme (FRGS) with reference code number FRGS/1/2021/STG04/USM/02/4 and account code number 203.CIPPT.6711973. A.E. Ibukun would like to acknowledge Tertiary Education Trust, TETFUND Nigeria (TETF/ES/ POLY/KWARA/TSAS/2020), and The Federal Polytechnic Offa, Kwara State, Nigeria, for the sponsorship.

References

1. Štěpánová S, Kašička V (2019) Recent developments and applications of capillary and microchip electrophoresis in proteomics and peptidomics (2015–mid 2018). J Sep Sci 42:398–414
2. Kristoff CJ, Bwanali L, Veltri LM, Gautam GP, Rutto PK, Newton EO, Holland LA (2019) Challenging bioanalyses with capillary electrophoresis. Anal Chem 92:49–66
3. Zhang Z, Qu Y, Dovichi NJ (2018) Capillary zone electrophoresis-mass spectrometry for bottom-up proteomics. TrAC Trends Anal Chem 108:23–37
4. Agregán R, Echegaray N, López-Pedrouso M, Kharabsheh R, Franco D, Lorenzo JM (2021) Proteomic advances in milk and dairy products. Molecules 26:3832
5. Shen X, Yang Z, McCool EN, Lubeckyj RA, Chen D, Sun L (2019) Capillary zone electrophoresis-mass spectrometry for top-down proteomics. TrAC Trends Anal Chem 120:115644
6. Štěpánová S, Kašička V (2023) Recent developments and applications of capillary and microchip electrophoresis in proteomics and peptidomics (mid-2018–2022). J Sep Sci 46. https://doi.org/10.1002/JSSC.202300043

7. DeLaney K, Sauer CS, Vu NQ, Li L (2019) Recent advances and new perspectives in capillary electrophoresis-mass spectrometry for single cell "omics". Molecules 24:10042. https://doi.org/10.3390/molecules24010042
8. Bagwe K, Gould N, Johnson KR, Ivanov AR (2023) Single-cell omic molecular profiling using capillary electrophoresis-mass spectrometry. TrAC Trends Anal Chem 165:117117. https://doi.org/10.1016/j.trac.2023.117117
9. Mantovani V, Galeotti F, Maccari F, Volpi N (2018) Recent advances in capillary electrophoresis separation of monosaccharides, oligosaccharides, and polysaccharides. Electrophoresis 39:179–189
10. Kun R, Jóna E, Guttman A (2021) Capillary electrophoresis-based N-glycosylation analysis in the biomedical and biopharmaceutical fields. Adv Exp Med Biol 1336:129–137
11. Portero EP, Nemes P (2019) Dual cationic–anionic profiling of metabolites in a single identified cell in a live Xenopus laevis embryo by microprobe CE-ESI-MS. Analyst 144:892–900
12. Choi SB, Polter AM, Nemes P (2022) Patch-clamp proteomics of single neurons in tissue using electrophysiology and subcellular capillary electrophoresis mass spectrometry. Anal Chem 94:1637–1644. https://doi.org/10.1021/ACS.ANALCHEM.1C03826
13. Foret F, Szökő E, Karger BL (1993) Trace analysis of proteins by capillary zone electrophoresis with on-column transient isotachophoretic pre-concentration. Electrophoresis 14:417–428
14. Torano JS, Ramautar R, de Jong G (2019) Advances in capillary electrophoresis for the life sciences. J Chromatogr B 1118:116–136
15. Mikšík I (2019) Coupling of CE-MS for protein and peptide analysis. J Sep Sci 42:385–397
16. Zhu Z, Lies M, Silzel J (2021) Native fluorescence detection with a laser driven light source for protein analysis in capillary electrophoresis. Anal Chim Acta 1183:338936
17. Li X, Yu L, Shi X, Rao C, Zhou Y (2021) Capillary isoelectric focusing with UV fluorescence imaging detection enables direct charge heterogeneity characterization of erythropoietin drug products. J Chromatogr A 1643:462043
18. Johnson KR, Greguš M, Ivanov AR (2022) Coupling high-field asymmetric ion mobility spectrometry with capillary electrophoresis-electrospray ionization-tandem mass spectrometry improves protein identifications in bottom-up proteomic analysis of low nanogram samples. J Proteome Res 21:2453–2461. https://doi.org/10.1021/ACS.JPROTEOME.2C00337
19. Couderc F, Ong-Meang V, Poinsot V (2017) Capillary electrophoresis hyphenated with UV-native-laser induced fluorescence detection (CE/UV-native-LIF). Electrophoresis 38:135–149
20. Štěpánová S, Acta AC (2016) Recent applications of capillary electromigration methods to separation and analysis of proteins. Elsevier, Amsterdam. Accessed https://www.sciencedirect.com/science/article/pii/S0003267016307681
21. Štěpánová S, Acta AC (2022) Applications of capillary electromigration methods for separation and analysis of proteins (2017–mid 2021)–a review. Elsevier, Amsterdam. Accessed https://www.sciencedirect.com/science/article/pii/S0003267022000186
22. Bhimwal R, Rustandi RR, Payne A, Dawod M (2022) Recent advances in capillary gel electrophoresis for the analysis of proteins. J Chromatogr A 1682:463453. https://doi.org/10.1016/j.chroma.2022.463453
23. Farmerie L, Rustandi RR, Loughney JW, Dawod M (2021) Recent advances in isoelectric focusing of proteins and peptides. J Chromatogr A 1651:462274
24. Raymond BY, Quirino JP (2022) Pseudophase microextraction for in-line sample concentration in micellar electrokinetic chromatography. Anal Chim Acta 1189:339219
25. Peng L-Q, Dong X, Zhen X-T, Yang J, Chen Y, Wang S-L, Xie T, Cao J (2020) Simultaneous separation and concentration of neutral analytes by cyclodextrin assisted sweeping-micellar electrokinetic chromatography. Anal Chim Acta 1105:224–230
26. Wang L, MacDonald D, Huang X, Chen DDY (2016) Capture efficiency of dynamic pH junction focusing in capillary electrophoresis. Electrophoresis 37:1143–1150
27. Zhu G, Sun L, Dovichi NJ (2016) Dynamic pH junction pre-concentration in capillary electrophoresis-electrospray ionization-mass spectrometry for proteomics analysis. Analyst 141:5216–5220

28. Malá Z, Gebauer P (2019) Recent progress in analytical capillary isotachophoresis. Electrophoresis 40:55–64
29. Yang Q, Chen B, He M, Hu B (2018) Sensitive determination of seven triazine herbicide in honey, tomato and environmental water samples by hollow fiber based liquid-liquid-liquid microextraction combined with sweeping micellar electrokinetic capillary chromatography. Talanta 186:88–96. https://doi.org/10.1016/J.TALANTA.2018.04.012
30. Khnouf R, Han C (2020) Isotachophoresis-enhanced immunoassays: challenges and opportunities. IEEE Nanotechnol Mag 14:6–17
31. Bahga SS, Santiago JG (2013) Coupling isotachophoresis and capillary electrophoresis: a review and comparison of methods. Analyst 138:735–754
32. Rogacs A, Marshall LA, Santiago JG (2014) Purification of nucleic acids using isotachophoresis. J Chromatogr A 1335:105–120
33. Eid C, Santiago JG (2018) Isotachophoresis applied to biomolecular reactions. Lab Chip 18:11–26
34. Grist SM, Mourdoukoutas AP, Herr AE (2020) 3D projection electrophoresis for single-cell immunoblotting. Nat Commun 11:6237
35. Yang X-M, Zhang S-H, Wang C, Wang Z (2013) Development of micellar to solvent stacking online pre-concentration technique in capillary electrophoresis. Chin J Anal Chem 41:1939–1946. https://doi.org/10.1016/S1872-2040(13)60698-2
36. Quirino JP, Haddad PR (2008) Online sample pre-concentration in capillary electrophoresis using analyte focusing by micelle collapse. Anal Chem 80:6824–6829. https://doi.org/10.1021/ac801258r
37. Quirino J, Terabe S (2000) Approaching a million. Fold sensitivity increase in capillary electrophoresis with direct ultraviolet detection: cation-selective exhaustive injection and sweeping. Anal Chem 72:1023–1030. https://doi.org/10.1021/ac990344b
38. Quirino JP, Terabe S (2000) Large volume sample stacking of positively chargeable analytes in capillary zone electrophoresis without polarity switching: use of low reversed electroosmotic flow induced by a cationic surfactant at acidic pH. Electrophoresis 21:355–359. https://doi.org/10.1002/(SICI)1522-2683(20000101)21:2<355::AID-ELPS355>3.0.CO;2-D
39. Ghiasvand A, Feng Z, Quirino JP (2018) Enrichment and separation of cationic, neutral, and chiral analytes by micelle to cyclodextrin stacking–micellar electrokinetic chromatography. Anal Chem 91:1752–1757
40. Quirino JP, Grochocki W, Markuszewski MJ (2017) Sample concentration of charged small molecules and peptides in capillary electrophoresis by micelle to cyclodextrin stacking. Anal Chem 89:13422–13428
41. Jin H-F, Shi Y, Shen Q-X, Cao J, Ye L-H (2024) Online concentration and separation of multiple derivatized monosaccharides from edible fruit with cyclodextrin-encapsulated sweeping by micellar electrokinetic chromatography. J Chromatogr A 1714:464563
42. Ma X-R, Wu Y-W, Shen Q-X, Jiao Y-H, Cao J, Ye L-H (2024) One-step admicelle to cyclodextrin sweeping of toxic aristolochic acids by capillary electrophoresis. Microchem J 200:110480
43. Zhen X-T, Chen Y, Yang J, Dong X, Zheng H, Cao J (2020) Online concentration of triazine herbicides in microemulsion electrokinetic chromatography by electrokinetic injection assisted micelle to cyclodextrin stacking. J Chromatogr A 1628:461438
44. Uzma HX (2025) Trends development and applications on electrophoresis techniques of slab gel, capillary, microchip/microfluidic capillary, and isotachophoresis. Talanta 293:128029. https://doi.org/10.1016/j.talanta.2025.128029
45. Staikos G, Dondos A (2009) Study of the sodium dodecyl sulphate–protein complexes: evidence of their wormlike conformation by treating them as random coil polymers. Colloid Polym Sci 287:1001–1004
46. Sänger CE (2019) CE-SDS method development, validation, and best practice—an overview. Electrophoresis 40:2361–2374
47. Dong Q, Guo X, Qu X, Bai S, You X, Cui H, Qin S, Gao L (2023) Chiral covalent organic framework-based open tubular capillary electrochromatography column for enantioseparation of selected amino acids and pesticides. Talanta 258:124415. https://doi.org/10.1016/j.talanta.2023.124415

48. Zheng X, Zhang Q, Ma Q, Li X, Zhao L, Sun X (2023) A chiral metal-organic framework {(HQA)(ZnCl2)(2.5H2O)}n for the enantioseparation of chiral amino acids and drugs. J Pharm Anal 13:421–429. https://doi.org/10.1016/j.jpha.2023.03.003
49. Maccarrone G, Contino A, Cucinotta V (2012) The study of solution equilibria in chiral capillary electrophoresis by the ligand-exchange mechanism. TrAC Trends Anal Chem 32:133–153. https://doi.org/10.1016/j.trac.2011.09.010
50. Sharaf El-Din MMK, Attia KAM, Nassar MWI, Kaddah MMY (2018) Development and validation of a copper ligand-exchange chromatographic method for the estimation of D-lactic acid in ringer-lactate solution. Talanta 189:86–91. https://doi.org/10.1016/j.talanta.2018.06.068
51. Yang S, Gu L, Wu F, Dai X, Xu F, Li Q, Fang X, Yu S, Ding C-F (2022) The chirality determination of amino acids by forming complexes with cyclodextrins and metal ions using ion mobility spectrometry, and a DFT calculation. Talanta 243:123363. https://doi.org/10.1016/j.talanta.2022.123363
52. Tůma P (2021) Determination of amino acids by capillary and microchip electrophoresis with contactless conductivity detection—theory, instrumentation and applications. Talanta 224:121922. https://doi.org/10.1016/j.talanta.2020.121922
53. Qi L, Qiao J (2022) Progress of chiral ligand-exchange capillary electrophoresis for enantioseparation. J Chromatogr A 1679:463381. https://doi.org/10.1016/j.chroma.2022.463381
54. Xue S, Ren S, Wang L, Zhang Q (2020) Evaluation of tetraalkylammonium amino acid ionic liquids as chiral ligands in ligand-exchange capillary electrophoresis. J Chromatogr A 1611:460579. https://doi.org/10.1016/j.chroma.2019.460579
55. Neequaye T, El Rassi Z (2023) Poly(carboxyethyl acrylate-co-ethylene glycol dimethacrylate) precursor monolith with bonded (S)-(−)-1-(2-naphthyl) ethylamine ligands for use in chiral and achiral separations by capillary electrochromatography. J Chromatogr A 1688:463713. https://doi.org/10.1016/j.chroma.2022.463713
56. Li Z, Mao Z, Zhou W, Chen Z (2020) γ-Cyclodextrin metal-organic framework supported by polydopamine as stationary phases for electrochromatographic enantioseparation. Talanta 218:121160. https://doi.org/10.1016/j.talanta.2020.121160
57. Lobato A, Pereira EA, Gonçalves LM (2021) Combining capillary electromigration with molecular imprinting techniques towards an optimal separation and determination. Talanta 221:121546. https://doi.org/10.1016/j.talanta.2020.121546
58. Schmid MG (2012) Chiral metal-ion complexes for enantioseparation by capillary electrophoresis and capillary electrochromatography: a selective review. J Chromatogr A 1267:10–16
59. Aydoğan C, Denizli A (2014) Chiral separation-based ligand exchange by open-tubular capillary electrochromatography. Anal Biochem 447:55–57
60. Gui Y, Ji B, Yi G, Li X, Zhang K, Fu Q (2021) Polydopamine-assisted rapid one-step immobilization of l-arginine in capillary as immobilized chiral ligands for enantioseparation of dansyl amino acids by chiral ligand exchange capillary electrochromatography. Molecules 26:1800. https://doi.org/10.3390/molecules26061800
61. Ali N, Liu Y, Wang F, Qi L (2025) Chiral ligand-exchange capillary electrochromatography with thermo-responsive poly(N,N-dimethylacrylamide) as coating for efficient separation of D,L-amino acids. Talanta 293:128090. https://doi.org/10.1016/j.talanta.2025.128090
62. Qi L, Qiao J (2022) Advances in stimuli-responsive polymeric coatings for open-tubular capillary electrochromatography. J Chromatogr A 1670:462957. https://doi.org/10.1016/j.chroma.2022.462957
63. Qi X, Wang S, Wu B, Qu H (2009) Determination of hepatotoxic pyrrolizidine alkaloids in gynura segetum by MEKC. Chromatographia 70:281–285
64. Yu L, Xu Y, Feng H, Li SFY (2005) Separation and determination of toxic pyrrolizidine alkaloids in traditional Chinese herbal medicines by micellar electrokinetic chromatography with organic modifier. Electrophoresis 26:3397–3404
65. Carbonell-Rozas L, Lara FJ, del Olmo Iruela M, García-Campaña AM (2020) Micellar electrokinetic chromatography as efficient alternative for the multiresidue determination of seven neonicotinoids and 6-chloronicotinic acid in environmental samples. Anal Bioanal Chem 412:6231–6240

66. Manmana Y, Liu C, Koino H, Sueyoshi K, Kitagawa F, Kubo T, Otsuka K (2022) Development of transient trapping micellar electrokinetic chromatography coupled with mass spectrometry for steroids analysis. Chirality 34:1328–1337
67. van der Burg D, Wätzig H, Sänger CE (2023) Design of experiments for micellar electrokinetic chromatography method development for the monitoring of water-soluble vitamins in cell culture medium. Electrophoresis 44:1548–1558
68. Belal F, El-Razeq SA, Fouad M, Zayed S, Fouad F (2018) Simultaneous determination of five coccidiostats in veterinary powders, feed premixes, and baby food by micellar electrokinetic chromatography: application to chicken tissues and liver. Food Anal Methods 11:3531–3541
69. Jin H-F, Shi Y, Jiao Y-H, Fei T-H, Cao J, Ye L-H (2023) Two-step pressure injection-assisted online enrichment of herbicides from foods with affinity micelle sweeping by micellar electrokinetic chromatography. J Chromatogr A 1706:464258
70. Abo Gabal R, Osama S, Hanafy N, Oraby A (2023) Micellization thermodynamics as a function of the temperature of a cationic zwitterionic dodecyl phosphocholine and anionic sodium dodecyl sulfate mixed micelles with fluorometry. Appl Phys A Mater Sci Process 129:201
71. Fan G, Xiao-Fei W (2019) Research and application progress of micellar electrokinetic chromatography in separation of proteins. Chinese J Anal Chem 47:805–813
72. Krehic M, Avenel-Audran M (2009) Allergic contact dermatitis from decyl glucoside in an antiseptic lotion. Contact Dermat 61:349–350
73. Liu S, Sang R, Hong S, Cai Y, Wang H (2013) A novel type of highly effective non-ionic Gemini alkyl O-glucoside surfactants: a versatile strategy of design. Langmuir 29:8511–8516
74. Nowicki J, Mościpan M, Nowakowska-Bogdan E, Woch J (2019) Micellar catalysis in fischer glycosidation: en route to diverse functional glycosides. ChemistrySelect 4:13841–13845
75. Rodriguez J, Contento AM, Castaneda G, Munoz L, Berciano MA (2012) Determination of morphine, codeine, and paclitaxel in human serum and plasma by micellar electrokinetic chromatography. J Sep Sci 35:2297–2306
76. Lu H, Pezron I, Gaudin T, Drelich A (2018) Non-equilibrium micelles formed by sugar-based surfactants under their Krafft temperature. Colloids Surf A Physicochem Eng Asp 540:167–176
77. Costa JL, Morrone AR, Resende RR, da Matta Chasin AA, Tavares MFM (2014) Development of a method for the analysis of drugs of abuse in vitreous humor by capillary electrophoresis with diode array detection (CE–DAD). J Chromatogr B 945:84–91
78. Mabuchi M, Takatsuka S, Matsuoka M, Tagawa K (2004) Determination of morphine, morphine-3-glucuronide and morphine-6-glucuronide in monkey and dog plasma by high-performance liquid chromatography-electrospray ionization tandem mass spectrometry. J Pharm Biomed Anal 35:563–573. https://doi.org/10.1016/j.jpba.2004.01.026
79. Alnajjar A, Butcher JA, McCord B (2004) Determination of multiple drugs of abuse in human urine using capillary electrophoresis with fluorescence detection. Electrophoresis 25:1592–1600
80. Mi J, Zhang X, Chang W (2004) Determination of morphine by capillary zone electrophoresis immunoassay combined with laser-induced fluorescence detection. J Immunoass Immunochem 25:57–70
81. Zhang X-X, Li J, Gao J, Sun L, Chang W-B (2000) Determination of morphine by capillary electrophoresis immunoassay in thermally reversible hydrogel-modified buffer and laser-induced fluorescence detection. J Chromatogr A 895:1–7
82. Breadmore MC (2009) Electrokinetic and hydrodynamic injection: making the right choice for capillary electrophoresis. Bioanalysis 1:889–894
83. Morani M, Taverna M, Mai TD (2019) A fresh look into background electrolyte selection for capillary electrophoresis-laser induced fluorescence of peptides and proteins. Electrophoresis 40:2618–2624
84. Breadmore MC, Grochocki W, Kalsoom U, Alves MN, Phung SC, Rokh MT, Cabot JM, Ghiasvand A, Li F, Shallan AI (2019) Recent advances in enhancing the sensitivity of electrophoresis and electrochromatography in capillaries and microchips (2016–2018). Electrophoresis 40:17–39

The manufacturer's authorised representative in the EU is Springer Nature Customer Service Centre GmbH, Europaplatz 3, 69115 Heidelberg, Germany. If you have any concerns regarding our products, please contact ProductSafety@springernature.com

Printed and bound by CPI Group (UK) Ltd, Croydon, CR0 4YY
21/09/2026
02219985-0018